International Conference on Applied Machine Learning and Data Science (ICAMLDS 2019)

Journal of Physics: Conference Series Volume 1423

Chengdu, China
15 - 17 November 2019

Editors:

Liangquan Ge
Xiaonan Xiao

ISBN: 978-1-7138-1327-9
ISSN: 1742-6588

Printed from e-media with permission by:

Curran Associates, Inc.
57 Morehouse Lane
Red Hook, NY 12571

Some format issues inherent in the e-media version may also appear in this print version.

Institute of Physics
Dirac House, Temple Back
Bristol BS1 6BE UK

Phone: 44 1 17 929 7481
Fax: 44 1 17 920 0979

techtracking@iop.org

Additional copies of this publication are available from:

Curran Associates, Inc.
57 Morehouse Lane
Red Hook, NY 12571 USA
Phone: 845-758-0400
Fax: 845-758-2633
Email: curran@proceedings.com
Web: www.proceedings.com

TABLE OF CONTENTS

DATA SCIENCE

APPLICATIONS OF MACHINE LEARNING

ARTIFICIAL INTELLIGENCE

Preface

The organizing committee of ICAMLDS 2019 is proud to present the proceedings of the 2019 International Conference on Applied Machine Learning and Data Science (ICAMLDS 2019: http://www.icamlds.net), held in Chengdu, China from 15 to 17 November, 2019. We warmly welcome prospected authors submit new research papers to ICAMLDS 2019, and share the valuable experiences with the scientists and scholars around the world.

The conference is an international forum for the presentation of technological advances and research results in the fields of applied machine learning and data sciences such as databases, big data, data mining, applied machine learning, and other aspects of information systems and technology involving advanced applications of artificial intelligence and data. The conference will bring together leading researchers, engineers and scientists in the domain of interest from around the world.

ICAMLDS 2019 received more than 148 manuscripts, and 68 submissions have been accepted by our reviewers. By submitting a paper to ICAMLDS 2019, the authors understood and agreed that papers would undergo a rigorous peer-review process. Manuscripts were reviewed by at least two independent, qualified experts in the field selected by the conference committee, who took detailed comments and the authors would submit a revised version in which these feedback taken into consideration. All papers were reviewed using a double-blind review process: authors declared their names and affiliations in the manuscript for the reviewers to see, but reviewers did not know each other's identities, nor did the authors receive information about who had reviewed their manuscript.

The committees of ICAMLDS 2019 invested great efforts in reviewing the papers submitted to the conference and organizing the sessions to enable the participants to gain maximum benefit. We sincerely appreciate your participation and do believe our joint efforts would make our conference indeed fruitful and successful.

Editors
Prof. Liangquan Ge
Prof. Xiaonan Xiao

Committees

Conference Chairs

Prof. Xiaonan Xiao, Xiamen University, China

Prof. Liangquan Ge, Chengdu University of Technology, China

Program Chairs

Dr. Fei Li, Chengdu University of Technology, China

Program Committee

Professor Kecheng Liu, University of Reading, United Kingdom

Dr. Yasir Faiz, Pakistan Institute of Nuclear Science and Technology (PINSTECH), Islamabad, Pakistan

Prof. Jiqian Fang, Sun Yat-sen University University, China

Prof. Shirshov, E. V. Northern Arctic Federal University, Russia

Prof. Guoqiang Zeng, Chengdu University of Technology, China

Prof. Shitong Ye, Huali College Guangdong University of Technology, China

Prof. Yunli Cheng, Guangzhou NanYang Polytechnic College, China

Dr. Zhenming Xu, Xiamen University Tan Kah Kee College, China

Dr. Yanling Zhou, Hefei University, China

Prof. Qingxian Zhang, Chengdu University of Technology, China

Shahrum Abdullah, University Kebangsaan Malaysia, Malaysia

Dr. Jinhuan Wang, Intelligent Science & Information Engineering College, Xi'an Peihua University, China

Associate Prof. Peijiang Chen, Linyi University, China

Prof. Bachir ACHOUR, University of Biskra, Algeria

Dr. Chao Xiong, East China University of Technology, China

Dr. Jiankun Zhao, East China University of Technology, China

Journal of Physics: Conference Series **1423** (2019) 011003 doi:10.1088/1742-6596/1423/1/011003

Peer review statement

All papers published in this volume of *Journal of Physics: Conference Series* have been peer reviewed through processes administered by the proceedings Editors. Reviews were conducted by expert referees to the professional and scientific standards expected of a proceedings journal published by IOP Publishing.

Journal of Physics: Conference Series **1423** (2019) 012001 doi:10.1088/1742-6596/1423/1/012001

Deterministic Dual-stack Query Tree RFID tag anti-collision algorithm

Yanling Zhou[1], Thomas Clemen[2]

[1]Artificial Intelligence and Big Data College, Hefei University, Hefei 230601, China
[2]Department of Computer Science, Hamburg University of Applied Sciences, Berliner Tor 7, 20099 Hamburg, Germany

Corresponding author and e-mail: Yanling Zhou, zhouyanling1006@163.com

Abstract. Tag collision problems are a major issue affecting the performance of RFID systems. The probabilistic tag anti-collision algorithm has tag starvation and cannot identify some tags. This paper proposes a deterministic dual-stack query tree algorithm (DDQT). It successfully implements the anti-collision algorithm of tags by generating new query codes and tag identification functions. At the same time, the DDQT algorithm greatly saves memory space by adding a prefix code stack and a suffix code stack. The DDQT algorithm is simple, and there is no limit on the length of the electronic tag to be queried. Compared with the previous query-tree algorithms, the DDQT algorithm has obvious improvements in query comparison times, empty query times, and occupied memory space. It can effectively remove idle nodes, shorten recognition time and save memory resources, thus improving the speed and accuracy of label recognition to some extent.

1. Introduction

RFID (Radio Frequency Identification) technology plays a new type of automatic identification technology. In recent years, it has been widely used at home and abroad, mainly in the following areas, including logistics management, identification, anti-counterfeiting applications, and traffic management. RFID is an automated identification technology that is easy and fast to use. The RFID system is mainly composed of three parts: reader, label and information system. If there are multiple tags in the recognition range of a reader and the tags simultaneously transmit information to the reader, the reader will detect the conflict. The conflict is called "collision" which causes the tag recognition to fail. Tags are widely used in radio frequency identification, and tag collision problems have become a bottleneck limiting the efficiency of RFID systems. The tag collision avoidance problem belongs to the multi-access problem in the communication system [1]. The RFID tag anti-collision algorithm based on time division multiple accesses does not require hardware changes, and can be implemented at the algorithm level. The principle is simple and widely used. Commonly used RFID tag anti-collision algorithms fall into two categories. One is based on the Aloha algorithm, also known as the probabilistic algorithm. Since the number of idle time slots and collision time slots cannot be determined, some tags may not be recognized because of tag starvation. Tree-based algorithms that are also known as deterministic algorithms can avoid starvation and ensure that each tag is accurately identified. Since it is visited according to the binary combination rule, the time of reading the tag is extended. When the number of tags is large, the efficiency of the algorithm is significantly reduced. The tree-based anti-collision algorithm needs to group the IDs of the tags according to a certain length. When the packet length is 1, it is a binary tree algorithm. When the packet length is 2, it is a quad-tree algorithm. When the packet length is 3, it is the Octree algorithm.

2. RFID tag anti-collision algorithm based on query tree

According to the number of tree branches, it can be divided into binary query tree, quad-check tree, eight-check tree, etc. The binary tree anti-collision algorithm is the most widely used algorithm in RFID system. Its basic principle is: the reader sends a bit of the query code Q (0 or 1 for the tag, which is equivalent to forming two subtrees, first querying the 0 subtree and then querying the 1 subtree). Each tag in the reader response range determines whether its ID starts with Q, and if so, its own ID is transmitted to the reader. At this time, there may be three cases: identification (only one tag starting with Q), collision (two or more tags starting with Q), and idle (five labels begin with Q) [1]. If a collision occurs, add 0 and 1 respectively after the previous query code to form two new query codes (equivalent to splitting into two subtrees). At first, it sends a new query code with 0 added at the end to the label and queries the left subtree. Then it sends a new query code with 1 added at the end to the label and queries the right subtree. If a collision occurs again during the query, the above operation is repeated until all the tags are successfully recognized [2]. In the tree-based anti-collision algorithm, the nodes of the whole tree can be divided into four types: initial node, identification node, collision node and idle node [4]. There are one and only one initial node, and the number of identification nodes is equal to the number of tags. The number of the initial and identification nodes is determined. Therefore, to improve recognition efficiency, the key method is to reduce the number of collision nodes and idle nodes.

In the tree-based RFID anti-collision algorithm, in addition to the binary tree anti-collision algorithm [3,5,6], there is a quadtree anti-collision algorithm [7,8,9]. In the quadtree anti-collision algorithm, the tag group length is 2, and the node coding has four combinations. And a tree with four branches is formed in the querying process. Comparing the quadtree anti-collision algorithm with the binary tree anti-collision algorithm, it can be found that the number of the initial node and the identification nodes are the same, and the number of the collision nodes and the idle nodes are the difference. In the quadtree anti-collision algorithm or hybrid query tree algorithm [7], the number of the collision nodes is less than that in the binary tree collision algorithm, but there are more idle nodes. If the encoding algorithm of the quadtree's query code can be improved and the number of idle nodes is reduced, the recognition efficiency will be greatly improved.

3. The description forming the query code of DDQT algorithm

The DDQT algorithm is a RFID tag anti-collision deterministic dual-stack query tree algorithm. The querying codes are formed by encodeing the highest bit and the lowest bit of the collision bit. When collision occurs, it generates four new query codes at a time, removes idle nodes and shortens the recognition time, thereby improving Identify efficiency. The whole algorithm consists of two parts: generating new query code and tag identification. The new query code is generated by the quadtree dual-stack coding algorithm. The identification of the tag is implemented using a quadtree anti-collision algorithm.

The dual-stack algorithm is proposed to solve the problem of memory waste. The code used in this paper is a dual-stack query code, where the query code includes two parts, the plain code and the secret code, as shown in Figure 1 below. When the reader issues the inquiry code, the electronic tag compares with the corresponding code bit according to the plain code in the inquiry code. If the comparing result is the same, the tag sends the entire code to the reader. If the reader can correctly read the electronic tag code, it means there is no collision. When the reader cannot correctly read the electronic tag code, it means that two or more electronic tags are the same as the plain code bits of the query code, and a tag collision occurs.

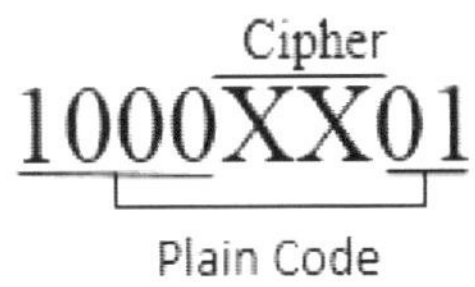

Figure 1. Example of new query code

When a collision occurs, the reader encodes the highest collision bit and the previous plain code according to the collision bit. The highest bit can take values 0 and 1, generating two new first front half of the plain code. The reader encodes the lowest bit and the following plain code. The lowest collision bit can take values of 0 and 1, generating two new second rear half of the plain code. By means of encoding the high bit and low bit, the two new plain codes of the front half are pushed into the prefix code stack, and the two new plain codes of the rear half are pressed into the suffix code stack. Figure 2 below shows an example of the new contents of the prefix-code stack and the suffix-code stack when the first collision occurred.

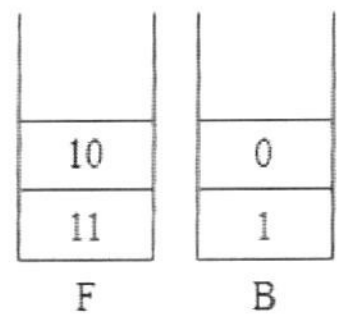

Figure 2. Example of prefix-code stack and suffix-code stack

The formation of the new query code is done with a prefix-code stack and a suffix- code stack. The suffix code stack works with the prefix code stack. Since each prefix code corresponds to two suffix codes, the same two suffix codes are used for every two prefix codes. Each time a new query code is generated by combing the prefix code popped from the prefix-code stack and the suffix code popped from the suffix stack.

The two consecutive popped-up suffix codes and the same prefix code are formed two new query codes. The two suffix codes are pushed into the stack again. The next prefix code popped up from the prefix-code stack and the two consecutive popped-up suffix codes is formed two new query codes. The two suffix codes do not need to be pushed into the stack. When the prefix code stack is empty, it means that all tags have been successfully identified. Figure 3 below is a flow chart of the new query code formed by the dual stack.

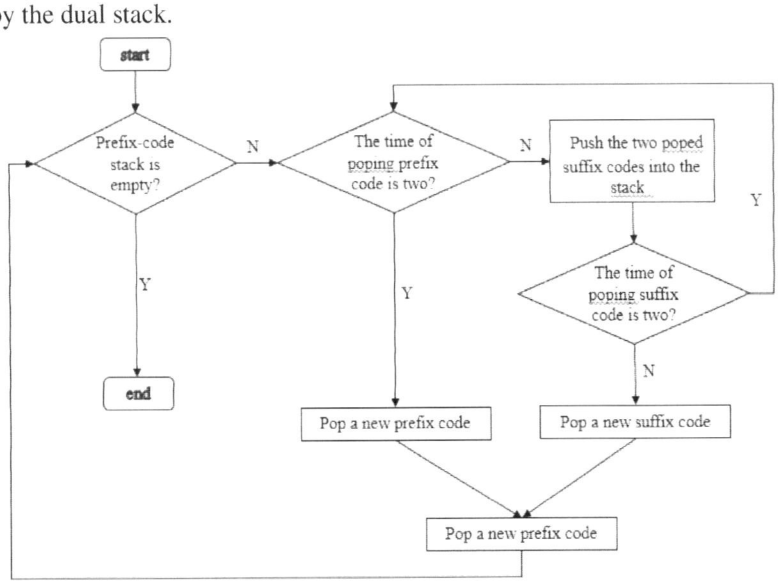

Figure 3. Flow chart of forming a new query code by dual stack

The query process of the new query code can include three operations: idle operation, collision operation, and identification tag operation. The idle operation and the identification tag operation are relatively simple. When the collision operation occurs, four new query codes are formed based on the highest and lowest bits of the collision bit. The four query codes are pushed into the stack. Then, continue to pop up from the stacks and form a new query code. Proceed to the next round of inquiry until the stack is empty and the query process is over. At this time, At this point, all the tags are identified. In the whole process, all coding operations are done by the reader, thus the reader should have the ability to store and encode. Fortunately, all current readers have these capabilities.

4. Recognizing tags process of DDQT algorithm

In the process of tag identification, the electronic tag may be in three states [4], which are activation state, waiting state, and sleep state.

Activation state: When the reader sends the query code to the tag in the response range, all tags enter the activation state. In addition, when the tag code is consistent with the request code sent by the reader, the tag will also be activated.

Waiting state: When the tag code is different from the query code currently issued by the reader, the tag will temporarily enter the waiting state and wait for the next activation.

Sleep state: When a tag has been identified, it will go to sleep, exit the communication connection, and no longer respond to the reader until the entire recognition process is completed. The specific identification process flow chart is shown in Figure 4 below:

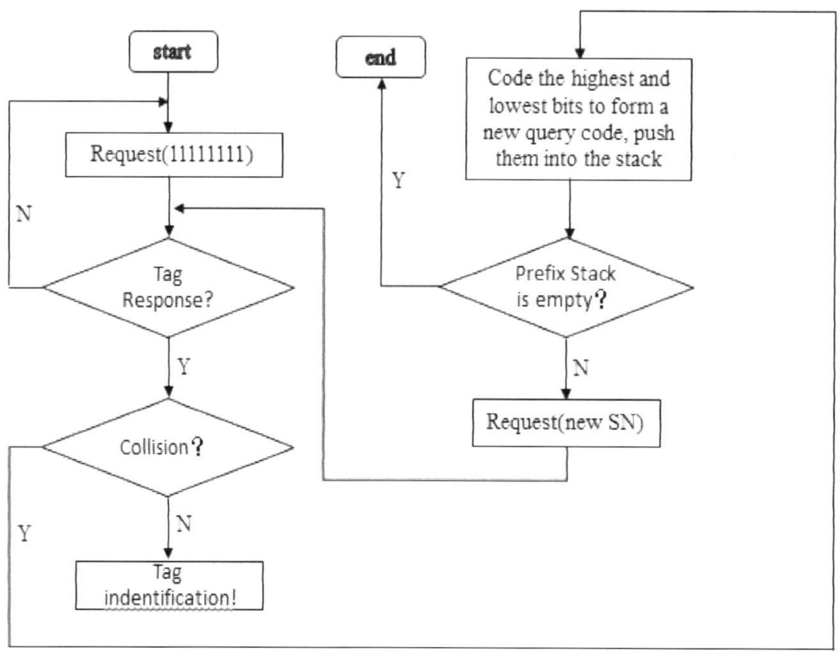

Figure 4. Flow chart of tag identification process

The following is a specific example of the algorithm execution process. Assume that there are 7 tags A, B, C, D, E, F, and G to be identified in the reader response range, and their IDs are 10100011, 11001100, 10101001, 10101100, 100011101, 11001001, 11101001. Here defines the left side of the ID code is high. The identification process is as following Table 1.

Table 1. Identification process of RFID tags under DDQT algorithm

Times	Query code sent	Query result	Collision code	Prefix code stack	Suffix code stack	New query codes
1	11111111	collision	1XXXXXXX	10 11	0 1	10XXXXX0 10XXXXX1 11XXXXX0 11XXXXX1
2	10XXXXX0	Tag D				
3	10XXXXX1	collision	10X0XXX1	1000 1010	01 11	1000XX01 1000XX11 1010XX01 1010XX11
4	1000XX01	Tag E				
5	1000XX11	NULL				
6	1010XX01	Tag C				
7	1010XX11	Tag A				
8	11XXXXX0	Tag B				
9	11XXXXX1	collision	11X01001	11001001 11101001		11001001 11101001
10	11001001	Tag F				
11	11101001	Tag G				

According to the case, the execution process of the DDQT algorithm is described in detail.

(1) The query code 11111111 is sent for the first time, and the collision code 1XXXXXXX is obtained. The highest and lowest bits are selected according to the collision code, and then the codes of the highest and lowest bits are 0.....0, 0...1, 1.......0, 1..1, the query codes are sequentially formed according to the quadtree dual stack code;

(2) Forming a new query code 10XXXXX0. According to the new query code 10XXXXX0, the tag D can be identified, at which time the RFID and the tag D perform normal communication. The communication ends, and the tag D enters a sleep state;

(3) Forming a new query code 10XXXXX1. According to the new query code 10XXXXX1, a collision occurs and generates a collision code 10X0XXX1. It will continue to select the highest and lowest bits according to the collision code, and then the code of the highest and lowest bits is 0.....0, 0...1, 1.......0,1.......1, the query code is sequentially formed according to the quadtree dual stack code;

(4) Forming a new query code 1000XX01, according to the new query code 1000XX01, the tag E can be identified. At this time, the RFID and the tag E perform normal communication. The communication ends, and the tag E enters a sleep state;

(5) Forming a new query code 1000XX11, according to the new query code 1000XX11, no response, NULL;

(6) Forming a new query code 1010XX01, according to the new query code 1010XX01, the tag C can be identified. At this time, the RFID and the tag C perform normal communication. The communication ends, and the tag C enters a sleep state;

(7) Forming a new query code 1010XX11, according to the new query code 1010XX11, the tag A can be identified, at this time the RFID and the tag A perform normal communication. The communication ends, and the tag A enters a sleep state;

(8) Forming a new query code 11XXXXX0, according to the new query code 11XXXXX0, the tag B can be identified, at this time the RFID and the tag B perform normal communication. The communication ends, and the tag B enters a sleep state;

(9) Forming a new query code 11XXXXX1, according to the new query code 11XXXXX1, a collision occurs, and a new collision code 11X01001 is generated. Since there is only one collision bit, a new search code is formed: 11001001, 11101001, and the prefix stack code is pushed up from the stack. The suffix code stack is empty;

(10) A new query code 11001001 is formed. According to the new query code 11001001, the F tag can be identified. At this time, the RFID and the tag F perform normal communication. The communication ends, and the tag F enters a sleep state;

(11) A new query code 11101001 is formed. According to the new query code 11101001, the tag G can be identified. At this time, the RFID and the tag G perform normal communication, the communication ends, and the tag G enters a sleep state; the entire communication process ends, and all tags are identified.

(12) The prefix code stack is empty and the communication ends.

The query code generated during the whole communication process is as following figure 5. Figure 6 is a change diagram of the dual stack in the producing the query code process and Figure 7 is an explanatory diagram of the symbols of Figure 6.

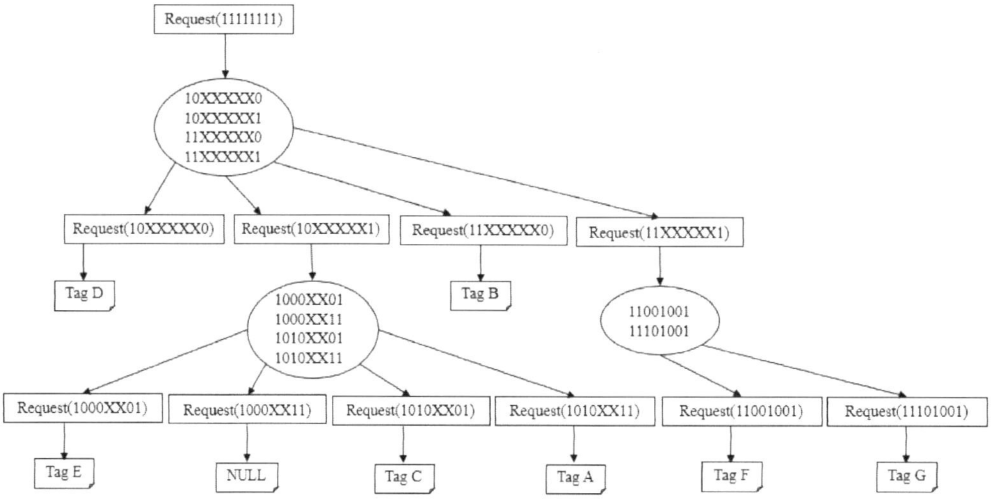

Figure 5. Query code tree based on DDQT algorithm

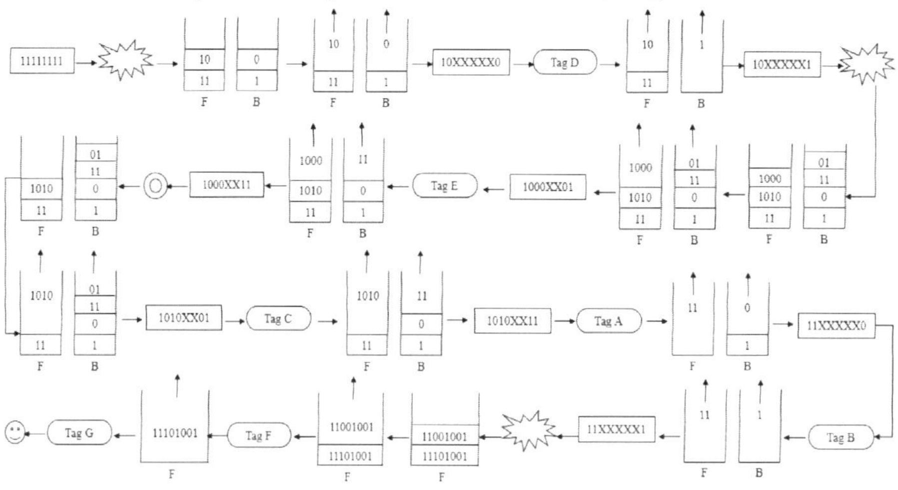

Figure 6. Change chart of dual stacks based on DDQT algorithm

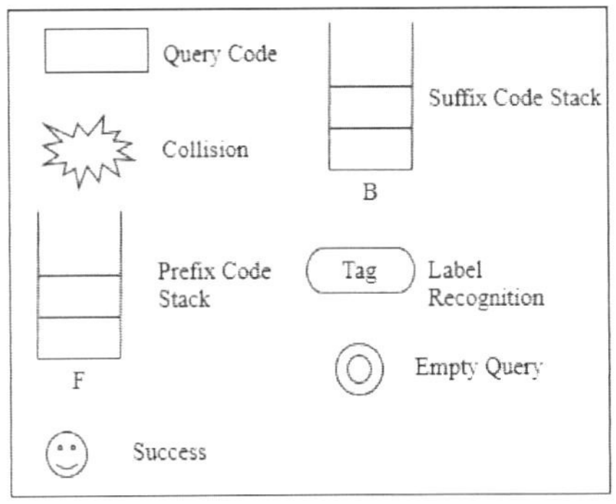

Figure 7. An explanatory diagram of the symbols of Figure 6

5. Performance analysis and simulation

In the RFID tag anti-collision algorithm, factors affecting performance include the number of query, the number of collisions generated during the query process, the number of the null query generated during the query process, and the size of the memory occupied by the generating query code.

In this paper, seven electronic tags are queried in the reading range of the reader, and four different tag anti-collision algorithms are compared. The four anti-collision algorithms include binary query tree, quadtree, bybrid query tree, and dual-stack query tree. The influencing factors of the four algorithms are shown in Table 2 below.

Table 2. Comparison of performance parameters of different anti-collision algorithms

Query tree type	Query Number	Collision Number	Empty Query Number	Memory Size (bit)
Binary Tree	18	7	3	68
Quad Tree	11	3	1	52
Hybrid Query Tree	11	3	0	48
Dual Stack Query Tree	11	3	1	34

In the RFID query tree tag anti-collision algorithm, the memory space is mainly affected by the number of search codes, and the number of search codes is affected by the number of collisions. When the number of collisions is more, the number of search codes generated is larger. According to the analysis, the number of collisions of the binary query tree is twice of the quadtree query tree. However, the binary query tree generates two query codes in the process of one collision, and the quad-tree query algotithm generates four query codes in the process of one collision. Therefore, there is no significant improvement in the occupation size of the memory space. The DDQT algorithm combines different pop-up codes into different query codes. In the process of collision, the two parts of the query code are stored in the prefix-code stack and the suffix-code stacks in a separate manner to achieve different prefix and suffix codes to store in the same memory space. In the DDQT algorithm, the generation of the query code is dynamic. The process of the traditional query tree code generation is a static process, and the query codes always occupy memory space after generation. By comparison, the DDQT algorithm greatly improves the utilization of memory space compared to the other three algorithms.

6. Conclusion

The DDQT algorithm greatly saves memory space by adding a prefix-code stack and a suffix-code stack. And the algorithm is simple and easy to implement. There is no limit to the length of the electronic tag which needs to query. Compared with the binary query tree algorithm, quadtree query tree algorithm and hybrid query tree algorithm, the querying number of DDQT algorithm is the same as that of the quadtree query tree algorithm and the hybrid query tree algorithm, which is nearly half of that of the binary query tree algorithm. The number of collisions is the same as that of the quadtree query tree algorithm and the hybrid query tree algorithm, which is significantly less than that of the binary query tree algorithm. On the null query, it is better than the binary query tree algorithm, which is not much different from the quadtree query tree algorithm and the hybrid query tree algorithm. In occupation of the memory space, it is obviously smaller than the other three algorithms. The binary tree anti-collision algorithm has the largest memory space occupancy rate, which is twice of the DDQT algorithm. The memory-occupancy rate of DDQT algorithm is significantly reduced by half compared to the binary tree query algorithm and significantly improved compared to the quadtree query tree algorithm and the hybrid query tree algorithm.

Acknowledgments

This paper is supported by the Excellent Talents Training Funded Project of Universities of Anhui Province (No. gxgwfx2019065) and Provincial Humanities and Social Science Research Project of Anhui Colleges (No.SK2018A0605).

References

[1] M.K. Yeh, Y.L. Lai, J.R.Jiang. An Efficient Query Tree Protocol for RFID Tag Anti-Collision[C]. IEEE 2015: 346-352.
[2] Kuo-Hui Yeh, N.W.Lo. An efficient tree-based tag identification protocol for RFID systems[C]. 22nd International Conference on Advanced Information Networking and Applications, WAINA, 2008,220:966-970.
[3] Yun Tian,Gongliang Chen, jianhua Li. Bi-slotted binary tree algorithm with stack for radio frequency identifi cation tag anti-collision[J]. Journal of Shanghai Jiaotong University (Science), 2013,18(2):173-179.
[4] Wu J. Henxin J and etc. An improved Query Tree Anti-collision Algorithm in RFID System [J]. Computer technology and development. 2015, 25(2): 86-89.
[5] Zhiguo X. Yigang H. Zhouguo H. Binary-tree bit-detecting RFID tag anti-collision algorithm[J]. Computer Engineering and Applications, 2010,46(20):245-248.
[6] T.Wang. Enhanced binary search with cut-through operation for anti-collision in RFID systems. IEEE Commun.Lett, Vol.10,No.4,April 2006: 236-238.
[7] J.Ryu,H.Lee,Y.Seok etc. A Hybrid Query Tree Protocol for Tag Collision Arbitration in RFID system. IEEE ICC,Jun 2007: 981-986.
[8] Y. Kim, S.Kim, S.Lee and K.Ahn. Improved 4-ary Query Tree Algorithm for Anti-Collision in RFID System[C]. IEEE Computer Science. 2009:699-704.
[9] T.F. Shin, C.H. Teng. The Design of a Stride Query Tree Algorithm for RFID Systems[C]. IEEE computer Science. 2011:108-113.

Journal of Physics: Conference Series **1423** (2019) 012002 doi:10.1088/1742-6596/1423/1/012002

Evaluation on least square method applied to gamma spectrum de-noising

Fei Li[a, b], Xinle Lang[b], Yi Chen[b], Liangquan Ge[a, b,*], Li Feng[b], Siwei Li [c]

[a] Applied Nuclear Technology in Geosciences Key Laboratory of Sichuan Province, China
[b] College of Nuclear Technology and Automation Engineering, Chengdu University of Technology, China
[c] Chengdu Shude foreign language school

* Corresponding author and e-mail: Liangquan Ge, glq@cdut.edu.cn

Abstract. Gamma energy spectrum is a way to confirm the type and content of radioactive nuclides in substances. In order to obtain effective data of gamma spectrum, it is necessary to analyze the data of gamma spectrum obtained by the detector. Nevertheless, due to the influence of statistical fluctuation and gamma generated by the environment itself during the measurement of gamma spectrum, the obtained gamma spectrum is not accurate enough as a result of needing spectral analysis of gamma spectrum, and noise reduction can make the spectral data smoother. Therefore, this paper studies the noise reduction algorithm for gamma spectrum analysis, analyzes the design and application of different algorithms in the process of noise reduction, and summarizes the advantages and disadvantages of this algorithm and others.

1. Introduction

Gamma spectrometry is regarded as an intuitive instrumental analytical technique for its fast, reliable and non-destructive qualitative and quantitative analysis of radionuclides. The analysis systems generally consist of scintillators, photomultiplier tubes, electronic circuits and back-end displays. The principle is that gamma rays released during the process of gamma decay of nuclides interact with the scintillator in the instrument in terms of photoelectric effect, Compton effect, electron pair effect and so on, which are finally output, collected, recorded and analyzed in the form of voltage pulses. Continuous development of detector technology has improved accuracy and precision, resulting in rapid progress in gamma spectrum analysis technology. Quantitative analysis of gamma spectrum data lead to errors due to the inherent statistical fluctuations, electronic system noise, measurement time and other factors. In order to obtain accurate radioactivity of various nuclides from gamma spectrum and keep the data feature of spectrum data as far as possible, it is necessary to reduce the influence of various noise[1][2]. The denoising algorithm is to eliminate the interference information in the original spectrum and form a reasonable smooth curve by means of statistical mathematics. At present, there are many smoothing methods for gamma spectrum, including sliding average method [3], barycenter method [4], wavelet transform method [5-6], Gauss method and least square method [7].

The least squares method is one of the most widely used curve fitting methods, which use linear regression to transform a non-linear problem into a linear problem and establish a fitting function, taking the optimal solution of the function as the root of the equation system. Barton et al selected Savitzky-

Golay smoothing method for data processing, which adopted the least square method for fitting and noise reduction at the window to improve the accuracy of the results to be measured [8]. Experimental results show that this algorithm has superior noise reduction capability and the SNR is improved by 100%. Zaiyu Duan et al. obtained gamma spectrum data by using NaI(Tl) scintillation detector, programmed on MATLAB to achieve smoothing processing, and improved the least square method by using multi-point smoothing. The method basically suppresses the statistical fluctuations, and the weak peaks are also evident, which shows that the least square method and the barycenter method have a good smoothing effect on energy spectrum processing in MATLAB [4]. Guobing Yu introduced the principle and algorithm of weighted least square method (WLSQ) in the study of NaI (Tl) scintillation detector, and programmed it on MATLAB. Experimental results showed that the processing error of weighted least square method was lower than 10% compared with HPGe spectrometer [9]. Liang Chen proposed a nuclide recognition algorithm for portable gamma spectrometer. In the process of smoothing energy spectrum data, the time-domain least square method was used for noise reduction, which could prepare for peak searching and spectrum resolution in the following[10]. Ming Hu proposed the design principle of the least square vector machine to reduce the noise of airborne gamma spectrum. Based on the principle of statistics, the method is applied to segment regression fitting noise reduction of airborne gamma spectrum based on the distribution of energy channel addresses. It is concluded that the segmented LS-SVM noise reduction method can significantly reduce the statistical fluctuation of airborne gamma ray at the segmented region, and has good adaptability and generalization [7].

Nevertheless, the definition of optimal solution often has different interpretations according to the experimental object and the measured data. In this paper, noise reduction output of second, third and fourth order polynomials under different conditions is analyzed based on the least square method for gamma spectrum data. Data processing of various cases is compared and summarized, and the fitting effect of least squares polynomials is evaluated accurately. By using MATLAB platform for noise reduction fitting, different energy windows of gamma spectrum were respectively processed to evaluate the noise reduction effect of least square method in different energy windows, which provided important reference basis and practical analysis for the research on noise reduction technology of gamma spectrum.

2. Theory

2.1. Algorithms and principles of least square method

When the sum of squares of all parameter points and fitting points (assuming that the error of plane figure is the distance between data points and fitting points) is the minimum, that is, the sum of squared errors is the smallest. Then the energy spectrum curve fitted by the least square method can be obtained. The mathematical principle is as follows:

Assuming a given set of known data: { ($x_{i,}y_i$) (i=1,2,3......m) }, if the mathematical model of fitting curve is y=f(x), then the distance from the ith error is f(x_i)- y_i,so the sum of the squares of all point is $\sum_1^m[f(x_i)-y_i]^2$. The minimum value of $\sum_1^m[f(x_i)-y_i]^2$ and the corresponding parameters are obtained by calculating the zero point of derivative of $\sum_1^m[f(x_i)-y_i]^2$ as a function. Finally, the fitting function curve y = f(x) is obtained.

If there are known data points {($x_{i,}y_i$) (i=1, 2, 3......m)},φ denotes a class of functions consisting of polynomials of no more than m, thus the solution of $f_k(x_i)=\sum_{k=0}^n a_k x^k \notin \varphi$ needs to satisfy:

$$\text{i}=\sum_1^m[f(x_i)-y_i]^2=\sum_{i=0}^m\left(\sum_{k=0}^n a_k\,x^k-y_i\right)^2=min \tag{1}$$

Formula (1) is a polynomial function, so its condition of minimum value existing and satisfying is that its corresponding partial derivative value is zero. Therefore, there are two equations:

$$\frac{\partial \text{I}}{\partial a_j}=2\sum_{i=0}^m\left(\sum_{k=0}^n a_k x_i^k-y_i\right)x_i^j=0, j=0,1\dots n \tag{2}$$

$$\sum_{k=0}^n\left(\sum_{i=0}^m x^{j+k}\right)a_k=\sum_{i=0}^m x_i^j y_i\,, j=0,1,2\dots n \tag{3}$$

The fitting function can be calculated as follows:

$$f_k(x_i)=\sum_{k=0}^n a_k x_i^k,\ \begin{cases}k=1\\k>1\end{cases} \tag{4}$$

a_k (k=0, 1, 2...n) is the only solution to the equation.

2.2. Experimental data

In order to make the energy spectrum data more reliable, the Monte Carlo software Geant4 open source particle transport development platform is used to simulate the 40K energy peak data in the gamma spectrum. The experimental models are all taken from actual physics laboratory benches and equipment to ensure that the simulation data are highly consistent with the experimental results.

3. Experimental results and discussion

In this paper, aiming at the gamma spectrum data, the noise reduction effect of the least square method based on the second-order, third-order and fourth-order polynomials in different situations was compared and concluded, so as to accurately evaluate the optimal fitting effect of the least square polynomial. Gamma spectrum data was substituted into MATLAB software to obtain the fitting function parameters of various polynomials, and then the noise reduction parameters signal-to-noise ratio (SNR), RMS error (RMSE) and Pearson correlation coefficient (PCC) were calculated.

3.1. Second-order fitting function

The denoising effect parameters are obtained by denoising the address spectrum data selected by multiple gamma spectra (Table 1). It can be clearly seen that when the fitting function is f(x)=ax2+b and f(x)=ax2, RMSE is greater than 1.3, SNR is less than 3 and PCC is less than 0, so the fitting effect is very poor and it is not suitable to use this method for noise reduction. When the fitting function is f(x) =ax2+bx+c and f(x) =ax2+bx, RMSE is less than 0.33, SNR is greater than 14, PCC is close to 0.9, and the fitting works well. Compared with the first two groups of methods, when the fitting function is f(x) =ax2+bx+c, RMSE is relatively small and PPC is relatively large. Although the signal-noise is relatively small, due to the small sample data sets adopted in this experiment, accidental error occurs. Conclusion: according to the following table, when the least square fitting of order 2 is performed, the noise reduction effect is most significant when the fitting function is f(x) =ax2+bx+c. (Figure 1)

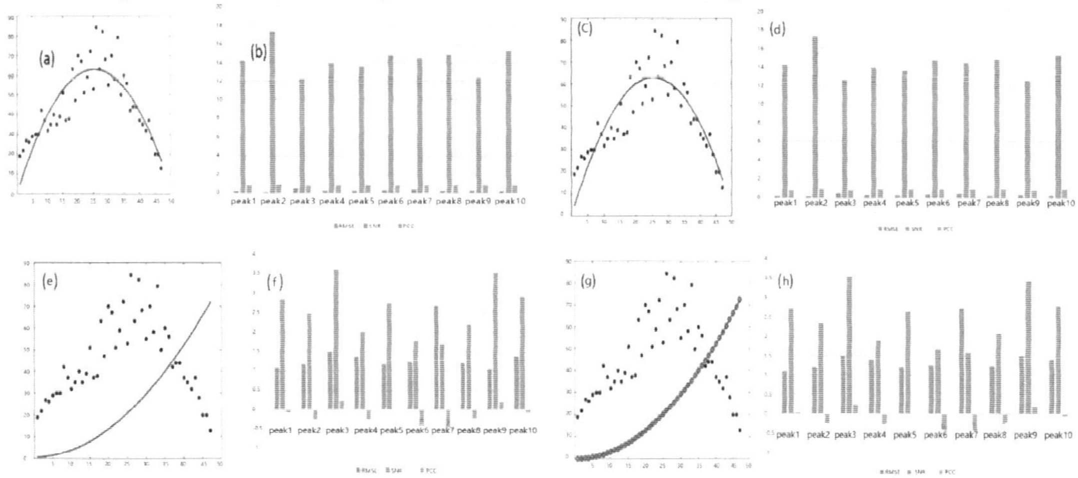

Figure 1. Comparison diagram of the fitting curve of the second-order fitting function and the noise reduction effect

Table 1. The noise reduction effect when the fitting function is the second order

Serial number	Function	RMSE (average value)	SNR/db (average value)	PCC (average value)
1	$f(x) = ax^2 + bx + c$	0.3275	14.3162	0.8938
2	$f(x) = ax^2 + bx$	0.3277	14.3503	0.8814
3	$f(x) = ax^2 + b$	1.3853	2.5781	-0.1226
4	$f(x) = ax^2$	1.4488	2.4688	-0.1128

3.2. Third-order fitting function

The denoising effect parameters are obtained by denoising the address spectrum data selected by multiple gamma spectra (Table 2). It can be clearly seen that when the fitting function is f(x)=ax3+b and f(x)=ax3, RMSE is greater than 1.3, SNR is less than 2 and PCC is less than 0, so the fitting effect is very poor and it is not suitable to use this method for noise reduction. When the fitting function is f (x) = ax3 + bx2 + c, f (x) = ax3+ bx2, f (x) = ax3 + bx + c, and f (x) = ax2 + bx, the fitting effect is better. When the fitting function is f(x)=ax3+bx2+cx+d and f(x)=ax3+bx2+cx RMSE is less than 0.33, SNR is close to 15, PCC is greater than 0.9, and the fitting works well. Compared the two groups of methods with better results, when the fitting function is f(x)=ax3+bx2+cx+d, the RMSE value is relatively small and the SNR is relatively high. Although the PPC is small, due to the small sample data set adopted in this experiment, accidental errors may occur. According to the above table, when the least square fitting of order 3 is performed, the noise reduction effect is most significant when the fitting function is f(x)=ax3+bx2+cx+d. (Figure 2)

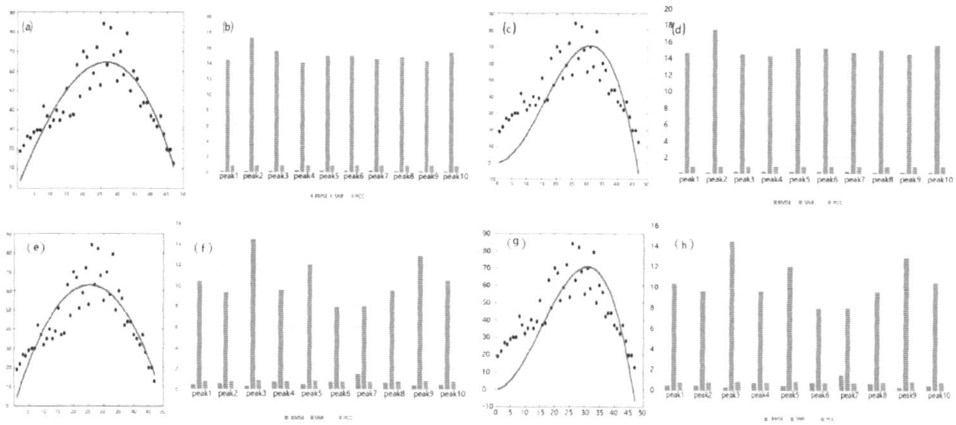

Figure 2. Fitting function $f(x)=ax^3+bx^2+cx+d$(a)(b), $f(x)=ax^3+bx^2+cx$(c)(d), $f(x) = ax^3 +bx^2+c$(e)(f) and $f(x)=ax^3+bx^2$(g)(h) fitting curve and contrast diagram of noise reduction effect.

Table 2. The noise reduction effect when the fitting function is the third order

Serial number	Function	RMSE (average value)	SNR/db (average value)	PCC (average value)
1	$f(x) = ax^3 + bx^2 + cx + d$	0.2841	15.0918	0.9042
2	$f(x) = ax^3 + bx^2 + cx$	0.2925	14.9619	0.9050
3	$f(x) = ax^3 + bx^2 + c$	0.6396	10.4212	0.8377
4	$f(x) = ax^3 + bx^2$	0.6438	10.40491	0.8370
5	$f(x) = ax^3 + bx + c$	0.389	14.1432	0.8919
6	$f(x) = ax^3 + bx$	0.3896	14.1348	0.8919
7	$f(x) = ax^3 + b$	1.3261	1.6167	-0.2622
8	$f(x) = ax^3$	1.3509	1.4651	-0.2622

3.3. Fourth-order fitting function

The denoising effect parameters are obtained by denoising the address spectrum data selected by multiple gamma spectra (Table 3). It can be clearly seen that when the fitting function is $f(x) = ax^4 + bx^3 + c$, $f(x) = ax^4 + bx^3$, $f(x) = ax^4 + bx^2 + c$, $f(x) = ax^4 + bx^2$, $f(x) = ax^4 + b$ and $f(x) = ax^4$, RMSE is greater than 0.7, SNR is less than 10 and PCC is less than 0.8, so the fitting effect is very poor and it is not suitable to use this method for noise reduction. When the fitting function is $f(x) = ax^4 + bx^3 + cx^2 + d$, $f(x) = ax^4 + bx^3 + cx^2$, $f(x) = ax^4 + bx + c$ and $f(x) = ax^4 + bx$, RMSE is small, SNR is large, PPC is great and fitting effect is general. When the fitting function is $f(x) = ax^4 + bx^3 + cx^2 + dx + e$, $f(x) = ax^4 + bx^3 + cx^2 + dx$, $f(x) = ax^4 + bx^3 + cx + d$, $f(x) = ax^4 + bx^3 + cx$, $f(x) = ax^4 + bx^2 + cx + d$, and $f(x) = ax^4 + bx^2 + cx$, RMSE is less than 0.3, SNR is greater than 15, PCC is greater than 0.9. Among the six methods with good noise reduction effect, when the fitting function is $f(x) = ax^4 + bx^3 + cx^2 + dx + e$,RMSE is the minimum, SNR is the maximum, PCC is the maximum, and the noise reduction effect is the best. (Figure 3)

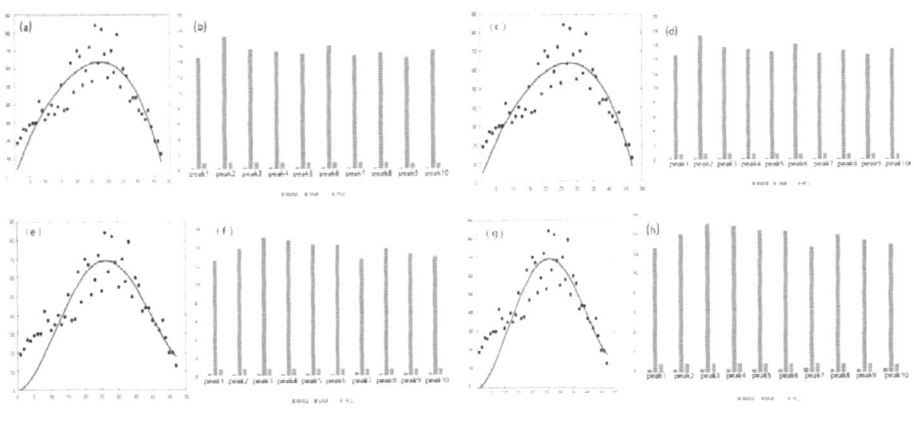

Figure 3. Fitting function $f(x) = ax^4 + bx^3 + cx^2 + dx + e$(a) (b), $f(x) = ax^4 + bx^3 + cx^2 + dx$(c) (d), $f(x) = ax^4 + bx^3 + cx^2 + d$(e) (f) and $f(x) = ax^4 + bx^3 + cx^2$(g)(h) fitting curve and contrast diagram of noise reduction effect.

Table 3. The noise reduction effect when the fitting function is the fourth order

Serial number	Function	RMSE (average value)	SNR/db (average value)	PCC (average value)
1	$f(x) = ax^4 + bx^3 + cx^2 + dx + e$	0.2511	15.4255	0.9119
2	$f(x) = ax^4 + bx^3 + cx^2 + dx$	0.2513	15.4220	0.9118
3	$f(x) = ax^4 + bx^3 + cx^2 + d$	0.3604	13.8960	0.9087
4	$f(x) = ax^4 + bx^3 + cx^2$	0.3622	13.8709	0.9085
5	$f(x) = ax^4 + bx^3 + cx + d$	0.2606	15.2490	0.9092
6	$f(x) = ax^4 + bx^3 + cx$	0.2608	15.2490	0.9092
7	$f(x) = ax^4 + bx^3 + c$	0.7947	6.5093	0.6700
8	$f(x) = ax^4 + bx^3$	0.8944	6.4163	0.6650
9	$f(x) = ax^4 + bx^2 + cx + d$	0.2746	15.1109	0.9042
10	$f(x) = ax^4 + bx^2 + cx$	0.2748	15.1084	0.9041
11	$f(x) = ax^4 + bx^2 + c$	0.7411	9.0944	0.7771
12	$f(x) = ax^4 + bx^2$	0.7470	8.7698	0.8736
13	$f(x) = ax^4 + bx + c$	0.4641	13.1441	0.8736
14	$f(x) = ax^4 + bx$	0.4641	13.1474	0.8734
15	$f(x) = ax^4 + b$	1.2515	1.1422	-0.2309
16	$f(x) = ax^4$	1.2498	0.9571	-0.3503

According to the results of the above program fitting noise reduction, we can know that the following functions have significant noise reduction effects, including the second order $f(x) = ax^2 + bx + c$, the third order $f(x) = ax^3 + bx^2 + cx + d$, and the fourth order $f(x) = ax^4 + bx^3 + cx^2 + dx + e$. When the fitting function is equal to $f(x) = ax^4 + bx^3 + cx^2 + dx + e$, the RMSE value is the minimum, the SNR is the maximum, PCC is the maximum, and the noise reduction effect is the best. Similarly, the analysis results of 233Th window and 238U window are consistent with the results of 40K window.

4. Conclusion

In this paper, the least square method is used to reduce the noise of gamma spectrum data, and MATLAB software is utilized to complete the development and design of the program. Moreover, according to the RMSE, PCC and SNR, the noise reduction effect of each fitting function is evaluated and analyzed, and finally the function with the most significant noise reduction effect is obtained by comparison. When the fitting function is the fourth order, the evaluation effect is obviously better than that of the second and third order. The results show that the least squares method has remarkable noise reduction effect in processing gamma spectrum, especially for high frequency channel data.

Acknowledgement

This work was funded by the National Natural Science Foundation of China (No.41774147), the National Key R&D Project (No.2017YFC0602100) and Science and technology activity project for overseas personnel in Sichuan province.

References

[1] Knežević, D., Jovančević, N., Krmar, M.1;Petrović, J. Modeling of neutron spectrum in the gamma spectroscopy measurements with Ge-detectors. NUCLEAR INSTRUMENTS & METHODS IN PHYSICS RESEARCH SECTION A-ACCELERATORS SPECTROMETERS DETECTORS AND ASSOCIATED EQUIPMENT. 2016,833:23-26.

[2] Fei Li, Zhixing Gu, Liangquan Ge, Hui Li, Xinyu Tang, Xinle Lang, Bo Hu. Review of recent gamma spectrum unfolding algorithms and their application. Results in Physics. 2019,13: 102211.

[3] Junsong Ren. The Study of γ Spectrum Analysis and Radioactive Nuclide Identification Method. Southwest university of science and technology.2017.

[4] Zaiyu Duan, Jianhua Chen, Guixin Zhang, Junjun Gong. Disposal of Smooth γ Spectrum on Matlab Process. Nuclear Power Engineering .2007,28(3):125-128.

[5] Lunhui LI, Jian-feng He, Qin Wang, Yuan Wang. Study of γ Energy Spectrum Denoising Based on Improved Wavelet Threshold Method. Atomic Energy Science and Technology.2016, 50(7):1279-1283.

[6] Dongsheng Shi, Yuming Di, Chunlin Zhou. Comparative study on γ energy spectrum denoise by fourier and wavelet transforms. Nuclear Electronics & Detection Technology.2006,6(26):134-137.

[7] Ming Hu. Based on Least Square Support Vector Machine Airborne Gamma Ray Energy Spectrum Segmentation Denoising Method Research. East China university of technology.2016

[8] Boris Ryabko.2013.A confidence-set approach to signal denoising [J].Statistical Methodology, 15(2013):115-120.

[9] Guobing Yu, Shan Jiang, Chuankai Chen. Study on Efficiency of Detection in Different Geometry Sample Cup in γ-ray Spectrometer Analyze. Journal of Anhui Agricultural Sciences.2009,37(24):11641-11642.

[10] Liang Chen. Research on the Nuclide Identification Algorithm and Digital Spectra Acquisition System. Tsinghua University.2009.

Journal of Physics: Conference Series 1423 (2019) 012003 doi:10.1088/1742-6596/1423/1/012003

The satisfaction of entrepreneur instructors and higher vocational students majoring in economic and business administration with the cooperative instruction in the work post practice

Xifeng Liao[1]

School of Economic and Business Administration, Hunan College of Information, Changsha Hunan, 410200, China

Corresponding author and e-mail: Xifeng Liao, martha463@126.com

Abstract. From the perspective of both instructors from entrepreneur and students, this research studied the satisfaction level with the cooperative instruction in the Work Post Practice, a very important course for senior students at higher vocational colleges in China. The method of descriptive analysis was adopted. The statistic results illustrated that the practice posts of students didn't fit well with students' majors, and students responded that it did improve their professional abilities and qualities soundly. Business instructors demonstrated a higher satisfaction level on students' work and qualities shown during the practice than students' self evaluations. Suggestions like using the internet platfom, establishing a stable team of entrepreneur instructors, allocating salary and other subsidies to entrepreneur instructors were proposed to improve instruction quality of the work post practice.

1. Introduction

The work post practice is a very important course for Chinese tertiary vocational and technical education, which aims to enhance the senior students' professional skills and career abilities in the real work environment. In order to complete this course, students are required to work fulltime in the entrpreneur and receive internship salary and other subsidies as the contract stated. The course is mainly instructed by the entreprenear worker and a college instructor work as the assisted instructor. At present, it is the main form of coopertion between Chinese vocational and technical colleges and entrpreneurs, and usually last for 6 months. It is a very important way to cultivate technical talents for the fast developing economy and technology. Therefore, both administors and instructors at vocational colleges in China put close attention on this course.

When examined the literature on vocational students' work post practice, the researcher found 10 papers in Chinese academic databases and only 4 were published in the Chinese Science Citation Database. Most of these papers adopted quantitative research methods, and the samples were narrowed in a certain major or a small area (Bai, 2017; Chi & Zhang, 2018; He & Xu, 2017; Tang & Zhang,

[1] This research was supported under the 2019 Vocational Teaching and Research Reformation Program Fund of Hunan Province of China: The Implementation of Big Data Supported Work Post Practice of Vocational College Students in Enterprises (Project Reference: ZJGB2019026).

2013; Wu, 2015; Yang, 2018). The popular methods used were descriptive analysis, logistic regression, factor analysis and T test. The research contents centered on the entrepreneur working environment, students' subsidies, and student administration. The satisfaction researches were all done from students' position and didn't take the instructors satisfaction into account.

2. Research Objectives

In order to know better about the quality of corporative instruction and examine how students and entrepreneur instructors were satisfied with the corporation between the college instructor and the entrepreneur instructor. This research creatively used the descriptive quantitative analysis to study the satisfaction level of this two groups and made a comparison between them. It aimed to answer the following questions:

Q1: How are students satisfied with the cooprative instruction?
Q2: How are students satisfied with their work post practice experience?
Q3: How are entrepreneur instructors satisfied with the cooperation with the college instructors?
Q4: How are entrepreneur instructors satisfied with the performance of students?

3. Method

3.1. Samples

The target population in this research are students who studies in the Department of Economic and Business Management and instructors that tutoring students work post practice in the entrepreneur. The questionair was sent to students and entrepreneur instructors right after students completed their 6-months practice and 55 valid student data and 8 instructor data was collected. Data showed that there were more female students (69.1%) than male students, most students (83.6%) completed their practice in private enterprises, of which most of them (69.2%) are in small and micro enterprises in which has less than 200 workers, and 26.3% students in enterprises in which has more than 500 workers. The demographic features showed that the respondents represent students majoring in Economic and Business Adminstration very well. The descriptive data also showed that students find their practice position by different means. A big part of students (23.6%) found their position via job hunting web, but the biggest part (36.3%) get the position by friends', families' or relatives' recommendation. A small part (3.6%) of them found their position via on campus employment, i.e. enterprises came to the campus to hold interview for the employment.

The 8 instructor respondents are all female and 87.5% are working in private enterprises. Half of them got three-year college degree, which are equal to the associate degree in some western countries, and half got bachlors degrees. All instructors had at least one-year work experience, and 75% of them had 3-9 years experience, some (12.5%) had more than 10 years. Most of instructors (62.5%) had at least one-year experience in tutoring students, and some (25%) had 3-5 years. But no respondents reported having experience which was longer than 5 years.

3.2. Analysis

The collected data was analyzed by using SPSS 24.0, and descriptive methods was mainly adopted to compare the percentages of different level of satisfaction. Pie charts and tables are used to illustrate the differences in satisfaction level in each sector. When comparing the overall satisfaction of students and instructors, the means of variables and the standard deviation was taken into account to measure the difference.

4. Results

4.1. Students' satisfaction with their work post practice

Percentages were counted to measure the satisfaction of students' with the wok post practice. Most students reported that this course improved their professional skills and the required career abilities. Some students (29.1%) reported extremely satisfied that the practice improved their professional skills, 47.3% reported somewhat satisfied, and very small part (3.63%) reported somewhat dissatisfied and extremely dissatisified (1.81%), the percentages can be seen in Figure 1. On the aspect of improving required career abilities, 41.8% students reported extremely satisfied and a very tiny part (3.6%) reported extremely dissatisfied. Data can be seen in Figure 2. Students reported a good satisfaction level to their overall practice experience.

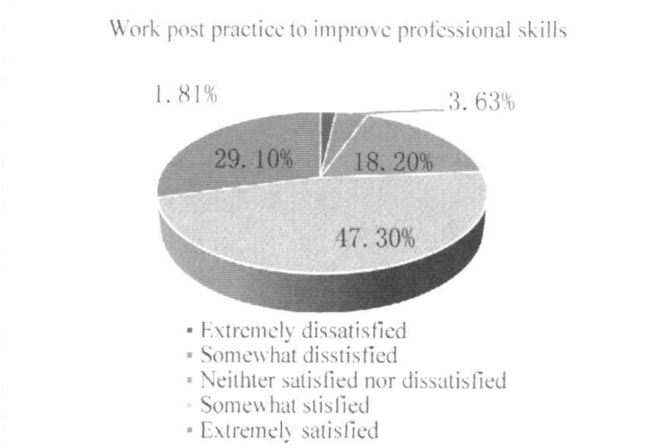

Figure 1. Students' self reports of the work post practice to improve their professional skills

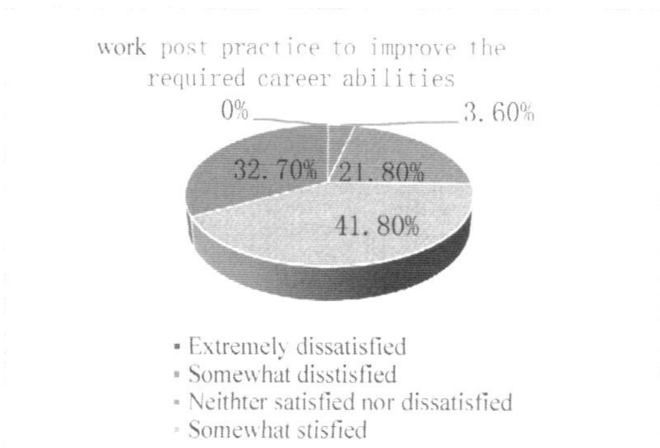

Figure 2. Students' self report of the work post practice to improve the required career abilities

4.2. Students' satisfaction with the cooperative instruction

The college instructors work is mainly on giving directions to students to submit some required paper work, such as signing the contract with the enterprise and the college, writing the practice diary, writing practice plan and report etc. The data reflected that a good satisfaction level was shown on the college instructors work. Compared to the college instructor, the enterprise instructor received lower satisfaction percentage. The detailed data can be seen in Table 1.

Table 1. Students' evaluation to the coopertive instruction

		Extremely dissatisfied (%)	Somewhat dissatisified (%)	Neither Dissatisfied nor satisified (%)	Satisified (%)	Extremely satisfied (%)
College instructor	Sing the contract and submit online	3.6	3.6	12.7	38.2	41.8
	Write practice diary	5.5	0	7.3	36.4	50.9
	Write practice plan	5.5	0	7.3	43.6	43.6
	Write practice report	1.8	1.8	10.9	32.7	52.7
Entrepreneur instructor	On daily work	1.8	5.5	23.6	45.5	23.6
	On professional skills	3.6	5.5	21.8	45.5	23.6
	On required career abilities	5.5	5.5	23.6	45.5	20.0

4.3. Entrepreneurs' satisfaction to students' performance on posts

The researcher compared the means of satisfaction with the overall practice experience of students with that of entrepreneur. The results showed that students' self evlation was lower than the evaluation from entrepreneur instructors. Entrepreneur instructors reported either somewhat satisified and extremely satisfied. See Table 2.

Table 2. Students' and enterprises' satisfaction with the overall work post practice of students

	Extremely dissatisfied	Somewhat dissatisified	Neither Dissatisfied nor satisified	Satisified	Extremely satisfied	Mean	SD
Students' satisfaction	0%	2%	29%	44%	26%	3.93	0.79
Entrepreneurs' satisfaction	0%	0%	0%	50%	50%	4.50	0.53

When asked how likely the entrepreneur instructors were to hiring vocational practice students on the post, half respondents reported they were very likely to hire students in the future and a small part of them expressed not sure about the decision. Data can be seen in Figure 3.

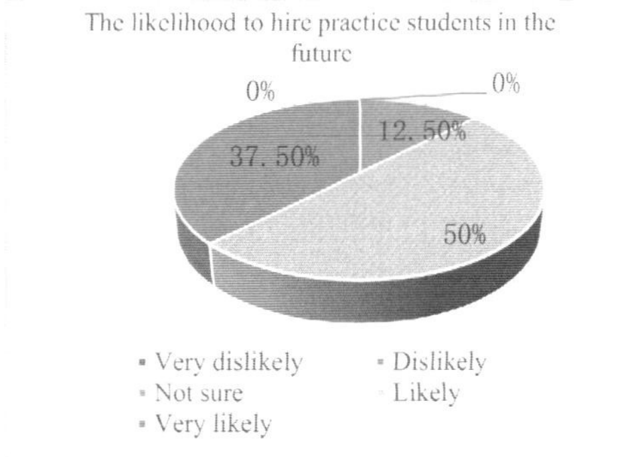

Figure 3. The likelihood to hire practice students in the future

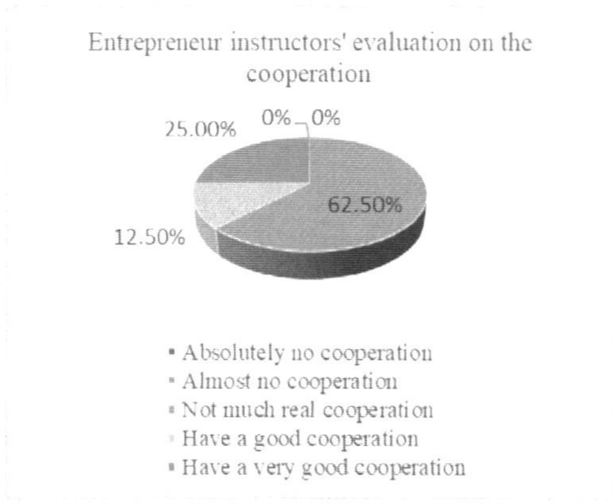

Figure 4. Entrepreneur instructors' evaluation on the cooperation with college instructors

4.4. Entrepreneurs' satisfaction with the cooperation

The results showed that more than half (62.5%) entrepreneur instructors reported there was not much real cooperation between college and entrepreneur instructors. Only 25% reported a very good cooperation. See figure 4. But a contradictory satisfaction level as Figure 5 shows, 37.5% responded that they were extremely satisified and 50% reported somewhat satisfied with the cooperation.

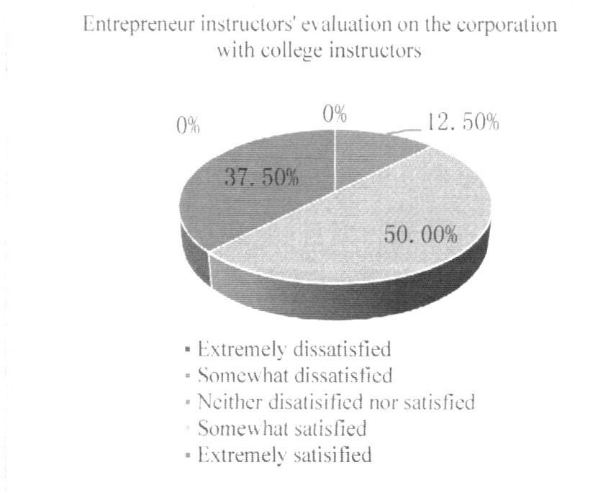

Figure 5. Entrepreneur instructors' evaluation on the cooperation with college instructors

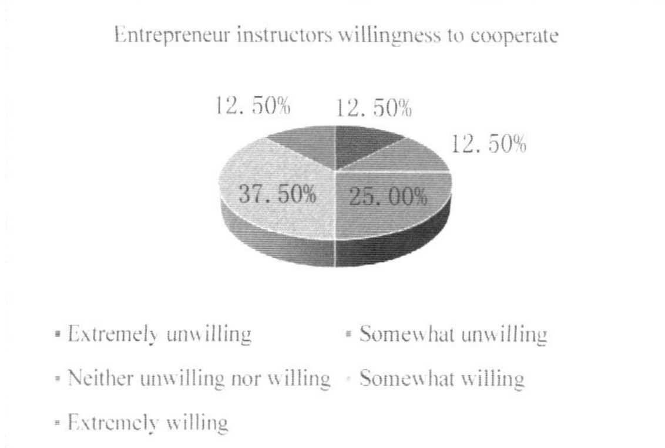

Figure 6. Entrepreneur instructors' willingness to cooperate with college instructors with the support of internet technology.

Journal of Physics: Conference Series **1423** (2019) 012003 doi:10.1088/1742-6596/1423/1/012003

When entrepreneur instructor were asked whether willing to get involved in a more closed cooperation with the support of internet technology, such as some social media like Wechat, QQ or some professional intruction platform, entrepreneur instructors presented a comparatively low willingness. Only 12.5% responedents reported extremely willing to cooperate, and 37.5% were somewhat willing to cooperate. Other data can be seen in Figure 6.

5. Discussion

5.1. Apply internet technology in finding work post position

The survey results showed that the main way students finding their practice positions is quite traditional, such as by recommendation of families and friends or by attending on campus face to face interviews. The platform named "Internet and accurate employment" comes into being due to the development of big data and is adopted widely among colleges in Hunan province of China. Its operation concept is "Internet + big data + intelligent matching" (Yang, 2018). In the era of big data, teachers and students should make full use of the convenience brought by the internet technology. Before the work post practice starts, college staff should encourage students and entrepreneurs to use this platform. By using this platform students can search hire information, and enterprises can release recruitment information. At the same time, the further development of the platform should be continuously strengthened to enhance the function of intelligent job matching. For example, the platform function of submitting resumes of students is expected a further develop, by which the platform can automatically recommend some students to the employer according to their work intentions and majors.

5.2. Enhance social responsibility of enterprises to promote cooperation

At present, most enterprise instructor are lack of enthusiasm and motivation in college-enterprise cooperation in guiding students. One of the important reasons is that they fail to realize that the cooperation is beneficial to the enterprise development, and they also fail to realize that cultivating students is the social responsibility and obligation that enterprises should fulfill. In particular, some medium, small and micro enterprises cannot realize that by participating in the cooperation they can reserve needed technology talents and promote the long-term development of enterprises and help enterprises achieve their strategic goals. Although the most basic goal of enterprises is to obtain benefits, human resources can be obtained as well in the process of cultivating students jointly by schools and enterprises. Resérchers also pointed that the cooperation can support the enterprises to obtain economic benefits and promote its own brand (Ma & Chen, 2018). A number of foreign studies have proved that enterprises' participation in school-enterprise cooperative teaching and involve in apprenticeship education have considerable economic benefits (Dionisius et al., 2018). Therefore, as a benefitiary in the cooperation, enterprises should firstley approve that the college- enterprise education is beneting students, industry and society, then they will be willing to cooperate with vocational colleges to give students the opportunity to practice , to help students better adapt to the working environment, and take cultivating technological talents as their own social responsibility (Ma & Chen, 2018). At the same time, fulfilling corporate social responsibility also needs the promotion of social morality and legal supervision. To establish corporate social responsibility standards with Chinese characteristics and to standardize the enterprise qualification specifications, and to make specific regulations for different cooperating entities, industries and enterprises at different development stage are very important. An evaulation to the quality and efficacy of ccoperation with vocational colleges carried out by a third party entitiy is also a vital process for the cooperative instruction. Because it will improve the credibility of the cooperate social responsibility report and provide evidence and reference when colleges choosing the cooperation partner and when enterprises apply for government tax reduction after involving vocational education.

5.3. Improve qualifications of enterprise instructors

The data analysis of this study also showed that only 25% of corporate instructors have a bachelor degree. Some other Chinese scholars also pointed out that there are many problems exist in current corporate instruction. On one hand, the quality of teachers is uneven, some instructors do not have professional certificates or professional technical positions, and most enterprise instructors do not have a teaching certificates (Ma & Chen, 2018). In order to ensure the quality of school-enterprise cooperative teaching for work post practice, it is very important to establish a long-term and stable enterprise instructor team. Some scholars proposed that government at all levels, colleges and enterprises should attach great importance to the role of enterprise teachers in work post practice instruction. Therefore, all vocational colleges, enterprises, business and educaitonal departments should take improving the enterprise instructors qualification as an important task, and should set up relevant policies to let the whole society care about and develop the enterprise instruction team (He & Xu, 2017).

Reference

[1] Bai, L. (2017). The Analysis on Factors Affect Higher Vocational Students' Satisfaction on The Work Post Practice Based on ISM. *Journal of Vocational Education*(22), 38-45.

[2] Chi, W., & Zhang, J. X. (2018). On Satisfaction Degree of Students toward Post Practice of Business English Specialty in Higher Vocational Education. *Vocational and Technical Education, 897*(39), 75-79.

[3] Dionisius, R., Muehlemann, S., Pfeifer, H., Walden, G., Wenzelmann, F., & Wolter, S. C. (2018). Cost and Benefit of Apprenticeship Training:A Comparison of Germany and Switzerland. *IZA DP*(3465), 1-33.

[4] He, X., & Xu, X. Q. (2017). Higher Vocational College Students' Satisfaction With the Work Post Study Based on SEM. *Journal of Vocational Education*(21), 76-80.

[5] Ma, Y. H., & Chen, D. (2018). The Research on The Motivation of Cooperative Edcuation Between Entrepreneur and Educational Institute——Based on the Economic Benefits and Corporate Social Responsibility *Higher Education Exploration*(3), 5-13.

[6] Tang, F., & Zhang, Z. H. (2013). Research on Factors That Affect Students Satisfaction Accounting *Journal of Vocational Education*(27), 82-84.

[7] Wu, R. (2015). On the Satisfaction of Students Majoring Hotel Management With Their Work Post Practice. *Journal of Vocational Education*(15), 75-78.

[8] Yang, W. T. (2018). On the Application of Big Data in Accurate Employment Guidance in Higher Vocational Edcuation. *Journal of Fujian Institute of Education*(7), 16-19.

Journal of Physics: Conference Series **1423** (2019) 012004 doi:10.1088/1742-6596/1423/1/012004

Effects of anesthetic eugenol on respiration and excretion of *Sinogastromyzon szechuanensis*

Xiaojiang Chen[1], Liling Xiong, Jiangu Wu[2], Quan Wang[3*]

[1, 2, 3] Jiangsu Agri-animal Husbandry Vocational College, Jiangsu, China

*Corresponding author and e-mail: Quan Wang, 153791449@qq.com

Abstract. In order to select a more suitable anesthetic for the breeding and transportation of *Sinogastromyzon szechuanensis*, we used different concentrations of anesthetic eugenol (0, 6, 12,18,24,30 mg•L^{-1}) to test the effects of the respiratory and metabolic on *Sinogastromyzon szechuanensis* with an average mass of (2.029±0.26) g under the water temperature (18±0.5) ℃. Based on regression analysis method, the regression equations of oxygen consumption rate, ammonia removal rate and temperature were calculated. (R^2> 0.99). Oxygen consumption rate and ammonia removal rate were calculated and compared by One-Way ANOVA (Turkey method) using SPSS 25.0 software. The multiple comparison results revealed that when the mass concentration of eugenol reached 24 mg•L^{-1}, it could significantly reduce the oxygen consumption rate and ammonia excretion rate ($p<0.05$). With the increase of eugenol concentration, the oxygen consumption rate and ammonia excretion rate of the *Sinogastromyzon szechuanensis* first increased and then decreased. Under the influence of eugenol, the oxygen consumption rate and ammonia excretion rate of *Sinogastromyzon szechuanensis* were increased with the increase of temperature ($p<0.01$). When the water temperature is 14-26℃, the oxygen-nitrogen ratio is 8.0-19.4, and the temperature is positively correlated with the oxygen-nitrogen ratio. It is concluded that eugenol with a mass concentration of 24 mg•L^{-1} can make *Sinogastromyzon szechuanensis* enter the stage of anaesthesia II (sedation), so this concentration is recommended as the anesthetic dosage for transportation, and the anesthetic effect at low temperature is more obvious.

1. Introduction

There are many types of anesthetics used in the transportation and production of aquatic animals, of which MS-222 and eugenol are most commonly used. Its mechanism is: it passes through the gill or body surface into the fish body, inhibits the cerebellar cortex, then acts on the basal ganglia and cerebellum, and finally reaches the medulla spinalis [1]. Studies have shown that MS-222, eugenol and other anesthetics can reduce the oxygen consumption and ammonia nitrogen emissions of aquatic animals, resulting in changes in oxygen consumption and ammonia excretion rates of aquatic animals. The tolerance level of different aquatic animals to anesthetics is also different. [2].

The *Sinogastromyzon szechuanensis* live in the tributaries of the upper reaches of the Yangtze River in China, mainly distributed in the gravel beaches of the swift flowing river. They swim swirly, feeding on algae, chironomid larvae and other aquatic invertebrates [3]. The body color is beautiful and it is delicious, so it is popular. Human destruction of its habitat and overfishing have led to a sharp decline in its population resources. It has been listed as a local protected fish in some province in China. There published literatures related to *Sinogastromyzon szechuanensis* mainly focusing on resource investigation, early development, mitochondrial genome and phylogenetic analysis [4]. In this study, the effects of eugenol on oxygen consumption rate, ammonia excretion rate and oxygen-nitrogen ratio of the *Sinogastromyzon szechuanensis* were studied for the purpose of selecting suitable anesthetics for reducing the damage in production and transportation and improving their survival rate. It also provides basic information for the study of respiratory physiology of the *Sinogastromyzon szechuanensis*.

2. Materials and methods

2.1. Materials

The experimental fish were collected from Neijiang city of Sichuan Tuojiang River System, the sampling point was in N 29° 36'41.93", E 105°01'47.67". After a month of indoor domestication culture, 135 healthy *Sinogastromyzon szechuanensis* with a uniform size, whose average body weight was (2.029±0.26) g and the average total length was (5.01±0.15) cm, were chosen. The experimental water was fully-aerated tap water. The initial dissolved oxygen amount was 11.18±0.1 mg•L^{-1}, and the water temperature was (18±0.5) °C (except for different temperature tests). And the pH is 7.6±0.1.

We choose a 2.5L glass triangle bottle as the respiration chamber, supplemented by plastic wrap and rubber band to make a sealed hydrostatic breathing apparatus. The anesthetic eugenol are purchased from the Jiangxi Jin Yuan Natural Perfume Co., Ltd.. Eugenol and ethanol are mixed by mass ratio 1:5 and then transferred to volume container and then add pure water to the scale line. After the mother liquor is prepared, it is refrigerated in the refrigerator, and prepare the liquid is temporarily before the experiment is carried out.

2.2. Methods

2.2.1. Effects of eugenol on oxygen consumption rate and ammonia excretion rate of Sinogastromyzon szechuanensis under the same temperature.

We set the mass concentration of eugenol to be 0,6,12,18,24,30 mg•L^{-1} on the basis of pre-experiment,. In each cone bottle, put 5 *Sinogastromyzon szechuanensis*, and set three parallel for each concentration, and add a blank control group (respiration chamber without fish), and the water temperature is (18±0.5) °C.

Let the fish starve for 24 hours before the test. We place the respiratory chamber in a constant temperature water tank. The breathing experiment lasts for 2 hours (9:00 am-11:00 am). At the end of the experiment, the bottle was upside down and blended, then we siphon water sample from bottom of triangle bottle. We use Winkler iodometry to determine the dissolved oxygen and use nesslerization to determine ammonia (ammonium) nitrogen. After taking the water sample, anaesthetize the fish quickly then weigh the test fish without water (accurate to 0.01g), and measure the body length with foot gauge and ruler (accurate to 0.1cm), and then put the fish into the fresh water for recovery.

2.2.2. Effects of eugenol on oxygen consumption rate and ammonia excretion rate of Sinogastromyzon szechuanensis at different temperatures.

According to the anesthetic effect in 1.2.1, eugenol 24 mg•L^{-1} was selected to carry out experiments on the effects of different temperatures on oxygen consumption rate and ammonia excretion rate of *Sinogastromyzon szechuanensis*. Set up tanks with temperature gradient of 14°C, 20°C, 26°C, Nine 2.5L conical flasks (respiratory chamber) are placed in each water tank, and among them, there are respectively three eugenol and no-anesthetic which are used as contrast. In addition, set a blank control (no fish, and it is used to measure initial dissolved oxygen at different temperatures). Place 5 *Sinogastromyzon szechuanensis* in each of them (Note: Transfer temperature difference is no more than 2 °C), Start experiment immediately after adding anesthetic. The other methods are the same as 1.2.1.

2.2.3. Measurement and calculation of each index.

$$R_{oc}=\frac{1000\times(C_0-C_1)\times V}{W\times T}$$

$$R_{am}=\frac{1000\times(C_0'-C_1')\times V}{W\times T}$$

$$O/N= R_{oc}/R_{am}$$

R_{oc} is the oxygen consumption rate /(μg•g^{-1}•h^{-1}); R_{am} is the ammonia excretion rate /(μg•g^{-1}•h^{-1}); C_0 and C_1 represent the dissolved oxygen concentration (mg•L^{-1}) of the control group and the experimental group after the experiment, C_0' and C_1' represent the dissolved oxygen concentration (mg•L^{-1}) of the control group and the experimental group after the experiment, V is the actual water volume (L) of the respiratory chamber (conical bottle); W represent the quality of the test fish (g); T represent the test time (h). O/N represent the oxygen-nitrogen ratio.

2.2.4. Statistics and analysis of data.
The data were processed by Excel and SPSS Statistics 25 software. Duncan's was used to compare the differences of groups. If one of the above parameters is the same in the same column, there is no significant difference between the groups ($p > 0.05$). The statistical values are expressed by average value± standard deviation. Regression analysis of oxygen consumption rate and ammonia discharge rate with temperature by Excel software.

3. Result and analysis

3.1. Effects of eugenol with different concentrations on oxygen consumption rate, ammonia excretion rate and oxygen-nitrogen ratio of the Sinogastromyzon szechuanensis
The effects of eugenol on oxygen consumption rate, ammonia excretion rate and oxygen-nitrogen ratio of *Sinogastromyzon szechuanensis* are shown in Table 1. With the increase of eugenol concentration, the oxygen consumption rate and ammonia excretion rate increase first and then decrease. One-way ANOVA shows that when the concentration of eugenol is 12 mg•L^{-1}, the oxygen consumption rate and ammonia excretion rate of the *Sinogastromyzon szechuanensis* are significantly higher than those of the control group ($p<0.05$). The sucker of fish is absorbed on the glass wall or bottom of the bottle, and the breathing rate is slightly higher than that of the control group, and the difference is significant ($p<0.05$). When the concentration of eugenol reaches 24 mg•L^{-1}, the oxygen consumption rate and ammonia excretion rate are significantly decreased, and the breathing rate ratio is significantly lower than that of the control group ($p<0.05$), and when concentration continues to rise, the oxygen consumption rate and ammonia excretion rate have no significant difference. Observing their behavior, we find that abdomen of some fish is upward, facing upwards on the bottom of the bottle, and muscles are flabby. The oxygen-nitrogen ratio is between 11.30 and 13.00, and there is no significant difference among the groups ($p>0.05$).

Table 1. Effects of different concentrations of eugenol on oxygen consumption rate, ammonia excretion rate and oxygen-nitrogen ratio of *Sinogastromyzon szechuanensis*

Anesthetic	Concentrations (mg•L^{-1})	R_{oc}/(μg•g^{-1}•h^{-1}) (n=90, $\bar{x}$±SD)	R_{am}/(μg•g^{-1}•h^{-1}) (n=90, $\bar{x}$±SD)	O/N
Eugenol	0	457.8 ± 28.7^{b}	38.3 ± 4.9^{b}	11.50
	6	485.3 ± 31.2^{ab}	40.9 ± 2.3^{b}	12.10
	12	543.2 ± 23.5^{a}	49.5 ± 8.0^{a}	11.30
	18	470.9 ± 40.8^{ab}	37.3 ± 2.0^{b}	12.40
	24	387.4 ± 40.1^{c}	33.8 ± 4.4^{c}	11.90
	30	383.3 ± 35.5^{c}	30.0 ± 5.9^{c}	13.00

3.2. Effects of eugenol on oxygen consumption rate, ammonia excretion rate and oxygen-nitrogen ratio of Sinogastromyzon szechuanensis at different temperatures
The control group and eugenol (24 mg•L^{-1}) group, the oxygen consumption rate and ammonia excretion rate of the Sinogastromyzon *szechuanensis* increase significantly with the increase of temperature, and there is a significant correlation (Fig 1, $R^2>0.99$). One-way ANOVA analysis showed that there are significant differences in oxygen consumption rate and ammonia excretion rate among the three temperature gradients of different test groups ($p<0.01$), The oxygen-nitrogen ratios of anesthetic group and blank control group increase with the

increase of temperature. At the same temperature, the oxygen-nitrogen ratio of anesthetic group is lower than that of control group (Table 2).

Table 2. Effects of eugenol on oxygen consumption rate, ammonia excretion rate and oxygen-nitrogen ratio of *Sinogastromyzon szechuanensis* at different temperatures

Treatment group	Temperature ℃	$R_{oc}/(\mu g \cdot g^{-1} \cdot h^{-1})$ (n=45, $\bar{x}$±SD)	$R_{am}/(\mu g \cdot g^{-1} \cdot h^{-1})$ (n=45, $\bar{x}$±SD)	O/N
Control group	14	135.9±20.0[a]	13.9±1.3[a]	9.8
	20	340.3±42.5[b]	19.5±2.2[b]	17.4
	26	476.9±49.0[c]	24.6±3.7[c]	19.4
Eugenol 24 mg•L[-1]	14	115.2±24.8[a]	11.3±0.4[a]	10.2
	20	208.2±28.8[b]	15.7±1.8[b]	13.3
	26	293.4±28.9[c]	20.1±1.7[c]	14.6

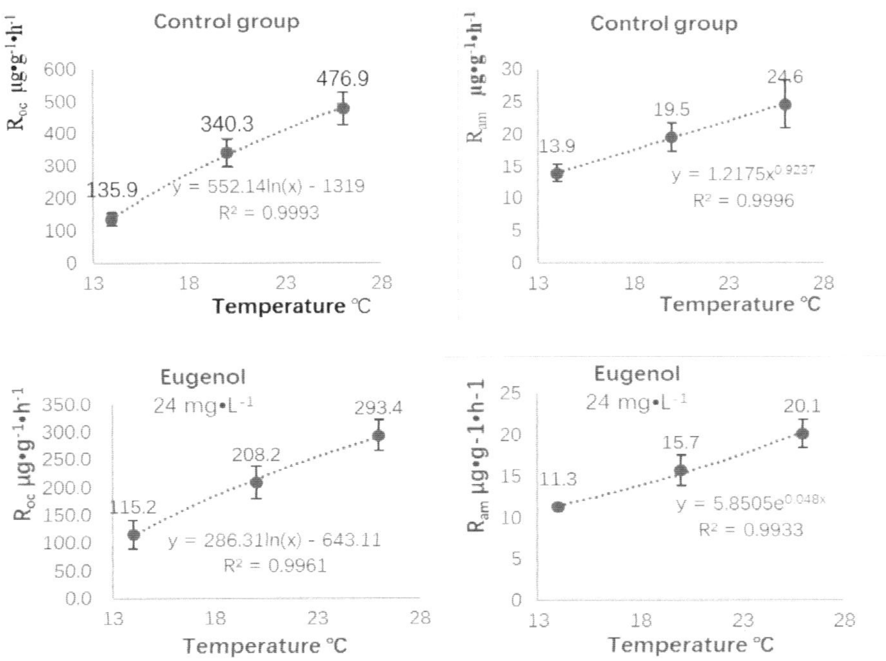

Figure 1. Effects of eugenol on oxygen consumption rate and ammonia excretion rate of the *Sinogastromyzon szechuanensis* at different temperatures

4. Discussion

There are many factors affecting the anesthetic effect of anesthetics, including the anesthetics type, concentration, temperature, dissolved oxygen and other water quality indicators, as well as the tolerance of different varieties to anesthetics [5]. After the tested aquatic animals are anesthetized, we can use oxygen consumption rate, ammonia excretion rate and breathing rate to judge the depth of anesthesia. According to the behavioral characteristics of the *Sinogastromyzon szechuanensis* in the anesthetic solution, the anesthetic process was divided into five stages (Table 3):

Table 3. Stages of anaesthesia in the *Sinogastromyzon szechuanensis*

Stages of anaesthesia	Behavior characteristics
I—Stress stage	The response to stimulation was obvious, swimming was active, scurrying around, and breathing frequency increased.
II—Sedation stage	The response to stimulus was reduced, adsorbed on the bottle wall, sometimes the head reached to the surface of the water, and the respiratory rate decreased slightly.
III—Mild anesthetic stage	The tactile response was slow, lying on the bottom of the bottle, and the respiratory rate was further reduced.
IV—Deep anesthetic stage	Complete loss of swimming ability, no response to touch, lying on the bottom of the bottle, breathing rate dropped sharply, gill moved slightly. It can be resuscitated about 1 minute after being transferred to clean water.
V—Death stage	The gill completely stopped opening and it could not recover after it was transferred to fresh water.

4.1. Effects of different concentrations of eugenol on oxygen consumption rate and ammonia excretion rate of Sinogastromyzon szechuanensis

The metabolism of aquatic animals usually takes oxygen and then releases carbon dioxide and nitrogen, so the oxygen consumption rate and ammonia excretion rate can directly or indirectly reflect the metabolic state and physiological condition. In production and transportation, the use of a certain number of anesthetics can reduce oxygen consumption, reduce carbon dioxide and ammonia nitrogen emissions, reduce stress injury, and help to improve survival in production and transportation. Different kinds of aquatic animals, and different individual sizes, water temperatures and dissolved oxygen all influence the anesthetic effect.

When *Sinogastromyzon szechuanensis* were anesthetized with eugenol at different concentrations, their oxygen consumption rate and ammonia excretion rate decreased significantly. The oxygen consumption rate and ammonia excretion rate of eugenol group were higher than those of control group when the concentration of eugenol was 12 mg•L^{-1}. The results of this study were consistent with that of Zhuang. The main reason was that the stress effect of anesthetics increased respiratory rate and metabolism. When concentration of eugenol was 24 mg•L^{-1}, it could lead to the stage II of anaesthesia, reduce the oxygen consumption rate and ammonia excretion rate significantly. Among them, eugenol reduced its oxygen consumption rate by 15.4%, ammonia excretion rate by 11.7%. The anesthetic effect of eugenol on *Procambarus clarkii* was similar to that of this study. With the increase of concentration, the oxygen consumption rate and ammonia excretion rate increased first and then decreased. When the concentration was 40 mg•L^{-1}, the oxygen consumption rate and ammonia excretion rate both decreased significantly [5]. When eugenol anesthetized the *Acipenser baeri* with an average body weight of 1.44g, the oxygen consumption rate of 2-4 mg•L^{-1} group was significantly higher than that of the control group, and 8-30 mg•L^{-1} was significantly lower than that of the control group [6]. Wang (2018) simulated the transport of 130g juvenile largemouth bass. When concentration of eugenol was 10 mg•L^{-1}, the juveniles remained in the second stage of anesthesia.

4.2. Effects of eugenol on oxygen consumption rate, ammonia excretion rate and oxygen-nitrogen ratio of Sinogastromyzon szechuanensis at different temperatures

With the increase of temperature, the basal metabolism and active metabolism become stronger, resulting in the increase of oxygen consumption rate and ammonia excretion rate. Oxygen consumption rate and ammonia excretion rate of most shellfish increase with the increase of temperature. After exceeding the tolerance temperature range, the physiological function of shellfish is disturbed, and the oxygen consumption rate and ammonia excretion rate decrease [7].

Regardless of whether the *Sinogastromyzon szechuanensis* were anesthetized, the oxygen consumption rate and ammonia excretion rate increased with the increase of temperature at the temperature of 14-26°C ($p<0.05$). The results are similar to those of juvenile *Acipenser sinensis*, *Oplegnathus punctatus*, juvenile *Lucioperca lucioperca*, *Monopterus albus* eel [8-10]. From Figure 1, although the anesthetic agent can reduce the oxygen consumption and ammonia excretion rate of *Sinogastromyzon szechuanensis*, the effect of temperature on the oxygen consumption rate and ammonia excretion rate of *Sinogastromyzon szechuanensis* is extremely significant. The eugenol has better anesthetic effect at low temperature.

Oxygen-nitrogen ratio (O/N) is an important physiological index in the process of animal metabolism, which can predict the chemical nature of energy substances consumed in the metabolism of organisms. When energy is supplied entirely by protein, the oxygen-nitrogen ratio is about 7 [11]. When energy is supplied by protein and fat, the oxygen-nitrogen ratio is about 24 [12]. When energy is supplied by fat or carbohydrates, the oxygen-nitrogen ratio becomes infinite [13]. In this paper, the ratio of oxygen-nitrogen ratio of eugenol groups with different concentrations ranged from 11.30 to 13.00. However, there was no significant difference among the groups, which indicated that anesthetics eugenol did not significantly change the main energy-supplying substance—protein. Zhuang (2009) pointed out that the oxygen-nitrogen ratio decreased with the increase of anesthetic concentration in the anesthesia test of juvenile *Acipenser sinensis*, which was inconsistent with the results of ours experiment. Zhang (2010) considered that the oxygen-nitrogen ratio was not positively correlated with the concentration of anesthetic in the narcotic experiment of Nile tilapia, and the results are consistent with the viewpoints of this study. In the temperature range of 14-26°C, the oxygen-nitrogen ratio of *Sinogastromyzon szechuanensis* ranged from 8.0 to 19.4, and it increased with the increase of temperature. At the beginning, energy was mainly protein, then the protein utilization rate decreased, and the utilization of fat and carbohydrates increased.

5. Conclusion

The results showed that anaesthetic eugenol had significant effects on oxygen consumption rate and ammonia excretion rate of *Sinogastromyzon szechuanensis* ($p<0.05$). With the increase of anesthetic concentration, anesthetic eugenol with a concentration of 24 mg•L-1 can make the *Sinogastromyzon szechuanensis* enter the stage of anaesthesia II, and this concentration meets the requirements of live *Sinogastromyzon szechuanensis* transportation, so it can be used as the recommended anesthetic dosage for transportion, and the anesthetic effect at low temperature is more obvious.

Acknowledgment

This research was financially supported by “Qing Lan Project of Jiangsu Province”. This work was supported by “311 Talents Project in Taizhou”, Research project of the Jiangsu Agri-animal Husbandry Vocational College (Grant No. NSF201711).

Reference

[1] Liu Changlin, He Li, Chen Siqing, et al. Review of fish anesthesia research [J]. Fisheries Modernization, 2007, 34(5): 21-25. (In Chinese)

[2] Wang Wenhao, Dong Hongbiao, Sun Yongxu, et al. Anesthetic effect of MS-222 and eugenol in simulated transport of juvenile largemouth bass [J]. Southern Fisheries Science, 2018, 14(6): 52-58.

[3] Yang Jun, GUO Yanxi. A new species of the genus Sinensis in Sichuan Province, China (Scorpio, Climbing) [Z]. Zoological Systematics, 2013, 38(4): 895-900. (In Chinese)

[4] Zou, Y. , Zhang, J. , Zhang, T. , Xie, M. , & Wu, T. . (0). Sequence identification and phylogenetic analysis of the mitochondrial genome of microphysogobio kiatingensis (cypriniformes: cyprinidae). [J]. Mitochondrial DNA Part B, 2018, 3(1).

[5] SD Coyle, RM Durborow, JH Tidwell, et al. Anesthetics in Aquaculture [R]. SRAC Publication, 2004: 3900-3095

[6] Feng Guangpeng, Zhuang Ping, Zhang Longzhen, et al. Effects of anesthetic eugenol on oxygen consumption rate of Acipenser baeri Brandt larvae and blood biochemical parameters of juveniles[J]. Journal of Dalian Ocean University, 2010, 25(2): 113-118.

[7] NIE Hongtao, JIANG Liwen, WANG Hailun, et al. Effects of different temperature and salinity on oxygen consumption rate and ammonia excretion rate of Solen grandis [J]. Ocean Science, 2017, 41(4): 44-50. (In Chinese)

[8] Song Xiefa, Cheng Yawei, Xing Daochao, et al. Effects of temperature and body weight on oxygen consumption rate and ammonia excretion rate of zebrafish and circadian rhythm changes[J]. Journal of Ocean University of China (Natural Science Edition), 2017,47(9): 21-29. (In Chinese)

[9] FU Jian-gui, ZHANG Zhen-zao, LI Cai-juan, et al. Effects of temperature on oxygen consumption rate, ammonia excretion rate and asphyxiation point of juveniles of different masses of P. chinensis [J]. Aquaculture, 2018, 39(10): 22-26. (In Chinese)
[10] YUAN Quan, Lü Wei,. Effects of temperature and body mass on oxygen consumption rate and ammonia excretion rate of Astragalus membranaceus[J]. Journal of Jiangxi Agricultural University, 2018, 30(11): 80-85. (In Chinese)
[11] Mayzaud P. Respiration and nitrogen excretion of zooplankton. IV. The influence of starvation on the metabolism and the biochemical composition of some species [J]. Marine Biology, 1976, 37(1): 47-58.
[12] Ikeda T. Nutritional ecology of marine zooplankton [J]. Memoirs of the Faculty of Fisheries Hokkaido University, 1974, 22(1): 1-97.
[13] Dall W, Smith D M. Oxygen consumption and ammonia-N excretion in fed and starved tiger prawns, Penaeus esculentus Haswell [J]. Aquaculture, 1986, 55(1): 23-33.
[14] Zhuang Ping, Xu Bin, Zhang Longzhen, et al. Effects of MS-222 and eugenol on oxygen consumption rate and ammonia excretion rate of juvenile Chinese sturgeon[J]. Chinese Journal of Fisheries Science, 2009, 16(4): 612-618. (In Chinese)
[15] Zhang Jiandong, Chen Gang, Tang Baogui, et al. Effects of anesthetic MS-222 on oxygen consumption rate and ammonia excretion rate of juvenile Nile tilapia[J]. Fisheries Modernization, 2010(4): 35-39. (In Chinese)

Research on gamma spectrum semi-quantitative analysis based on convolutional neural network

Fei Li[a, b], Jing Wang[b], Liangquan Ge[a, b,*], Fangyan Hu[b], Feng Cheng[b], Kun Sun[b]

[a] Applied Nuclear Technology in Geosciences Key Laboratory of Sichuan Province, China

[b] College of Nuclear Technology and Automation Engineering, Chengdu University of Technology, China

* Corresponding author and e-mail: Liangquan Ge, glq@cdut.edu.cn

Abstract. In the field of practical application of nuclear technology, one of the vital steps is the interpretation of gamma energy spectrum, which can obtain the type and content of radionuclides and achieve further analyzation. Convolutional neural network is one the most basic and effective algorithm structures in deep learning because it has local receptive field, weight sharing and down sampling structure, it solves the problem of parameter expansion caused by full connection and reduces the number of weights. That can be used to study the resolution of gamma-ray spectra of convolutional neural networks. This paper introduces the spectral forming principle of gamma spectra and the convolutional neural network, and USES the convolutional neural network to study the spectral decomposition of gamma spectra. In this paper, a multi-layer convolution neural network model is built based on C# software. The convolution neural network is applied to gamma-ray spectrum decomposition, and the U, Th, K nuclides are identified and semi-quantitatively calculated. By identifying and analysing different energy spectrum data, it is shown that the modified model structure can be applied to the spectral decomposition of gamma spectrum.

1. Introduction

With the rapid development of computer technology. In order to make machine equipment have intelligent thinking ability when dealing with complex data, artificial neural network (ANN) is formed and has been continuously developing over last three decades. Gamma spectrum analysis is regarded as a fast, reliable and non-destructive technology on determining the type and intensity of radionuclides [1]. The concepts of convolution neural network (CNN) and convolution kernel came into being. At present, the neural network has developed to a certain stage. There are not only types in structure, but also breakthroughs in algorithm. Therefore, the application of convolutional neural network in gamma spectrum is feasible.

For nuclear science and technology, it is also natural to apply ANN to gamma-ray spectroscopy. As early as last century, scholars began to study the use of neural networks for gamma-ray spectrum analysis. Olmos put forward the neural network in 1992 and Vigneron proposed in 1996 to use the neural network to carry out statistical modeling in gamma-ray spectrum analysis [2]. Because the neural network has the functions of non-linear mapping, fast parallel distribution processing and recognition of spectral features for energy spectrum analysis [3], the results of nuclide recognition can be obtained quickly and accurately in the analysis and processing of gamma spectrum. In recent years, the research of using neural networks to solve gamma-ray spectra has also begun to increase gradually in China. Jianwen Huo used probabilistic neural networks for rapid identification of radionuclides in 2005 [4]. Chongjie Wang used RBF artificial neural networks to analyze gamma-ray spectra in 2016. [5]. Jianping He used the neural network model to train the fast nuclide recognition based on Walsh-Hadamard transform and stack self-coding published in 2017 [6].

The application of convolution neural network in gamma-ray spectrum decomposition analysis is essentially to identify the spectrum image, so as to find the characteristic peaks of the spectrum, and ultimately get the spectrum information through analysis. The decomposition gamma spectrum of convolution neural network is different from that of traditional gamma spectrum. It extracts features directly from the spectrum data, and then obtains the categories and contents of elements through appropriate weights.

This paper mainly studies the qualitative evaluation of K, U and Th elements by using multi-layer convolution method based on C# platform. C# platform can greatly reduce the number of parameters of multi-layer convolutional neural network and locally optimize the structure of the neural network.

2. Material and method

2.1. Convolution principle

Convolution is divided into two categories, one is discrete convolution, referred to as convolution sum, such as formula 1; the other is continuous convolution, referred to as convolution integral, such as formula 2.

$$f[x]*g[x]=\sum_{k=-\infty}^{\infty}f[k]g[x-k] \tag{1}$$

$$f(x)*g(x)=\int_{-\infty}^{\infty}f(k)g(x-k)dk \tag{2}$$

Where, $x=K+(x-k)$, this means that every input data has a corresponding weight [7]. All the methods adopted in this paper are based on discrete convolution, and the matrix is convoluted by discrete convolution.

2.2. Principle of Convolutional Neural Network

CNN can automatically learn training weights in supervised mode, and constantly extract features from sample data. Traditional pattern recognition is generally not very effective and convolutional neural network has good effect on image recognition. The structure of convolutional neural network mainly includes sparse connection and value sharing. It has three forms of constraints: feature extraction, feature mapping and sub-sampling.

2.3. Convolutional neural network structure

Convolutional neural network is mainly composed of convolution layer, sub-sampling layer and full connection layer. After convolution operation, the features will be mapped to the next layer, and then the data features will be extracted by using the layer. The convolution layer and the sampling layer alternately form the feature extraction layer of an input layer, and then the data after feature extraction is used as the input layer to continue feature extraction. Finally, the entire connection is carried out to classify the output layer and obtain the output result [8]. Figure 1 shows the technical route in details.

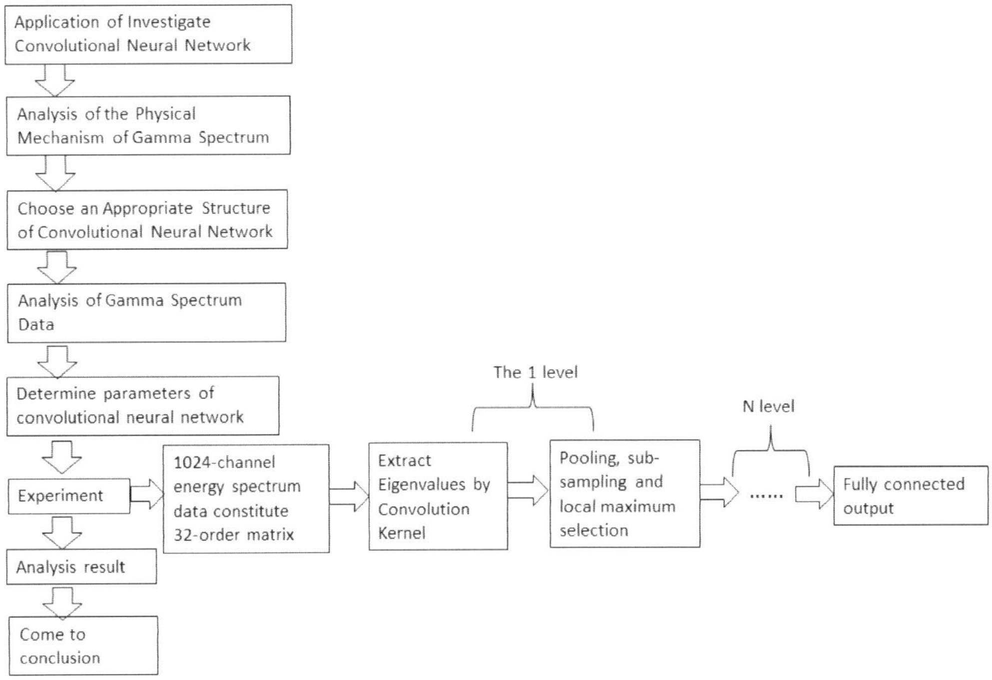

Figure 1. Technical route diagram

3. Results and discussion

3.1. Experimental Selection of CNN Structural Quantity Method

Through theoretical study and analysis, the number of convolution cores selected is 6. The convolution layer and the pooling layer have two layers respectively. Finally, there is one layer of full connection layer. Firstly, 1024-channel gamma-ray spectroscopy data are input to form a 32*32 two-dimensional matrix. The energy spectrum data are arranged horizontally into a two-dimensional matrix, as shown in Figure 2. The positions of the energy peaks of U, Th and K nuclides are generally different, and the positions of all energy peaks are basically kept within a range. Therefore, the positions of nuclide corresponding matrices of U, Th and K are basically unchanged in the 32 *32 two-dimensional matrixes. The dimension matrix is regarded as a pixel image, and the feature pixels of U, Th and K nuclides in the image are recognized by grasping the feature information of the image, so as to realize the possible spectral recognition.

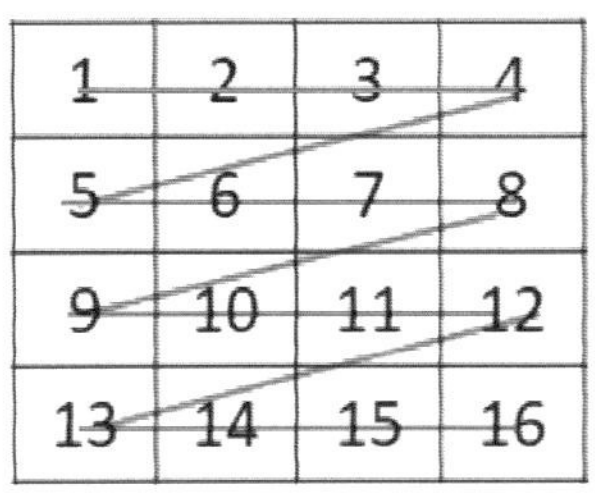

Figure 2. The arrangement of the input energy spectrum data to form a matrix

The selection of convolution kernel is the key. Six 5*5 convolution kernels are selected in this experiment. In this experiment, the choice of convolution kernels is mainly around 0.5 and 12 parameters. Each convolution core does not affect each other. After convolution, six feature mapping images are obtained.

The first convolution layer is to convolute the input layer data according to different convolution kernels. The convolution core moves in the same way as the two-dimensional matrix of 32*32 composed of energy spectrum data, as shown in Figure 2. The first pooling layer is to pool six feature mapping matrices obtained from the first convolution layer and get six 14*14 feature extraction matrices. In this experiment, the maximum pooling method is used to select the maximum value in the 2*2 region of each feature mapping matrix and each maximum value is taken as a value in the feature extraction matrix. The pooling method is the same as the two-dimensional matrix of energy spectrum data, which not only reduces the dimension, but also makes the feature data more prominent and concentrated.

Because a "convolution-pooling" process can not accurately extract features, we need to construct another layer and finally get six 5*5 feature extraction matrices. After full connection, the mapping feature real numbers are converted to be between [0,1]. The final function value can be used to judge whether there are U, Th and K nuclides in gamma spectrum and the content of U, Th and K nuclides can be calculated according to the output results.

3.2. Evaluation of experimental results

The energy spectrum data used in this experiment are obtained from a geological exploration survey and detected by gamma ray spectrometer. Among them, the range of K-peak is 458-470, U-peak is 555-564, and U-peak is 826-858. In this paper, 80 sets of energy spectrum data are obtained. Fifty-six of them are used as training samples according to the proportion of 7:3. The remaining 24 sets of energy spectrum data are used as experimental group to test the effect of neural network model.

The two-dimensional matrix of the spectral line yields the following initial bottom convolution data. If its two-dimensional matrix is represented by gray scale graph, figure 3 can be obtained.

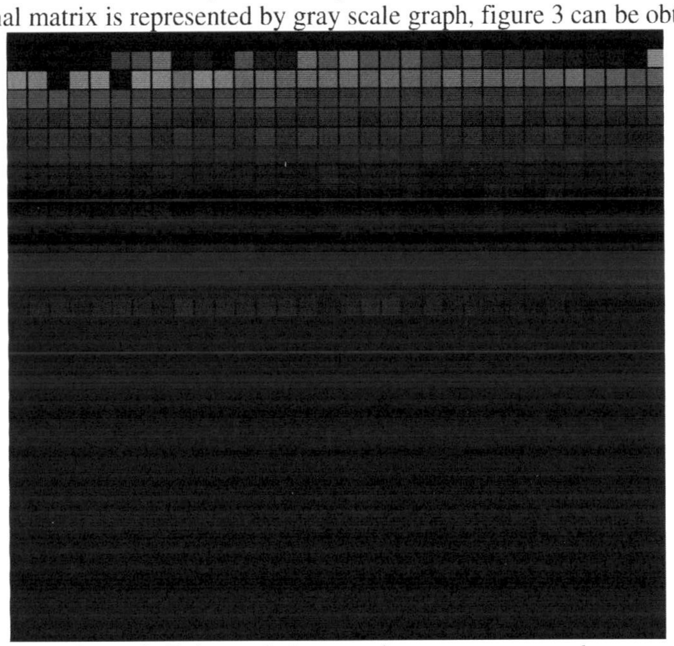

Figure 3. Color scale image of energy spectrum data

The initial convolution graph is input into the neural network model, and a series of operations such as convolution, pooling and full connection are carried out. The output result is 0.5473.

In the experiment, a group of gamma-ray spectra data was used as standard reference spectra. It is known that there are U, Th and K nuclides in the gamma-ray spectra, and the contents are shown in Table 1.

Table 1. K, U, Th Contents in Standard Energy Spectrum

K content /%	U content /ppm	Th content /ppm
2.76	2.21	9.4

The standard gamma-ray spectrum data is input into the convolutional neural network model built on C# platform and the output is 0.5477. Then other energy spectrum data are input into the program in order to make the result more accurate (see figure 4).

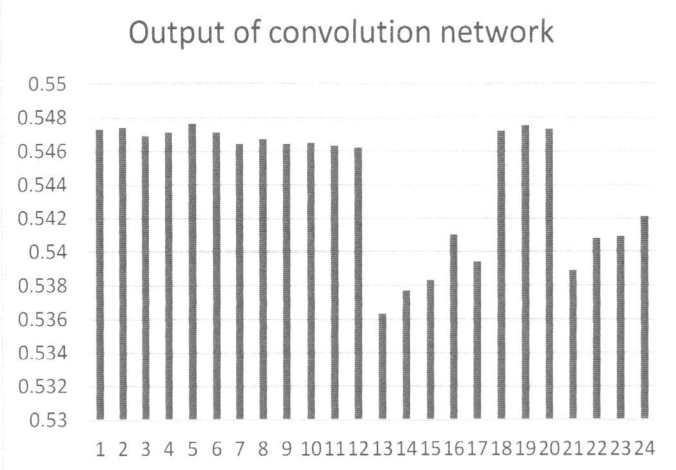

Figure 4. Sample data output in convolution neural network

After recognition by the convolution neural network program, it can be found that U, Th, K nuclides exist in each group from the chart. Compared with the actual results, the results are correct.

Table 2. Relevant data of K measured by experiment

Serial number	K content (μg • g^{-1})	Maximum counting of experimental data	Maximum count for each set of data	K peak channel	K peak count
1	2.62	1051	907	458~470	339
2	2.53	1051	925	458~470	357
3	2.64	1051	883	458~470	342
4	2.69	1051	1051	458~470	328
5	2.67	1051	999	458~470	350
6	2.67	1051	973	458~470	348
7	2.48	1051	986	458~470	327
8	2.65	1051	980	458~470	331
9	2.54	1051	987	458~470	334
10	2.63	1051	1005	458~470	332
11	2.72	1051	990	458~470	372
12	2.73	1051	946	458~470	324
13	2.04	1051	976	458~470	316
14	2.32	1051	981	458~470	345

Table 2, cont

15	2.01	1051	989	458~470	318
16	1.94	1051	725	458~470	278
17	1.92	1051	752	458~470	276
18	2.6	1051	776	458~470	295
19	2.73	1051	871	458~470	270
20	2.54	1051	820	458~470	275
21	2.08	1051	·751	458~470	240
22	1.89	1051	848	458~470	255
23	2.38	1051	784	458~470	320
24	2.51	1051	750	458~470	298

The experiment was quantitatively calculated based on standard energy spectrum. The experimental group selected K content for quantitative analysis (see table 2). According to the normalized results and standard gamma-ray spectrum K content to predict the experimental group of gamma-ray spectrum K content (see figure 5).

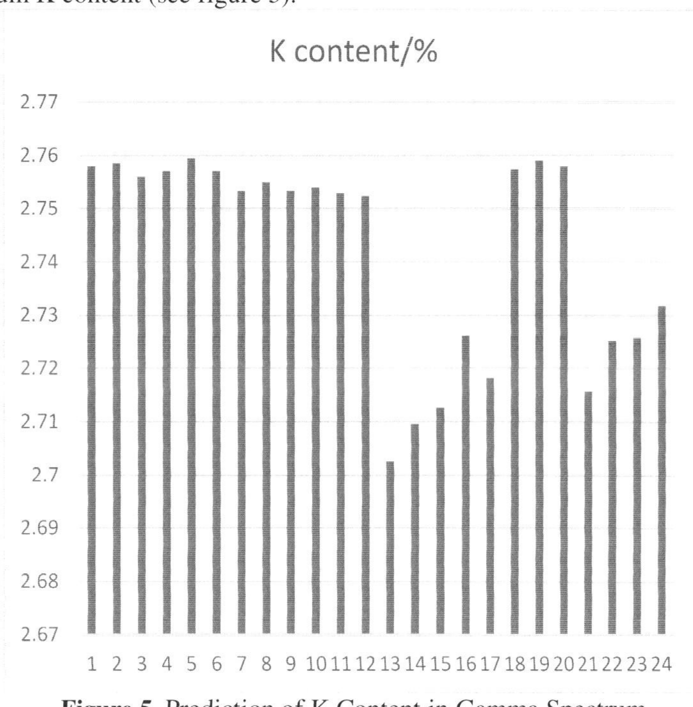

Figure 5. Prediction of K Content in Gamma Spectrum

Relative errors are calculated according to the predicted and real values of potassium content in experimental samples. Compared with the actual results, the absolute error of gamma-ray spectrum K content calculated by convolution neural network model is not large. The minimum and maximum relative errors are 0.82 and 44.19 respectively. By calculating the relative error with 5% of the initial value, the qualified rate is 37.5%. If the relative error is calculated by 10%, the qualified rate is 62.5%. It basically meets the expected requirements.

4. Conclusion

This design chooses convolution neural network as the experimental model and optimizes the model of convolution neural network. Finally, the convolution neural network model is applied to the gamma-ray spectrum decomposition to identify K, U, Th nuclides and calculate the content. The feasibility of convolution neural network model is proved by comparing experiment with practice. However, the comparison results show that there are some deviations between the experimental results and the standard, which indicates that there is still room for further optimization in the model structure and parameter selection of convolutional neural network. The application of convolution neural network in gamma spectrum resolution is effective and has great potential, which can be further studied.

Acknowledgement

This work was funded by the National Natural Science Foundation of China (No.41774147), the National Key R&D Project (No.2017YFC0602100) and Science and technology activity project for overseas personnel in Sichuan province.

References

[1] Fei Li, Zhixing Gu, Liangquan Ge, Hui Li, Xinyu Tang, Xinle Lang, Bo Hu. Review of recent gamma spectrum unfolding algorithms and their application. Results in Physics. 2019,13: 102211.
[2] V. Vigneron, J. Morelb, M.C. Lkpyb, J.M. Martineza. Statistical modelling of neural networks in y-spectrometry. Nuclear Instruments and Methods in Physics Research. 1996,369:642-647.
[3] Donsheng Shi. Application of Particle Swarm Optimization in Recognition of Gamma Spectrum by Neural Network. Nuclear Technology. 2007, (7): 615-618.
[4] Chongjie Wang. Gamma Spectrum Analysis Based on RBF Artificial Neural Network. Nuclear Electronics & Detection Technology. 2016, (1): 56-59.
[5] Jianwen Huo. Research on Rapid Recognition of Radionuclides Based on Probabilistic Neural Network. Nuclear Electronics & Detection Technology. 2015, (3): 253-258.
[6] Jianping He. Fast Nuclide Recognition Based on Walsh-Hadamard Transform and Stack Self-Coding. Nuclear Electronics & Detection Technology. 2017, (8): 834-840.
[7] Jian CHENG,Pei-song WANG,Gang LI,Qing-hao HU,Han-qing LU. Recent advances in efficient computation of deep convolutional neural networks. Frontiers of Information Technology & Electronic Engineering. 2018, 19(01):64-77.
[8] Dan Li. An Improved Convolutional Neural Network Based on Lenet-5. Computer Age. 2016(08):4-6.

Journal of Physics: Conference Series **1423** (2019) 012006 doi:10.1088/1742-6596/1423/1/012006

Seeking for optimum culture medium for chlorella pyrenoidosa and analysis of the amino acid component

Wei Li [1], Xiaojiang Chen[2*], Mengmeng Zheng[3], Zhicheng Li[4], Haiyang Li[5]

[1,2,4,5] Jiangsu Agri-animal Husbandry Vocational College
[3]Jiangyan County Agriculture Commission of Taizhou Prefecture

*Corresponding author and e-mail: Xiaojiang Chen, cq_cxj@126.com

Abstract. *Chlorella Pyrenoidosa* is one of the widely used algae in aquaculture. In order to find out a proper culture medium for high-density culturing to meet the demand of fish reproduction and water quality control, five normal algae culture media were test via multiple comparisons. The densities of the five groups, with six parallel samples each, which were cultured in corresponding medium, were determined by blood counting chamber on the beginning (0d), the 3rd day (3d), the 6th day (6d) and 9th day (9d). Then the growth curves were described and the growth radio were calculated and compared by One-Way ANOVA (Turkey method) using the SPSS 21.0 software. The multiple comparisons results revealed that HGZ group could reproduced a rapid and stable speed at high density, while SE group and Aquatic 4# reproduced faster than the HGZ group. Amino acid component results revealed that C. Pyrenoidosa contains 18 kinds of amino acid (42.19 ~ 43.14% of the body mass). There are significant differences on total amino acid, fish essential amino acid and non-essential amino acid between groups based on the multiple comparisons. However, the amino acid could not meet the demand of fish or shellfish. Thus, extra amino acid, such as tryptophan and methionine should be added and the HGZ medium was the best choice for producing *C. Pyrenoidosa* to satisfy the aquaculture.

1. Introduction

Chlorella Pyrenoidosa is unicellular green algae, who belongs to Chlorella and be of with different cell sizes, goblet-shaped chromatophore. It has been suggested that the consumption of natural "whole foods" rich in macronutrients has many healthful benefits. Besides of rich in protein, vitamins and minerals, *C. Pyrenoidosa*are of high reproductive ability via their special autospore method [1]. Besides of their nutrition, *C. Pyrenoidosa*are also the ideal diet or feed supplement for aquatic larvae for their ability in controlling of water quality or for their small size in initial feed application. Because of rich nutrition, small shape and no pollution to water quality, it is difficult to replace them with artificial diet [1]. They are widely used in aquatic animal breeding and water quality control [2, 3]. Especially in the process of rotifer cultivation and reproduction of shellfish.

However, there are some barriers in aquicultureapplication. Such as, it is difficult to control the quality of different batches and to obtain algae in large numbers in a short time to match up the requirements of production. In this article, five different commonalgaeculture media were investigated to find the ideal medium for culturing the *C. Pyrenoidosa*in a high concentration and speed to meet the requirements of larva.

2. Materials and methods

2.1. Materials

Species: *Chlorella Pyrenoidosa* (Provide by Qingdao Microalgae Biotechnology Co., Ltd., China)

Main *Instruments*: Artificial climate Chamber (Shanghai Dutco Scientific Instruments Co., Ltd., GZP-150)

Medium: Basal, Aquatic Medium 4# [4], SE (Brostol's solution) [5], HGZ-145[6] and WAV medium [7]

2.2. Cultivation Methods and Procedure

Prepare thirty clean conical flask (50ml); Add 15 mL algae medium of *C. Pyrenoidosa* and 35 mL *corresponding* medium in five groups (Basal, Aquatic 4#, SE, HGZ and WAV), with 6 *repeated*trials.

The algae medium cultured in artificial climate Chamber (28℃, 10 000 Lux, photoperiod of 16L:8D). *Shake* the flask twice a day to guarantee the homogeneity. The concentration of *C. Pyrenoidosa*were counted via blood counting chamber.

2.3. Growth ratio formula

Growth ratio represents the average relative growth rate (or reproduction rate) of *C. Pyrenoidosa*of different periods.

$$\mu = \frac{\ln C_2 - \ln C_1}{t_2 - t_1}$$

Where

μ=*growth* ratio of *C. Pyrenoidosa*
C_1 = *concentration* of *C. Pyrenoidosa*at t_1 (10^5ind/L)
C_2=*concentration* of *C. Pyrenoidosa*at t_2 (10^5ind/L);
t_1=*start* point of the test period;
t_2=*termination* point of the test period.

2.4. Analysis of the Amino Acid Component

C.Pyrenoidosa of Basal group, Aquatic 4# and HGZ were harvested at the 9th day, with 3 repeat trials. Then, *C. Pyrenoidosa* were vacuum frozen drying at low temperature for amino acid component analysis.

3. Results

3.1. Growth ratio and curve of 5 different culture media

The concentrations of *C. Pyrenoidosa* of different cultural medium were presented in table 1 and figure 1at 0th d, 3rd d, 6th d, and 9th d. At 0th d, there is no difference between the groups. At 3rd d, group SE is primary high concentration, while group Aquatic 4# secondary. At 6th d, group Aquatic 4# came up to be the primary. At 9th d, HGZ showed **highest concentrationandgrowth radio** which were significant *higher*than the other groups (Fig 1 tab1) (Turkey method).

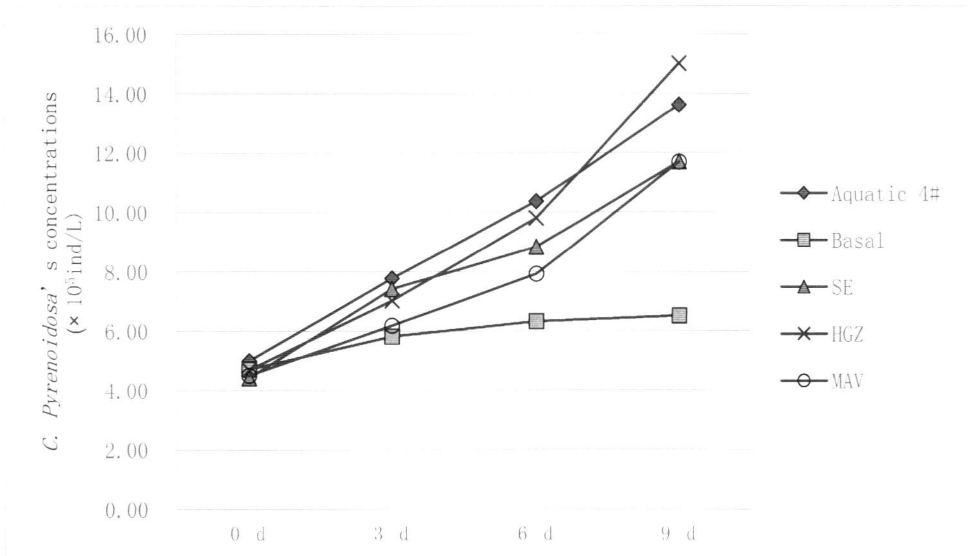

Figure 1. Line chart of *C. Pyrenoidosa*'s concentrations of groups at stages

Table 1. Growth ratio of *C. Pyrenoidosa*of groups at different periods

Medium	Growth ratio of *C. Pyrenoidosa*		
	0-3 d	3-6 d	6-9 d
Aquatic 4#	0.15±0.00^{c,m}	0.10±0.00^{cd,l}	0.09±0.00^{b,l}
Basal	0.07±0.00^{a,n}	0.03±0.01^{a,m}	0.01±0.00^{a,l}
SE	0.17±0.01^{d,n}	0.06±0.00^{b,l}	0.09±0.00^{b,m}
HGZ	0.13±0.01^{c,m}	0.11±0.00^{d,l}	0.14±0.00^{c,m}
MAV	0.11±0.02^{b,m}	0.08±0.01^{c,m}	0.13±0.01^{c,l}

Tips: a, b, c, d and e showed the significant difference between groups in a column; l, m and n showed the *significant* difference between groups in a line;different letter stands for significant difference between *groups* or periods, while same letter means no significant difference in a column or line.

3.2. Analysis of the Amino Acid Component

According to the results (Tab. 2), *C. Pyrenoidosa*consisted of 18 amino acid (42.19 ~ 43.14%), which contained 8 human essential amino acids (hEAA) with a high concentration (16.12 ~ 16.44%). While the *concentration* of fish essential amino acid (fEAA) was 18.36 ~ 19.67%. Essential amino acid (EAA) is animportantnutrition index, the EAA of *C. Pyrenoidosa* were higher than the other plant feedstuff. Such as defatted soybean cake (12.64%), defatted peanut cake (9.81%) and alfalfa meal (5.81%). Thus, *C. Pyrenoidosa*is a good resource for feedstuff.

Post Hoc Multiple Comparisons of different groups results revealed that the medium could affect the amino acid component of *C. Pyrenoidosa.*

Table 2. Amino Acid Component of different groups

Amino Acid	Component(g/100g)		
	HGZ	Aquatic 4#	Basal
Total Amino Acid	43.14 ± 0.11^{A}	43.10 ± 0.35^{B}	42.19 ± 0.15^{B}
human essential amino acid (hEAA[1])	16.44 ± 0.09^{a}	15.71 ± 0.18^{c}	16.12 ± 0.13^{b}
fish essential amino acid (fEAA)	19.67 ± 0.06^{M}	18.36 ± 0.28^{O}	18.86 ± 0.04^{N}
fish non-essential (fNEAA)	23.47 ± 0.11^{m}	24.74 ± 0.08^{m}	23.32 ± 0.17^{n}
fEAA/TAA	0.46	0.43	0.45
fEAA/fNEAA	0.84	0.74	0.81

tips: A and B showed the difference of total amino acid content in each group.
a, b and c showed the difference of human essential amino acids in each group.
M, N and O showed the difference of essential amino acids in fish in each group.
M and n showed the difference of nonessential amino acids in fish in each group.

4. Conclusion

C. Pyrenoidosa is rich in amino acids and high ratio of EAA/TAA. For their high protein quality, *C. Pyrenoidosa* could satisfy the basic growth needs of animals and reach the FAO "protein model" standard. All fish essential amino acids could be detected, and the concentration of several fEAA, such as leucine, lysine, valine, phenylalanine, arginine and threonine, were higher than 5%. According to Yuping Deng's and Kai Lu's researches, leucine can significantly improve the fish quality of grass carp and the intestinal antioxidant condition, who would improve immunity of fish and promote production [8-9]. Most fish's lysine requirement was 3.3-8.5% of diet. Lacking of lysine, fish would hesitate to intake fodder and be easy to get sick even to lost their lives.

Besides high nutrition, *C. Pyrenoidosa* is a kind of high-quality initial feeding for fish and shellfish. Unlike other fodders used in aquiculture, C. *Pyrenoidosa* could also improve the water quality and avoid fodder corruption in water. However, methionine (0.35%) and tryptophan (0.38%) could be the primary and secondary limiting amino acid for most fish for their low concentration. It is said by JIA Peng and Luo that methionine content of fodder was positively correlated with fish quality (protein and fat) [10-11]. Therefore, it is necessary to pay attention to the supplement of methionine, cysteine and other amino acids. In addition, *C. Pyrenoidosa* also contains a lot of nutrient material, such as vitamin, carotene, which could play a positive role in improving immunity.

The growth ratio of *C. Pyrenoidosa* varied significantly among different culture media and conditions. The variation of amino acids components was relative modest. Neither HGZ nor Aquatic No. 4 can meet all the amino acids needs of fish. Extra ones would be added into the fodder. Therefore, the application of *C. Pyrenoidosa* as feeding or additive should take more consideration of its growth rate and reproducing stability (at high density) rather than the components of amino acid. **Thus, HGZ would be the best choice for aquaculture application.**

Furthermore, *C. Pyrenoidosa* is also widely used in water quality and algal bloom control [12-13]. However, *there* are many problems in the application of aquaculture, such as unstable algae quality. Therefore, the research on how to standardize the chlorella culture system, concentration, storage and transportation should be carried out to make greater and wider usage in aquiculture.

Acknowledgment
This research was financially supported by "Qing Lan Project of Jiangsu Province". This work was supported by "311 Talents Project in Taizhou", Research projects of the Jiangsu agri-animal husbandry vocational college（No. NSF201711; NSFHP201910）andInvestigation and evaluation on culture system of Chlorella Pyrenoidosa（11710119010）

References

[1] Yang, et al., *Growth of Chlorella pyrenoidosa in wastewater from cassava ethanol fermentation.* World Journal of Microbiology & Biotechnology, 2008. **24**(12): p. 2919-2925.
[2] Shen, N.N. and LI, *Application of Chlorella pyrenoidosa and Bacillus lichenifirmis for water quality control in shrimp culture.* Marine Fisheries Research, 2008.
[3] Jiang, D.S., et al., *Acute toxicity of three typical pollutants to aquatic organisms and their water quality criteria.* Environmental Science, 2014. **35**(1): p. 279-285.
[4] Wang, J., *Effects of transgenic Bt rice on several trypical aquatic organisms.* 2014, Yangzhou University.
[5] LIFei, GUOJian-lin, and ZHANGAi-ju, *Effects of three culture media on the growth of Chlorella and its application in culture of freshwater rotifers.* Journal of Biology, 2016. **33**(1): p. 109-111.
[6] Ma, X., W. Xue, and Y. Wei, *Effeets of Different Mediums and Shake on the Growth of Mieoreystis Aeruginosa and Produetion of Mieorcystins.* Journal of Shanxi Agricultural University(Natural Science Edition), 2009. **29**(2): p. 157-160.
[7] Gao, X., et al., *A comparative study on total lipid and fatty acid composition of five newly isolated marine diatoms.* Journal of Biology, 2014. **31**(1): p. 60-63.
[8] Deng, Y., *Effects of graded levels of dietary leucine on growth, flesh quality, intestinal immunity and antioxidant statusin young grass carp (Ctenopharyngodon idella).* 2014, Sichuan Agricultural University.
[9] Zhou, F., *Study on effects of dietary lysine and arginine on grouth performance and the arginine/lysin antagonism mechanism in juvenile black sea bream, Acanthopagrus schlegelii.* 2011, Zhejiang University.
[10] JIAPeng, et al., *Effects of dietary methionine levels on the growth performance of juvenile gibel carp (Carassius Auratus gibelio).* Acta Hydrobiologica Sinica, 2013. **37**(2): p. 217-226.
[11] Zhi, L., et al., *Dietary l -methionine requirement of juvenile grouper Epinephelus coioides at a constant dietary cystine level.* Aquaculture, 2005. **249**(1): p. 409-418.
[12] Xinjian, Z., et al., *Study on Chlorella pyrenoidosa and photosynthetic bacteria as water quality modulator in turbot larvae culture.* MARINE FISHERIES RESEARCH, 2008. **29**(6): p. 116-121.
[13] Olguín, et al., *Comparative sensitivity of Scenedesmus Acutus and Chlorella Pyrenoidosa as Sentinel Organisms for Aquatic Ecotoxicity Assessment: studies on a highly polluted urban river.* Environmental Toxicology, 2015. **15**(1): p. 14-22.

Research on the acute toxicity and teratogenicity of the feeding L-tryptophan

Fuxing Hao[1], Li Liu[1], Yaya Xu[1], Guilin Han[1], Yumei Zhang[2, 3*]

[1] Jiangsu Agri-animal Husbandry Vocational College, Taizhou, Jiangsu Province, 225300

[2] College of Veterinary Medicine Yangzhou University, Yangzhou, Jiangsu Province, 225009

[3] Collaborative Innovation Center for prevention and control of important animal diseases and zoonoses in Jiangsu Province, Yangzhou, Jiangsu Province, 225009

*Corresponding author and e-mail: Yumei Zhang, dwddwd98@126.com

Abstract. In order to evaluate the safety of the feeding L-tryptophan, we conducted acute toxicity test and teratogenicity test in rats. Rats were fed with L-tryptophan by gavage. The oral LD_{50} was more than 20.0g/kg b.w. In the teratogenicity test, the doses of L-tryptophan were 5.0g/Kg b.w., 2.5g/Kg b.w. and 1.25g/Kg b.w. We set up a normal control group. The rats were fed by gavage once a day from the 7th to 17th day of pregnancy. At the 20th day of pregnancy, we counted and measured the litter size, placenta weight, live baby weight, body length, tail length, fetus malformation rate, live fetus average malformation rate, maternal fetus malformation rate, fetus death rate, absorptive fetus rate and other indicators. Then we use SPSS12.0 software to carry out data statistical analysis, and One-Way ANOVA and t-test for comparison between groups. The results showed that, when each dose group compared with the control group, there was no significant difference in average body weight, body length, tail length, placenta weight, fetus live rate, fetus death rate and absorptive fetus rate when compared to the normal control group($P>0.05$), and there was no significant difference in the common deformity indexes, such as fetus malformation rate, live fetus average malformation rate and maternal fetus malformation rate between each dose group and the normal control group($P>0.05$). The oral LD_{50} of L-tryptophan to Wistar rats was more than 20.0g/kg b.w., and the traditional teratogenicity test was negative, indicating that the feed additive belongs to the actually non-toxic level and has no obvious teratogenicity.

1. Introduction

Tryptophan is an absolutely necessary amino acid for the growth, development and metabolism of animals and human body, which has a variety of physiological functions [1]. L-tryptophan, following after methionine and lysine, is another important amino acid added to feed, which can promote the growth of livestock and poultry, anti-disease and anti-high density feeding [2-4].

L-tryptophan is one of the necessary limiting amino acids in the body, and its synthesis in the body is less, so it needs to be taken from the outside. In vivo, starting from L-tryptophan, 5-hydroxytryptophan, indoleacetic acid, niacin, pigment, alkaloid, coenzyme and other biologically active substance can be synthesized. At present, it has been used as a food additive in many places in

China to enhance the utilization of protein, which plays an important role for fortified food to enhance the utilization of plant protein. In addition, L-tryptophan also have functions of preventing mildew, disinfection and preventing oxidation, which can be used as fish preservative.

L-tryptophan, through blood-brain barrier, can be transformed into 5-Hydroxytryptamine in vivo, and the final product is excreted from urine. In recent years, it has been found that L-tryptophan is not only used for intravenous nutrition infusion, but also has good effects on depression, insomnia, hypertension and pain relief, so it has been used in clinical practice in some European and American countries [5].

L-tryptophan, following after methionine and lysine, is the third largest amino acids added to feed. By adding L-tryptophan, we can adjust the balance of amino acids in the diet and promote the growth of livestock and poultry [6-7]. In addition, the L-tryptophan metabolite, 5-hydroxytryptamine, has the effects of anti high density, weaning and other stress effects in animals, and can enhance the resistance of animals. Because L-tryptophan production cost is gradually reducing, it has been used in pig feed in Japan and Europe and America.

China is a country lacking in protein resources, so it is of great potential value to study how to reduce the protein level of the diet and how to reasonably allocate the low protein diet on the basis of the existing research. On the other hand, the low protein diet can significantly reduce the nitrogen emissions and the odor of excreta in the process of animal production, and reduce the environmental pollution caused by livestock breeding. Therefore, the research and application of tryptophan and other amino acids are of great significance to the conservation of protein resources, the reduction of environmental pollution and the development of livestock breeding. In this paper, L-amino acids produced by a biotechnology company were used to carry out acute toxicity and traditional teratogenicity tests for rats, so as to preliminarily evaluate the application safety of L-amino acids.

2. Materials and methods

2.1. Materials

2.1.1 Drugs and experimental animals

The feeding L-tryptophan, L-tryptophan>20%, it is yellowish granule, the batch number is 20170610. Provided by a biotechnology company.

Preparation of test substance solution: after grinding of test substance L-tryptophan, use 1% sodium carboxymethylcellulose solution to prepare the test substance suspension with.

Wistar rats (clean grade), 10 weeks old, 80 females and 80 males, provided by Yangzhou University Comparative Medicine Center. Animal production license No.: SCXK (Su) 2017-0004. Use license No.: SYXK (Su) 2017-0044. Male and female rats were caged at 1:1. It is the first day of pregnancy when we can see the vaginal plug, which is used in traditional teratogenicity test.

2.1.2 Reagent

Ethanol fixation fluid: 80% ethanol;

Bouin's Fluid: take 750ml of saturated picric acid solution, 250ml of 40% formaldehyde and 50ml of glacial acetic acid, and mix them.

Potassium hydroxide solution: prepare 1% concentration and 2% concentration potassium hydroxide respectively.

Alizarin red stock solution: add alizarin red into the following mixture until it is saturated. Glacial acetic acid 5ml, pure glycerin 10ml, 1% chloral hydrate 60ml.

Alizarin red application solution: take 1ml of Alizarin Red stock solution, add 1% potassium hydroxide solution until 1000ml, and prepare it temporarily before use.

Dehydrated transparent liquid: I glycerol 20mL, 2% Potassium hydroxide solution 3mL,add distilled water to 100mL

II glycerol 50mL, 2% Potassium hydroxide solution 3mL, add distilled water to 100mL

III glycerol 75mL， add distilled water to 100mL.

2.2. Methods

2.2.1 Acute toxicity test

According to the toxicity test process of Han Shiqun [8], the rats were given 5.0 and 15.0g/kg by gavage respectively, and each dose was given to two male and female rats respectively. They did not die within one week. It shows that the toxicity of L-tryptophan in the feed of the tested sample is relatively small, and we can not measure the specific value of LD_{50}. According to the technical specification of acute toxicity test, the maximum dose test was carried out for the tested sample.

Take 20 rats, half of which are male and half of which are female, and carry out gavage twice within 24 hours with the operable highest concentration and the largest dosage volume (0.8 mL/100g b.w.). The interval of dosing was 6 hours, and the total dosage was20.0 g/kg b.w.. Observe for 2 weeks and record the poisoning symptoms and death of animals.

2.2.2 Teratogenicity test

(1)Acquisition and treatment of pregnant rats

Cage the female and male rats according to 1:1 and let them stay overnight. The next day, check whether there is a vaginal suppository. The day of discovery of the vaginal suppository is the first day of conception. Weigh, number and randomly group the pregnant rats that day. The experimental doses of L-tryptophan were 5.0g/kg b.w., 2.5g/kg b.w. and 1.25g/kg b.w., and distilled water was used as the normal control group. On the 7th to 17th day of pregnancy, dose them according to 1mL / 100g by gavage once a day, and weigh them on the 1st, 7th, 11th, 15th and 20th day of pregnancy, so as to adjust the dosage and observe the weight change of pregnant rats. Record the pregnant rats at any time. On the 20th day of pregnancy, let the cervical vertebrae of pregnant rats dislocate and die. Immediately cut the abdominal cavity along the midline and take out the uterus. Examine and record the number of corpus luteum on the left and right, the litter weight (conjoined rats in uterus), the weight of uterus, the number of live fetus, the number of dead fetus and the number of absorbed fetus.

(2)Fetal rat measurement and appearance abnormality examination

Cleaned the fetus, weigh the live fetus's body weight and placenta weight one by one, measur the body length and tail length, and identify the gender. Then we examined the appearance of fetal rat. Pay attention to the location, color, shape, size, quantity of each anatomical part and whether there are abnormal incisions or grooves.

2.2.3 Examination of skeletal deformity in fetal rats

First, make bone samples: put 2/3 of the fetal rats to be tested in 75%~90% ethanol, and fix them for more than 48 hours. Transfer fetal rats into 1% potassium hydroxide solution for 2~7d, during which we change the solution every 1~2d. After the invasion, take out the fetal rats and staine them. Transfer fetal rats into alizarin red solution for 2~3d, shake them 2-3 times a day, and dye their bones red. The colored specimens were transferred into the dehydrated transparent liquid I, II and III for 1d respectively. At this time, the color on the soft tissue gradually faded, and the bone was bright dark red.

Bone examination: pour the specimen and transparent solution into the plate, first observe the whole-body bone with the dissecting microscope, and then carry out the local examination for the abnormal parts with a larger magnification.

2.2.4 Examination of internal organ malformation in fetal rats

Put 1/3 of the fetal rats to be examined in Bouin's Fluid, fix them for 10-15 days, wash them with tap water, put them on the wax plate, cut off the limbs and tail, slice each part of the fetal rats with a blade, and examine the internal organs of the fetal rats.

Journal of Physics: Conference Series **1423** (2019) 012007 doi:10.1088/1742-6596/1423/1/012007

2.2.5 Calculation of deformity index

Calculate the total number of malformation fetus, the total number of deformities, the average deformity rate of live fetuses and the maternal deformity rate according to different dose groups. When we calculate the total number of malformations, each living fetus is recorded as a malformation fetus as long as it has one or more malformations. When we calculate the total number of malformations, the corresponding number should be included in the total number if a live fetus shows several kinds of malformations.

2.2.6 Data analysis

In this test, the data are expressed as mean± standard deviation. We used SPSS12.0 software for data statistical analysis, and use One-Way ANOVA and t-test for group comparison (test level α=0.05).

3. Result

3.1. Acute toxicity test

When the total amount of the feeding L-tryptophan was 0.0 g/kg b.w., no rats died. During the 2-week observation period after dosing, the animal's mental state was good, the hair color was bright and clean, the movement was free, the breath, appetite and feces were normal, there was no abnormal secretion in the mouth and nose, and we did not find any abnormal reaction related to dosing. After 14 days, all the animals were examined by pathology, and no obvious pathological changes and abnormalities were found in any organ.

The oral LD50 of rats fed with the feeding L-tryptophan was more than 20.0g/kg b.w.

3.2. Teratogenicity test

3.2.1 Weight changes of pregnant rats

The number of pregnant rats in each experimental group was: 11 in the normal group, the low dose group and the high dose group respectively, and 10 in the middle dose group. The behavior of pregnant rats during pregnancy was normal, without poisoning or death. Comparing the total average weight gain of each group, we can see that there is no significant difference between the pregnant rats in the experimental group and the control group (Table 1).

Table 1. Effects of the feeding L-tryptophan on the weight gain (g / rat / day) of pregnant rats

Dose group	Number of pregnant rats	Weight gain of pregnant rats				
		1-6days	7-11days	12-15days	16-20days	Total average weight gain
Normal control	11	2.18±0.68	3.55±0.75	1.66±2.66	7.07±2.03	3.64±0.69
Low dose group	11	2.70±0.72	2.11±1.46*	3.30±1.36	4.22±2.17*	3.05±0.58
Medium dose group	10	2.10±1.38	2.88±1.18	4.38±0.83*	4.88±1.74	3.45±0.75
High dose group	11	0.62±1.94*	4.93±4.09	5.36±1.38*	6.58±1.55	3.76±0.83

Note: In the same row values mark with "*" superscripts mean there has significant difference when compare with the control group ($p<0.05$)

3.3. Effect on reproductive function of rats

In Table 2, the mean litter weight of the high dose group is significantly better than that of the normal control group. The number of corpus luteum, number of implantation and number of live litter in low dose group and high dose group are significantly better than those in control group ($P<0.05$). There is no statistical difference in the average litter weight, number of corpus luteum, number of implantation and number of live litter between the medium dose group and the normal control group, and there is no statistical difference in the number of absorbed fetus between each dose group and the normal control group($P>0.05$).

Table 2. Effects of the feeding L-tryptophan on the reproductive function of rats

Dose group	Number of pregnant rats	Litter weight (g)	The number of corpus luteum	Implantation number	Number of live litters	Number of absorbed fetus
Normal control	11	53.23±10.16	9.55±1.31	9.00±1.64	8.64±1.92	0.36±0.46
Low dose group	11	56.41±5.78	11.09±1.39*	10.82±1.22*	10.36±1.19*	0.45±0.58
Medium dose group	10	48.54±11.12	10.10±1.30	10.00±1.20	9.80±1.40	0.20±0.32
High dose group	11	61.63±5.98*	11.18±0.96*	11.27±1.07*	10.82±1.07*	0.27±0.40

Note: In the same row values mark with "*" superscripts mean there has significant difference when compare with the control group ($p<0.05$)

3.4. Toxicity to rat embryos

In Table 3 and table 4, there is no significant difference in the number of dead fetus, the number of absorbed fetus, absorbed fetus rate, fetus death rate, fetus live rate, the number of litter, the littler average body weight, body length, tail length and placental weight between each dose group and the normal control group($P>0.05$).

Table 3. Toxicity of the feeding L-tryptophan to the rat embryos

Dose group	Number of pregnant rats	Total litter number	Number of live litters	Number of death litter	Number of absorbed fetus	Absorbed fetus rate %	Fetus deat h rate %	Fetus live rate %
Normal control	11	95	95	0	4	3.06	0	95.96
Low dose group	11	114	114	0	5	3.39	0	95.80
Medium dose	10	98	98	0	2	2.00	0	98.00
High dose group	11	119	119	2	3	2.42	1.65	95.97

Note: fetus live rate (%) = number of live fetus / numbers of implantation

Note: if there is No * marked on the shoulder of the same column of data in the table, it means that there is no significant difference when compared with the control group ($P>0.05$).

Journal of Physics: Conference Series **1423** (2019) 012007 doi:10.1088/1742-6596/1423/1/012007

Table 4. Effects of the feeding L-tryptophan on the growth of fetal rats

Dose group	Number of pregnant rats	Litter rat / nest	Weight (g)	Body length (cm)	Tail length (c,)	Placenta weight (g / litter)
Normal control	11	8.6±1.9	3.94±0.32	3.63±0.17	1.28±0.09	0.73±0.10
Low dose group	11	8.9±1.2	3.71±0.37	3.56±0.18	1.25±0.09	0.66±0.14
Medium dose group	10	9.8±1.4	3.8±0.45	3.49±0.21	1.21±0.08	0.66±0.12
High dose group	11	9.7±1.1	4.12±0.32	3.67±0.16	1.30±0.09	0.65±0.08

Note: if there is No * marked on the shoulder of the same column of data in the table, it means that there is no significant difference when compared with the control group (P>0.05).

3.5. Effect on fetal rat deformity

There is no significant difference(P>0.05) between the test group fed with the feeding L-tryptophan and the normal control group in the appearance deformity rate, skeleton deformity and internal organ deformity, such as malformation fetus rate, live fetus malformation rate and maternal malformation fetus rate (Table 5).

Table 5. Teratogenic effect of the feeding L-tryptophan on rats

Dose group	Number of pregnant rats	Number of live litters	Number of malformation fetus	Number of maternal malformation fetus	Live fetus malformation average rate (%)	Maternal malformation fetus rate (%)	Malformation		
							Appearance	Skeleton	Internal organs
Normal control	11	95	1	1	1.05	9.09	0	1	0
Low dose group	11	114	1	1	0.88	9.09	0	1	0
Medium dose group	10	98	0	0	0	0	0	0	0
High dose group	11	119	1	1	0.84	9.09	0	1	0

Note: if there is No * marked on the shoulder of the same column of data in the table, it means that there is no significant difference when compared with the control group (P>0.05).

4. Discussion

In the study of teratogenesis, it can be divided into traditional teratogenesis test and feeding reproduction teratogenesis test. It is mainly due to the different time to give the test substance. The traditional teratogenesis test is to give the test substance within the time of organ formation during embryo development, and the feeding reproduction teratogenesis test is divided into first-generation, second-generation or third-generation feeding reproduction test. In the first generation feeding reproduction test, the test substance is given during the formation of sperm or egg until the fetus is delivered, and then investigate the teratogenicity of the test substance. In the second or third generation reproduction teratogenesis test, we continue to give the test substance after the weaning of the offspring until the F2 or F3 generation is delivered, and investigate the teratogenicity to the offspring after a long time of contact of the test substance.

In teratogenesis, the most sensitive stage to teratogens is the organogenetic period, which is generally called the critical or dangerous stage. The traditional teratogenesis test is to give the test substance in the organogenetic period to let us understand the teratogenesis of the test substance to the embryo. In common experimental animals, the organogenetic period of rats, calculated from the date of fertilization, is about 9-17 days, small mice is 7.5-16 days, and rabbits is 11-20 days. In organogenetic period, the contact between teratogen and embryo may cause abnormal morphology and structure. However, if the contact between teratogen and embryo occurs in the blastocyst formation stage before implantation, embryo often dies and there are few deformities. The blastocyst formation period of rats, calculated from the date of fertilization before implantation, is about 3-4 days, and the time of implantation beginning is about 5.5-6 days; small mice is 3-4 days and 4.5-5 days; rabbits is 3-4 days and 7 days [9].

L-tryptophan is one of the three amino acid feed additives except tyrosine and methionine. If the L-tryptophan supply in animal body is insufficient, there will be a series of maladjustment symptoms in animals, such as slow accumulation of fat and slow growth of the body. In male animals, testicular dysplasia sometimes occurs [10]. With the continuous improvement of fermentation technology, L-tryptophan, as a feed additive, its application prospect will be more and more broad.

5. Conclusion

In the acute toxicity experiment, the oral LD_{50} of rats fed with the feeding L-tryptophan was more than 20.0g/kg b.w. According to the WHO acute toxicity classification standard, the non-toxic level is that the LD50 of rats is more than 15.0g/kg b.w., which indicates that L-tryptophan is non-toxic. The traditional teratogenicity test of the feeding L-tryptophan for the test substance is negative, which shows that it has no obvious teratogenesis effect on rats, and even in the number of corpus luteum, the number of implantation, the number of live litter and other aspects, L-tryptophan is significantly better than the normal control, indicating that L-tryptophan is safe in teratogenesis when used as feed additive.

Acknowledgment

This research was financially supported by "Qing Lan Project of Jiangsu Province". This work was supported by "Training program of innovation and entrepreneurship for college students in Jiangsu Province (201912806096y)" and Jiangsu Agri-animal Husbandry Vocational College (Grant No. NSFPT201828), (Grant No. NSFPT201825).

Reference

[1] Cui Qin, Cui Shan. Research progress of tryptophan nutrition [J]. Chinese feed, 2003, 15:20-22.
[2] Li Jianxin, Zhang Xumei, Xu Qishou. The physiology and biochemistry of tryptophan and its research [J]. Amino acids and biological resources, 2005, 27(3):58-62.
[3] Xu Guomao, Wang Zheng, Wei Hua, etc. Action mechanism of tryptophan and its application in pig production [J]. Chinese pig industry, 2009, 2:54-56.
[4] Ettle T, Ruth F X. Specific dietary selection for tryptophan by the pig-let [J]. J.Ani.Sci., 2004, 82(4):1115-1121.
[5] Song Wenxia, Wang Ruiming. Study on L-tryptophan [J]. Agricultural products processing, 2005, 34(3):18-21.
[6] Yu Dongyou, Zhou Bin, Rao Wei, etc. Effects of the feed tryptophan levels on performance and egg quality of laying hens [J]. 2010, 22(5):1265-1270.
[7] Ma Yu'e, Zhan xiu'an, Zhu Qiaoming, etc. Effects of the feed tryptophan levels on performance, antioxidant function and serum biochemical indexes of yellow-feathered broilers [J]. 2011(12):122-126.
[8] Han Shiqun, Xu Yudi, Zhou Qing, etc. Studies on the composition of acidolysis products of cyanobacteria and its feeding safety [J]. Journal of animal nutrition, 2019.31(4):1842-1849.
[9] Hao Fuxing, Zhou zewei, Zheng Lingmin, etc. Teratogenicity test of new microecological feed additive [J]. Jiangsu agricultural science, 2015, 43(7):218-219.
[10] Zhang Weiguo, Qian He. Production technology and application of amino acids. Beijing: China Light Industry Press,1997.

Empirical analysis on intra-industry trade in textile between China and Vietnam

Lei Yao[1] and Zheng Yang[2]

[1]School of Business, Beijing Institute of Fashion Technology, Beijing, 100029, P.R. China
[2]School of Business, Beijing Institute of Fashion Technology, Beijing, 100029, P.R. China

Corresponding author and e-mail: Lei Yao, sxyyl@bift.edu.cn

Abstract. The total volume of import and export trades between China and Vietnam exceeded 723 billion yuan in 2018. A comprehensive analysis of the Sino-Vietnamese intra-industry trade in textile is presented. By calculating G-L index and Bruelhart index, this paper analyzed the current situation of intra-industry trade of textile between China and Vietnam. A regression econometric model was established to explore factors affecting the intra-industry trade of textile. The study shows that adjusting textile industry structure and optimizing trade environment between China and Vietnam, as well as strengthening trade liberalization, are of great significance for reference, and are beneficial to promoting coordinated and sustainable development of economic and trade between China and Vietnam.

1. Introduction

In recent years, trade cooperation between China and Vietnam has continued to deepen. As of 2018, China as Vietnam's largest trading partner has continued to make efforts to promote trade between the two countries for 14 consecutive years. According to the statistics of the Vietnam Textile Association, there are currently more than 4,000 textile enterprises in Vietnam, and the scale of the industry has expanded year by year. Therefore, textile trade plays a very important role in the economies of both China and Vietnam. At present, China's textile industry needs to seek new economic and trade growth points through transformation. During this critical period, China needs to strengthen cooperation with other countries to obtain more textile raw materials and products to meet what it needs. This paper believes that analyzing and studying the intra-industry trade in textile industry between China and Vietnam can not only promote the economic and trade development of the two countries, but also have certain reference significance for strengthening further cooperation between the two countries in other areas.

2. Analysis of the status quo of Sino-Vietnamese intra-industry trade in textile

China's overall import and export of textiles to Vietnam has shown a growing trend. During 2008-2017, the trade balance between the two countries has been positive, and the gap is getting bigger and bigger. It shows that China's export trade to Vietnam is greater than that of China's imports from Vietnam, that is trade is in an unbalanced state. Based on the product trade structure data, the

G-L index and Bruelhart index are used to analyze the current status of Sino-Vietnamese textile intra-industry trade.

2.1. Sino-Vietnamese main textile G-L index

According to the data obtained in the UNCOMTRADE statistical database, the G-L index calculation and analysis of different types of textile trade between China and Vietnam were carried out, and the following results were obtained (Table 1).

Table 1. G-L Index of Sino-Vietnamese Major Textile Trade from 2007 to 2017.

category	2007	2008	2009	2010	2011	2012	2013	2014	2015	2016	2017
651	0.427	0.832	0.748	0.878	0.918	0.900	0.885	0.797	0.643	0.576	0.562
652	0.012	0.020	0.007	0.004	0.004	0.005	0.004	0.004	0.010	0.008	0.042
653	0.072	0.070	0.064	0.065	0.054	0.040	0.029	0.017	0.012	0.012	0.018
654	0.053	0.042	0.015	0.008	0.003	0.004	0.003	0.002	0.004	0.011	0.006
655	0.022	0.016	0.016	0.037	0.051	0.039	0.033	0.032	0.065	0.106	0.130
656	0.103	0.044	0.050	0.032	0.033	0.046	0.043	0.034	0.033	0.028	0.021
657	0.054	0.064	0.074	0.054	0.070	0.072	0.070	0.080	0.057	0.074	0.077
658	0.307	0.245	0.091	0.059	0.070	0.066	0.080	0.073	0.092	0.124	0.148

According to the results, there are inter-industry and intra-industry trades in the Sino-Vietnamese textile industry, but most of the textiles are traded between China and Vietnam based on the inter-industry trade model. Only textile yarns such as 651 yarn, wool yarn, cotton yarn and hemp yarn are mainly based on intra-industry trade, showing a trend of polarization. In general, the difference in the level of intra-industry trade between yarn-based textiles and other types of textiles is still very high. The level of intra-trade trade in raw materials is maintained at a relatively high level while that in intermediate and finished textiles is low. This can explain to a certain extent that China's textiles mainly participate in the Sino-Vietnamese textile trade with the comparative advantage of lower labor costs. Although the import and export volume is large, the products are still mainly low-end.

2.2. Analysis of Sino-Vietnamese Textile Bruelhart Index

The difference between the Bruelhart index and the G-L index is that the Bruelhart index measures the dynamics of intra-industry trade based on trade flows, which makes up for the defects of the G-L index [1]. According to the marginal intra-industry trade index formula, the marginal intra-industry trade index of China and Vietnam textiles from 2007 to 2017 is calculated. From a dynamic perspective, it reflects the extent of the changes in the intra-industry trade of textiles at different times. The results calculated from the previous data are shown in Figure 1.

As shown in the chart, from a dynamic perspective, the trend of intra-industry trade in China and Vietnam has increased and decreased, and the fluctuations are more intense. The years of China-Vietnam textiles' overall marginal intra-industry trade index greater than 0.5 are 2008-2009, 2011-2012, 2015-2016, 2016-2017, indicating that in these time spans, the increase in Sino-Vietnamese textile trade is mainly caused by intra-industry trade. The marginal intra-industry trade index for the remaining six time periods is less than 0.5, indicating that the increase in Sino-Vietnamese textile trade is mainly caused by inter-industry trade. As can be seen from the figure, since 2013, the Sino-Vietnamese textile marginal intra-industry trade index has generally grown steadily.

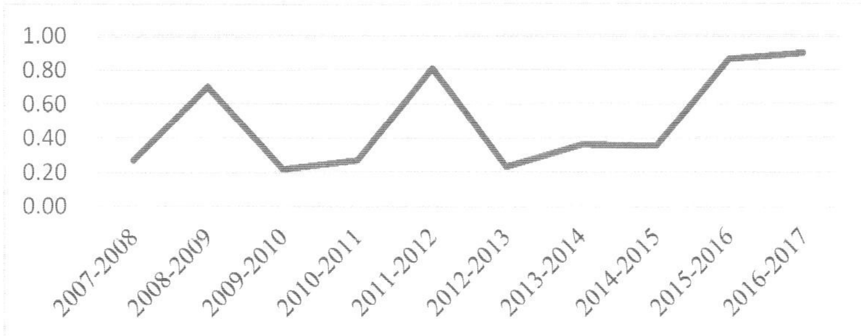

Figure 1.Bruelhart Index of Sino-Vietnamese Textile in 2007-2017

3. Empirical Analysis on Influencing Factors of Sino-Vietnamese Textile Industry Trade

3.1. Theoretical assumptions and variable selection

There are many factors affecting intra-industry trade. According to whether the various factors are quantifiable and whether sufficient sample size can be obtained, this paper finally selects and quantifies five influencing factors to measure the intra-textile trade between China and Vietnam.
(1)Differences in per capita income: The higher the per capita income level, the more obvious consumers feel about the differentiation of textiles, which will make the division of labor more refined.
(2) Trade imbalance: Therefore, this paper believes that the trade imbalance is negatively related to the level of Sino-Vietnamese intra-industry trade in textiles. This article uses China's Vietnam import and export textile trade balance (China's textile trade volume to Vietnam - China's textile trade volume from Vietnam) to reflect the textile trade balance between the two countries [2]. (3) China's direct investment in Vietnam: This paper assumes that China's FDI to Vietnam is positively related to the level of intra-industry trade in textiles. This article uses China's FDI data for Vietnam to reflect China's investment in the Vietnamese textile industry. (4) Market size: This paper assumes that the scale of marketization is positively related to the level of intra-industry trade in textiles of the two countries. This paper uses the average market size (average of China and Vietnam's GDP) to reflect market size. (5)Official launch of the China-ASEAN Free Trade Area: The China-ASEAN Free Trade Area was officially completed on January 1, 2010. Therefore, it is assumed that the official launch of the China-ASEAN Free Trade Area is positively related to the level of Sino-Vietnamese intra-industry trade in textiles. This article sets this variable as a dummy variable [3].

3.2. Model construction and the results

3.2.1 Model setting

Based on the data of 18 years from 2000 to 2017, the G-L index of textiles in the two countries is taken as the explained variable. Creating a double logarithmic model with the following model [4]:

$$\ln y = C + \beta_1 \ln x_1 + \beta_2 \ln x_2 + \beta_3 \ln x_3 + \beta_4 \ln x_4 + \beta_5 D_5 + \mu \qquad (1)$$

y is the index of Sino-Vietnamese intra-industry trade in textiles, $x_1$is the absolute value of the Per capita GNI difference between China and Vietnam.x_2 is the absolute value of the textile trade balance, x_3 represents China's stock of direct investment in Vietnam, $x_4$represents the average GDP of China and Vietnam, and D_5indicates whether the China-ASEAN Free Trade Area has been launched (dummy variable).

3.2.2 The results of models

Finally, a regression result with good fitting in all aspects is obtained, as shown in table 2.

Dependent Variable: LNY

Sample: 2000 2017

Included observations: 18

Table 2. Model regression results

Variable	Coefficient	Std. Error	t-Statistic	Prob.
C	-26.12974	3.873203	-6.746286	0.0000
LNX1	2.380001	0.602858	3.947863	0.0017
LNX2	-1.204772	0.174087	-6.920533	0.0000
LNX4	0.864724	0.474261	1.823308	0.0913
D1	0.516973	0.215358	2.400526	0.0321

R-squared	0.912299	Mean dependent var	-1.698138
Adjusted R-squared	0.885314	S.D. dependent var	0.514196
S.E. of regression	0.174134	Akaike info criterion	-0.427851
Sum squared resid	0.394194	Schwarz criterion	-0.180526
Log likelihood	8.850663	Hannan-Quinn criter.	-0.393749
F-statistic	33.80781	Durbin-Watson stat	1.369907
Prob(F-statistic)	0.000001		

The regression results of the model show that the remaining four variables have passed the economic meaning test. The regression estimation results show that the R value of the model is 0.912299, the P value is 0.000001, and the P values of the four variables are all less than 0.1, so the model has been tested and the fitting effect of the equation is better. It shows that the difference of per capita income between China and Vietnam, trade balance, market size and the official launch of China-ASEAN free trade area have a significant impact on the index of Sino-Vietnamese intra-industry trade in textiles.

3.3. Model Analysis

(1) The difference of variable per capita income (X1) is positively correlated with the index of Sino-Vietnamese intra-industry trade in textiles, which is in line with theoretical assumptions that the degree of trade balance is negatively correlated withthe index of Sino-Vietnamese intra-industry trade in textiles. According to the results of the regression model analysis, the index of Sino-Vietnamese intra-industry trade in textiles will increase by 2.38 percentage points for every 1 percentage point increase in per capita income under the same other conditions. Therefore, increasing the difference in per capita income between China and Vietnam will promote the development of intra-industry trade between the two countries [5].

(2) The trade balance (X2) is negatively correlated with the index of Sino-Vietnamese intra-industry trade in textiles, which accords with the theoretical assumption. According to the results of the regression model analysis, the index of Sino-Vietnamese intra-industry trade in textiles will fall by 1.20 percentage points for every 1% increase in the textile import and export balance between China and Vietnam. Therefore, in order to improve the level of textile trade development between China and Vietnam, it is necessary to reduce the trade balance between the two countries [6].

(3) The market size (X4) is positively correlated with the index of Sino-Vietnamese intra-industry trade in textiles, which is in line with theoretical assumptions. According to the results of the

regression model analysis, the index of Sino-Vietnamese intra-industry trade in textiles will increase by 0.86 percentage points for every 1% increase in the average GDP of China and Vietnam. With the expansion of market scale, people have higher requirements for textiles, so as to promote the differential production of textiles and improve the level of domestic trade development in the textile industry between China and Vietnam [7].

(4) The official launch of the China-ASEAN Free Trade Area (D1) is positively related to the index of Sino-Vietnamese intra-industry trade in textiles, that is to say, the comprehensive completion of the China-ASEAN Free Trade Area is conducive to the improvement of the internal trade level of the textile industry in China and Vietnam.

4. Conclusion

China should implement a product differentiation strategy to optimize the product structure of trade textiles between the two countries. China's textile industry is in a very critical stage of transformation and development. It is undergoing an industrial structure upgrade. China has gradually lost its competitive edge in terms of price [8]. In order to improve the economic benefits brought by China's textile industry, China needs to accelerate the adjustment and upgrading of the industrial structure of the textile industry. According to the principle of comparative interests, both China and Vietnam can make reasonable investments, especially pay attention to intra-industry investment, and gradually form a reasonable division of labor in textile production and export, which can stimulate the healthy and sustainable development of intra-industry trade between China and Vietnam.

Acknowledgement

(1) This project is supported by Research on the Cultivation of Beijing Residents' Awareness of the Green Consumption Concept and Related Educational Approaches (SZ20171001209).

(2) This project is supported by Cooperative Development Research on Used Clothing Recycling and the Resource Utilization System in the Beijing-Tianjin-Hebei Region (PXM2016_014216_000022).

References

[1] Grubel H G, Lloyd P J, Intra-industry trade: the theory and measurement ofinternational trade in differentiated products, Journal of International Economics. 1975, 85(339):312-314.

[2] He Fangnan, Research on trade competitiveness of textiles and garments in Vietnam, East China University of Science and Technology. 2017.

[3] Cheng Lu, Ning Ling, Research on Trade Enhancement between China and Vietnam, Journal of Guangdong Ocean University. 2016, 36(02): 39-44.

[4] Wang Na, Empirical Analysis of Intra-industry Trade in China's Textile Industry, Journal of Shandong Institute of Business and Technology. 2013, 27(02): 11-16.

[5] Liu Jianbo, "The theory of intra-industry trade and the structure of China's foreign trade model" [D], Fudan University. 2006.

[6] Balassa B, The Determinants of Intra-industry Specialization in United States Trade, Oxford Economic Papers.1986,38(2):220-33.

[7] XuPanpan, The status quo of textile industry in Vietnam, Textile Machinery. 2017 (09): 46-47.

[8] Cao Fang, An Empirical Study on the Status Quo and Determinants of China's Textile and Apparel Industry-Based on Panel Data Analysis from 2002 to 2010, International Business Research. 2012, 33(05):23-34.

Research on the guidance learning and monitoring system of autonomous learning for College English

Lu Wen

College of Foreign Studies, Guilin University of Electronic Technology, Guilin, China

Corresponding author and e-mail: Lu Wen, 10810489@qq.com

Abstract. This paper aims to find out the influence of guidance learning and monitoring on autonomous learning for College English. By applying the qualitative and quantitative research approach, we tested the efficiency of the guidance learning and monitoring system of autonomous learning by a semester-long teaching practice. We compared the final score with that of the previous semester. After analyzing the data, we concluded that compared with the previous semester, the grades of performance scores of the experimental semester were significantly improved as well as the final scores. This shows that the effect of implementation of guidance learning and monitoring system to autonomous learning in College English is obvious. Finally, this paper provides some strategies to improve the effect of autonomous learning, and points out that there is a harmony relationship between autonomy and monitoring.

1. Introduction

The information revolution has brought about the continuous change of teaching and learning methods of College English, which has made the deep integration of information technology and teaching. Various teaching resources have emerged in an endless stream, teaching methods have been constantly updated, and teaching platforms have been increased. College English Curriculum Requirements (2017) point out that classroom teaching should combine with autonomous learning in college English curriculum in order to stimulate personalized learning strategies and promote the development of students' autonomous learning ability. Holec (1981) considers that autonomous learning means students are responsible for their own learning. For learners, with the help of teachers, it is necessary to determine the learning objectives, contents and methods, and then determine the learning time, place and progress, and finally evaluate their own learning effect. Dickinson (1995) believes that autonomous learning is "the attitude towards learning and the ability to learn independently".

2. Literature review

Holec (1981) integrated autonomous learning into foreign language teaching. Since the beginning of the 21st century, the study of autonomous learning has been favored by foreign language experts and scholars. Under the influence of computer network technology, in order to cultivate students'autonomous learning ability, a new college English teaching model supported by computer network technology has been established (Wang Shouren 2011). In order to meet students' individual needs, many researchers have explored the blended online and offline teaching model of College

English, aiming to cultivate good habits of autonomous learning, improve autonomous learning ability of English, and finally improve students' comprehensive ability of using English. According to Chen Jian-lin, Wang Jing (2016), this combination has become the dominant mode of College English teaching in China. Chen Dong-chun (2011) regards more specific teaching mode should be implemented face-to-face teaching of oral English in small classes plus reading and writing and translation courses in classroom teaching, while Network autonomous learning is audio-visual and oral autonomous learning. It can be seen that the current College English teaching has been deeply integrated and inseparable with information technology, which also shows that autonomous learning plays a vital important role in English learning. Therefore, many researchers prove the importance of autonomous learning from the perspectives of learning motivations, learning strategies, sense of self-efficacy, sense of social support and sense of belonging to the class and so on.

However, there are many factors that determine the level of students' autonomous learning ability, and teachers cannot achieve the expected teaching effect by implementing blended teaching mode. Professor Xu Jin-fen (2014) pointed out that although College English teachers have attached great importance to cultivating students' autonomous learning ability in the past ten years, the overall situation is still not ideal, and students' autonomous learning ability will not improve with the increased grade. In the process of autonomous learning practice, there are some unpleasing phenomena (Chen Dong-chun 2006). Without clear learning objectives and corresponding strategies, many students cannot effectively manage their own time by showing no specific learning plan or poor execution of the plan. To solve these problems, teachers and teaching administrators need to monitor, intervene and help them (Hou Xiao-hong, 2013). Therefore, cultivating students' autonomous learning ability requires teachers to explore and guide the students inner potential.

According to Metacognition Theory, cognitive subjects possess the abilities of self consciousness, selfmonitoring and self adjusting cognitive activity. Therefore, the effective integration of Metacognitive monitoring theory into teaching practice, especially students' autonomous learning, will achieve good results. The cognitive subjects are students, and the cognitive activity is autonomous learning. Students have self-understanding, self-monitoring and self-adjustment after a stage of autonomous learning. Finally, students' self-monitoring ability should be strengthened (Fu Zheng, 2014).

Therefore, some scholars proposed to monitor students' autonomous learning in order to improve the effect of autonomous learning. Some guided students to monitor their learning process and adjust their learning strategies. Some proposed the guiding principle and monitoring system of English autonomous learning to promote Metacognition. He Ming-xia (2012), starting from the autonomous learning monitoring, analyzed the current situation and necessity of the autonomous learning monitoring of College English in China, and then analyzed the objectives, types and principles of the autonomous learning monitoring, and finally focused on the contents and the monitoring strategies. Yin Hua-dong (2014) pointed out that the relationship between autonomous learning ability and academic performance was overemphasized.

In conclusion, a current prominent problem in the autonomous learning of College English is that students lack the awareness of self-management and the ability of self-decision-making. Many students do not have clear learning objectives and lack effective monitoring from teachers and classmates in the learning process, because of the low autonomous learning ability. Therefore, in order to improve students' autonomous learning ability and comprehensive application of English language, students must first understand its purpose and create a complete and multi-channel monitoring system. Therefore, it is particularly urgent and important to construct the guidance learning and monitoring system of College English autonomous learning and apply it to the practical teaching.

3. The construction of guidance learning and monitoring system of autonomous learning

3.1. The construction of the guidance learning system of College English

In this study, Book 4 of New HorizonCollege English--Viewing, Listening &Speaking (second edition) is selected to design the course guidance and unit guidance. The overall objective of the curriculum is to cultivate international talents. Through systematic teaching activities, this course takes satisfying students' personality and needs as the core of its work, helps students develop autonomous learning habits and abilities, and improves their English language proficiency and practical ability. According to the specific requirements of College English, students should achieve the following goals. From the listening perspective, students should master some basic listening skills, understand the teacher's English teaching and daily English conversation.At the same time, when they listen tosome English broadcasts (such as VOA Special English), they need to grasp the main idea and the key points. From the oral English perspective, students should master the basic oral communication skills, and can answer the teacher's questions accurately in English in class.And they can make a speech about 3 minutes on a specific topic discussionclearly and fluently.

The teaching schedule is 14 weeks in a semester and 28 class periods aiming at finishing five units of the teaching material, ie., Book 4 of New HorizonCollege English--Viewing, Listening &Speaking (second edition). Finally, we will take the formative assessment, that is, the total final assessment score is equal to 50% of the final exam score plus by 50% of the performance score. By the way, the performance score includes 10% performance score of the reading and writing class, 10% of the reading and writing autonomous learning, 10% of the Viewing, Listening &Speaking autonomous learning, and 10% of final oral examination. Unit guidance learning is based on the unit theme, including the learning tasks, questions, key points and evaluation criteria of the unit.

3.2. The construction of autonomous learning monitoring system of College English

When we refer to autonomous learning, it is by no means learning without teachers' participation (Benson Voller, 1997:63). In the network environment, students' autonomous learning of College English does not mean that students can learn without any constraints. Both the administrators and English teachers should carry out multi-directional monitoring on students' autonomous learning, including the Director of Teaching and Research Section, the Director of the Autonomous Learning Center and the other administrators.

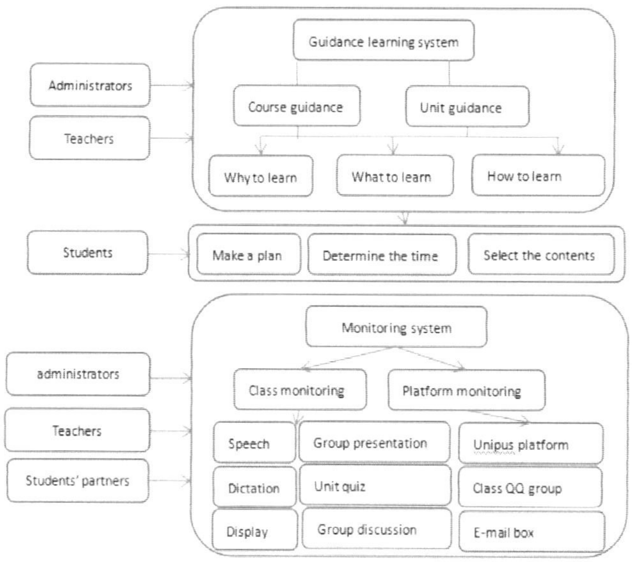

Figure 1. The guidance learning and monitoring system of autonomous learning

The monitoring system includes two aspects, one is the monitoring of the network autonomous learning platform and the other is the classroom monitoring. Platform monitoring mainly refers to the monitoring of students' completion of tasks on the platform by managers or teachers, including online hours, the number of completed assignments and discussions published, and finally the effect of autonomous learning. We release homework reminder and the latest platform learning feedback to students through the class QQ group, making students know their learning progress and effect in a timely manner. At the same time, the teaching assistant of the class helps the teacher to remind and supervise the students who have not finished the homework with high quality and quantity. When referring to the classroom monitoring, it mainly includes a series of classroom activities such as the theme speech, pair discussion, group discussion and in-class quiz and so on. Finally, a "four-in-one" monitoring system is formed among individual students, teachers, administrators and students' partners which will be shown as follows.

4. A practical study on the guidance learning and monitoring system of autonomous learning

4.1. Teaching practice of guidance learning and monitoring system

The object of this study is a total of 176 students in three classes of Grade 2014, School of Mechanical and Electrical Engineering, Guilin University of Electronic and Technology. The teaching experiment lasts for one semester, with a total of 14 weeks and 28 hours of the Viewing, Listening & Speaking courses. There are five units are taught in class, among which 4 class periods are the autonomous learning hours, and teachers just guide students' autonomous learning. The last 2 class periods are for oral examination, and the remaining 22 class periods are for classroom teaching. The effect of students' autonomous learning is tested in class and on the Unipus Platform. Autonomous learning began with the course guidance learning. In the first class, the teacher introduced the learning objectives, progress, contents of each unit, time arrangement of online autonomous learning, specific content and assessment approach of the course to the students in detail. After receiving the tasks, students began to make their own learning plans, determine the learning time and choose the contents of the learning sequence. After that, teachers and students discussed the suitable learning strategies.

Unit guidance learning is carried out in the classroom and on the Unipus Platform respectively. The Platform is used to release assignments and check the effect of autonomous learning. Specific learning objectives and contents are defined in class and listening contents are discussed in depth. Let's take unit 1 *Enjoy Your Feelings* of New HorizonCollege English--Viewing, Listening &speaking (second edition) as an example. There are three learning objectives, that is, learn how to describe different emotions, the listening skills of inferring, and the speaking skills of promising and refusing to help others. In order to have enough time for discussion in class, we chose Task1 and Task3 of the Basic Listening Practice part in class, and let the students finish the rest in the Unipus Platform. At the same time, in order to make students fully understand the joys and sorrows in life, we arranged the topic *How do you understand happiness* for oral speech after class and so we will discuss the topic in depth in class. When students gave a speech in class, the rest of the students need to listen carefully and fill in a form, which includes the topic of the speech, the general points and supporting details, and finally, they need to make an objective evaluation of the speech. This is a big challenge for students, because they have to understand the presentation and write down some points. Every week, there will be a topic related to the unit theme. During the semester, students' critical thinking abilities are greatly trained. In order to consolidate the knowledge they have learned, students also need to complete the vocabulary test on the UnipusPlatform after the unit learning. We believe that autonomous learning does not mean learning by the students themselves. It must be combined with classroom teaching. Only through monitoring between teaching administrators, teachers and students' partners can students effectively complete autonomous learning content and obtain good learning effect.

4.2. Teaching effect

Afterthe teaching experiments for one semester, we compared the performance scores, final assessment scores and total assessment scores in one of the classes. The statistical softwareSPSS22.0 is applied to calculate the mean and standard deviation, and Pairs Samples Test is conducted, which is shown as follows.

Table 1. Paired Samples Test of Class 1

		Paired Differences					t	df	Sig. (2-tailed)
					95% Confidence Interval of the Difference				
		Mean	Std. Deviation	Std. Error Mean	Lower	Upper			
Pair 1	Performance score1 - Performance score2	-13.73770	13.57375	1.73794	-17.21410	-10.26131	-7.905	60	.000
Pair 2	Test scores1 - Test scores2	1.35246	11.44055	1.46481	-1.57760	4.28252	.923	60	.360
Pair 3	Total score1 - Total score2	-6.63934	8.95923	1.14711	-8.93391	-4.34478	-5.788	60	.000

As it can be seen from Table 1, the mean values of performance scores and total scores of class 1 are-13.73770 and-6.63934 respectively, and the standard deviations are 13.57375 and 8.95923 respectively. At the same time, all their P values are 0.000, which are less than 0.05. That is, if the P values are in the 95% of the confidence interval, it indicates that there is a significant difference of the average score between the performance score and the total score of the class 1 in the two semesters. The experimental results show that the guidance learning and monitoring of College English autonomous learning is conducive to the improvement of students' performance scores and total evaluation scores. However, language learning is a long process and it is impossible to get immediate results in a short time, so the improvement of students' English proficiency cannot be reflected in a short time.

4.3. Implications for College English teaching

Although the implementation of online autonomous learning guidance and monitoring of College English has made great progress in students' performance and final grades, some problems have also been exposed. Therefore, we also put forward targeted solutions to these problems.

4.3.1. Enrich the content of autonomous learning.

From the point of view of the learning time of students, some students just spend less than 5 hours on the Unipus Platform, which means that these students only participate in autonomous learning for the purpose of obtaining autonomous learning scores. In addition to their passive status, many students even download the answers from the Internet then copy and paste them. The reason for this is that there are many factors that interfere with autonomous learning. A large part of the reason is that students only need to complete the listening which is matched with the textbook, and the content is not colorful. Some students regard their listening skills are poor, so they have to copy and paste their answers in order to get high marks. Therefore, in addition to the descriptions in the China's Standards of English Language Ability, we can construct more materials fitted for students at different levels. There should also be a variety of forms, such as tests, exercises, high-quality audio, video, MOOCs and micro-lectures, which are both interesting and related to culture.

4.3.2. Improve network platform services.

As the platform provided for students is the Unipus Platform, which is a supporting platform by Foreign Language Teaching and Research Press. So, students can log in only via the campus network. Students are forced to go to libraries, Foreign Language Autonomous Learning Center or places with campus networks for their autonomous learning, which causes great trouble to students. In addition, during the peak period, the system often broke down or could not save the data when they finished the tasks. Sometimes, even if it was safe to save the data, there was still something wrong with the scores. As a result, students had to do it again and again, wasting a lot of time with no harvest. Therefore, the service of the Unipus Platform should be improved, and the convenience and mobility of online autonomous learning should be given full play to meet the needs of students at anytime and anywhere.

4.3.3. Refine the evaluation index of autonomous learning.
At present, there is only one index to evaluate the autonomous score, that is the marks of the given teaching material in the Platform. This gave a lot of students to take advantage of the opportunity to search the answers directly from the Internet, which leads to a wicked phenomenon that the students who got the answers from the Internet gained higher grades than the students did the listening by themselves. Therefore, in addition to the achievement of the textbook content, we should also integrate the learning time length and participation in various tests and discussions into the autonomous learning, which account for a large proportion. In this way, even if students get the answers from the Internet, they can only get part of the score, which will decrease their opportunities of copying. And all teachers must be informed of the evaluation indicators, so that they must strictly follow this standard of teaching.

5. Conclusion
The online autonomous learning of College English has become an indispensable part of college English. Teachers should provide more learning strategies and methods, and curriculum guidance learning and unit guidance learning are very necessary for students to know their specific tasks. At the same time, as Littlewood (1999) believed that autonomous learning is not the same as individual learning or personal learning. It is necessary to strengthen the monitoring of students' autonomous learning in an all-round way, not only by using the learning platform, but also by combining with classroom teaching, so as to timely detect the effect of their learning. Autonomy and monitoring are not contradictory, but a relationship of mutual coordination, promotion and improvement.

Acknowledgement
This work is supported by Construction of Autonomous Learning Guidance and Monitoring System of College English (Grant No. JGW201606), Guilin University of Electronic and Technology Teaching Reform Project, by Research on Construction of Ecological System of English Smart Learning (Grant No. 2019KY0243), Basic Ability Improvement Project of Scientific Research for Young and Middle-aged Teachers in Universities of Guangxi, by Research on the Teaching Mode of College English 'Integrated Teaching' from the Perspective of the Cultivation of International Engineering Application -oriented Talents(Grant No. 2018JGB177), Guangxi Higher Education Undergraduate Teaching Reform Project, by Research and Practice on Smart Flipped Classroom Teaching Mode in College English Reading Class (Grant No. 2018JGB187), Guangxi Higher Education Undergraduate Teaching Reform Project, and by Research on the Long-term Mechanism and Implementation of Guangxi Minority Culture Inheritance Education with Cooperation Between University and Secondary Vocational School.

References
[1] Benson P&VollerP (eds.) 1997*Autonomy and Independence in Language Learning*. (Harlow: Longman), p63.
[2] Dickinson L 1995*Autonomy and Motivation: a Literature Review*, 23/2, p165-174.
[3] HolecH 1981 *AutonomyandForeignLanguageLearning* (Oxford: Pergamon Press).
[4] LittlewoodW 1999 Defining and developing autonomy in East Asian contexts. *Applied Linguistics*, p71-94.
[5] Chen Dong-chun 2006 on the orientation of autonomous learning in college English teaching in China. *Foreign Language World*, No.3, p 32-37.
[6] Chen Jian-lin, Wang Jing 2016 on the development of "Normal" in IT-based foreign language

education. *Computer-assisted Foreign Language Education*, No. 4, p3-9.
[7] College English Curriculum Requirements 2017 https://wenku.baidu.com/view/05b52209a36925c52cc58bd63186bceb19e8ed2e.html.
[8] Fu Zheng 2014 on the management of autonomous learning in network English teaching. *Journal of Inner Mongolia Normal University* (Educational Science), No.27 (02), p124-126.
[9] He Ming-xia 2012 *the monitoring of College English self-related learning based on network environment from theory to practice*. A dissertation of Shanghai International Studies University
[10] Hou Xiao-hong 2013 Research and practice of self-directed learning (SDL) in the Internet-based autonomous learning of College English viewing, listening and speaking. *Foreign Languages in China*, No.6 p60-64.
[11] Wang Shou-ren 2011 Some Thoughts on College English Teaching in China. *Foreign Language Learning Theory and Practice*, No.1 p1-5.
[12] Xu Jin-fen the current situation and reflection on Chinese college students' autonomous English learning abilities. *Language Education*, No. 2(04), P2-7.
[13] Yin Hua-dong 2014 A Survey on Studies of Autonomous Foreign Language Learning in Mainland China. *Foreign Language Teaching*, No.35 (01), p64-67+103.

Study on the surface ceramization of aluminum alloy

Zhifu Wu[1] and Liyun Hu[2]

[1]School of Materials Science and Engineering, Baise University, PR China 533000
[2]Library of Baise University, PR China 533000

Corresponding author and e-mail: Zhifu Wu, 1781177844@qq.com

Abstract. Aluminum alloy has been widely used as a metal material. However, it is usually vulnerable to corrosion, so the protection of aluminum alloy surface is particularly important. At present, the commonly used methods of surface protection of aluminium alloys are anodic oxidation, plasma micro-arc oxidation, chemical oxidation and so on. In this experiment, chemical protective coatings were prepared by chemical oxidation of Aluminium 1100 and 6063 treated with H2O2 or KMnO4 solution. The results showed that the oxidation capacity of H2O2 solution from 3% concentration to 15% concentration to Al1100 was successively enhanced. Moreover, the ceramic coating produced when the concentration of H2O2 is 12%-15% has a better effect, and the structures are cauliflower, concave and convex, and hydrophobic. It is expected that the surface has lotus leaf effect.

1. Introduction

Aluminum element is the most abundant metal element in the crust, accounting for about 7.73 percent of the total, and widely distributed, aluminum material in production and life in the application range after steel, is the second largest application material. At present, pure aluminum is rarely used in the production of practical structural materials, but the alloy formed by the combination of pure aluminum and other elements is a very good composite material [1].With the progress and development of the economy, aluminum alloy materials play a very important role in various fields, and its surface decorative [2] and other performance requirements are increasingly high. Aluminum alloy materials have the characteristics of small relative density, light mass, non-magnetic, high specific strength, stable structure, good conductivity, and no pollution to the environment [3], which are widely used in many industrial sectors. However, the surface hardness of aluminum alloy is soft and easy to be worn, and it is easy to be corroded in the atmospheric environment, damaging the surface gloss and various surface properties, and it is difficult to meet the requirements of higher-level products in various fields [4].Although the aluminum alloy itself will generate a layer of oxide film covering the surface, which has some protective effect on the alloy itself, the natural oxide film of aluminum alloy can only provide limited protection to the matrix, and the moisture and salt in the air will accelerate the corrosion of the aluminum matrix. Preparation of a protective coating on the surface of aluminum alloy can improve its surface hardness, corrosion resistance, high temperature resistance, hydro phobicity and other properties [5]. Aluminum alloy surface is usually treated by electroplating, chemical conversion, anodic oxidation and micro-arc oxidation [6].Recently, the surface ceramization technology of aluminum alloy has attracted more and more attention in various fields, and it is the main research method of surface modification technology of aluminum alloy. This method is that aluminum alloy is the matrix material, and a layer of ceramic film is prepared on the surface of the matrix, so that the surface of the matrix can have the properties of ceramic material, and can combine and cooperate with each other. Ceramic coating hardness, resistance to high temperature, wear and corrosion resistance and cheap manufacturing cost, has some excellent properties of ceramics. The excellent properties of the metal material and combined the excellent properties of ceramic materials, ceramic coating adhesive on the metal material surface does not affect the performance of metal materials, and the composite materials with special performance of metal materials and ceramic materials, superior comprehensive performance [7], aluminum matrix ceramic coating in the shuttle, electrical energy, the machine is widely applied in aspects of chemistry, is suitable for the single metal or ceramic materials can apply environment, so is the current solution is not stable, easy to corrode metal material surface an important technology, such as the defects by attention and concern. Among the preparation technologies of ceramic coating on aluminum alloy surface, several technologies are widely used: anodic oxidation, plasma micro-arc oxidation, chemical oxidation, laser cladding and sol-gel [8-10].

Anodic oxidation technology on the surface of the metal material: put metal materials as anode materials in electrolyte solution, then the cathode material is a kind of corrosion resistance of the conductive material, pass into the current in the current under the action of the anode materials generated on the surface of a layer of good wear-resisting caustic resistance, corrosion resistance and other functional or decorative ceramic layers of coating process [11].Anodic oxidation mainly deals with some light metal materials, especially aluminum and aluminum alloy, magnesium alloy, titanium and titanium alloy. The coating prepared by anodic oxidation technology is thicker, stronger and has good density, hardness, abrasion resistance, corrosion resistance, absorption and insulation [12].The anodic oxidation coating thickness of aluminum alloy can be large or small, and it has a special honeycomb structure. The porosity of the film layer is also affected by the different abilities of electrolyte and the generation speed of coating [13].The type of electrolyte, process design and pretreatment will affect the structure and performance of the film. However, anodic oxidation also has some disadvantages, such as the substrate surface needs to have a better quality, flat and clean. Therefore, before the preparation of the membrane, a series of complicated pre-treatment procedures must be carried out first, which is time-consuming and time-consuming. Moreover, the equipment required in the preparation of anodic oxidation film is more expensive, the technology is more complex, the energy consumption is more, and the quality of the film layer is not easy to control, especially for the coating of large parts, the operation is more difficult [14].

Micro-arc oxidation is a new technology of material surface modification. In this technology, aluminum, magnesium, titanium, zirconium, niobium, tantalum and other metals are immersed in the corresponding electrolyte, and a higher voltage is applied on the surface of the material to generate local instantaneous high-temperature and high-pressure oxidation of the material and generate a ceramic coating mainly composed of the matrix material oxide and containing the elements contained in the electrolyte [15].Micro-arc oxidation technology is simple to operate, green environmental protection, low production cost, wide range of applications. Micro-arc oxidation technology can greatly enhance the hardness, wear resistance and damage resistance of the surface of materials. Its micro hardness is equivalent to that of cemented carbide.The coating prepared by anodic oxidation technology has better corrosion resistance. It can make many metal materials in the application process to enhance the corrosion resistance of the material itself, so that in the actual application of the material selection has more consideration. The film has a good heat resistance, short time to withstand the high temperature of about 1000℃ and the film is not destroyed. Compared with anodic oxidation, chemical oxidation is easier to operate, does not require expensive and complex equipment, and the production process is relatively clean. The ceramic coating prepared by chemical oxidation method can enhance the adsorption and wet ability of the material and ceramic layer, and has the advantages of uncomplicated preparation process, simple operation, cheap production price, excellent and stable coating structure, etc. In this paper, ceramic coating on aluminum alloy surface by chemical oxidation is studied.

Chemical oxidation method is to put the metal material in the corresponding solution with the same temperature, make the metal material react with the solution, make the metal ions and oxygen in the oxidation solution combine with each other, and create a protective oxide film layer on the surface of the metal material . Chemical oxidation can also be used to prepare a ceramic coating on the surface of aluminum alloy as a protective layer to protect the substrate. This ceramic coating can protect the surface of the substrate and can also be used as the bottom layer of paint to enhance the adsorption .The chemical oxide ceramic layer endows the aluminum alloy with some properties that it does not have and effectively broadens the applicable range of aluminum alloy. Chemical oxidation method is used to prepare ceramic layer on the surface of aluminum alloy. In addition to the fact that the existence of the film layer does not affect the original properties of aluminum alloy, the ceramic layer has dense structure, strong wear and corrosion resistance, bright appearance, and tight combination, which does not affect the properties of the substrate itself. Compared with other preparation methods, chemical oxidation method can be used for coating some work piece with uneven, uneven and irregular surface morphology, and the combination is better. Although chemical oxidation is more environmentally friendly than anodic oxidation, there are still some pollutants discharged, causing damage to our living environment. The chemical oxidation process in the paper may solve these problems.

2. Experimental

2.1. Reagents and instruments

Hydrogen peroxide 30% (AR pure, Tianjin komeo chemical reagent co., LTD.), potassium permanganate (AR pure, Sinopharmaceutical group chemical reagent Beijing Co., LTD.); 1100 aluminum plate, 6063 aluminum bar.

SU-5000 Field emission scanning electron microscope (Hitachi, Japan), MX-6RT type optical metallographic microscope (Ningbo shun-yu instrument Co., LTD.), X8-600 hand-held alloy analyzer (Condensing technology Co., LTD.).

2.2. Experimental process

2.2.1. H2O2 oxidation 1100 aluminum sheet

The hand-held alloy analyzer was used to test the aluminum sheet in the package shell and the purchased aluminum bar, to determine the alloy components, and to cut a number of small pieces with a length of about $2mm^2$.With 30% H_2O_2 solution, 10ml of H_2O_2 solution with concentration of 3%, 6%, 9%, 12% and 15% were prepared. The samples of equal mass were put into the beaker of H_2O_2 solution with the above concentration, soaked for 24 hours, taken out and dried, and then the surface morphology of the samples was observed by SEM.

2.2.2. KMnO4 oxidation 6063 aluminum rod

Using a 10cm diameter aluminum strip, cut the sample with a growth of about 15cm.Use metallographic sandpaper to smooth the surface.The composition of the sample aluminum bar was determined as 6063 aluminum alloy by the hand-held alloy analyzer.Solid $KMnO_4$ was taken and $KMnO_4$ solution with concentration of 6%, 9%, 12% and 15% was prepared. The prepared samples were immersed in different concentrations of $KMnO_4$ solution for 24 hours, then taken out and dried.

3. Results and dicussion

3.1. Alloy analyzer analysis results

Table 1. Chemical Constituents of Aluminum Sheets for Pharmaceutical Packaging

composition	content%	+/-[*2]
Al	98.658	0.859
Fe	0.650	0.047
Si	0.600	0.066
Co	0.0041	0.032

Table 2. Chemical Constituents of Aluminum Alloy Bars Used in the Experiment

composition	%	+/-[*2]
Al	98.424	0.914
Mg	0.749	0.225
Si	0.596	0.063
Fe	0.103	0.023

According to the analysis of the alloy composition of the aluminum sheet of the handheld alloy analyzer, Table 1 shows that the content of Al accounts for 98.658% and contains a small amount of Fe, Si and other impurities. It can be determined from table 1 that the aluminum sheet is Al1100.According to the analysis of the alloy composition of the aluminum bar by the hand-held alloy analyzer, table 2 shows that the content of Al accounts for 98.424%, and contains a small amount of Mg, Si, Fe and other impurities. It can be determined from table 2 that the aluminum bar is 6063 aluminum alloy.

3.2. SEM analysis

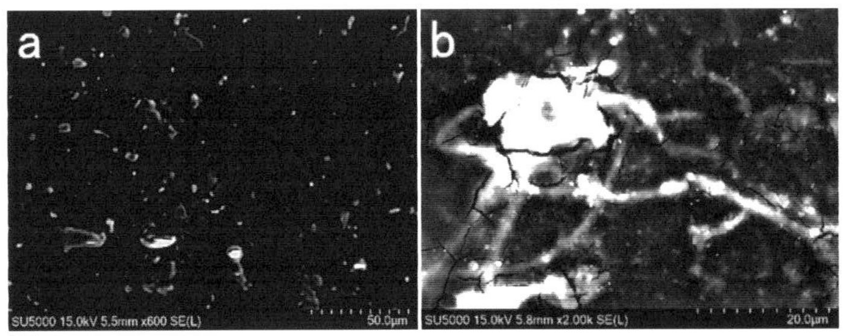

Figure 1. SEM images of aluminium sheets treated with 3% H_2O_2 solution
(a) 600 times amplification (b) 2000 times amplification

It can be seen from figure 1 that when the concentration of H_2O_2 solution is 3%, the oxidized part on the surface of the sample is sparse and distributed in small spots, with uneven distribution.

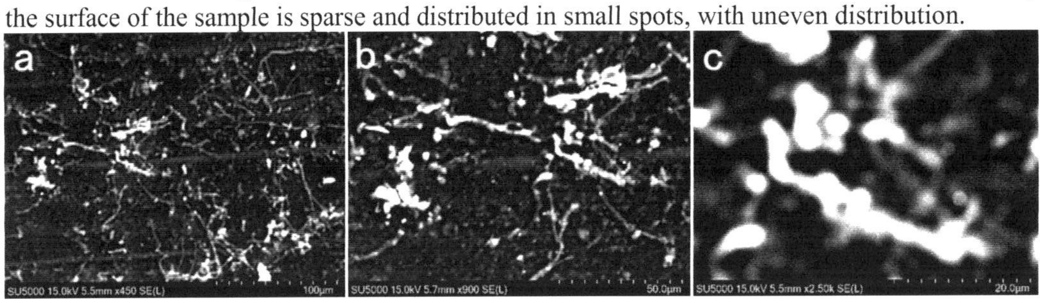

Figure 2. SEM images of aluminium sheets treated with 6% H_2O_2 solution
(a) 450 times amplification (b) 900 times amplification (c) 2500 times amplification

Journal of Physics: Conference Series **1423** (2019) 012010 doi:10.1088/1742-6596/1423/1/012010

It can be seen from figure 2 that when the concentration of H_2O_2 solution is 6%, the oxidized part on the surface of the sample presents a linear distribution, but the coverage is relatively small, and many places are still in the state of no coating.

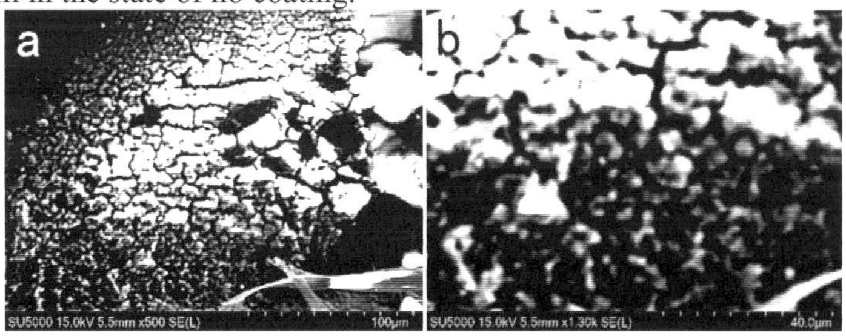

Figure 3. SEM images of aluminium sheets treated with 9% H_2O_2 solution
(a) 500-times amplification (b) 1300-times amplification

It can be seen from figure 3 that when the concentration of H_2O_2 solution is 9%, there are obvious and dense oxide layers on the sample surface. But it still doesn't completely cover the surface of the sample.

Figure 4. SEM images of aluminium sheets treated with 12% H_2O_2 solution
(a) 2200 times amplification (b) 4000 times amplification

It can be seen from figure 4 that when the concentration of H2O2 solution is 12%, there is a layer of oxidation products on the surface of the sample, and the surface organization of the oxidation products is dense, concave and convex, which is in the shape of cauliflower.

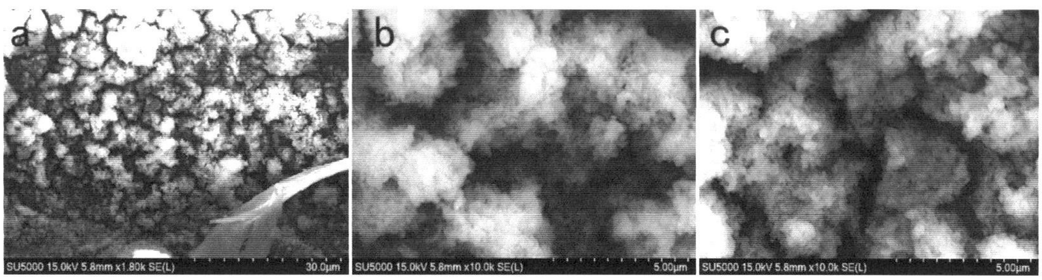

Figure 5. SEM images of aluminium sheets treated with 12% H_2O_2 solution
(a) 1800 times amplification (b) 10000 times amplification (c) 10000 times amplification

It can be seen from figure 5 that when the concentration of H_2O_2 solution is 15%, there is a layer of dense oxidation products on the surface of the sample, and the oxidation products are evenly distributed, with good organizational structure and concave and convex surface, in the shape of cauliflower.

3.3. Metallographic analysis

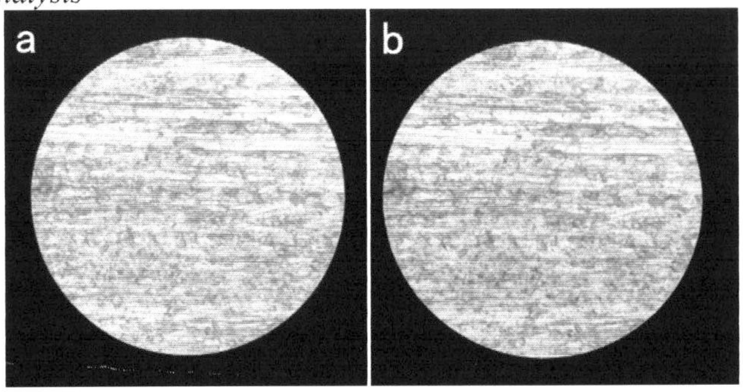

Figure 6. Metallographic micrograph of aluminum bar treated without KMnO4 solution (1000 times amplification)

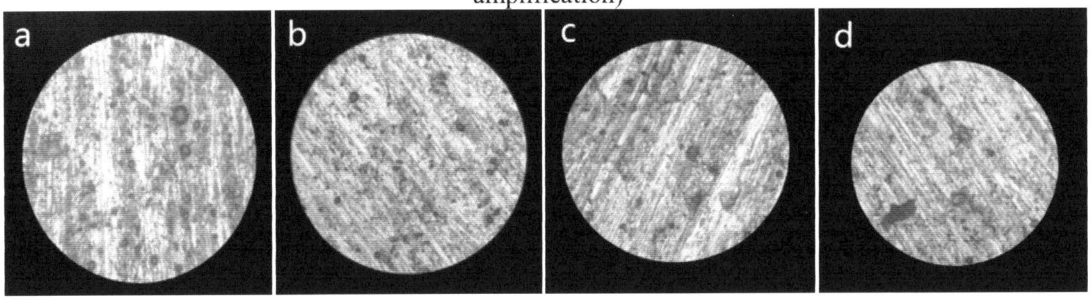

Figure 7. Metallographic micrograph of aluminum bar treated with different concentrations of $KMnO_4$ solution (1000 times amplification)
Concentration is (a): 6%, (b): 9%, (c): 12%, (d):15%

After looking at the metallographic photographs in FIG. 6 (a), FIG. 6 (b), FIG. 7 (a), FIG. 7 (b), FIG. 7 (c) and FIG. 7 (d), it was found that the surface of the sample was loose and not dense, and some special structures on the surface were difficult to be seen.After treating the sample with $KMnO_4$ solution, the product may be manganese dioxide, potassium manganate, etc.And potassium permanganate solution itself has color, compared to H_2O_2 is not easy to observe the resolution.Therefore, it is difficult to tell whether the ceramic coating was formed after the sample was treated with $KMnO_4$ solution.

4. Conclusion

Compared with the experimental process and result analysis of sample treatment with H_2O_2 solution and sample treatment with $KMnO_4$ solution, it shows that the use effect and preparation effect of H_2O_2 solution are better than that of $KMnO_4$ solution, so the preparation technology of ceramic coating with H_2O_2 solution is superior. From 3% concentration to 15% concentration, the oxidation capacity of H_2O_2 solution to Al1100 was successively enhanced, moreover, the ceramic coating produced when the concentration of H_2O_2 is 12%-15% has a better effect, and the structures are cauliflower, concave and convex, and hydrophobic. It is expected that the surface has lotus leaf effect.

Acknowledgement
Material Physics and Chemistry Open Project of Guangxi University Key Subject, Baise University Aluminum Matrix Composite Characteristic Research Team Project , Funds for the Construction of Master's Degree Granting Units in Guangxi Zhuang Autonomous Region in 2019.

References

[1] Smitha V S , Syamili S S , Peer Mohamed A. 2018, New J. Chem., 42, 10337-10347.
[2] Yan G J, Chen G D, Sun K. 2014. RSC Advances.4.6840-6844.
[3] WuYH,ZhaoW J,WangW R.2016. RSC Advances .6.5100-5110.

[4] Zhuang J L, 2018. Study on Technology of Aluminium Alloy/Ceramic Gradient Material by Laser Deposition. Shenyang Aerospace University Thesis for Masters Degree.
[5] He X,2013, Ordnance Material Science and Engineering. 36(1):39-42.
[6] Wang M, Dong X P, Yang F, Zheng X Y, 2015, Special Casting & Nonferrous Alloys. 35(04):427-429.
[7] Wang M, 2015, Study on Preparation Processes of Ceramic Coating on Aluminum Surface.Huazhong University of Science and Technology.
[8] Wu Y H , Zhao W J , WangW R. 2016, RSC Advances.6, 94074-94084.
[9] Gao J H,Li F H, Gong Y L,Ren S,2018, Plating & Finishing. 40, 18-23.
[10] Qian J C, Zou H Q, Fang M, Xu B, 2019 , Equipment Environmental Engineering, 16(3):58-62.
[11] Wang Y M, Zou Y C, Wang S Q, Chen G L, Ou Y J H, Wei D Q, Jia D C, Zhou Y , 2018, China Surface Engineering, 31(4):20-45.
[12] Ling H, 2015. Effect of Anodic Oxidation Film on Formation of Micro-arc Oxidation Ceramic Coatings on Aluminum Alloy. Xihua University.
[13] Wu H W, 2015.The Study of technology on cast aluminum alloy micro arc oxidation coatingand and its wear resistance. Harbin Institute of Technology.
[14] Tian L,Song Q L,Wang A R , 2013, Plating & Finishing, 35(11):5-8.
[15] Tian Y, Zhao Y G, Liu C W, Wang Y L, 2014, Aeronautical Science & Technology, 25(1):51-53.

Journal of Physics: Conference Series **1423** (2019) 012011 doi:10.1088/1742-6596/1423/1/012011

Analysis and empirical research about the satisfaction related factors of business administration specialty practice teaching in Chinese Medicine Universities

Xiao Cheng [1], Cuiling Guan [1], Xiaoyi Pan [1], Yongjian Wang[2]

[1] Hubei University of Chinese Medicine, Management School, Hubei Provincial Chinese Medicine Development Research Center, Hubei Wuhan 430065;
[2] West Chester Business School, Pennsylvania, United States of America 19383

Corresponding author and e-mail: Cheng Xiao, 48899293@qq.com

Abstract. To study the effects of pre-control, central control and post-test management programs on student satisfaction in management practice teaching in TCM University, 57 students in the experimental class of medical marketing were investigated by test tools like questionnaire and self-made satisfaction scale. There is no statistical difference in student satisfaction ($P>0.05$), no statistical difference in the sense of knowledge ($P>0.05$). There are statistical differences in the understanding of content, and understanding is higher than unknown satisfaction ($P< 0.05$). There are statistical differences in the number of preparatory projects before the internship. The more preparatory projects, the higher the satisfaction ($P<0.05$). There are statistical differences in the number of internship questions. The more preparatory questions, the higher the satisfaction ($P<0.05$). Quality and satisfaction of enterprise lecturers were correlated ($P<0.05$). The factors that affect the satisfaction of practice activities of management practice in Chinese medicine colleges include the content of pre-internship, the number of projects and questions for preparation, and the quality of enterprise lecturers.

1. Introduction

Since the 1990s, some medical colleges and universities have successively opened management majors to meet the needs of the market. According to statistics, about 20 Medical colleges and universities in China have established management majors [1].Management majors have become an indispensable part of medical colleges and universities. Like medical majors, management majors require a lot of practical practice teaching, but the form and content are different [2].

Practice teaching is a necessary means for modern colleges and universities to cultivate complex innovative talents. It can be divided into two types: outside school and inside school. In the courses of management majors in foreign universities, the proportion of internship practice hours can be as high as 70 % -100 % [1], In 2012, the Ministry of Education issued the "Notice on the Construction of the" Undergraduate Teaching Project "College Students'Practical Education Base" and decided to build college student practical education base to undertake the task of college practice education [2]. In 2018, the "National Standards for Teaching Quality in Business Administration"(hereinafter referred to as "National Standards") promulgated by the Ministry of Education[3] , "Pay attention to theory and practice, aiming to cultivate and train students' processing skills and decision-making ability" and have "knowledge acquisition ability, knowledge application ability, and innovation and entrepreneurship ability". The practice credits in the teaching plans of institutions of higher learning must reach more than 15 %. Practical teaching forms include "actual training experiments, internships, and social practices" [3].The practice of college management major in China mainly uses simulation laboratory for simulation experiments and simulation experiments. Some colleges and universities have logistics practice bases. Outside the school are mainly to visit the internship base and practice on the top post [4-6].

The quality evaluation of practical teaching in management specialty has always been the most intractable problem in universities [4]. The practice teaching in colleges and universities generally does not attach importance to student feedback and lacks pre-control, central control and post-test. The evaluation standards for the quality of practical teaching are not scientific and the quality monitoring system is not perfect enough, making practical teaching mostly in the form. College students are still unable to meet social requirements [1].

Regarding the research on practical teaching, the existing literature are generally inductive, comparative, and work summary. The research methods are relatively single, and there is a lack of quantitative and empirical research. This makes the research conclusions unconvincing and can not be widely used [5-9].

Medical marketing major in our school is a pilot program of JingChu Excellent Management Talents Program in Hubei Province. This plan aims to explore the talents training model of combining science and education, integration of production and education, cooperation between schools and enterprises, and uniting well-known pharmaceutical enterprises in our province. To construct a long-term mechanism for the joint training of high quality management talents in universities and enterprises, and to train innovative application talents who understand medical marketing. The practice of internship is the most important part of this plan. In order to improve the quality of personnel training, explore suitable practice forms, and provide decision-making basis for the "National Standard". This study targets students in experimental classes. The effects of pre-internship preparation and internship process on internship satisfaction were studied by combining qualitative and quantitative methods.

2. Object and methodology

2.1. Objects of study

On March 1, 2019, after the completion of the visit to the pharmaceutical enterprises, the students of the 17th medical marketing major were surveyed by sending a link to the online questionnaire. A total of 57 questionnaires were distributed and 57 valid questionnaires were received. The recovery rate was 100 %. The average age of students is 19.86 ± 0.693 years old, all of whom are college students.

2.2. Research Tools

This questionnaire is mainly to investigate the student's satisfaction evaluation of various factors before, during and after visiting the internship. It includes four parts: (1) Population variables gender, age, grade, and basic questions. (2) Pre-internship preparation section (pre-control): Including 4 questions: Do you know if the visit is part of the Hubei Province Outstanding Talent Program, if the teacher explained the purpose and content of the visit before the visit, and if you did it before the visit Which preparation (multiple selection), how many questions did you prepare before the visit, all classified variables. (3) Internship Process Part (Central Control): The internship process involves generally the internship process, internship content, and internship lecturer, etc.. It is the main aspect that affects student internship process experience [8-10]. Therefore, the evaluation of this visit activity includes three questions: whether you are satisfied with the design of the visit process, whether you are satisfied with the content of the visit, and whether you are satisfied with the corporate lecturer, all of which are Licket's Level 5 scales, from very unsatisfactory to very satisfied, 1-5 points. The higher the score, the more satisfied it is. (4) Post-internship experience section (post-test): Including overall

Journal of Physics: Conference Series **1423** (2019) 012011 doi:10.1088/1742-6596/1423/1/012011

satisfaction, you are most impressed with the visit activities (multiple choices), the impact of this visit on your future learning life is (multiple choices). And an open question: What do you think and suggest about this visit?

2.3. Methods of Statistical Analysis

The statistical software SPSS24 .0 was used to process the data, and ANOVA was used to analyze the differences in the satisfaction of demographic variables and pre-internship preparation variables. Using VIF value (variance expansion factor) [11] Conlinear diagnosis of independent variables; Multiple linear regression was used to explore the effect of various variables of the practice process on satisfaction. The hypothesis of this paper is that: H1 Gender and pre-internship preparation variables have statistical differences in satisfaction; H2 Internship process, internship content, corporate lecturer and satisfaction are linearly related.

3. Outcomes

3.1. Degree of satisfaction of various factors in preparation for internship

There is no statistical difference in student satisfaction. There is no statistical difference in the meaning of knowledge; there are statistical differences in understanding content, and understanding is higher than unknown satisfaction; there are statistical differences in the number of preparatory projects before the internship. The more preparatory projects, the higher the satisfaction. There are statistical differences in the number of internship questions. The more preparatory questions, the higher the satisfaction. (Table 1) The study assumes that H1 holds.

Table 1. Results of factors satisfaction of demographic variables and pre-internship preparation

program	gender			meaning knows			content knows			
	male (n=21)	female (n=36)	F	yes (n=36)	no (n=21)	F	yes (n=52)	no (n=0)	not clear (n=5)	F
satisfaction	3.7±1.1	4.0±0.7	1.83	4.0±0.7	3.7±1.0	1.83	4.0±0.8	0	3.2±0.8	4.05*

program	number of preparation job before visit					number of questions before visit			
	1 (n=17)	2 (n=17)	3 (n=14)	0 (n=9)	F	0 (n=16)	1-3 (n=39)	>3 (n=2)	F
satisfaction	3.9±0.5	3.9±0.8	4.5±0.7	3.1±1.2	6.29**	3.5±0.9	4.0±0.8	5.0±0.0	4.31*

*P<0.05 **P<0.01

3.2. Impact of the internship process on overall satisfaction

This paper further discusses the impact of the internship process, the internship content and the enterprise lecturer on the satisfaction. Taking the overall satisfaction as the dependent variable, taking the internship process, the internship content, and the enterprise lecturer as the independent variables, multiple linear regression analysis was carried out, and the common linear diagnosis results VIF<5, indicating that there is no colinearity between independent variables[11]. Eventually, business lecturers entered the model and were able to explain 58.8 % of the total variation. (Table 2) The study assumes that H2 holds.

Table 2. Multiple linear regression analysis results based on satisfaction variables

independant	β	SE	t	P	standedβ	R^2	R^2_{adj}
process	0.251	0.166	1.511	0.137	0.262	0.606	0.548
content	0.281	0.196	1.433	0.158	0.267		
lecturer	0.373	0.170	2.200	0.032	0.310		

3.3. Post-internship experience

When asked if you are most impressed with your internship activities(multiple choices), students are most impressed with the work environment of the company, followed by quality management, production arrangements, human resources management and corporate discipline. (Figure 1)

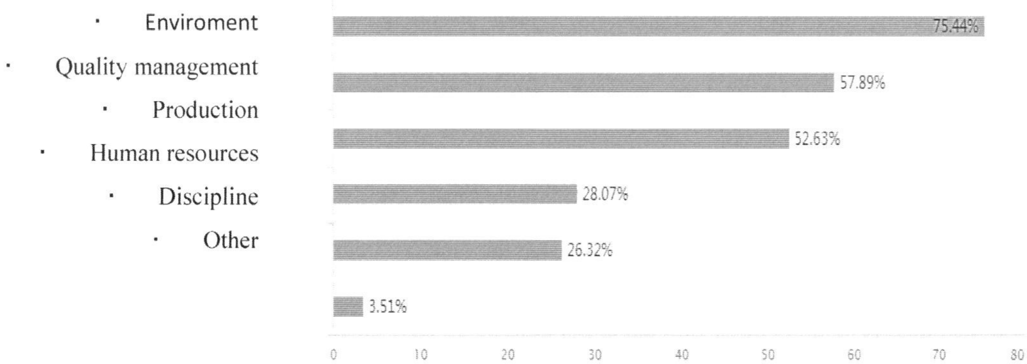

Figure 1. Most impressive internship projects

When asked about the impact of this internship on your future learning life (multiple choices), the most important choice for students is to inspire and improve their own knowledge structure. The second is to motivate me to study hard professional knowledge, motivate me to improve their language expression and communication skills, correct their own learning attitude. (Figure 2).

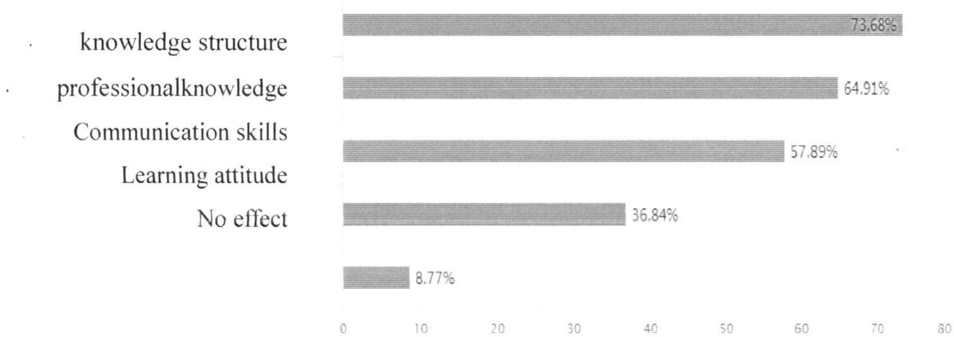

Figure 2. Impact of this internship on future learning life

When asked how you feel and suggest about this visit, through NVIVO word frequency analysis, students feel deeply about the "environment" of the company, are very interested in the knowledge of drug production, hope to organize such activities, and also hope to practice themselves.

4. Conclusions and recommendations

4.1. Pre-internship preparation work is related to internship satisfaction

This study found that understanding the internship content is higher than understanding the satisfaction, the more the number of pre-internship preparation projects, the higher the satisfaction, the more pre-internship preparation questions, the higher the satisfaction. In particular, there is a high degree of satisfaction between the number of pre-internship preparatory work projects. Pre-internship preparation includes understanding of the company's situation, understanding of the relevant content of the company's brand, and understanding of knowledge related to medicine production. The result is same to Zheng Hong-ling [5] and Zhang Fen [7], it suggests that the understanding of the internship before the internship will help to improve the internship satisfaction. Understanding the internship unit before the internship helps students to improve their learning expectations, stimulate their curiosity, improve participation, so that students have more sense of acquisition.

The "National Standards" stipulates that the practice teaching norms should include the preparation of internship activities, the evaluation and summary of internship achievements, but in fact, the standards of various universities and colleges on how to evaluate the quality of internship and how to score are inconsistent, and many do not even evaluate the quality of internship. Examination Form 1 turned in [1] [4]. At this visit, the teacher mentored the students to carry out preparatory work guidance before the internship, including the preparation topics for the students to refer to the week before the internship (to understand the situation of the company, the content of the company's brand, and the knowledge of medicine production). On the previous day, students 'preparation for the project and the preparation of questions were sampled.

4.2. Excellent corporate lecturers(mentors) are related to internship satisfaction

In this study, the internship process is related to satisfaction, and the satisfaction of corporate lecturers can significantly affect overall satisfaction. This is consistent with the results of previous studies. The quality of the instructor of the internship objects will make students feel the value and significance of internship activities. Put more psychological capital into learning. Stimulate their stronger learning motivation and more active learning behavior [10].

The "National Standard" stipulates that industry and commerce management majors should hire "people with practical experience to undertake some teaching tasks." "Encourage the employment of well-known scientists, entrepreneurial successes, entrepreneurs, venture capitalists and other outstanding talents from all walks of life as professional courses, innovative entrepreneurship courses or guidance teachers"[3].There is no quality requirement for external enterprise lecturers(mentors), but according to the results of this survey and long-term school enterprise experience, corporate lecturers(mentors) are the most important factors that affect student internship effectiveness and satisfaction. Excellent corporate mentors generally have the competence of outstanding employees in the company. They mainly include professional experience, solid integrity, dedication, outstanding performance, and smooth expression [5] [7-9]. Before practice, the school should cooperate with the company to select employees or managers with the above qualities in order to achieve a multiplier effect.

4.3. Relationship between short-term internship and long-term practice

Compared with other social sciences, medical marketing majors have high requirements for students 'practical ability and require a lot of hands-on experience of classroom knowledge. As educators, they need to systematically embed these practices into teaching plans while learning and practicing [1]. Practical teaching is an important link that promotes the cross-combination of knowledge structure, ability structure and scientific thinking. It is the key to cultivate practical and innovative talents and is an important experience learning. Students acquire emotional experiences that are not achieved by symbolic learning (books, online learning) through operational learning, communicative learning, observational learning, and reflective learning, and then internalize into comprehensive abilities [10]. Intern practice should also be conducted in accordance with "from basic to advanced, from individual to comprehensive, and perceptual recognition of experience and innovation"[3].The company's visit activities belong to the "cognitive internship", acquire perceptual knowledge in the professional field, consolidate the theory learned, provide a cognitive basis for the "professional internship" in the later grades, and use professional knowledge to carry out practical operations. Understand the basic rules of the main content of management activities in various fields, use professional knowledge to conduct comprehensive research on real problems, and propose solutions to real problems during formal off-campus internships in senior year. Therefore, visiting internships is a compulsory project for the management of the lower grades of majors.

The "National Standard" clearly stated that undergraduate education should be transformed into a "student-centered and competency-oriented talent training model", with special emphasis on quality assurance monitoring to be "dynamic and effective" [3] .This study is a useful exploration of the monitoring mechanism. In the future, a short-term evaluation and long-term monitoring system will be established.

There are three shortcomings in this study: the first is the choice of students 'majors, age is relatively single, which is not conducive to the comparison of these factors; The second is that the survey was only aimed at one project in the internship, and the students did not feel sustained enough. Future research will combine short-term internship surveys with long-term surveys, compare the differences in students 'perceptions of internship activities at different ages and majors, and integrate variables to add learning burnout and career adaptability to make research more practical. To form a more effective quality evaluation index system.

Acknowledgement

Fund projects: 2018 "Jingchu Outstanding Talents" Collaborative Education Program Project (project number: Egao Letter 2018-20-46); 2019 Hubei University Outstanding Youth Science and Technology Innovation Team Project (T201910)

Reference

[1] Cheng Xiao, Wang Yong-Jian, Guan Cui-Ling. Problems and Countermeasures of the Construction of Practical Education Base in Colleges and Universities: Taking Medicine Marketing Specialty in Chinese Medicine Colleges as an Example[J] .Shi Zhen Chinese medicine, 2016, 27(02): 447-449. (In Chinese)
[2] Ministry of Education of the People's Republic of China. Notice of the Ministry of Education on the Construction of "Undergraduate Teaching Project" College Students 'Practical Education Base. Http://www.moe.gov.cn(2012-03-02). (In Chinese)
[3] People's Republic of China Ministry of Education. National Standard for Teaching Quality of Undergraduate Specialty in Colleges and Universities. Http://www.moe.gov.cn(2018-1-30). (In Chinese)
[4] Zheng Heng-Bin. Problems and Countermeasures of Practice Teaching in Local Colleges and Universities [J] .Educational Exploration, 2012(11): 69-70. (In Chinese)
[5] Zheng Hong-Hing, Song Yan-Hua. Research on the Theory and Practice of Internship for Students with Major in Administration [J]. Education and occupation, 2016(20):107-109. (In Chinese)
[6] Wang Dong. Research on the Construction of Multi-professional Simulation Comprehensive Practice Platform [J]. Experimental Science and Technology, 2014, 12(3): 129-132. (In Chinese)
[7] Zhang Fen. Investigation and Analysis of Top Class Practice of Administrative Students in Xinjiang Higher Vocational Colleges: Taking Xinjiang Light Industry Vocational and Technical College as an Example [J].Introduction to Economic Research, 2018, 380(30): 157-158. (In Chinese)
[8] Jia Xian-Min. Construction of practical teaching system of applied undergraduate management [J] .Laboratory Research and Exploration, 2013, 212(10): 217-220. (In Chinese)
[9] Luo Ming-Zhong, Huang Sha-Sha, Duan Yu. On the Reform of Curriculum and Teaching Content in Colleges and Universities from the Perspective of Employment Orientation [J]. Higher agricultural education, 2013(8): 66-70. (In Chinese)
[10] Chen You-Qing.Relationship and Interaction between Symbolic Learning and Empirical Learning in Student Development [J].Journal of East China Normal University(Education Science Edition), 2010, 28(02): 24-32. (In Chinese)
[11] Zhang Xin-Min, Shen Jie, Yu Shun-Zhang.Discussion on the method of studying colinearity of independent variables in multivariate regression analysis [J].China Health Statistics, 1995(1): 62-63. (In Chinese)

Study on the influence of the gas superficial velocity on the radon adsorption of activated carbon

Meng Wang [1], Yi Gu [1 2], Fei Li [1 2], Liangquan Ge[1 2], Heng Lu [1], Liang Luo [3]

[1] College of Nuclear Technology and Automation Engineering, Chengdu University of Technology, Chengdu 610059, China
[2] Key Laboratory of Applied Nuclear Techniques in Geosciences Sichuan, Chengdu University of Technology, Chengdu 610059, China
[3] Chongqing Jian an Instrument Co Ltd, Chongqing 400060, China

Corresponding author and e-mail: Yi Gu, guyi10@cdut.cn

Abstract. In order to improve the radon adsorption efficiency of activated carbon and provide a reference for reducing the cost of radon purification, the radon adsorption experiment of activated carbon under different gas superficial velocities under the active adsorption mode was studied to measure the radon adsorption quantity of activated carbon. The radon adsorption quantity of apricot shell activated carbon and coal-based activated carbon did not change significantly with the increase of the gas superficial velocity. When the gas superficial velocity increased from 0.9cm/s to 2.0cm/s in turn, the radon adsorption quantity of coconut shell activated carbon did not change significantly. The radon adsorption quantity of coconut shell activated carbon increased first and then decreased when the gas superficial velocity increased from 2.0cm/s to 3.1cm/s. When the gas superficial velocity was about 2.7cm/s, the radon adsorption quantity of coconut shell activated carbon reached its peak value. 2.7cm/s can be selected as the optimum gas superficial velocity for the radon adsorption of coconut shell activated carbon.

1. Introduction

As a radioactive gas, radon can be absorbed into the human body and forms the internal radiation. Radon exists widely in the closed environment, such as underground mine tunnels, nuclear fuel factories or related laboratories. Therefore, it is necessary to take corresponding measures to reduce and eliminate radon with relatively high radon concentration or potential radon threat. The activated carbon adsorption method is one of the most widely use methods for radon reduction and radon removal [1-4]. At present, the research on the radon adsorption by activated carbon mainly focuses on: (1) the influence of the structure of activated carbon on the radon adsorption [5-7]; (2) the influence of external conditions on the radon adsorption by activated carbon [8]. While in the research on the external conditions, it mainly focused on either the adsorption effect of inert gases, such as krypton, xenon, etc. [9-10], or the influence of temperature, humidity, pressure and other factors on the adsorption performance [11-12]. There are relatively few references on the factors affecting the gas superficial velocity in the process of radon adsorption of activated carbon. In our work, based on previous studies, the influence of different gas superficial velocities on the radon adsorption of activated carbon is further discussed by referring to relevant researches and designing the experiment.

2. Experimental platforms

2.1. Adsorption platform

The experimental adsorption platform consists of the activated carbon adsorption bed, electronic flow meter, diaphragm pump and multi-functional radon self-controlled chamber. The radon extracted from the radon chamber by the diaphragm pump forms an internal circulation through the activated carbon adsorption bed and the flow meter, as shown in Figure 1 and the electronic flow meter has been verified by National Institute of Measurement and Testing Technology. The gas flow rate of the diaphragm pump is controlled by the working voltage to provide different gas superficial velocity for the experiment. The volume of the multi-functional radon self-controlled chamber is 3.8m^3, and the radon concentration regulation range is 0-5000Bq/ m^3. The volume of the radon chamber is much larger than the volume of the radon gas extracted from the radon chamber (The maximum extraction gas volume is 300L). Therefore, the influence of the experiment about radon adsorption of activated carbon on the radon concentration in the radon chamber can be neglected.

The activated carbon materials used in the experiment were three kinds of granular activated carbon materials; these include coconut shell activated carbon, coal-based activated carbon and apricot shell activated carbon. The particle sizes of activated carbons were between 1mm and 2mm, and their iodine values were between 850 and 1050. Before experiment, the deionized water was used to remove the ash on the surface of activated carbon and the gas adsorbed from the air. Then they were put into the drying oven and dried for 4 hours at 120°C. After drying, they were placed in the activated carbon adsorption bed as shown in Figure 2.

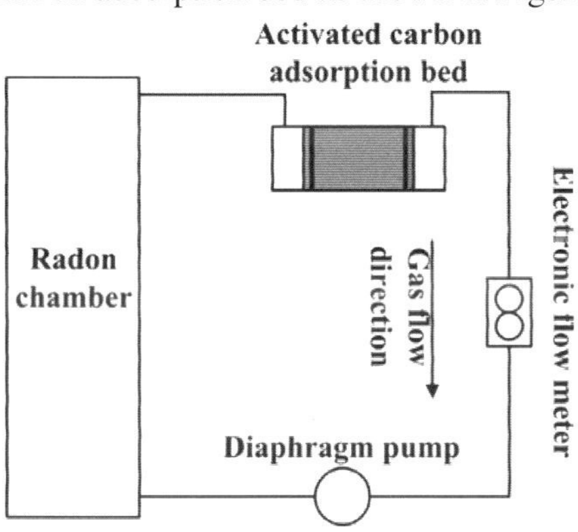

Figure 1. Schematic diagram of the adsorption experimental platform

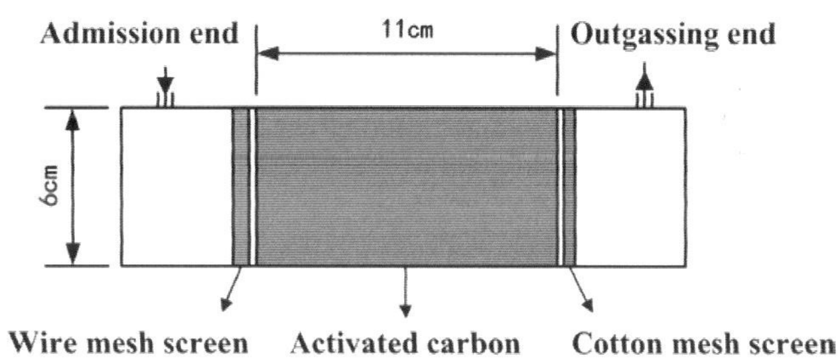

Figure 2. Schematic diagram of the structure of the activated carbon adsorption bed

2.2. Measurement platform

After the adsorption experiment was completed, the activated carbon was taken 200g from the gas admission end of the activated carbon adsorption bed, and then it was filled in polyethylene holder (φ7.5cm×7.5cm). The activated carbon samples were measured by HPGe γ-ray spectrometer, and the γ-ray spectrometer with high energy resolution was produced by ORTEC. A lead shield with the thickness of 4mm was used in the measuring system to minimize the environmental background. The γ-ray spectrum of both the environmental background and the activated carbon included radon and its daughters were measured separately. The radon in the radon chamber reached the radioactive equilibrium before the adsorption experiment. According to the radioactive equilibrium theory, ^{214}Bi which is the decayed daughter of radon had higher net counts of 0.609 MeV peak inγ-ray spectrum than other decayed daughters, therefore it was selected to analyze the radon adsorption quantity of activated carbon samples [13]. And the radiation background of coconut shell activated carbon, apricot shell activated carbon and coal-based activated carbon were $349h^{-1}$, $239h^{-1}$ and $228h^{-1}$ about net counts of 0.609 MeV peak.

3. Determination of adsorption time about activated carbon

The radon concentration in radon chamber was set to 1400 Bq/m^3, and the temperature and relative humidity were stable at 18℃ and 50% (the following experiments were all under the conditions of temperature and relative humidity), and the gas superficial velocity was set to 1cm/s. After the coconut shell activated carbon adsorption bed adsorbed for 45, 60, 75 and 90 min separately, and then the activated carbon samples were measured byγ-ray spectrometer. The changes of 0.609 MeV peak counts with measurement time are shown in Figure 3.

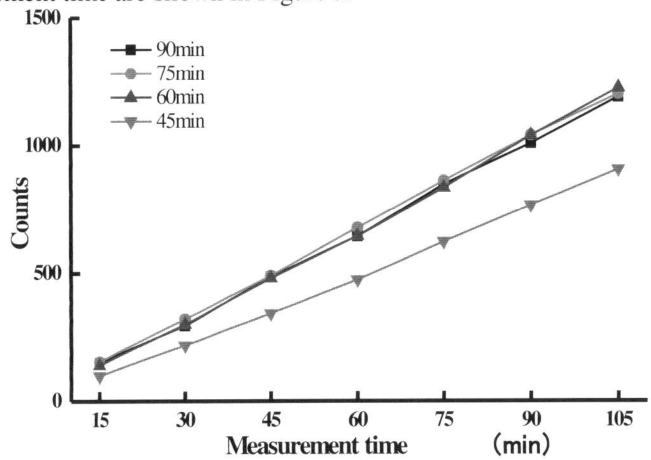

Figure 3. Measurement results of the activated carbon at different adsorption times

According to the growth rate of 0.609 MeV peak counts of the activated carbon samples at different adsorption times, the adsorption depth of activated carbon at different adsorption times can be judged. When the adsorption time is between 60 min and 90 min, the changes of 0.609 MeV peak counts at the same measurement time are very weak, which can almost be ignored. Therefore, when the adsorption time is 60min, the adsorption depth reaches 100%. When the gas superficial velocity is about 1cm/s or greater than 1cm/s, the adsorption depth of the carbon bed can reach 100% within the adsorption time of 60min.

4. Analysis of the activity of radon adsorbed by activated carbon

The detection efficiency of the energy spectrometer is obtained by measuring the standard material, when the activated carbon samples to be detected and the standard material are in the radioactive equilibrium, the 0.609MeV peak counts can be obtained by the γ-ray energy spectrum of activated carbon, and the activity of radon (i.e., the radon adsorption quantity) adsorbed by activated carbon can be calculated by the formula (1):

$$A_s = A_{Ra} \times (N_s / N_{Ra}) \tag{1}$$

In the formula: A_s is the activity of radon, A_{Ra} is the activity of the standard material (^{226}Ra,312Bq), N_{Ra} is the net counts of 0.609MeV peak of the standard material, N_s is the net counts of 0.609MeV peak of activated carbon after it adsorbs the radon.

In order to obtain the net counts of 0.609Mev peak, the background of gamma spectrum is subtracted by the algorithm of Sensitive nonlinear iterative peak (SNIP). Before applying SNIP algorithm, the LLS (twice log operators plus square-root operator) operator is used to transform each energy spectrum count with the formula (2):

$$V(i) = \log[\log(\sqrt{y(i) + 1}) + 1] \tag{2}$$

Where i represents the channel, y(i) are the energy counts in the channel i, V(i) is the saving vector after y(i) is transformed by LLS operator, then V1(i), V2(i) are calculated up to Vm(i) step by step, and the parameter of m is generally the FWHM. The new value in the channel i in the p-th iteration is obtained by the formula (3):

$$V_p(i) = \min\left\{1/2 \cdot \left[V_{p-1}(i-p) + V_{p-1}(i+p)\right], V_{p-1}(i)\right\} \quad (3)$$

Where p grows iteratively from 1 to a given m value with step size 1. When Vm (i) is calculated, the net value of the spectrum can be obtained by performing the inverse calculation of LLS operator.

After background subtraction, the spectrum needs to be smoothed and denoised before calculating the activity of radon. And the spectrum is denoised by the denoising method of Least squares support vector machine (LS-SVM). In the method, the spectrum was segmented according to the principle of three-window method, and the method can be described by the formula (4):

$$Y(i) = \begin{cases} y_1(i); & 1 \le i < 34 - t \\ y_2(i); & 34 + t < i < 114 - t \\ y_3(i); & 114 + t < i < 256 \\ k_j \cdot y_1(i) + \overline{k_j} \cdot y_2(i); & 34 - t \le i \le 34 + t \\ k_j \cdot y_2(i) + \overline{k_j} \cdot y_3(i); & 114 - t \le i \le 114 + t \end{cases} \quad (4)$$

Where i is the spectrum channel, y_1,y_2, and y_3 are the results of spectrum three-section regression fitting, the channel distribution of the three-energy segment is (1～34+t), (34-t～114+t), and (114-t～256), respectively. The overlapping channel address of the spectrum is 2t+1. k_j and —k—k_j are the smoothing weighting factors and their mathematical expressions are described in formula (5):

$$\begin{aligned} k_j &= \frac{1}{2} + \frac{1}{2}\sin\left(\frac{j-1}{2t}\pi + \frac{\pi}{2}\right) j = 1,2,\ldots,2t+1 \\ \overline{k_j} &= \frac{1}{2} + \frac{1}{2}\sin\left(\frac{j-1}{2t}\pi + \frac{\pi}{2}\right) j = 1,2,\ldots,2t+1 \end{aligned} \quad (5)$$

Where j states the position number of the spectrum in the overlapped data segment. The net counts of 0.609 Mev can be obtained by the spectrum that was eliminated noise, and then the activity of radon can be calculated by the net counts.

5. Results and discussion

When Guo Liangtian et al. [14] explored the influence of the gas superficial velocity on the adsorption coefficient of krypton of activated carbon, the gas superficial velocity was controlled in the range of 0.1cm/s-5cm/s when this gas flowed through the activated carbon adsorption bed. Radon and krypton belong to the same family elements, and their physical properties are similar. When studying the influence of the gas superficial velocity on the radon adsorption of activated carbon, based on the experimental conditions of predecessors and the existing experimental instruments, the experimentally selected gas superficial velocity is controlled in the range of 1cm/s-3cm/s, the radon concentration in radon chamber is set to 1000Bq/m^3, the adsorption time of activated carbon is 60min, and the measurement time is 60min. The influence of the gas superficial velocity on the radon adsorption of activated carbon is analyzed mainly by the radon adsorption quantity of the activated carbon bed. The radon adsorption experiments were carried out on the coconut shell, apricot shell and coal-based activated carbon under the adsorption conditions of different gas superficial velocities respectively, and the radon adsorption quantity was calculated. The results are shown in Table 1.

Table 1. Radon adsorption measurement results of coconut shell activated carbon at different gas superficial velocities

Gas superficial velocity /cm·s^{-1}			The radioactivity of radon /Bq		
Coconut shell	Apricot shell	Coal-based	Coconut shell	Apricot shell	Coal-based
0.98	1.10	0.91	57.01	66.01	16.12
1.10	1.26	1.01	77.11	65.19	22.43
1.27	1.35	1.28	76.52	61.22	19.51
1.55	1.46	1.64	45.80	51.41	17.41
1.94	1.63	1.69	75.71	42.18	17.06
2.10	1.93	1.91	86.80	61.34	19.63
2.35	2.12	2.15	151.76	60.99	20.33
2.69	2.45	2.24	166.02	80.03	18.11
2.83	2.64	2.49	160.88	70.57	23.37
2.90	3.10	2.72	118.93	71.38	17.76
2.98	——	2.74	145.34	——	18.69
3.01	——	2.79	121.27	——	19.16
——	——	3.08	——	——	18.34

The radon adsorption quantity of three kinds of activated carbon under different gas superficial velocities is shown in Figure 4. When the gas superficial velocity is in the range of 0.9cm/s-2.0cm/s, the changes of the radon adsorption quantity of coconut shell activated carbon and apricot shell activated carbon are roughly the same and at the same level. When the gas superficial velocity is in the range of 2.0cm/s-3.1cm/s, the radon adsorption quantity of coconut shell activated carbon is significantly higher than other two kinds of activated carbon. The radon adsorption quantity of coconut shell activated carbon reaches its peak when the gas superficial velocity is 2.7cm/s. The radon adsorption quantity of coal-based activated carbon has no obvious change under different gas superficial velocity, and the radon adsorption quantity is much smaller than the others.

When the gas superficial velocity is more than 2.7cm/s, the radon adsorption quantity of coconut shell activated carbon shows a downward trend. The reason for this change may be that after the gas superficial velocity increases to a certain value, the resistance of external diffusion material begins to decrease, and the adsorbed molecules are separated from the active center of the surface of the adsorbent, causing a part of the radon atoms adsorbed by activated carbon to be desorbed [15], and resulting in a decrease in the radon adsorption quantity.

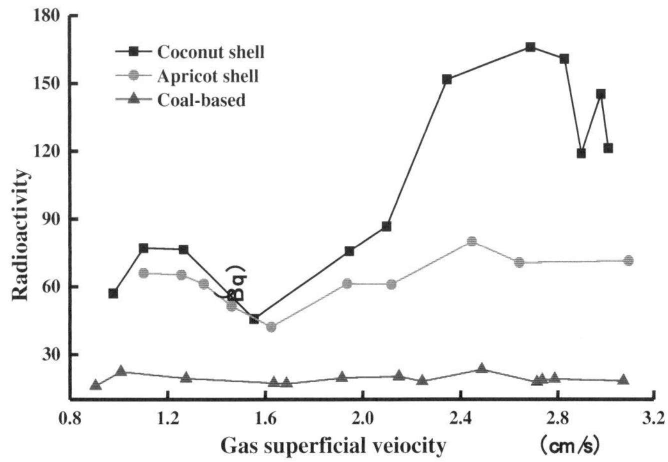

Figure 4. Radon adsorption quantity of activated carbon at different gas flow rates

6. Conclusion

Through radon adsorption experiments of the activated carbon bed in radon chamber and the measurement of activated carbon samples by the HPGe γ-ray spectrometer with high resolution, the conclusions of the analysis are as follows: (1) When the gas superficial velocity is 0.9cm/s-3.1cm/s, the radon adsorption capacity of three kinds of activated carbon is: coconut shell activated carbon > apricot shell activated carbon > coal-based activated carbon, the radon adsorption capacity of coal-based activated carbon is low, so it is not recommended to use. (2) Among the three kinds of activated carbon, the radon adsorption capacity of coconut shell activated carbon is greatly affected by the gas superficial velocity, and the radon adsorption capacity of coal-based activated carbon is basically not affected by the gas superficial velocity. (3) When the gas superficial velocity is in the range of 0.9cm/s-3.1cm/s, the optimal gas superficial velocity that the coconut shell activated carbon absorbs radon is 2.7cm/s. The above conclusions have certain reference significance for the selection of activated carbon and the setting of the gas superficial velocity in the practical application of the radon adsorption of activated carbon.

When the gas superficial velocity is between 1.5cm/s and 1.8cm/s, the radon adsorption quantity of the three kinds of activated carbons presents the V-shaped fluctuation in varying degrees, and the measurement results of repeated experiments are the same as those of the previous ones at this gas superficial velocity, the abnormal results caused by improper operation in the experiment are excluded. There is no clear explanation for the fluctuation of the radon adsorption quantity of activated carbon at the gas superficial velocity, which needs to be further explored.

Acknowledgement

The authors acknowledge the financial support by the National Key R&D Program of China (Project No.2017YFC0602105), the National Science Foundation of China (Project No. 41774190) and the Key R&D Projects of Sichuan Province (Project No. 2018GZ0004).

References

[1] National Research Council. Committee on health risks of exposure to radon (BEIRV 1) health effects of exposure to radon [M]. Washington: National Academy Press, 1999.

[2] Yingxia Shi, Tianliang Guo, Ruiyun Wang, et al. Study of Adsorption of Radioactive noble gases by Activated carbon under Engineering Conditions [J]. Radiation Protection, 2007, 27(5):302-307.

[3] Wenyan Ma, Chunlin Zhou, Guiying Chen, et al. Study on sorption action of activated carbon to radon [J]. Nuclear Electronics & Detection Technology, 2002, 22(1):84-87

[4] Weiwu Jiang. The Effect on the Activated Charcoal Measuring Method for Indoor Radon Concentration[J]. Environmental Monitoring in China, 2007, 23(3):26-27
[5] Qingbo Wang, Wengkai Zhu, Jingyuan Qu, et al. Characterization of Pore Structure of Several Activated Carbons With Different Radon Adsorption Capabilities[J]. Atomic Energy Science & Technology, 2011, 45(11):1382-1387.
[6] Geng Wang, Liheng Liu. Effect of Modification methods on the structure, properties and adsortion capacity of activated carbons [C]. Environmental Engineering, 2017, 2.
[7] Xiaowa Wang, Jun Wan, Yu Tu. Study on radon absorption characteristics of structurally modified activated carbon[J]. Radiation Protection, 2015, 35(4):227-231.
[8] Zikovsky L. Temperature dependence of adsorption coefficients of 222Rn on activated charcoal determined by adsorption- desorption method [J]. Health Physics,2001,80(2):175-176.
[9] Guopu Qu, Qiu Ling, Yinglan Guo, et al. The research of 85Kr adsorption by activated carbons [J]. Nuclear Electronics & Detection Technology, 2005, 25(3):268-270.
[10] Yongguo Li, Fei Liang, Yuan Zhang, et al. Research status on activated carbon absorption of radioactive noble gases and their influencing factors [J]. Radiation Protection, 2015, 35(2):112-116.
[11] Xu Feng, Detao Xiao, Shoukang Qiu, et al. Research on High Pressure Adsorption of Radon on Activated Carbon [J]. Atomic Energy Science & Technology, 2016, 50(4):763-768.
[12] Defu Xie. Study on the Influence of Radon Dynamic Adsorption in Activated Carbon [D]. University of South China, 2011.
[13] Yang Zhang, Chunyu Liu, Guohui Yin, et al. Study of Detection Efficiency of 222Rn Samples Obtained by Active Activated Charcoal by HPGe γ Spectrometry [J]. Atomic Energy Science & Technology, 2014, 48(1):142-146.
[14] Tiangliang Guo, Yingxia Shi, Ruiyun Wang, et al. Study on the Charcoal Delay bed for Radioactive noble gases [J]. Radiation Protection, 1994, 14(1):15-24.
[15] Changxian Li, Tingwu Qin. Coal-based activated carbon [M]. China Coal Industry Publishing House, Beijing, 1993.

Journal of Physics: Conference Series **1423** (2019) 012013 doi:10.1088/1742-6596/1423/1/012013

Analysis and evaluation of nutritional components in muscle of *sinogastromyzon szechuanensis*

Lijia Yue, Xiaojiang Chen*, Wei Li

Jiangsu Agri-animal Husbandry Vocational College, Taizhou 225300, Jiangsu, China

* Corresponding author and e-mail: Xiaojiang Chen, cq_cxj@126.com

Abstract. Based on routine biochemical analysis methods, the nutritional composition in muscle of *Sinogastromyzon szechuanensis* were calculated using the Excel 2019 and SPSS 25.0 software. The results showed that the rates of moisture, crude protein, crude fat and ash in muscles of *S. szechuanensis* are (77.80±0.70) %, (18.38±0.21) %, (2.00±0.10) % and (1.55±0.08) % respectively. 16 kinds of amino acid were detected. The total content of amino acids (TAA), essential amino acids (TEAA) and delicious amino acids (TDAA) were 17.80%, 6.83% and 7.37% respectively. Amino acid score (AAS), chemical score (CS) and essential amino acid index (EAAI) were calculated using Excel 2019 software. The results revealed that the first and second limited acids were Ile and Val respectively. The essential amino acid index (EAAI) was 89.65. Four kinds of delicious amino acids (TDAA) accounted for (7.37±0.32) %, with the 41.43% of the total amino acids. The ratio of branched chain amino acid to aromatic amino acid (F value) were 2.31±0.04. It revealed that *S. szechuanensis* was a kind of delicious and high-quality fish with high-protein and low-fat composition.

1. Introduction

Sinogastromyzon szechuanensis belongs to Cypriniformes and Homalopteridae, and is a common characteristic fish in the mainstream and tributaries of the upper Yangtze River. At present, there are few studies on *S. szechuanensis*. Those studies mainly focus on the early development of fish and resource survey [1-2]. However, there is no report on the nutritional components of *S. szechuanensis*. In this study, the main nutrient components in the muscle of *S. szechuanensis*weredetermined and their nutritional value is evaluated comprehensively so as to provide references for the nutritional research of *S. szechuanensis* and the development and utilization of its resources.

2. Materials and methods

2.1. Materials

The material fishwere collected from the Neijiang section of Sichuan Province in the Tuojiang River system in October 2018, with a total of 90 fish. The body weight was (2.62±1.12) g and the length was (5.13±0.55) cm. After peeling each fish, the muscles on both sides of the spinal column behind the head of the fish were removed, and the muscles of each 30 fish were crushed and mixed evenly, as a sample, and stored at low temperature for reserve. The samples were divided into two parts, one for the determination of general nutrients and the other for the determination of amino acid content after freeze-dried and crushed by freeze-drying machine.

2.2. Methods

Determination of general nutrients: moisture: 105 ℃ drying andconstant-weight method (GB/T 5009.3-2003); ash: 550℃ muffle furnace burning method (GB/T 5009.4-2003); crude protein: Kjeldahl method (GB/T 5009.5-2003); crude fat: methanol-chloroform method [3]. Amino acid content determination: determined by JSYKO-S-433D automatic amino acid analyzer. The dates were analyzed using the Excel 2019 and SPSS 25.0 software.

2.3. Quality evaluation of nutritional components

According to FAO/WHO Amino Acid Scoring Model [4] andWholeEgg Protein ScoringModel [5], calculate AAS (Amino acid score), CS (Chemical score), and EAAI (Essential amino acid index) of *S. szechuanensis* using Excel 2019 software. The calculation formula were as follows:

$$\mathrm{AAS} = \frac{aa}{AA_{(FAO/WHO)}}$$

$$\mathrm{CS} = \frac{aa}{AA_{(Egg)}}$$

$$\mathrm{EAAI} = \sqrt[n]{\frac{100\mathrm{A}}{\mathrm{AE}} \times \frac{100\mathrm{B}}{\mathrm{BE}} \times \frac{100\mathrm{C}}{\mathrm{CE}} \times \cdots \times \frac{100\mathrm{F}}{\mathrm{FE}}}$$

Note:aa: Content of Amino Acid in sample to be Measured[mg/g N]; $AA_{(FAO/WHO)}$: Content of the Same Amino Acid in FAO/WHO Amino Acids Scoring Model[mg/g N]; $AA_{(Egg)}$:Content of the Same Amino Acid in Whole Egg Protein Scoring Model[mg/g N]; A, B, C,... F represents the essential amino acid content of the test sample; AE, BE, CE,... FE represents the essential amino acid content of whole egg protein.

$$\text{Content of amino acid to be measured [mg/g N]} = \frac{\text{content of amino acid in the sample\%(fresh)}}{\text{content of crude protein in the sample\%(fresh)}} \times 6.25 \times 1000;$$

F: It represents the ratio of branched chain amino acids (BCAA) to aromatic amino acids (AAA).

$$F = \frac{\mathrm{Leu} + \mathrm{Ile} + \mathrm{Val}}{\mathrm{Phe} + \mathrm{Tyr}}$$

3. Results

3.1. General nutritional components

According to the test, the moisture content in the muscle of *S. szechuanensis* was 77.80%, the crude protein was18.38%, thecrude fat was 2.00% and the ash was1.55%.

Table 1. The main biochemical compositions in muscle of *S. szechuanensis*

mean ± SD, n =3, %, fresh weight

Components	Moisture	Crude protein	Crude fat	Crude Ash
wet weight	77.80±0.70	18.38±0.21	2.00±0.10	1.55±0.08

3.2. Composition and content of amino acids

The composition and content of amino acids in the muscle of *S. szechuanensis* are shown in Table 2. Totally 16 amino acids were detected (tryptophan and cystine were not detected), including 7 essential amino acids (Thr, Val, Met, Ile, Leu, Phe, Lys), 2half-essential amino acids (Arg, His) and 7 non-essential amino acids (Ser, Pro, Tyr, Asp, Ala, Gly, Glu). Among them, the content of Glu (2.92%) was the highest, followed by Asp, Lys, Ala and Leu. The content of His is the lowest. The total contents of amino acids (TAA), essential amino acids (TEAA), non-essential amino acids (TNEAA) and delicious amino acids (TDAA) in muscle of *S. szechuanensis* were 17.80%, 6.83%, 9.40% and

7.37%, respectively. The ratio of branched chain amino acids (Leu, Ile, Val) to aromatic amino acids (Phe, Tyr) in muscle of *S. szechuanensis* is 2.31(*F* value).

Table 2. Amino acids composition in muscle of *S. szechuanensis*

mean ± SD, *n* =3, %, fresh weight

Amino acids	Content	Amino acids	Content
Asp	1.82±0.07	Gly	1.13±0.04
Thr	0.78±0.02	Glu	2.92±0.14
Val	0.92±0.02	TAA	17.80±0.75
Met	0.47±0.03	TEAA	6.83±0.24
Ile	0.73±0.02	THEAA	1.56±0.09
Leu	1.49±0.05	TNEAA	9.40±0.44
Phe	0.77±0.03	TDAA	7.37±0.32
Lys	1.67±0.07	BCAA	3.14±0.09
Ser	0.70±0.04	AAA	1.36±0.06
Pro	0.74±0.05	TEAA/TAA (%)	38.40±0.32
Arg	1.12±0.05	TEAA/TNEAA (%)	72.70±1.08
Tyr	0.59±0.03	TDAA/TAA (%)	41.43±0.42
His	0.44±0.03	*F*(BCAA/AAA)	2.31±0.04
Ala	1.50±0.08		

Note: TAA: total amino acids; TEAA: total essential amino acids; THEAA: total half-essential amino acids; TNEAA: total non-essential amino acids; TDAA: total delicious amino acids; BCAA: branched chain amino acid; AAA: aromatic amino acid.

3.3. Evaluation of muscle nutrition value

After converting the essential amino acid content data in Table 2, the results were compared respectivelywith the amino acid scoring criteria model recommended by FAO/WHO and the egg protein amino acid scoring model. The amino acid score (AAS), chemical score (CS) and theessential amino acid index (EAAI) are shown in Table 3. According to AAS and CS scoring criteria, the AAS (1.67) and CS (1.29) of Lys were all the highest.Thefirst limiting amino acid isIle and the second limiting amino acid isVal of *S. szechuanensis*. The essential amino acid index (EAAI) of *S. szechuanensis* is 89.65.

Table 3. Evaluation of essential amino acids composition in muscle of *S. szechuanensis*

mg/g N

Essential amino acids	*S. szechuanensis*	FAO/WHO	TEPAA	AAS	CS
Ile	248	250	331	0.99	0.75
Leu	508	440	534	1.15	0.95
Lys	567	340	441	1.67	1.29
Thr	265	250	292	1.06	0.91
Val	313	310	410	1.01	0.76
Phe+Tyr	462	380	565	1.22	0.82
EAAI (Essential amino acids index)	89.65				

4. Discussion

4.1. The comparison of muscle composition of S. szechuanensis with that of other fishes

Nutritional composition of fish muscle is related to the species, age, size, sampling season and other factors of fish. From Table 4, we can see that the moisture content in muscle of *S. szechuanensis* is lower than that of *Paramisgurnus dabryanus*, *Mastacembelus armatus* and *Triplophysa siluroides*, slightly higher than that of other fishes. The protein content in muscle of S. szechuanensisis18.38%, which is higher than that of Para*misgurnus dabryanus*, *Misgurnus anguillicaudatus*, *Mastacembelus armatus, Triplophysa siluroides and Triplophysa dalaica.* The ash content is lower than that of *Parabotia fasciata Dabry*, *Misgurnus anguillicaudatus* and *Mastacembelus armatus*, and the fat content was lower than that of *Paramisgurnus dabryanus*, *Mastacembelus aculeatus*, *Triplophysa siluroides* and *Ctenopharyngodon idellus.* The results indicated that the muscle of *S. szechuanensis* has high water content, low ash content, making the fish taste tender, and itbelongs to relatively high-protein and low-fat food, which meets the dietary needs of people in modern society.

Table 4. Comparison of approximate composition in muscle of *S. szechuanensis* with other fishes

%, fresh weight

Species	Moisture	Crude protein	Crude fat	Crude Ash
S. szechuanensis	77.80	18.38	2.00	1.55
Parabotia fasciata Dabry[6]	76.25	20.73	1.25	1.62
Paramisgurnus dabryanus[7]	78.80	17.40	2.57	1.13
Misgurnus anguillicaudatus[7]	76.60	17.90	2.00	1.80
Mastacembelus aculeatus[8]	76.74	21.42	2.66	1.49
Mastacembelus armatus[8]	79.54	17.72	1.63	1.64
Sinibotia superciliaris[9]	76.65	20.05	1.56	1.34
Triplophysa siluroides[10]	81.04	16.98	2.14	1.13
Triplophysa dalaica[11]	78.30	17.77	1.77	1.14
Carassius auratus[12]	77.80	20.50	1.20	1.40
Ctenopharyngodon idella[12]	76.60	19.80	4.20	1.20

4.2. Composition of amino acids and evaluation of nutritional value of S. szechuanensis

The composition and content of amino acids is one of the important indicators for evaluating the nutritional composition of food. Proteins containing many kinds of essential amino acids and high content of amino acids have relatively high nutritional value. According to the ideal FAO/WHO model, when the TEAA/TAA value is about 40% and the TEAA/TNEAA value is more than 60% of a protein, itcan be considered a high-quality protein. In this study, seven kinds of essential amino acids were detected in the muscle of *S. szechuanensis*, and the types of essential amino acids are abundant. The TEAA/TAA and TEAA/TNEAA values are 38.40% and 72.70%, respectively. Therefore, they can be considered to behigh-quality protein. In the amino acid composition of *S. szechuanensis*, the top three amino acids are Glu, Asp and Lys, which are similar to those of *Mastacembelus armatus* [8] and *Triplophysa dalaica* [11].

According to AAS and CS scores, we can see that the content of Lys is the highest in the essential amino acids contented in the muscle of *S. szechuanensis*, and the first and secondlimiting amino acids are Ile and Val, respectively. The AAS scores of all kinds of essential amino acids in *S. szechuanensis* are all greater than 0.9, and CS scores allgreater than 0.7, which indicates that the content of essential amino acids in *S. szechuanensis* is rich and its composition is relatively balanced. The essential amino acid index (EAAI) in muscles of *S.szechuanensis* is 89.65, which is higher than that of *Sinibotiareevesae* [14], *Mastacembelusaculeatus* [8], *Mastacembelusarmatus* [8], *Parabotiafasciata-Dabry* [6] and *Leptobotiaelongata*. It shows that *S. szechuanensis* has high nutritional value.

The flavor of food is mainly related to the composition and content of delicious amino acids (Asp, Glu, Gly, Ala), especially Glu and Asp, which have the strongest flavor. The total content of four kinds of delicious amino acids in the muscles of *S. szechuanensis* is 7.37%, accounting for 41.43% of the total amino acids, which is higher than that of *Triplophysa dalaica*[11](38.40%), *Triplophysa siluroides*[10](39.56%), *Sinibotia superciliaris*[9] (36.73%), *Parabotia fasciata Dabry*[6] (41.41%) and *Mastacembelus armatus* (37.51%). The content of Glu in the muscle of *S. szechuanensis* is the highest, followed by Asp, which is consistent with that of *Triplophysadalaica* [11], *Sinibotiasuperciliaris* [9], *Mastacembelusaculeatus* [8] and *Mastacembelusarmatus* [8] and other kinds of fish. It is concluded that the content of delicious amino acids in *S. szechuanensis* is rich, which makes it have delicious taste.

Branchedchain amino acids in food can reduce cholesterol and protect the liver. In this study, the ratioof branchedchain amino acids to aromatic amino acids in muscle of *S. szechuanensis* is 2.31, which is higher than that of *Triplophysasiluroides* [10] (2.24), *Mastacembelusarmatus* [8] (2.06) and *Mastacembelusaculeatus* [8] (2.17), lower than that of*Triplophysadalaica* [11] (2.48) and *Sinibotiasuperciliaris* [9] (2.61), indicated that itcould be used as a supplementary source of branchedchain amino acids.

5. Conclusion

The muscle of *S. szechuanensis* has high moisture content, low ash content and tender meat. It also has the characteristics of high-protein and low-fat composition, which belongs to high-quality protein. The content of amino acids and delicious amino acids in muscle of *S. szechuanensis* are abundant, and its composition ratio is reasonable, which makes it have delicious flavor and high nutritional value, and its development and utilization prospects are broad. However, due to environmental pollution, deterioration of water quality, human activities and other factors, the living environment of *S. szechuanensis* is deteriorating, and its wild resources are declining. We suggest that we should strengthen the research on *S. szechuanensis*, carry out artificial domestication and reproduction, develop suitable artificial feed according to its nutrient characteristics, and rationally develop and utilize *S. szechuanensis* resources to promote the healthy development of its resources.

Acknowledgement

This research was financially supported by “Qing Lan Project of Jiangsu Province”. This work was supported by “311 Talents Project in Taizhou” and Jiangsu Agri-animal Husbandry Vocational College (Grant No. NSF201711).

Reference

[1] Jinming Wu, Qianqian Wang, Fei Li, et al. Early development of Sinogastromyzon szechuanensis in the Chishui River. Sichuan Journal of Zoology, 2011,30(4):527-536.

[2] Jinming Wu, Qianqian Wang, Fei Li, et al. Fish resources of early life in Chuishui section of the Chishui River. Resources and Environment in the Yang Basin,2010,19(11):1270-1276.

[3] Shaomei Wang, Shaolian Chen, Yibo Cui. On the procedures of Chloroform-Methanol extraction for the determination of lipid content of fish samples. Acta Hydrobiologica Sinica, 1993,17(2):193-196 .

[4] Pellett P L and Yong V R. Nutrition evaluation of protein foods [M] . Tokyo: The United National University Publishing Company,1980:26-29.

[5] Wenxun Fan, Zeying Li, Xuhe Zhao. Nutritional evaluation of protein foods [M] . Beijing: People's Health Press, 1984:42-44.

[6] Yuanan Wu, Zhangqiang Liang, Xiping Yuan, et al. Analysis and evaluation of nutritional components in the muscle of Parabotia fasciata Dabry. Freshwater Fisheries, 2011,41(1): 87-91.

[7] Jie Yin, Chuanxi Xiong, Shenghua Li, et al. Analysis of nutritional composition in tetraploid loach, diploid loach and Paramisgurnus dabryanus. Journal of Fujian Fisheries, 2009,(1):73-76.

[8] Yuanan Wu, Zhiqiang Liang, Chuanwu Li, et al. Analysis and evaluation of nutritional components in the muscle of two kinds of Mastacembelus. Acta Nutrimenta Sinica, 2010, 32(5):499-502.

[9] Bin He, Xianjun Chen, Zhihai Long, et al. Analysis and evaluation on nutritional components of the muscle of Botia superciliaris. Freshwater Fisheries, 2013,43(3):93-96.

[10] Hua Li. Comparison of Nutrient Components in Muscles of Wild and Artificial Cultured

Triplophysa siluroides. Sichuan Journal of Zoology, 2013,32(4):584-587.
[11] Pengfei Wu, Longwu Geng, Haifeng et al. Jiang,Analysis and Evaluation of Nutritional Components in Muscle of Plateau Fish Triplophysa dalaica. Chinese Journal of Fisheries, 2016,29(6):19-25.
[12] Haiyan Ding, Xiaojie Sun, Xiaofeng Sheng, et al. Comparison and analysis of the nutritional composition in muscle of some cultured fresh-water and marine-cultured fishes. Food Science and Technology, 2016,41(3):150-155.

Journal of Physics: Conference Series **1423** (2019) 012014 doi:10.1088/1742-6596/1423/1/012014

Study on identification and classification of hidden dangers in highway toll stations based on accident causation theory

Qiangqiang Jing [1], Yufeng Bi [2], Niantao Liu [3], Qiangmin Zhang [4] and Renshan Chen [5]

[1] Qilu Transportation Development Group, Jinan 250100, China
[2] Shandong provincial communication planning and design institute, Jinan 250031, China
[3] Qilu Transportation Development Group, Jinan 250100, China;

Corresponding author and e-mail: Yufeng Bi, 838786509@qq.com

Abstract. Aiming at high safety risk of expressway toll station, based on the theoretical model of accident cause, this paper makes a systematic analysis and correlation test on the unsafe state of object, personal unsafe behaviour, and the imperfection of safety management system within the jurisdiction of expressway toll stations. Through statistical analysis, the types and distribution areas of unsafe physical state, the types and main causes of unsafe behaviours within the jurisdiction of toll stations are identified. The results show that: education and training, risk management and control, hidden dangers, operating procedures, equipments and facilities management, emergency management, etc., have a great impact on the physical safety of highway toll stations; bad driving habits, trouble free mentality, poor safety awareness, lack of safety knowledge are the main reasons for unsafe behaviours. The results of the study have guiding significance for the risk management of toll station.

1. Introduction

Toll station is the basic unit of highway operation management and an important facility for collecting the cost of passing vehicles. Located at the entrance and exit of the highway, it has become a necessary passage for vehicles to enter and leave, a gathering area for people and vehicles and also an area with frequent traffic accidents [1]. Toll station accidents usually include frontal collision, rear-end collision, scratching and etc, and the occurrence of accidents has a great impact on subsequent vehicles. It easily causes large-scale traffic jams, and even rear-end collisions and chain collisions, which may lead to secondary accidents [2-3]. For example, on the evening of November 3, 2018, a 31-car continuous collision accident occurred at the Lanzhou South Toll Station, causing 15 deaths and 44 severe injuries [4]. In addition to causing direct casualties and vehicle damage, the toll station accidents often cause serious damage to the power, communication and other highway equipment and facilities. What's worse, it may even cause the operation of the entire highway to be interrupted, and bring a huge threat to the life of employees in the organization. Therefore, the accident prevention work in the highway toll station jurisdiction is particularly important.

The occurrence of toll station accidents is closely related with the driver's behaviors outside the organization, the performance of vehicle, the location of toll stations, the management of equipment and facilities, and the management of operating organizations [5-6]. Previous studies have proved that unsafe behaviors such as fatigue driving, speeding, illegal parking, and seizing lanes are the main causes of accidents [3]. Therefore, from the perspective of highway operators, if they can strengthen daily safety management and standardize passing vehicles' order, it can better prevent accidents in toll stations. In addition, the service and management of toll station will directly affect the driver's behavior outside the organization and also their choices for the lanes. If the hidden dangers that affect the driving safety in operation areas can be timely identified and eliminated, it will effectively avoid traffic accidents or reduce accident losses, and will be very helpful for the emergency disposal work after the accidents.

Strengthening toll stations' hidden danger elimination and risk management from the highway enterprise level is an effective measure to prevent accidents in the jurisdiction. At present, the investigation and elimination of hidden dangers in toll stations is mainly conducted from the following aspects, including the function of vehicles, the safety of employees, the safety of tolls, and the safety of toll equipment and facilities [7-8]. However, for the problems like what are the hidden dangers affecting the operation safety of toll stations, what methods should be used to identify those hidden dangers, and how to ensure the comprehensiveness and independence of accidents classification still remain to be the key issues that needs to be solved urgently. Based on study of accident causes, this paper firstly determined the influencing factors of accidents in the toll station jurisdiction, and then investigated the relevant contents involved in the hidden dangers, and then classified and analyzed the hidden danger data according to the accident causation model, and finally further put forward targeted governance measures to solve those problems.

2. Accident Causation Theory

Hidden dangers refer to the violation of safe production laws, regulations, rules, standards, regulations and management systems by producing and operating units, or the potential dangers caused by other factors like product's dangerous state, operator's unsafe behaviors and management deficiencies during production and operation [9]. These three factors are the main causes of accidents, and they are also the key points for the identification and classification of hidden dangers in the highway toll station jurisdiction. The accident causation model explains various causes and their logical relationships, and thus provides an effective tool for the identification and classification of accidents.

At present, there are many typical accident causation models both at home and abroad, and this paper applied the accident-causing "2-4"model [10-11] as the tool for identifying and classifying the hidden dangers in the highway toll station because of its clear definition, logical clarity, practical operation and wide application. The model was shown in Figure 1. The accident causation "2-4" model firstly classified the accident causes into the internal and external two parts of the organization in which the accident was triggered. On one hand, external supervision, nature, family, and political, economic, cultural, legal and other factors are collectively referred to as organizational external causes. On the other hand, the internal organizational causes are divided into two phases: safety culture and safety management system. Safety culture is a collection of safety concepts and a guiding ideology for organizing safety work, and its imperfections are the original cause of accidents. The safety management system includes safety policy, organizational structure, safety management procedures and etc., and it could be fully implemented during running. Its deficiency is the root cause of accidents. The individual internal causes within organization are divided into habitual behavior and two stages of one-time behavior and physical state, among which the unsafe habits are considered as the indirect cause of accidents. It involves five aspects of deficiencies in knowledge, consciousness, habits, physiology, and psychology. The one-time behavior, that is, unsafe act, is a direct cause of an accident, including both the unsafe acts of the driver directly involved in the accident, and also the unsafe acts associated with the accident by other personnel in the organization. Unsafe physical state includes the missing or imperfection of safety guards and equipment, unsafe state of site environment and operating objects, which may be existingat the time of the accident, or may be caused by unsafe acts or habits.

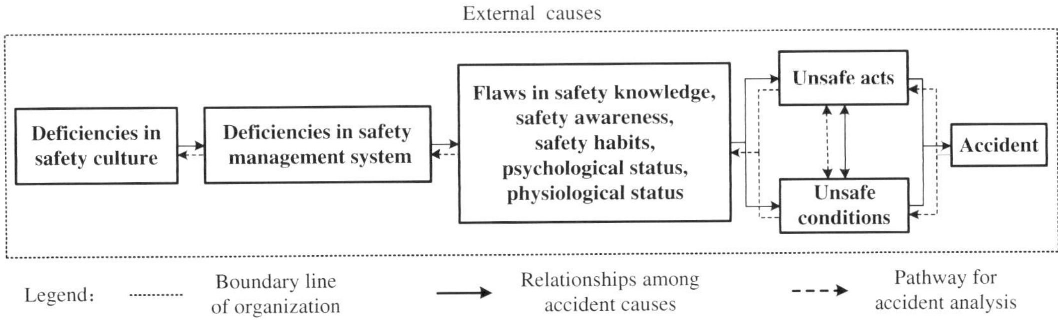

Figure 1. Accident causation "2-4" model [10, 11]

According to the accident causation model, various hidden dangers that easily cause accidents in the highway toll station can be identified one by one. The content contained in the hidden danger is refined on the basis of the original definition: one is the direct cause of the accident, that is, unsafe physical state and unsafe acts; the other is the deficiency of individual knowledge, consciousness, habits and psychology that causes the unsafe acts to generate or activate unsafe physical state. The third one is the factor imperfection of organizational management system which leads to unsafe behavior and physical state, and this one is the deep cause of accidents. In addition, the identification of organizational safety culture deficiencies in the model is difficult to determine, and it is mainly obtained by the measurement of scales, so no further study is conducted here.

3. Accident Causes Identification and Classification

The selected object of this study is the highway toll operating jurisdiction under the Qilu Transportation Development Company with altogether 52 toll stations. The study has shown that if unsafe physical state and acts near accident area, it can directly lead to accidents, and it is a key point for identifying the hidden dangers. However, in order to fundamentally eliminate these two hidden dangers, it is necessary to determine the organizational deficiencies which cause their emergence. Therefore, the identification and classification of hidden dangers in the toll station should firstly start from the unsafe acts of the operator and the unsafe state of the equipment and facilities, and then based on the results obtained, it can be inferred the organizational safety management system deficiency, and finally correlation tests can be used to determine the specific deficiency factors that must be focused on safety construction and improvement.

3.1. Unsafe physical state analysis

There are 25 types of unsafe physical states involved in the toll station jurisdiction, which are mainly distributed in four working places of toll sheds, toll booths, toll plazas and toll lanes. The names, descriptions and causes of hidden dangers are shown in Table 1 below. According to statistics, the types and frequencies of unsafe physical states appearing in the toll lanes are the most (as shown in Figure 2), which is the key and difficult point for the investigation and control of hidden dangers. The date from Classification of Enterprise Workers' Casualty Accidents have shown that besides the vehicles accidents (traffic accidents) outside the organization, the types of accidents such as electric shock, fire, object blow, and mechanical injury have also become the common hidden dangers in toll station jurisdiction, so it is imperative to strengthen risk control and emergency management.

The toll shed is the only large-scale construction facility in the entire toll station jurisdiction. Its hidden dangers involved are mainly distributed on the top of the toll shed, including the deficiencies in roofs, lights, signboards, lightning protection facilities and etc., which often cause the unexpected accidents like objects strike and collapse. The toll booth is the main area for charging transactions. One of the main problems involved is property security. The investigation have found that there are 35 toll booths with door and window damage, which can easily lead to public security incidents such as theft and robbery and then will bring economic losses to the organization. In addition, there are many hidden dangers of electrical equipment in toll booths where electric shocks and fire accidents frequently happen; and the hidden dangers of fire extinguishers failure and medium non-compliance in toll booths is widespread, which seriously delays the emergency disposal after the accident. For the toll plaza, it is the main gathering area of personnel and vehicles. Once there is any hidden danger, it will affect the normal operation of vehicles, and even cause large-scale congestion or traffic accidents. Therefore, besides strengthening the maintenance of various equipment and facilities in the toll plazas, emergency measures should also be taken for the gathering vehicles (especially transport vehicles with dangerous goods). The toll lane space is relatively narrow, but the transportation equipment involved is very complicated. The study have shown that the type and number of hidden dangers related to it are also the most statistically. It would seriously interfere with the normal stoppage of the contributory vehicles, so the organization should strengthen the routine maintenance of road equipment, such as speed bump, vehicle poles and traffic signs.

Table 1. The unsafe physical state and frequency statistics of the toll station jurisdiction

No.	Physical state name	Key descriptions	Frequencies	Causes
1	Deficiency in the roof of the toll sheds	Shedding, looseness, corrosion, water leakage	18	Object blow, electric shock, collapse
2	Deficiency in the lights of the toll sheds	Malfunction, looseness, aging	14	Vehicle damage, object blow, electric shock
3	Deficiency in the information board of the toll sheds	Looseness, corrosion, rupture	10	Object blow
4	Deficiency in the lighting protection equipment of the toll sheds	Falling off, electricity failure, broken parts	8	Electric shock, object blow
5	Honeycombs in the toll sheds	-	4	Others
6	Anti-theft door and window damage in toll booth	Door locks, window locks, etc.	35	Other (public security incident)
7	Defective socket in toll booth	Damage, looseness, aging	10	Electric shock, fire
8	Defective line in toll booth	Falling off, exposure	6	Electric shock, fire
9	Toll booth alarm failure	-	4	Other (public security incident)
10	Fire extinguisher failure	Being expired, under pressure, corrosion	20	-
11	Not compliant of fire extinguisher media	Carbon dioxide replace	3	-
12	Defect in the high pole light of the toll plaza	Missing wiring cover, aging line, loose lamp cover, lighting failure, skewed column, column covered with plants	17	Electric shock, fire, object blow, vehicle damage
13	Hazardous chemicals vehicles gathered in toll plaza	Early morning queues on high speed	8	Vehicle damage, fire, explosion
14	Damaged fence of toll plaza	Outsiders entering	4	Vehicle damage, other (public security incident)
15	Debris in the toll plaza	-	4	Vehicle damage, fire
16	Toll booth gantry loose	Screw corrosion	2	Collapse
17	Defective lane deceleration belt in toll lane	Abrasion, deficiency	19	Vehicle damage
18	Defect in the stop lane of toll lane	Damage, looseness, corrosion	13	Vehicle damage, mechanical damage
19	Defective signage of toll lane	Skewed traffic signs; skewed, dropped, and damaged speed limit signs	12	Vehicle damage, object blow

Table 1, cont				
20	Defects in toll lane roads	Water, ice, pits, loose slabs, debris	10	Vehicle damage, other (falling)
21	Defective lane markings	Damage, ageing, non-reflective film	10	Vehicle damage
22	Defective lane collision column	Skew, fracture, deformation	9	Vehicle damage
23	Defects in the safe lane of the toll lane	Damage, debris, exposed screws, exposed steel	7	Vehicle damage, mechanical damage
24	Defects in the toll lane isolation pier	Skew, damage	5	Vehicle damage
25	Wore toll lane markings	-	4	Vehicle damage

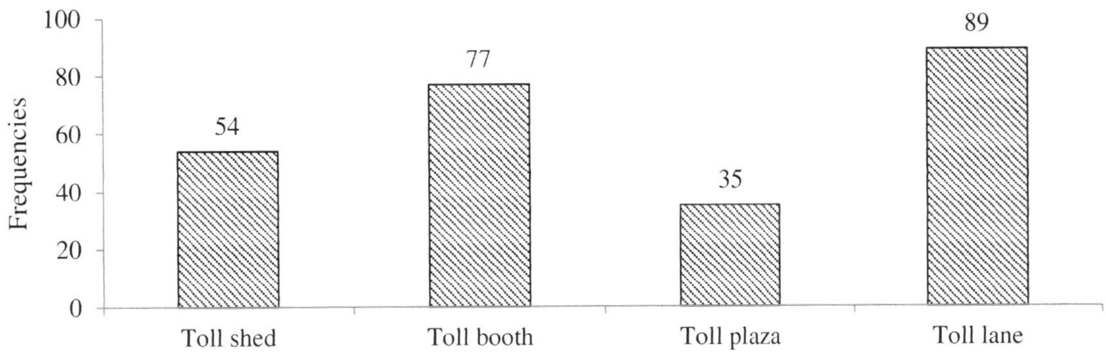

Figure 2. Frequency classification of unsafe physical states in different operating locations of toll stations

3.2. Unsafe behavior analysis

The identification of individual unsafe behaviors in the toll station jurisdiction mainly includes two parts: one is the unsafe acts that directly cause the accident, and the other is the habitual behavior deficiencies corresponding to the unsafe acts. The types of unsafe acts and their causes are shown in Table 2 below. Compared with the unsafe state, the categories and frequencies of individual unsafe acts involved in the toll station jurisdiction are relatively reduced, but it also affects the normal operation of the organization and thus should not be ignored. Considering the work site, the toll booth is the main operating area of the operator, and the involved individuals have the most unsafe acts, which mainly related to the layout of the pavilion and the maintenance and use of the fire extinguisher. The sender of the unsafe acts in the toll plaza is mainly involved with foreign intrusion personnel, field personnel, relevant operators, and etc., so the organization should strengthen the behavior control of the relevant responsible operators. The unsafe acts involved in the toll lanes are only one case, but the frequency of its occurrence still remain high reflecting with no safety sign for lane construction, so the organization should also strengthen the safety supervision for the relevant parties. And finally no unsafe acts related to the toll sheds were found in this study.

Table 2. The unsafe behavior and frequency statistics of the toll station jurisdiction

Number	Acts name	Key descriptions	Frequencies	Corresponding unsafe habits
1	Not standardized toll booth wiring	Messy, unused junction box, shedding	10	Insufficient knowledge, bad driving habits, energy-saving mentality
2	Insufficient fire extinguisher of toll booth	Toll booth	15	Insufficient knowledge, low safety consciousness, and luck mentality
3	No annual inspection of fire extinguishers	-	13	Insufficient knowledge, bad driving habits, energy-saving mentality, irrelevant psychology
4	Improper extinguisher positions	-	6	Insufficient knowledge and bad driving habits
5	No ability of using fire extinguisher	-	5	Insufficient knowledge
6	Untimely service vehicles	Traffic congestion	4	Physiological fatigue
7	Non-staff entering the toll plaza	-	8	Low safety consciousness, herd mentality
8	Not standardized staff dress	-	6	Bad driving habits, energy-saving mentality, herd mentality
9	Not standardized wiring of toll plaza	No pipe, wire, etc.	5	Insufficient knowledge, bad driving habits, energy-saving mentality
10	Private wiring in the construction site	No electric box, etc.	4	Low safety consciousness, energy saving mentality, and luck mentality
11	Lane closed without safety signs	Lane construction, temporary closure	11	Low safety consciousness, bad driving habits, energy-saving mentality, and paralysis

It can be seen from the accident causation model that unsafe acts are caused by the deficiencies of individual knowledge, consciousness, habits, physiology and psychology. Therefore, according to the unsafe acts and their specific performance, the corresponding unsafe habits can be derived. As shown in Figure 3 below, there are 9 types of unsafe habitual behaviors involved in the highway toll station jurisdiction. Among them, the individual's bad driving habits, their energy-saving mentality, lower safety awareness, and insufficient safety knowledge are most common, which become the main cause of unsafe acts. It is worth noting that the same unsafe act may be caused by multiple unsafe habitual behaviors, and the causes of unsafe acts in different working places and conditions may also be different.

Bad driving habits are manifested as a long-standing habitual violation, for example the drivers always installing and operating equipment and facilities according to their past cognitive methods, and they have been accustomed to it but without suffering any accident. It has a certain relationship with the deficiency of individual's safety knowledge, and its improvement method is mainly to strengthen safety training and supervision. The energy-saving mentality is a lazy psychology on the purpose of saving trouble and taking shortcuts, for example, the acts of wiring without following specific requirements, not using the established protective facilities, simplifying the dangerous operation process, and etc. If all those individual misconducts or illegal operation continue for a long time, it will form bad driving habits, and the effective way to improve this is also to strengthen safety training and supervision. The safety awareness deficiency is manifested by the individual's poor danger awareness in the workplace. They cannot realize the serious consequences of the illegal operation, and this is closely related with their poor individual safety knowledge. Insufficient safety knowledge means that individuals do not have the relevant skills like cannot use fire extinguishers; or they lack expertise or experience in certain areas, such as improper and irrational wiring. An effective way to both increase safety awareness and knowledge is to strengthen their safety training. In addition, for other unsafe habitual behaviors with relatively fewer frequencies, especially unsafe mentality, it should not be neglected. Staff training and psychological intervention must be strengthened, and the positions can be reasonably arranged through psychological assessment for employees.

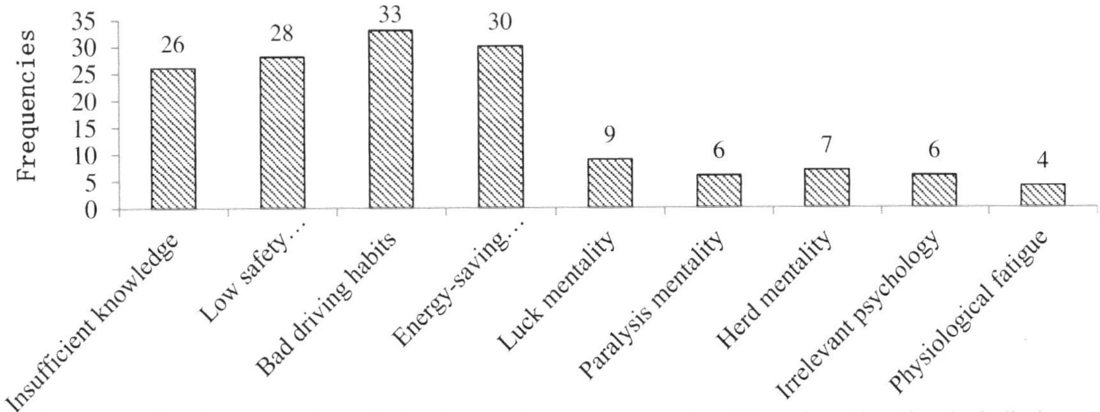

Figure 3. Classification of unsafe habitual behavior causes of accidents in toll station jurisdiction

3.3. Safety management system factors deficiency analysis

According to the principle of organizational behavior, organizational behavior influences and determines individual behavior, that is, the generation of individual unsafe behavior is ultimately caused by the management system factors deficiency at the organizational level. For the physical hidden dangers, it is closely related with the unsafe behavior or misconducts of the operators, and in essence, its deep causes still lie in the organization. Therefore, the organizational hidden dangers identification and risk control are of great significance in fundamentally preventing accidents. Based on the results of on-site investigations and combined with the relevant requirements of the Occupational Health and Safety Management System Requirements [12] andThe Basic Standards for Enterprise Safety Production Standardization [13], 14 types of safety management system factors in the highway toll stations jurisdictions were identified in this study, and these factors and their causes are shown in Table 3 below.

Table 3. The management system factors deficiency of the toll station

No.	System factors deficiency	Description of causes	Rank correlation coefficient	P-value
1	Institutions and responsibilities	Insufficient staffing; unclear personnel duty; not standardized duty records.	0.294*	0.029
2	Safety production investment	Insufficient equipment for emergency equipment; inadequate fire-fighting equipment; no account for security funds.	0.026	0.853
3	Operating procedure	Not standardized line layout; equipment facilities not being in place.	0.430**	0.001
4	Education and training	Incomplete training materials; not standardized training records; missing training contents (missing plans, new equipment).	0.602**	0.000
5	Equipment and facility management	Improper selection; inadequate configuration; not standardized installation; maintenance not being in place.	0.269*	0.047

Journal of Physics: Conference Series **1423** (2019) 012014 doi:10.1088/1742-6596/1423/1/012014

Table 3, cont

6	Operation safety	Temporary power management not being in place; workers not being dressed properly.	0.090	0.514
7	Related party management	Temporary power management not being in place; safety signs not being set properly.	0.041	0.767
8	Safety sign management	Missing safety signs; worn out safety signs.	0.171	0.211
10	Risk management and hidden danger elimination	Incomplete scope of hidden danger investigation; insufficient number of hidden danger investigations; not closed-loop management of hidden danger investigation; not standardized hidden danger investigation records.	0.586**	0.000
11	Emergency management	Insufficient emergency equipment configuration; poor management of fire protection facilities; defects in warning signs; imperfect emergency plans; not standardized emergency material records.	0.370*	0.038
12	Safety production information construction	Insufficient use of hidden danger investigation and rectification system; failure of timely reporting safety work summary.	N/A	N/A
14	Document management	Not organized and archived safety records; incomplete safety self-examination data; not standardized safety data records.	N/A	N/A

Note: ** Relevance is significant at the 0.01 level; * Relevance is significant at the 0.05 level.

In order to further determine the role relationship and cause path of the organizational imperfections on unsafe physical state and behavior, the correlation between the above statistics was analyzed in this study. Considering that the distribution of variables is non-normal, so the correlation is measured by the non-parametric rank correlation coefficient (Spearman), and the significance of the correlation coefficient is tested by t-test. And the studydetermines the p-value less than 0.05 as correlation and less than 0.01 as significant correlation. The rank correlation coefficient takes [-1, 1], and the positive result was expressed as positive correlation. If the result closer to 1, the positive correlation will be stronger. The negative result is expressed as negative correlation, and if the closer to -1, the negative correlation will be stronger. The Spearman rank correlation coefficient expression is as follows:

$$\rho_s(i,j) = 1 - \frac{6\sum_{k=1}^{n} d_i^2}{n^3 - n} \tag{1}$$

Where d_i is the difference between the rank of the independent variable X_i and the dependent variable Y_i.d_i = rg (Xi) - rg (Yi), where n is the total number of observations.

According to statistics, the factors deficiency of three management systems, including education and training, risk management and hidden danger elimination, and operational procedures, have a significant impact on the unsafe behaviors and physical states. So the organization should firstly add training contents, formulate targeted training programs, and make training records. At the same time, the organization should strengthen the investigation of hidden dangers in toll stations (expanding the scope and increasing the frequency), and then make full use of the information construction platform to achieve hidden danger's closed-loop management and hierarchical management. For each type of operation, specific operational procedures should be established to achieve safety standardization of daily operations. For other management system deficiency factors, such as the insufficient safety personnel, the unclear division of duties, the poor management of the selection, arrangement, installation and maintenance of the relevant transportation equipment and facilities, the imperfect emergency plan, the insufficient emergency reserve, and the irregular records, are also the main causes that can affect the unsafe physical state and behavior. For the insufficiency of the two safety production information construction and archive management, because it mainly acts on individual behavior and physical state through indirect means, so it is not easy to establish the correlation between individuals and organizations. Their accounting management deficiencies include the lower utilization rate of the hidden danger investigation and rectification system, the untimely compiled and archived' safety data, and the not standardized records (including safety inspection, safety training, safety activities, etc.). Due to its higher frequencies, it is also very important for identifying and controlling hidden dangers, and it should also be strengthened.

4. Conclusions

Based on the accident causation model, this paper identified accidents hidden dangers in the highway toll station jurisdiction, and conducted classification and correlation test on the causal factors from aspects of physical states, individual acts and habitual behaviors, and organizational safety management system.The main conclusions are as follows:

(1)The unsafe physical states involved in the toll station jurisdiction mainly distributed in four operating areas of toll sheds, toll booths, toll plazas and toll lanes. Among them, the types and frequencies of hidden dangers appearing in toll lanes are the most and the highest, which is the key and difficult area for hidden danger investigation and control.

(2) The injuries and deaths in toll stations mainly included vehicle injury accidents (traffic accidents), electric shocks, fires, object strikes, and other (safety incidents). So the organization should strengthen the control and emergency management of related risks based on this.

(3) The individual unsafe acts involved in the toll station jurisdiction with relatively fewer types and lower frequencies, so the organization should regulate the use and maintenance of toll booth fire-fighting facilities, and strengthen safety supervision of relevant parties in the jurisdiction. In addition, bad driving habits, energy-saving mentality, poorer safety awareness and insufficient safety knowledge are the main causes of unsafe acts, so the organization should strengthen the training, supervision and psychological intervention of operators.

(4) The deficiencies in safety management system factors, such as education and training, risk management and hidden dangers elimination, operational procedures, equipments and facilities management, and emergency management, have a significant impact on the unsafe physical states and unsafe acts in highway toll stations. It is the key point of safety management system construction.

Acknowledgment

Thispaperissupported by Shandong Transportation Science and TechnologyPlan2017B58, and by The Soft Science Research Plan2016RKB01453: Research on road safety countermeasures based on "Internet + bad weather".

References

[1] Yin Xiaoting, Qian Yongsheng, Liu Yufei. Research on Traffic Safety of Highway Toll Stations. Intelligent Communications. Vol. 18(2010) No. 1, p. 97-100.
[2] Kong Ying. Traffic Safety Management of Highway Toll Stations under Emergencies. Master's thesis of Jilin University, 2012.
[3] Wu Wanwu. Research on Traffic Safety of Expressway Toll Station. Master's Thesis of Chang'an University, 2004.
[4] Sina News. A Major Traffic Accident Occurred at a Toll Station in Lanzhou. http://news.sina.com.cn/o/2018-11-05/doc-ihnknmqw9830225.shtml.
[5] Liang Xiaocheng. Research on Traffic Safety Technology of Expressway Toll Station. Northern Communications. 2013, No. 3, p. 116-118.
[6] Zhang Hui. Research on the Management of Highway Toll Stations. Master's thesis of Hebei University, 2013.
[7] Teng Wei. Research on Security System Evaluation of Highway Operation Management.

Master's Thesis of Chang'an University, 2011.
[8] Zhang Min, Chen Hong, Wu Xiaowu. Safety Evaluation Model of Highway Toll Station. China Safety Science Journal. Vol. 19 (2009) No. 10, p. 139-144.
[9] GB 6441-1986. Classification Criteria for Casualties of Employees in Enterprises.
[10] Fu Gui. Safety Management - Behavior Control Method for Accident Prevention. Beijing: Science Press, 2013.
[11] Fu Gui, Fan Yunxiao, Zhai Ruipeng, et al. General Cause Accident Analysis Method (4th Edition). Journal of Accident Prevention. Vol. 2(2016) No. 1, p. 7-12.
[12] GB/T 28001-2011. Occupational Health and Safety Management System Requirements.
[13] GB/T 33000-2016. Basic specifications for Enterprise Safety Production Standardization.

The application and evaluation of "Production-oriented Approach" in College English Teaching

Ruiyun Zeng[1], Qianmei Li[2]

[1, 2] Guilin University of Electronic Technology, Guilin541004, Guangxi, China

Corresponding author and e-mail: Ruiyun Zeng, 602633037@qq.com

Abstract. This paper introduces a semester-long of experiment process with the production-oriented approach (POA) to test the effectiveness of POA in an ordinary university in China. The findings are as follows. First, POA has obvious effects on improving the overall English level of students. Second, with POA, there is an obvious increase in students' reading ability, a slight increase in their listening ability, and a slight decrease in their vocabulary ability. This paper analyzes the research results and puts forward corresponding solutions to the problems existing in the implementation process of POA in ordinary colleges and universities.

1. Introduction

In 2008, taking Swain's (1995, 2000) "output hypothesis" as the important theoretical basis, Professor Wen Qiufang put forward the "output driven hypothesis" based on the concept of "integration of learning and application" for English professional skills curriculum reform. Since then, some scholars and teachers begin to apply this hypothesis in English major teaching. In 2013, Professor Wen extended this hypothesis to college English teaching, making it more concerned by college English teachers and practiced in classes. In 2015, Professor Wen formally named it as "production-oriented approach" (POA). At present, many teaching experiments have provided practical classroom experience and lessons for the construction of the theoretical system. Some of these studies focus on a particular section of the teaching links of "motivating – enabling – assessing" (Zhang Wenjuan2016; Qiu 2017; Sun 2017), some on the teaching experiments on a single unit (Cao, 2017; Zhang Wenjuan2016), some on application of POA in a certain course (Qi 2016; Zhang Wenjuan 2017), and some on the effectiveness analysis on POA teaching (Zhang Lingli 2017). InAugust of 2017, Professor Wen andher team held an English teaching and training class to promote POA teaching to front-line teachers of college English from all over China. The author participated in this session and began to conduct POA teaching experiments in college English teaching in Guilin University of Electronic Technology (GUET).

Professor Wen (2015) pointed out that this approach is mainly aimed at middle and advanced foreign language learners. Applying the common European framework for reference, the target audience should be at least A2 or above. At present, the colleges and universities that apply this method are basically key universities with students of high level of English proficiency. The application and effectiveness analysis of POA in ordinary colleges and universities are till in the way. In this paper, the author attempts to share her experience and in POA teaching experiments in an ordinary university and test the effectiveness the approach in both the overallEnglish proficiency and some particular language skills.

2. Research design

In line with the situation and requirements of the national college English teaching reform, GUET has reduced teaching hours on college English course since September 2012, and implemented graded teaching. However, there is no significant change in teaching methods. College English teachers still adopt the "text-centered method", that is, the teaching mode of "warm-up practice –text learning – after-class practice", aiming at understanding the words, sentences and passages in the textbooks. The result is that the students show no strong interest in class learning, the classroom teaching efficiency is low, and the problem of "separation between learning and using" is relatively prominent. Therefore, how to stimulate students' initiative and creativity in English learning and comprehensively improve the teaching effect according to their different internal needs in a limited period of time is the key problem to be solved urgently. Based on the above background, this study aims to solve two problems. First, can POA improve students' overall English level? Second, which skills of listening, speaking, reading, writing and translating are most effective in POA?

2.1. Subjects of the experiment

The subjects of this study are471 sophomore students in Business College of GUET. The experimental group consist of 59 students in one of the class. In the university, all students are not allowed to register for College English Test – Band 4(CETt-4) until the third semester. Starting from 2017, the university has conducted an entrance test for ordinary undergraduates before the new semester begins. Those who have achieved a certain score can take the CET-4 in the first semester. At the same time, some students are allowed to take the CET-4 in the second semester according to their scores of the final examination in the first semester. The classes in the first and second semesters are arranged according to the students' professional major. At the third semester, those who have passed the CET-4 in the two semesters will be selected into the advanced classes, and the rest can freely choose any parallel class with a particular teacher assigned to this college. Therefore, in the third semester, among the ten classes of this college, there are two advanced classes with 87 students who have passed CET-4, and ten parallel classes with 471 students. The author chose one of the parallel classes as the experimental group, applying the teaching method of POA, with a total of 59 students. The controlled group is another parallel class of 65 students in another college under the same teacherwith the traditional "text-centered method". The two groups are analyzed against the 471 students in all parallel classes inBusiness College. The reason why the author did not adopt POA in the controlled group is that the English teaching in this collegeseparated listening and speaking course with reading and writing, so the overall POA teaching could not be adopted.

In addition, since the teaching and test contents of the advanced classesare different from those of the parallel classes in the third semester, the average data of the whole college, whether in the pre-test or the post-test are all excluding the data of the advanced classes.

2.2. Preparation for the experiment

POA teaching experiment lasted for 14 weeks, with 4 periods per week and 56 periods in total. All the students spend two weeks in the language laboratory for independent learning, completing the exercises off the designated teaching units. The next 12 weeks are spent in the classroom for the teaching of the fivedesignatedunits both of the textbooks. There are both communicative and language goal for each unit. The production tasks includes spoken task and written task.The former consists of group videos and group presentations or role plays, the latter English composition, translation, practical writing and the like. Each production task is decomposed into several subtasks. The whole teaching follows the three links of "motivating – enabling – assessing" (Wen, 2015), and the assessing link adopts the "teacher-student cooperative evaluation" (Wen, 2016). In order to effectively carry out POA teaching experiments, the author elaborated the overall evaluation scheme of college English courses to students at the beginning of the semester.The class openly elected a course representative, and set up an English learning QQ group for cooperative learning. The whole class is divided into 10 groups, with a leader for each group. At the same time, the learning information of each student is recorded as an important basis for formative evaluation of for theirCollege English course.

3. Research process

The research tools used in this experiment are pre- and post-tests. There are two pre-tests. The first is the English proficiency test, namely the entrance test mentioned above, which is consistent with the CET-4, including listening, reading, writing and translation. The entrance examination is conducted on the intranet of GUET, and the papers will be marked by machine immediate after the examination. The second is the final examination of the first semester. The post-test isthe final exam of the third semester. The two final exams are the same types of stage test, including listening, reading, vocabulary, translation and writing, among which listening, reading and vocabulary are marked by machine. Writing and translation are sealed and marked collectively within all the teachers in the whole grade. Therefore, there is no significant difference in marking standards of both the pre-tests and post-test. The author applies the SPSS21.0 software package in the quantitative data analysis of the differences between the two groups of students and the itemized skills.

3.1 Pre-test analysis

3.1.1 Comparison of admission level between experimental group and college average.

The experimental group are from the business college of GUET, a major with good enrollment. In order to clarify the English level of the students before the teaching experiment, the author compared the scores of the experimental group with the average score of the whole college, rather than with that of the whole university. The total full score of entrance test is 710 points, which was converted into a percentage system for the convenience of comparison.

In the pre-test, the average score of the whole college is 39.6, while the average score of the experimental group is 39.9729. The one-sample t test is as follows:

Table 1. One-sample t test of scores of experimental group in pre-test 1 (T0)

	Test Value = 39.6					
				Mean	95% Confidence intervals of difference	
	T	df	Sig. (2-tailed)	Difference	Lower	Upper
T0	.348	58	.729	.37288	-1.7738	2.5196

As can be seen from Table 1, although the average score of the experimental group is slightly higher than the average score in the college, the critical confidence (2-tailed) was 0.729, greater than 5%. Therefore, there is no significant difference between the scores of students in the entrance test class and the average scores of students in the whole college, which is 39.6 points. Therefore, the English proficiency of the experimental group was about the same as the average level of the whole college.

3.1.2 Comparison of admission level between the experimental group and the controlled group

After the entrance test, the students were divided into 10 teaching classes according to their major, not according to their score of the entrance examination.That is to say, 59 students of the experimental class were scattered in different classes and taught by five different teachers under the traditional teaching method. At this time, the author did not participate in the teaching of this college. After the final examination of the first semester, the author selected all the scores of the experimental group and all the students in parallel classes with the aid of Excel. The average score of the experimental group is 60.0169, while the average of the college is 57.5. Then the test result is as follows:

Table 2. One-sample t test of scores of experimental group in pre-test 2 (T1)

	Test Value = 57.6					
				Mean	95% Confidence Intervals of Difference	
	T	df	Sig. (2-tailed)	Difference	Lower	Upper
T1	1.278	58	.206	2.41695	-1.3683	6.2022

As can be seen from Table 2, after a semester of traditional college English teaching, in different classes and under different teachers, although the average score of the experimental group was slightly higher than the average level of the whole college, the significance (2-tailed) is 0.206, greater than 5%. Therefore, there is no significant difference between the performance in the first semester and the average score of 57.6 in the whole college. Therefore, under the same teaching method, the overall English level of the experimental group did not improve significantly.

3.2 POA teaching process

The teaching process of POA covers three stages, namely motivating, enabling, and assessing. During the three stages, the mediating effect of the teacher mainly consists of guiding, designing, scaffolding and the like.

3.2.1 Motivating

In the motivating phase, Professor Wen (2015) elaborates the steps and requirements. She said, When designing the communicative scenes for driving links, the author designs communicative scenes close to the learning and living environment of students, such as designing the "100-yuan challenge of one-day tour in Guilin", applying for the "world's most liveable city" for their hometown, and writing the "business plan" for the innovation and entrepreneurship project of college students. Students are interested in these production tasks, but they also find that the departure of pragmatic existing distance, will produce a sense of urgency and initiative to continue learning.

3.2.2 Enabling

In the facilitation stage, the successful completion of a production task requires at least the idea, language and discourse structure to express the content in language. Considering the actual level of students in GUET, the author focuses on the learning of English expressions and forms, including words, phrases and sentence patterns that can serve for the production tasks. Step by step, language facilitating activities are designed to help students transform "declarative knowledge" into "procedural knowledge" according to the path of "identification and understanding –output as required – open output". The third step is to extract the discourse structure required by the production task from the input. With the help of the scaffolding to facilitate the activities, students have built up the confidence of language expression and can successfully achieve the advanced level of language production.

3.2.3 Accessing

The production can be assessed immediately and afterclass. In the assessing stage, teachers should specify specific operable assessing criteria according to different forms and contents of production tasks. At the beginning of the semester, the author combined the writing standards of composition and professor Lucas's standards of speech (2010), designed detailed evaluation standards, and sent to students for self-evaluation and mutual evaluation.For example, if the production task was a written composition, the author the evaluation criteria could be broken down into the following, each of which accounted for 20 percent.

If the production task is oral, in addition to the above criteria in terms of content, teachers should also specify assessing criteria in terms of delivery, as shown in the following table:

It must be pointed out that the evaluation step is not a step after the completion of the motivating and enabling stage, but a step running through the entire POA teaching process. That is to say, in the process of students' selective learning and production task practice, teachers should evaluate students' learning effect and timely adjust teaching steps and teaching contents. In addition, the evaluation stage is time-consuming, but it cannot be handled hastily. Teachers can make use of the network communication platform to allow students participate in the assessment process and make pre-class and after-class assessment possible. Teachers can send evaluation forms to students in advance for students to make self-assessment, and conduct mutual assessment on the production works of other students uploaded. At the end, the teacher make targeted assessment in class.

4. Results and discussion

After a semester of teaching, the author tested the POA teaching effect of the experimental class based on the scores of the final examination of the third semester. The effectiveness of POA on overall English proficiency and individual English skills is reported below.

4.1 *Effect on the overall English proficiency*

4.1.1 Comparison of average score between the experimental group and the whole college

The final examination of the third semester is based on the same principle as the first semester. Both are periodic exam based on the textbooks. In this test, the average score of the experimental group is 67.7712. According to the single- sample t test, the comparison results are as follows:

Table 3. One-sample t test of scores of experimental group in post-test (T3)

	Test Value = 63.6					
					95% Confidence Interval of the Difference	
	t	df	Sig. (2-tailed)	Mean Difference	Lower	Upper
T3	3.296	58	.002	4.17119	1.6379	6.7044

As can be seen from Table 3, after a semester of POA teaching experiment, the overall average score of the experimental group has improved (4.17119) compared with the whole college. And the significance (2-tailed) is 0.002, much less than 5%. Therefore, it indicates that there is a significant difference between the grade of students between the experimental group and the college average, which is 63.6 points. It can be concluded that POA helps to improve the overall English proficiency of ordinary college students.

4.1.2 Comparison between the experimental group and the controlled group

It is proved that the average score of the experimental group was significantly higher than that of the college after a semester of POA teaching. In order to eliminate the differences between the teachers competence, the author compared the results of the experimental group and the control group again. In pre-test 1, we have found no significant difference between the experimental group and the control group, indicating that there is no significant difference between the two groups in English proficiency. After a semester of traditional teaching, in the second pretest, there was still no difference in the scores of the two groups, indicating that there is also no significant difference in the teachers' competence of the two groups. Now let's examine the changes in the English proficiency of the two groups of students after a semester of POA teaching.

Table 4. Comparison of post-testbetween the experimental group and the controlled group

Group	N	Average	Std. Deviation	Mean Difference	Sig. (2-tailed)
Experimental	59	67.7712	9.72081	8.90196	.000
Controlled	65	58.8692	12.23509		

In Table 4, the method adopted is the t-test of equal variance of the two samples. It can be seen that the significance (2-tailed) is 0.000, far less than 5%. Therefore, it indicates that there is a significant difference between the scores of the two groups in the third semester of the survey.At the same time, we can also find that after three semesters of traditional teaching, the English proficiency gap between the controlled group and the experimental group is getting larger and larger. This may be because the traditional English teaching fails to stimulate the learning initiative of students, especially the students in the middle and high grades. After the intensive high college English teaching, they find that the traditional college English teaching is no longer challenging and lose the interest in English learning, and their English level remains the same as in the high college.

4.2 Effect on individual skills

Now that we have proved that POA teaching has a significant effect on the overall English proficiency of students, we are going to prove which skills of English POA can promote. Due to the subjective factors in the marking of translation and writing, only three types of questions, namely listening, reading and vocabulary, are selected in this comparison.

Table 5. Comparison of individual skills between the experimental group and the controlled group

Item	Experimental group average	College average	Std. Deviation	Mean Difference	Sig. (2-tailed)
Listening 20%	15.9492	15.1612	2.91503	.7480	.053
Reading 30%	18.5085	16.7396	5.18078	1.8084	.010
Vocabulary 15%	6.1864	6.3079	2.12517	-.1135	.683

It can be seen from Table 5 that in the three language skill in comparison, the significance (2-tailed)of listening and vocabulary is more than 5%, which indicates that compared with the average level of the whole college, POA teaching has no prominent advantage in improving students' listening and vocabulary ability. The significance (2-tailed) for reading is 0.010, less than 5%. Therefore, it indicates that there is a significant difference between the reading score of this class and the average score of 16.7 points of all the students in the college. This shows that the POA teaching has significantly improved the students' reading ability. According to previous experiments of other teachers (Qi 2016, Zhang Lingli 2017), POA is more effective in improving writing and speaking abilities. Since Guet has not installed standardized speaking test platform at present, the author cannot conduct credible speaking level test and make persuasive effectiveness analysis. In terms of writing, the writing type of this test does not fit well with the actual communication scenes of students, which fails to reflect the obvious advantages of POA teaching method. However, in general, in the writing part of this test, the language and structure of the experimental group are more standardized. The influence of POA on writing needs to be further studied in the future with a more effective analysis design.

5. Conclusions

The POA teaching experiment has been implemented for half a year and needs to be continued. Teachers and students are gradually adapted to the "motivating - enabling - assessing" teaching mode which obtained significant teaching effects. Studies have shown that POA can effectively improve students' reading, writing and speaking abilities and promote their all-round development. However, due to subjective and objective reasons, the design of this study still has defects. First, the research cycle is too short, and changes and learning effects of students observed are limited. If the experiment can start since the enrollment, and the experiment tracking lasts for four semesters, it is believed that it will have a more obvious effect on the improvement of students' overall English proficiency and individual skills. Second, there is no well-designed method for measuring the production ability of speaking and writing, so the production ability of students in the experimental stage cannot be tracked in detail. At the same time, the traditional English proficiency test is not satisfactory in testing teaching effectiveness comprehensively and objectively. In addition, POA challenges students' learning initiative, especially in ordinary universities where students' admission English level is not so high. In the initial stage, students are resistant to participating in POA experiments when the requirements of the whole university are not unified. POA also requires teachers to be more committed and engaged. Textbooks used for College English teaching in GUET are not completely in accordance to the concept of POA principle. Therefore, a lot of time and energy is required for teachers in the design of communicative scenes, production tasks and enabling steps suited with students' actual English level. Of course, if POA can be comprehensively promoted in the whole university and teaching teams can be formed in and out of the college, it is believed that POA can give full play to its obvious advantages.

Acknowledgement
This work is supported by the Teaching Reform Project of Guangxi Higher Education (grant NO. 2018 JGB177, NO. 2018JGB187), Basic Ability Improvement Project of Scientific Research for Young and Middle-aged Teachers in Universities of Guangxi (Grant No. 2019KY0243), and Guangxi Higher Education Undergraduate Teaching Reform Project (Research on the Long-term Mechanism and Implementation of Guangxi Minority Culture Inheritance Education with Cooperation Between University and Secondary Vocational School).

References
[1] Cao Qiaozhen. Teacher Scaffolding in Production-oriented Approach [J]. Foreign language Education in China, 2017(10). (In Chinese)
[2] Lucas, Stephen E. *The Art of Public Speaking* (M). Beijing: Foreign Language Teaching and Research Press, 2010. (In Chinese)
[3] Qiu Lin. The Step-by-step Design of Language Activities in the Production-oriented Approach [J]. Modern ForeignLanguages, 2017 (3). (In Chinese)
[4] Sun Shuguang. Teacher-student Collaborative Assessment in Classroom Teaching: A Reflective Practice [J]. Modern Foreign Languages, 2017 (3). (In Chinese)
[5] Swain, M. The Output Hypothesis and Beyond: Mediating Acquisition through Collaborative Dialogue [A]. In J.P. Lantolf (Ed.), *Sociocultural Theory and Second Language Learning* [J]. Oxford: Oxford University Press, 2000. (In Chinese)
[6] Swain, M. Three Functions of Output in Second Language Learning [A].In G. Cook & B. Seidlhofer (eds.). *Principles and practicein applied linguistics: Studies in Honour of H.G. Widdowson* [J].Oxford: Oxford University Press, 1995. (In Chinese)
[7] Wen Qiufang. Application of the Output-driven Hypothesis in College English Teaching: Reflections and Suggestions [J]. Foreign Language World, 2013 (6). (In Chinese)
[8] Wen Qiufang. Developing a Theoretical System of Production-oriented Approach in Language Teaching [J]. Foreign Language Teaching and Research, 2015 (4). (In Chinese)
[9] Wen Qiufang. Output-driven Hypothesis and Curriculum Reform of English Professional Skills [J]. Foreign Language World, 2008 (2). (In Chinese)
[10] Zhang Lingli. An Experimental Study on the Effectiveness of the Production-oriented Approach [J]. Modern ForeignLanguages, 2017 (3). (In Chinese)
[11] Zhang Wenjuan. "Enabling" Students' Production to Enable Students' Language Development: Applying Production-oriented Approach to TEFL Classroom [J].Foreign Language Education in China, 2015 (4). (In Chinese)
[12] Zhang Wenjuan. An Experimental Study on the Effect of the Production-oriented Approach on College Students'English Writing Quality [J]. Modern ForeignLanguages, 2017 (3).
[13] Zhang Wenjuan. Applying Production-oriented Approach to College English Classrooms: A Teaching Experiment [J]. Foreign Language and Foreign Language Teaching, 2016 (2).

Research of Chinese Government debt sustainability and its effect on economic growth

Jifan Li

Southern University of Science and Technology, Shenzhen 518055, P.R. China.

Corresponding author and e-mail: Jifan Li, jeffery.li097@outlook.com

Abstract. This paper researches the appropriate scale of Chinese government debt from the perspective of debt sustainability and its effect on economic growth. First, we made a model based on fiscal response framework and introduced trade openness, savings ratio and so on as control variables. We use it to measure the upper limit of the sustainable debt ratio over the years. Then we did empirical research and found that government debt has a threshold effect on economic growth, with a turning point of about 33.3%. Besides, we also found that debt affects economy by affecting the formation of total society capital significantly, and excessive debt may inhibit the formation. According to our research, the government should release savings reasonably, improve its sensitivity to debt and use it more efficiently to enhance the turning point of debt impact. Government should also gradually reduce current debt ratio to optimize the relative scale of debt.

1. Introduction

After the financial crisis, many economies experienced adverse debt impacts. Economic recession, financial aid and expanded government spending have led to rapid increases in debt ratios in many countries. It has also aroused extensive discussions on the relationship between government debt sustainability, debt scale and economic growth in academic areas. There are two perspectives to study debt sustainability: theoretical model analysis and empirical indicator system. In the theoretical model, a mainstream thinking system was proposed by Bohn (2007). He put forward the idea that the current debt policy would respond to the previous period government debt ratio and constructed a framework of government debt ratio research. Later scholars such as Ostry (2010) and Ghosh (2013) made some compensation. They believed that when the debt scale is higher than the critical value, the government's ability to adjust fiscal surplus is limited (this situation is called fiscal fatigue), and the surplus is not enough to cover the actual interest expenditure of debt, which leads to the further expansion of the debt scale and eventually the outbreak of the crisis. On the other hand, the relationship between debt size and economic growth has many possibilities in theory. The mainstream view is that there exists a non-linear quadratic curve relationship. It was put forward by Checherita et al. (2012) firstly, which also accords with the theoretical analysis and empirical test. The phenomenon that the impact of government debt on economic growth has a turning point is called the threshold effect.

In China, since the financial crisis and the rapid accumulation of local government debt, the central government attached great importance to the debt problem. In 2011 and 2013, the official national debt audit result was published for the first time. Meantime, local debt limit management was also launched. Whether the debt of the Chinese government is in a sustainable stage and has a good role in promoting the economic growth ratio is a question worthy of study under the current situation. The famous 60% debt ratio warning line given by the EU Treaty of Maastricht is roughly based on the average of its member level at that time, which does not necessarily have wide applicability, especially for developing countries. For example, in some Latin American countries, the debt ratio was less than 60% when the debt crisis broke out, while Japan did not break out under the debt ratio of 200%. And the scale of debt that most promotes economic growth is quite different because of the different economic environment and the use of debt. The main work of this paper is to study the upper limit of the relative sustainable scale of Chinese government debt in recent years and observe the debt space. The second is to study the relationship between debt scale and economic growth in the context of debt sustainability.

The new contributions of this paper are as follows: Firstly, this paper combine the two perspectives of debt sustainability and the threshold effect of debt on economic growth. Previous studies including Ostry (2010), Ghosh (2013) or Checherita (2012), are based on a single perspective of sustainability. In this paper, sustainability is included in the research framework. We further sought the scale that can satisfy both conditions of debt sustainability and economic growth. Secondly, in terms of debt sustainability, this paper solves the financial response function of China to explore the debt ceiling from the perspective rarely used in Chinese literature, which provides evidence for the phenomenon of financial fatigue in China. The data processing has been improved, and more accurate ratio data of China's fiscal surplus has been obtained compared with the predecessors. For the first time, China's inflation ratio and trade openness are included in the Chinese literature, which provides a new empirical study on the factors affecting China's debt ceiling. Thirdly, in terms of debt threshold effect, this paper takes into account the variables that have an important impact on China's economic growth (such as total savings ratio, real interest ratio, trade openness, etc.) and enriches the study of the turning point factors that affect debt threshold effect. Fourthly, the empirical study further finds that the impact of debt on social capital formation is a significant channel for debt to affect economic growth, which enriches the research on how debt affects economic growth. Fifthly, in view of the fact that the current situation of the developed and developing economies is the main object of study in a large number of literature. In this paper, the historical data of China as the research object, more targeted.

2. Literature review

Research on government debt risk and sustainability has the following perspectives:

Comparing the real economic growth ratio with the real debt interest ratio, comparing current assets and liabilities from the view of balance sheet, quantitatively calculating default risk or building an index evaluation system to measure. The existing analysis mainly analyzed the probability of debt default risk and relevant research were abundant. However, there are some blanks, such as little research on the relative scale of the sustainable debt ceiling. For fiscal policy reference, it is intuitive and meaningful to calculate the ceiling ratio of debt.

Academic research models on the relative scale of debt sustainability have been developed first in foreign countries. The current mainstream ideas and methods originated from Bohn (1998, 2007) who constructed a new framework to test the sustainability of government debt by using the idea that current debt policy would respond to the previous debt ratio. Follow-up scholars continue to improve it. Ostry et al. (2010) argued that when a country's debt level is too high, it is difficult for the government to adjust its basic fiscal surplus to deal with debt. So a non-linear model was introduced to simulate the fiscal response function. Ghosh (2013) put forward the idea of debt space, using the current debt ratio and debt ceiling gap to measure the sustainability of government debt. The estimated debt space can not only more intuitively reflect the sustainability, but also quantitatively reveal the degree.

On the other area about the relationship between debt and economic growth, under the framework of neoclassical economics, debt affects economic growth by influencing capital formation. It can not only squeezes out private investment, but increase public investment, therefore causing an uncertain impact on total investment and economic growth ratio. The specific empirical results depend on the relative size of the two impacts, which are related to the unique economic environment, the relative scale of debt and even the efficiency of public investment. When the debt ratio is low, the effect of increasing public investment may be dominant, while when the debt ratio is high, the effect of crowding out private investment funds is significant. Therefore, we argue that the formation of social capital and the relationship between economic growth and debt ratio may be a quadratic curve, and there is a turning point in the effect of debt ratio on economic growth.At the empirical level, the literature on their relationship has been less. Up to recent years, Reinhart and Rogoff (2010) conducted descriptive studies using data from 44 countries and found that government debt has a threshold effect on economic growth, which aroused a sensation in the academic and political circles. Checherita et al. (2012) considered that there is a non-linear quadratic curve relationship between government debt and economic growth, which has become a widely accepted academic view.

These documents have the following characteristics. Literature on sustainability rarely directly showed the relative scale of debt from the perspective of fiscal response function. In the study of economic threshold effect, previous studies such as Checherita (2012) focused on developed countries and developing countries as a whole, rather than on a specific target like China. There are obvious differences between China and developed countries in capital formation, factor productivity and other important variables, so the group empirical conclusions are not necessarily suitable for China.In addition, more importantly, the connotation of moderatio debt scale in existing papers is almost only from the perspective of economic growth. This is not comprehensive. For example, Reinhart believes that the appropriate debt size of developing countries is about 90%, but the Treaty of Maastricht gives a reference value of 60% for the debt ratio ceiling. This shows that some countries are limited by sustainability and the debt scale can not reach the level that most promotes economic growth. This paper expands the connotation of moderatio scale in view of this problem.

3. Model and Data

3.1. Analysis Framework of Sustainable Government Debt

This paper mainly uses the analytical framework proposed by Ghosh et al. (2013) to establish the equation between the basic fiscal surplus (basic fiscal surplus = fiscal revenue - fiscal expenditure deducting debt expenditure) and the government debt ratio to solve the debt ceiling.

Since the gap in the basic fiscal surplus is compensated by debt financing, it can be concluded that:

$$D_t - (1 + r_t)D_{t-1} = E_t - R_t \tag{1}$$

Among them, D_t represents the balance of the debt at the end of the t period, R_t and E_t refers to the fiscal revenue and fiscal expenditure deducting debt expenditure respectively. After simplification, the equation is as follow:

$$D_t = (1 + r_t)D_{t-1} - S_t \tag{2}$$

S_t represents the basic fiscal surplus at t period. Because we care about the relative scale of these variables, the following equation is obtained by introducing GDP and simple calculation.

$$d_t - d_{t-1} = (r_t - g_t)d_{t-1} - s_t \tag{3}$$

Among them, d_t represents the government debt ratio (government debt/GDP) at the end of the t period. s_t represents the basic fiscal surplus ratio (basic fiscal surplus/GDP). r_t is the debt interest ratio and g_t is the economic growth ratio. $(r_t - g_t)d_{t-1}$ represents the ratio of actual interest expenditure in the t period.

To further study the upper limit of debt ratio through this equation, we need to estimate the fiscal response function, that is, the relationship between basic fiscal surplus ratio and debt ratio. In this paper, we adopt the conclusion that the relationship is non-linear put forward by Ostry (2010) and Ghosh (2013). They believe when the debt ratio starts to rise from a lower level, the government will increase the basic fiscal surplus to tackle it. But when the debt ratio exceeds a certain level, the basic fiscal surplus ratio will decrease with the increase of the debt ratio because of the limited ability. This paper constructs the following fiscal response function:

$$s_t = a_0 + f(d_{t-1}) + X_t + \varepsilon_t \tag{4}$$

$f(d_{t-1})$ is the cubic function of debt ratio. X_t represents other control variables that may affect the basic fiscal surplus. The significant ones from previous research results are inflation ratio, trade openness, output gap, government expenditure gap (Ghosh, etc.), economic growth ratio (proposed by Tang Wenjin), and savings level (proposed by Hoshi).

In the fiscal response function, the government debt ratio data comes from the IMF. As for the calculation of the basic fiscal surplus ratio, this paper calculates the accurate basic fiscal surplus ratio on the basis of China Fiscal Yearbook. Inflation ratio, nominal GDP comes from the National Bureau of Statistics. The degree of trade openness comes from China's Foreign Trade Statistics Yearbook. Referring to the research of Tang Wenjin (2014), the non-stationary economic growth ratio is used to replace the output gap and fiscal expenditure gap as explanatory variables. The data of economic growth ratio comes from the National Bureau of Statistics. The data of residents' savings deposit balance come from the National Bureau of Statistics and the People's Bank of China.

After calculating the fiscal response function, the debt ceiling can be obtained by substituting it into equation (3).

Analysis Framework of Government Debt and Economy Growth

Referring to Checherita's (2012) idea, this paper sets the government debt ratio as the main explanatory variable. The equation is as follows:

$$g_{t+1,t+5} = a_0 + h(d_t) + a_1 \ln gdp_t + X_t + \varepsilon_t \tag{5}$$

In the regression equation, the response variable is the medium-term economic growth ratio. We use the real average per capita GDP growth ratio in the next five years as an agent, which can eliminate the impact of economic fluctuations and endogenous problems.

d_t refers to government debt ratio. In economic theory, government debt has both positive and negative effects on economic growth. Government debt is represented by the government debt ratio. The data comes from IMF.

$h(d_t)$ is a function of debt ratio. We adopt the quadratic function form proposed by Checherita (2012) of it. In China, the specific form of this function can be further judged according to the image. $\ln gdp_t$ is the logarithm of per capita real GDP. X_t represents the control variables. This paper chooses Martin (2004) and Checherita (2012) theory and empirical results as reference to select the total savings ratio (sav_t), real interest ratio (rir_t), trade openness (tra_t), dependency ratio of the elderly population (old_t), population growth ratio (pop_t) as control variables. Data comes from National Bureau of Statistics, People's Bank of China, World Bank, China Foreign Trade Statistics Yearbook.

4. Empirical Analysis

4.1 Calculating the upper limit of Chinese government debt ratio

4.1.1 Estimating Fiscal Response Function

Fiscal response function is the basis of solving the dynamic equation of government debt. According to the latest research, this function is a non-linear (cubic) function because of financial fatigue. From the figure, this phenomenon does exist in China. Therefore, the financial response function is constructed as follows:

$$s_t = a_0 + f(d_{t-1}) + X_t + \varepsilon_t \tag{6}$$

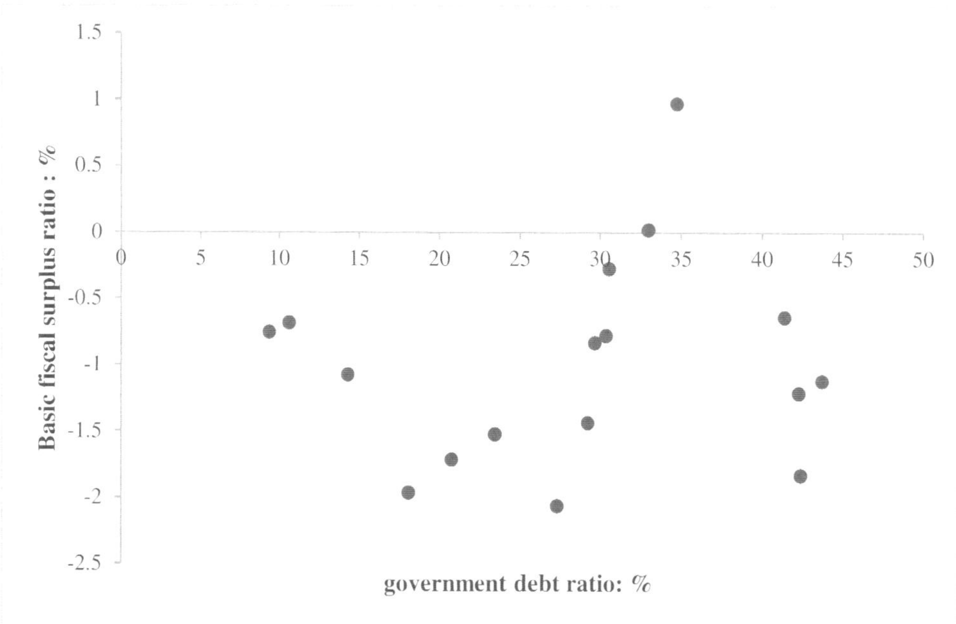

Figure 1. Spot chart of government debt ratio and basic fiscal surplus ratio

The unit root test of all variables shows that the first order difference of all variables passes the stationarity test at 1% significance level. This paper estimates the fiscal response function in four forms of the $f(d_{t-1})$, uses stepwise regression method to determine the proper specific form of the cubic function, and excludes the economic growth ratio in the explanatory variables. The regression results of the equation are shown in Table 1.

Table 1. Estimated results of fiscal response function

Variable	Correlation coefficient	Standard error	P-value
Constant term	0.0330	0.010	0.009
Debt Ratio Squared Term	1.5366	0.382	0.0008
Debt Ratio Cubic Term	-1.5380	0.405	0.001
Debt Ratio	-0.4491	0.121	0.001
Resident savings ratio	-0.0655	0.016	0.0005
Trade openness	0.0827	0.015	0.0000
Inflation ratio	-0.0690	0.020	0.008
R^2		0.846	
Adj R^2		0.797	
DW stat		1.970	
Prob(F)		0.000	

The explanatory power and significance of the equation are good, and the coefficients of variables are significant at 0.01 level. Residual items pass the EG co-integration test to exclude the possibility of "pseudo-regression". DW statistic shows that there is no autocorrelation in random error terms. From the result, we can see high inflation will have a negative impact on the adjustment of China's basic fiscal surplus. Increasing trade openness helps to reduce debt risk. It is worth mentioning that the saving ratio coefficient is significantly negative, proving our theory assumption that in China, excessive savings ratio will weaken the government's response and regulatory capacity to debt. The cubic-term coefficient of government debt ratio is significantly negative, which confirms the phenomenon of "fiscal fatigue" in China.

4.1.2 Calculating the upper limit of government debt ratio

The expression of fiscal response function is obtained and therefore we can solve equation (3). It means:

$$(r_t - g_t)d_{t-1} - s_t = 0 \tag{7}$$

Using equation (7) and data, we can get the upper limit of government debt ratio over the years, as shown in Figure 2. It can be seen that the upper limit of the debt ratio of the Chinese government varies from 53.8% (2016) to 70.3% (2007), which is also close to the "Maastricht Treaty". From 2003 to 2007, with the increase of trade openness, high economic growth and the release of savings ratio, the debt ceiling increased. After 2008, due to the impact of the financial crisis, economic growth slowed down gradually, while trade openness began to decline from the peak in 2006, and the savings ratio also maintained a high upward trend, which had a negative impact on the debt ceiling.

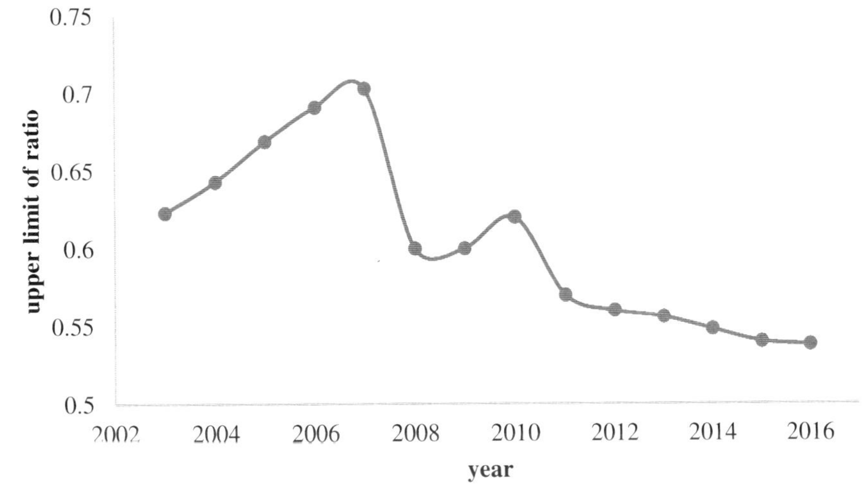

Figure 2. Chinese debt ceiling between 2003-2016

Measuring Turning Point of Government Debt Economic Effect

Referring to the analytical framework of Checherita et al. (2012), the model is set as follows:

$$g_{t+1,t+5} = a_0 + h(d_t) + a_1 \ln gdp_t + X_t + \varepsilon_t \tag{8}$$

Considering the scale of government debt ratio was very small before 1994, the sample period of this paper is 1995-2016. The specific form is judged by observing scatter plots. As can be seen from Figure 3, quadratic function is more suitable for simulation.

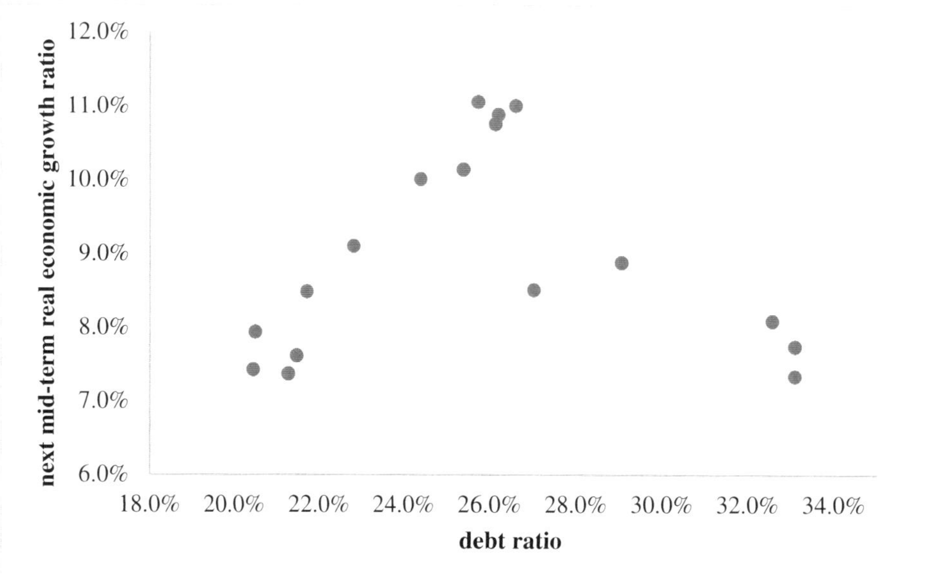

Figure 3. Scatter plot of government debt ratio and economic growth ratio

First, we did ADF test, and the first-order difference of all variables passed the stationarity test at 1% significance level. In this paper, we use the stepwise regression method and eliminate the control variable of population growth ratio. The regression results of the final equation are shown in Table 2.

Table 2. Estimation results of equation

Variable	correlation coefficient	Standard error	P-value
d_t	3.6960	0.69	0.0010
d_t^2	-5.5500	1.28	0.0017
$\ln gdp_t$	-0.0474	0.02	0.0403
rir_t	0.1340	0.05	0.0179
tra_t	0.0970	0.10	0.0143
old_t	1.6178	0.83	0.0818
sav_t	-0.1430	0.07	0.0403
Constant term	-2.086	0.10	0.0503
R2		0.9485	
Adj R2		0.9085	
DW stat		2.3400	
Prob(F)		0.0000	

The model is significant and has good explanatory power. The P value of cointegration test is less than 0.01, which excludes pseudo-regression. The DW value indicates that there is no autocorrelation. Heteroscedasticity White test shows that the model has no heteroscedasticity.

As per capita GDP rises, Chinese economic growth slowed down. Opening trade and rising real interest ratios could help stimulate economic growth. The coefficient of dependency ratio of the elderly population is positive, it is very interesting. Obviously, it is not because the increase of the proportion can directly promote economic growth, but because of China's "demographic dividend" period in the process of population aging in the past two decades. "Demographic dividend" means that there has been an increase in the proportion of working-age population and a period of abundant labor resources. The explanation is consistent with the fact that the proportion of working-age population (15-64 years old) in China rose in the past two decades (data from the National Bureau of Statistics).

The total savings ratio coefficient is negative. A reasonable explanation is that excessive savings may inhibit household consumption, thus inhibiting the enthusiasm of social investment in production, which makes the savings funds unable to effectively transform into productive capital.

From the estimation results, we can estimate that the turning point of government debt in China is about 33.3%. That is to say, when the relative scale of government debt exceeds this value, the marginal effect of increasing debt on economy will be negative. The critical value obtained in this paper is close to that in some famous literature. For example, Kumar's famous paper (2010) also found that when the government debt ratio was less than 30%, the impact of government debt on economic growth was positive. When the government debt ratio was between 30% and 90%, the overall impact of the increase of government debt ratio on economic growth was negative.

4.2 Exploring how government debt affects economic growth

Under the neoclassical framework, capital accumulation is most closely related to debt among the driving forces of economic growth ratio. Therefore, government debt affect economic growth through affecting total social investment or capital formation is an obvious theoretical channel. This paper will explore the assumption.

We use total capital formation ratios to measure the variable of capital accumulation (total capital formation ratio = total capital formation / GDP). Data comes from the World Bank web. The government debt ratio is taken as the main explanatory variable, and the sample period is 1994-2016. We refers to Fan Xiaoyun(Chinese NanKai University Finance Prof.)'s research(2014), and chose the lagging total capital formation ratio, lagging per capita GDP growth ratio(g), lag savings ratio and current real interest ratio as the control variables. The econometric model is constructed as follows:

$$Cap_t = a_0 + h(d_{t-1}) + Cap_{t-1} + X_t + \varepsilon_t \quad (9)$$

We did unit root test. It was found that the first-order difference of total capital formation ratio and its lag term passed the test at the 5% significant level, while the first-order difference of other variables passed the stationarity test at the 1% significant level. Using stepwise regression method, the measurement results are shown in Table 3.

Table 3. Estimation results of equation

Variable	Correlation coefficient	Standard error	P-value
Constant term	-0.198	0.061	0.007
d_{t-1}	2.557	0.502	0.000
d_{t-1}^2	-3.613	0.808	0.000
g_{t-1}	0.316	0.179	0.097
Cap_{t-1}	0.417	0.151	0.015
R2		0.965	
Adj R2		0.955	
DW stat		2.160	
Prob(F)		0.000	

The model is very significant and the explanatory is good. It also passed the EG cointegration test and we excluded the pseudo-regression. According to the empirical test results, we reasonably infer that in China, the impact of government debt on the formation of total capital is also a quadratic curve, with a turning point of about 35.4%. When the debt ratio exceeds 35.4%, the marginal effect of government debt on total capital is negative. This means that the crowding-out effect of government debt ratio on capital formation exceeds the effect of increase of public investment. This data is very similar to the turning point of government debt on economic growth ratio above. It can be reasonably inferred that government debt affects economic growth by affecting total social investment is reasonable and significant.

5. Conclusions and recommendations

This paper combines debt sustainability and debt threshold effect to study the Chinese government debt problem for the first time. In terms of debt sustainability, based on the analysis framework of Ghosh (2013), this paper introduces the savings ratio, trade openness and inflation ratio as control variables for the first time to study the situation of Chinese government debt, and calculates the upper limit of China's government sustainable debt ratio for consecutive years.

The conclusions are as follows. Firstly, there is indeed a phenomenon of "fiscal fatigue" in China. The adjustment ability of basic fiscal surplus is limited, and it will gradually fail when the debt ratio exceeds a point. Secondly, in China, high inflation will increase debt risk, while the improvement of trade openness will enhance fiscal adjustment ability and high savings ratio has a negative effect on fiscal surplus adjustment. Finally, at this stage, there is still a little room for Chinese government debt, which has gradually approached the upper limit. In the future, debt prevention and control needs to be monitored dynamically.

About the threshold effect of government debt on economic growth. For the first time, based on the idea of Checherita (2012), this paper conducts empirical research on China and finds that: firstly, the government debt ratio has a quadratic function effect on economic growth with a turning point of about 33.3%, so the marginal benefit is negative when the government debt ratio exceeds this point. Secondly, trade openness promotes economic growth, while high savings does not directly promote economic growth. It may be because high savings has inhibited the normal consumption of residents and the production investment of enterprises. In addition, we should use saving funds for more efficient investments, so as to transform saving into real productive capital.

In addition, this paper further analyses the mechanism of government debt's effect on economic growth, and concludes that the channel which government debt affects economic growth by affecting the formation of social total capital is reasonable and obvious. It is also find that previous research believe government debt can promote capital formation and society investment, but if it exceeds a certain limit, it will also inhibit the formation of total social capital. The turning point of government debt ratio on society capital formation is about 35.4%. After exceeding this scale, the crowding-out effect of government debt ratio on capital formation exceeds the effect of increase of public investment and finally reduces the formation of total social capital.

This paper holds that the appropriate scale of Chinese government debt should be around 33.3% and puts forward following opinions. Firstly, Chinese government should take measures to help to release the savings ratio reasonably to form social consumption and investment, improve its sensitivity to debt and monitor the debt ceiling and real scale dynamically to ensure that it does not exceed the ceiling. Secondly, Chinese government should gradually reduce relative proportion of debt to GDP in order to optimize government debt ratio. In addition, increasing the efficiency of public investment or transfer more government debt to useful public investment can not only increase the turning point (proper size) of debt, but also help to promote economic growth.

References

[1] Bohn. Are stationarity and cointegration restrictions really necessary for the intertemporal budget constraint?[J].Journal of Monetary Economics,2007,54(7):1835-1848.

[2] Bohn. The behavior of US public debt and deficits[J]. Quarterly Journal of Economics,1998,113(3):950-960.

[3] Checherita, Cristina, Philipp Rother. The Impact of High Government Debt on Economic Growth and Its Channels: An Empirical Investigation for Euro Area[J]. European Economic Review, 56(7), 2012.

[4] Ghosh A R, Kim J I, Mendoza E G,et al. Fiscal fatigue, fiscal space and debt

sustainability in advanced economies[J]. Economic Journal, 2013, 123(566): 1-20.

[5] Hoshi T,Ito T.Defying gravity:Can Japanese sovereign debt continue to increase without a crisis?[J]. Economic Policy,2014,29(77):5-14

[6] Kumar, Manmohan S. , Jaejoon Woo. Public Debt and Growth[J]. IMF Working Paper, July 2010.

[7] Manmohan S. Kumar, Jaejoon Woo. Public Debt and Growth[J].Policy Research Working Paper,2010,(July):8-17.

[8] Ostry J D, Ghosh A R, Kim J I, et al. Fiscal space[R].IMF Working Paper,2010.

[9] Reinhart,Carmen M., Kenneth S. Rogoff. Growth in a Time of Debt[J]. NBER Working Paper, January 2010.

Journal of Physics: Conference Series **1423** (2019) 012017 doi:10.1088/1742-6596/1423/1/012017

Decomposition and reconstruction of ore-caused anomalies information by airborne gamma-ray spectrum based on fractal singular value decomposition method

Kun Sun[1], Liangquan Ge[1,*], Fei Li[1], Qingxian Zhang[1], Yi Gu[1], and Wanchang Lai[1]

[1]The College of Nuclear Techonology and Autornation Engineering, Chengdu University of Techonglogy, Chengdu 610101, China

* Corresponding author and e-mail: Liangquan Ge, glq@cdut.edu.cn

Abstract. Identification method research on ore-caused anomalies information is one of the core technologies of airborne gamma-ray spectrum data post-processing. This paper proposed a decomposition and reconstruction method of anomaly information called fractal singular value decomposition (FSVD), aiming at the current ubiquitous truncation error and boundary effect problems in filter design in ore-caused anomaly information identification methods in airborne gamma-ray spectrum. Based on the multi-fractal theory, this method decomposes the singular value of the specific nucleus activity data of the airborne gamma-ray spectrum under time series, and quantitatively calculates the fractal singular value truncation parameter according to residual square sum minimum principle. According to the calculation result, the original data is partially reconstructed to rebuild the ore-caused anomalies information. This paper used the 1:50000 airborne gamma-ray spectrum trial production flight data of a certain area in Inner Mongolia (the surveying area contains proved metal ore occurrences) to verify the method. The results show that: after the original signals are decomposed by FSVD, the high, medium and low singular values can represent the background, anomaly, and noise signal, respectively. At the same time, the false anomaly signals caused by flight error exist in high singular values. After reconstructing the decomposed signals according to median singular values, background noises can be effectively separated, the ore-caused anomaly in the known ore-caused anomalies can be identified and band-like false anomalies caused by time domain batches can be removed.

1. Introduction

Airborne gamma-ray spectrum survey (AGS) can measure the distribution of radioactive elements in the primary halo of superficial soils. With fast measuring speed, large measuring area and ability to avoid unfavorable terrain, it has become an important part of China's mineral resources exploration technology [1]. Since radioactive elements can directly indicate the distribution of uranium deposits, areas with anomaly distribution of radioactive elements have higher metallogenic potential. The extraction effect of anomaly information in the processing of airborne gamma spectrum data directly affects the quality of the following work.

Nowadays, filtering original signals according to the nature of geological signal frequency domain has been widely recognized as one of the effective methods for extracting anomaly signals. Fourier transform has been widely used in petroleum exploration, seismic exploration, ground penetrating radar, geochemistry [2-4] and other fields. By filtering signals, the signal-to-noise ratio can be improved, and ore-caused anomalies can be divided. Although this method can effectively separate ore-caused anomaly signals, due to the mathematical principle of the Fourier transform (fundamental space is sine wave), truncation errors and boundary effects cannot be avoided.

Singular Value Decomposition (SVD) is an emerging method for analyzing signals in time series, and has been widely used in telecommunications, ecology, machinery industry, meteorology, geological exploration and other fields [5]. SVD uses the reduced-rank principle to achieve signal separation, which avoids truncation errors and boundary effects caused by Fourier transform. After being decomposed by SVD, original signals can form several singular values, and different singular values represent signals of different energies. By reconstructing part of the signals, the effect similar to filtering can be achieved. Fractal theory is an effective method to deal with nonlinear signals, which has been widely used in mathematics, materials science, ecology, geological exploration, computer information science and other fields [6]. Because the ore-forming process is affected by various elements such as different stratums and structures, the geologic body and corresponding geophysical field will exhibit spatial self-similarity characteristic. Fractal theory is an effective method to deal with information with this characteristic. In recent years, fractal theory has been applied to describe the distribution regularities with multiple measures in geophysics and geochemistry [6-8].

In this paper, a method for decomposing and reconstructing ore-caused anomaly information of airborne gamma-ray spectrum is proposed from the angle of energy domain in time series, namely FSVD method, which is an improved method for processing nonlinear signals, compared with SVD method. The method re-divides the original data from the angle of energy domain, processes the matrix formed by the interpolated original signals by SVD, piecewise fits the fractal curve of the singular values according to the principle of minimum residual square sum, separates the singular values corresponding to different energies, and reconstructs the singular values of each segment. In this way, errors caused by using naked eyes to determine critical singular values in the conventional SVD method can be avoided. FSVD separates the ore-caused anomaly from the background according to the difference and law of the critical singular value, and restores the geophysical field background at the energy scale, and extracts the ore-caused anomaly information of airborne gamma-ray spectrum.

2. Principle and Method

2.1. SVD (Singular Value Decomposition)

Assuming that the element content measured by airborne gamma-ray spectrometry can form matrix A ($m\times n$, m is the number of lines, n is the number of points measured by each line), rank of A is $r(r \le n)$, and there are orthogonal matrice V of size $n\times n$, and orthogonal matrice U of size $m\times m$,

$$U^T AV = B \tag{1}$$

In the formula (1), B is a non-negative diagonal matrix,

$$B = \begin{bmatrix} S & 0 \\ 0 & 0 \end{bmatrix}, \quad S = diag(\sigma_1, \sigma_2, \cdots, \sigma_r), \quad r \le \min(m,n) \tag{2}$$

$\sigma_1, \sigma_2, \cdots, \sigma_r$ is called the singular values of A, $\sigma_1 > \sigma_2 > \cdots > \sigma_r > 0$, and the column vectors μ_i, v_i of U and V is the left and right singular vectors of A respectively (μ_i is the *i-th* eigenvector of AA^T, and the *i-th* eigenvector of $A^T A$ is v_i).

Expression $A = UBV^T$ is called the singular value decomposition of A. The product of the matrix U and V^T constitutes an orthogonal complete space, denoted as $\phi_{(\sigma_k)} = U \otimes V^T$. The matrix product of the *k-th* column vector μ_k of U and the *k-th* row vector v_k of A is

$A_k = \begin{bmatrix} \mu_{1,k} \times v_{k,1} & \cdots & \mu_{1,k} \times v_{k,n} \\ \vdots & \ddots & \vdots \\ \mu_{m,k} \times v_{k,1} & \cdots & \mu_{m,k} \times v_{k,n} \end{bmatrix}$ k=1,2,…,r. A_k constitutes a complete 2D orthogonal space. The projection coefficient of matrix $A(m \times n)$ at A_k is singular value σ_k . σ_k(k=0, 1,···, r) forms the spectral density or feature space decomposition coefficient of matrix A, namely:

$$A(m \times n) = \sum_{k=1}^{r} \sigma_k \mu_k v_k \tag{3}$$

In formula (3), like the coefficients in the Fourier transform, σ_k represents the coefficients projected or mapped by matrix $A(m \times n)$ on the feature space $\mu_k v_k$. The accumulation of multiple characteristic matrices with weighting coefficients can be used to characterize the matrix A. The contribution of the singular value in matrix reconstruction is positively correlated with the magnitude of its corresponding singular value, and the larger singular value represents the main component; The energy of uranium geophysical field represented by matrix A is closely related to the singular value. The sum of eigenvalues ($\lambda_i = \sigma_i^2$) can reflect the sum of uranium geophysical field energy. Therefore, the greater the singular value (or eigenvalue) component, the greater the contribution power of the geophysical field signal. Taking k as abscissa and σ_k as ordinate, the energy density curve of A after SVD can be obtained. In the k-th eigenspace, the energy distribution percentage of the geophysical field of the original uranium element is as follows:

$$M_K = \frac{\sigma_k^2}{\sum_{i=1}^{r} \sigma_i^2}, \quad k=0, 1, \ldots, r \tag{4}$$

In formula (4), M_K is also called as the percentage of the spectral density of the matrix $A(m \times n)$ in the *k-th* feature space. Normally, the normalized energy integral curve is plotted with k as the abscissa and $M_A = \sum_{i=1}^{r} M_i$ (cumulative energy percentage) as the ordinate. When k=r, $M_A = 1$.

If $A(m \times n)$ is a real matrix, the range of p singular values is 0- $\lambda^{1/2}$. The singular value is defined as its spectral radius. Since the integral power spectral density at the spectral radius can be replaced by eigenvalues, the energy from the minimum spectral radius (0) to the current spectral radius $\lambda^{1/2}$ can be defined as the energy measure E_i in the sense of scale λ_i:

$$E_i = \sum_{k=i}^{P} \lambda_k, \quad 1 \le i \le P \tag{5}$$

Formula (5) is the energy corresponding to a spectral radius, which can also be understood as a energy measure that can be obtained at a spectral scale. If the rank of the matrix is P, a sequence pair with length P can be obtained. The energy measure of the sequence pair varies with the energy spectrum scale. Similar to the relationship between energy spectrum and frequency in Fourier transform, there is a fractal law between the scale defined in this way and the energy measure [9]. Using measure to characterize the energy distribution of ore-caused elements can quantitatively express complex geological mineralization in the form of energy redistribution. Therefore, the analysis of the measurement curve is a new way to quantitatively divide the energy boundary scale of mineralization.

2.2. Calculating Critical Singular Value Based on Multi-fractal Quantification

The multi-fractal model can qualitatively solve the problem of irregular nonlinear feature extraction related to mineralization [10]. In general, the critical singular value in airborne gamma-ray spectroscopy data can be determined by directly observing the singular value σ_i and the energy measurement curve E_i, but it has a large observation error which can be avoided by calculating the critical eigenvalue quantitatively by mathematical means. The distribution of singular value-energy measure curve has multi-fractal similarity under different observation scales. Critical singular value represents the quantization of different observation scales. The change of critical singular value can directly determine the scale of different energy signals being filtered. Therefore, when the multi-fractal model appears in the data studied, the geometric characteristics of the singular value-energy measurement curve will not change with the critical singular value. The curve with the above properties must obey the following power law distribution:

$$E_i = C ? s_i^{-D}(s_i \quad 0) \tag{6}$$

In formula (6), σ_i is a singular value; C is a non-negative constant, D is a fractal dimension; E_i is an energy measure corresponding to σ_i. Take logarithms on both sides of Formula (6), A simple linear equation between E_i, D and constant C can be obtained.

$$\log E_i = - \mathrm{D}\log(s_i) + \log(\mathrm{C}) \tag{7}$$

The slope D obtained by least square estimation is the fractal dimension.

Specific steps of FSVD:

(1) Perform Kriging interpolation on the Airborne gamma-ray spectrum data to form regular gridding data;

(2) Perform singular value decomposition on the gridding data to obtain a couple of non-zero singular values arranged in a large to small manner;

(3) Calculate the energy measure E_i corresponding to the first n singular values σ_i;

(4) Take σ_i as the horizontal axis and E_i as longitudinal axis, project scatter points on the X-Y plane coordinates.

(5) With the principle that the sum of residual squares is the smallest, fit scatter points into multiple linear segments, and singular values corresponding to intersections of the fitted segments are the critical singular values.

(6) Reconstruct the singular values corresponding to the ore-caused anomaly.

3. Experiment and Result

3.1. General Situation of the Survey Area

This paper selects the AGS863 airborne gamma spectroscopy system independently developed by Chengdu University of Technology as the detection tool, and selects the uranium content in the 1:50000 airborne radioactivity field test data within the scope of a metal mine in Inner Mongolia Autonomous Region as the data source. The location of the survey area is shown in Figure 1(A), and the geological map is shown in Figure 1(B). There are known lead-zinc mineralized spots in the survey area. The uranium element is high in lead-zinc ore, and target prospecting area can be identified by the anomaly area of uranium ore. Ground verification was carried out in lead-zinc mining area, and the geological map of the verification area is shown in Figure 1(C). AGS863 airborne gamma spectrometry detector consists of two boxes of crystals and each of them consists of 5 4L arrayed NaI crystal detectors (one of the crystals placed in the upper part is used to monitor the concentration of atmospheric radon and cosmic rays). The detector is installed on a Y-12 conveyor. The measurement platform is equipped with data acquisition system, radio altimeter, and global satellite positioning system, etc. With 21 design survey lines and 6705 survey points, the area of the survey area is about 300 km2. The average value of uranium content is 3.968 $\mu g/g$ with the highest value being 37.204 $\mu g/g$, the lowest value being 0.049 $\mu g/g$, and the standard deviation being 3.186 $\mu g/g$.

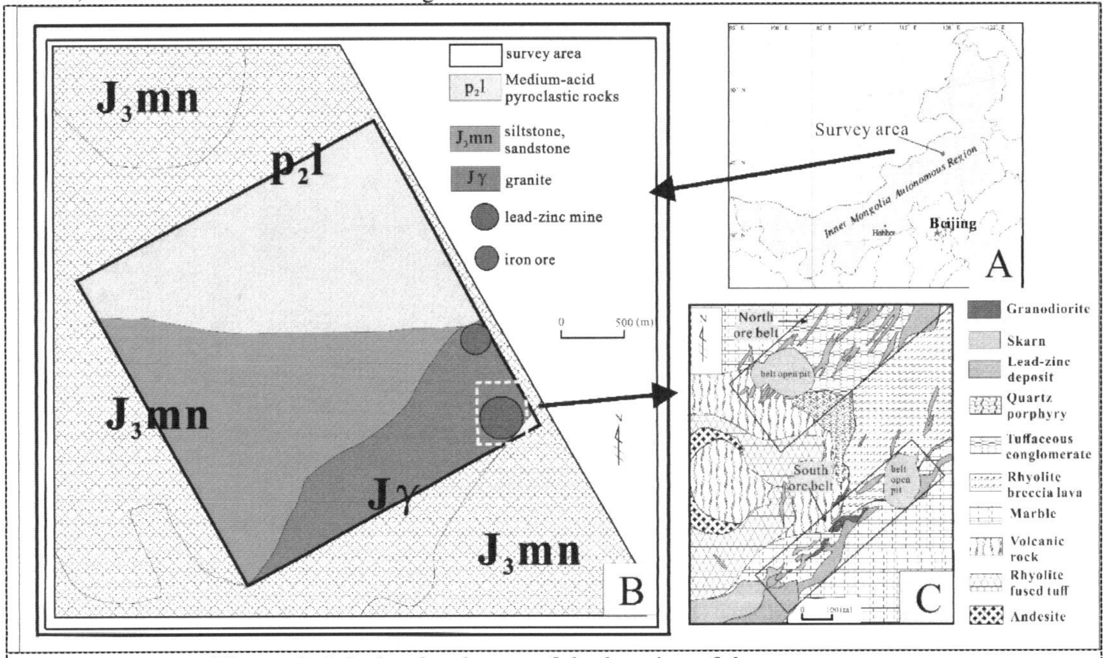

Figure 1. A is the sketch map of the location of the survey area;
B is the geological map of the survey area;
C is the geological map of the ground verification area

3.2. FSVD Anomaly Information Extraction

By formula (1), use SVD to interpolate the data of the survey area, calculate the relevant parameters and draw Table 1. It can be seen from Table 1 that the signals corresponding to the large singular values account for most of the energy of the total signals, and the small singular values account for a small proportion of energy. Only the first nine singular values contribute 93% of the total energy to the total energy.

Table 1. Singular Value Decomposition Information of Airborne Gamma Spectroscopy Uranium Content Data.

No.	Singular Value	Energy Density (%)	Cumulative Energy (%)
1	117.80	59.87	59.87
2	43.49	8.16	78.96
…	…	…	…
8	17.09	1.26	92.58
9	15.87	1.09	93.67
…	……	……	……
31	2.81	0.34	99.50
…	……	……	……

Journal of Physics: Conference Series **1423** (2019) 012017 doi:10.1088/1742-6596/1423/1/012017

By analyzing the relationship between the sequence of singular values and the cumulative energy percentage, it is found that when the singular value sequence is 31, the cumulative energy percentage reaches 99.5%. It can be considered that the first 31 singular values can reflect most of the energy of the total signals. At this point, linearly fit the singular value curve and quantitatively calculate critical singular values. The fitting curve is shown in Figure 2.

After calculation, σ_1=13.53, σ_2=49.52. In Figure 2, the fitting line segment A corresponds to the smallest singular value, and the corresponding energy is the lowest, which represents the perturbation anomaly caused by the high frequency noise in the survey area; segment B corresponds to relatively small singular values, and its corresponding energy is higher, which represents the anomaly information of element enrichment caused by hydrothermal mineralization and other geological processes in the survey area. segment C corresponds to the largest singular value and the highest energy, representing the stable geological background information in the survey area.

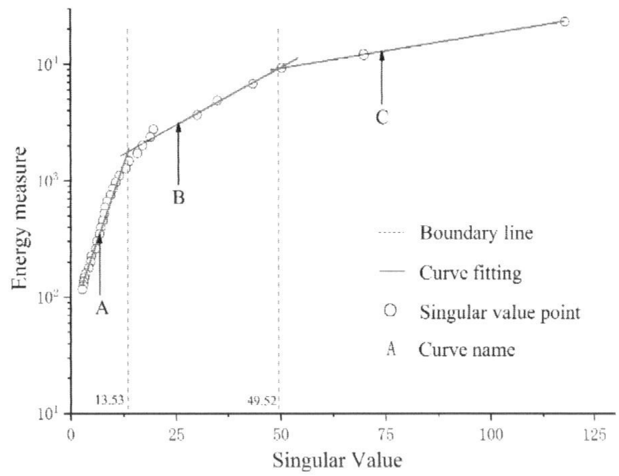

Figure 2. Figure with short caption (caption centred).

4. Result and Discussion

Using the singular values in segment B, the decomposed data is reconstructed, and the raw data of survey area and the anomaly data extracted by FSVD are respectively plotted into the contour map. The result is shown in Figure 3A and Figure 3B.

In Figure 3A, due to noise and different time domain batches, different degrees of irregular halo anomalies appear in the survey area. Among them, due to the metal mines in area *a*, the uranium content is extremely high, and thus a large number of irregular anomalies appear in the southeast corner of the survey area; there are banding anomalies along the route direction in areas *b* and *c*; large-scale weak anomalies exist in the J_3mn in area *d*; A step type of positive-negative anomaly crossover occurred in area *e* where the $J\gamma$ outcrop. The above anomalies are superimposed on ore-caused anomalies and interfering noises, causing interference to ore-caused anomaly identification and ground verification work.

In Figure 3B, after the raw data is processed by FSVD, the background is suppressed, negative anomaly area is reduced, and the ore-caused anomaly position of the ore occurrence is highlighted. A "crescent" anomaly occurs in area *a*, which coincides with the height of the ore shape in Figure 1B.The banding false anomalies along the route direction of the *b* and *c* areas are basically filtered out, showing a sporadic distribution of scatter-like anomalies; area *d* shows that there is a weak anomaly; the step type of positive-negative anomalies disappear and the weak anomalies are highlighted.

Comparing area a with ground verification map (Figure 3C), it is found that compared with the anomaly information presented by the raw data, the information about shape, location and scope of anomalies decomposed and reconstructed by are consistent with the ground verification result, and the anomaly intensity and background value are basically unchanged, which shows that FSVD method can filter out false anomaly information and eliminate interfering noise, suppress background, highlight anomalies, and accurately identify ore-caused anomalies from data doped with interference.

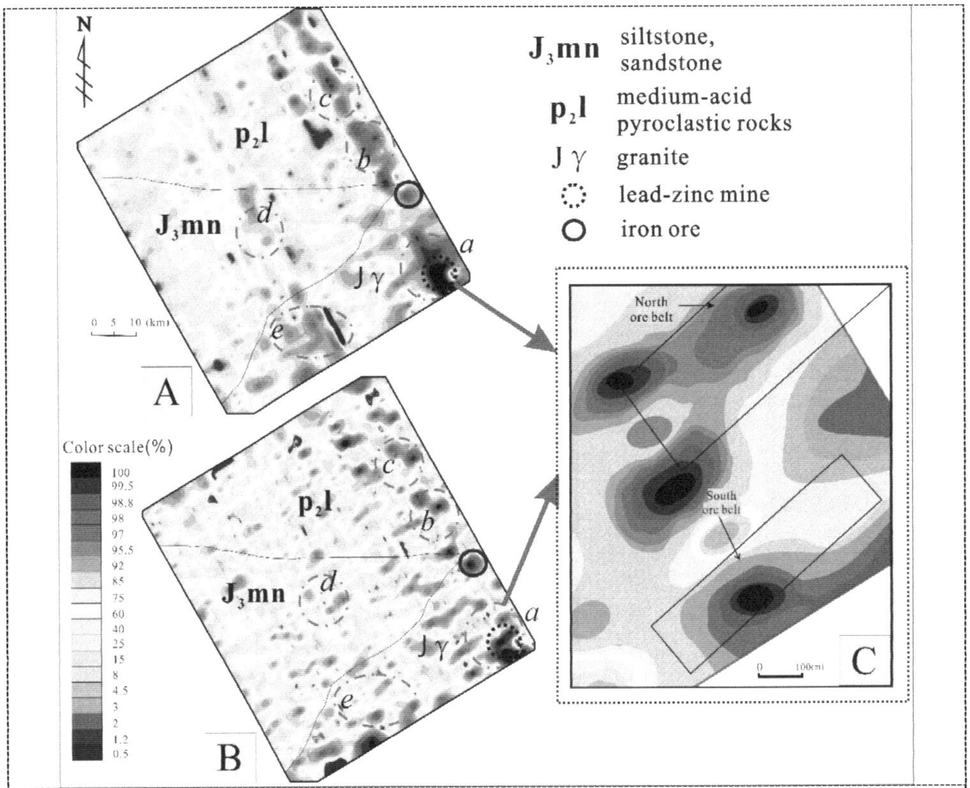

Figure 3. A Raw Data Contour Map; B FSVD Processing Effect contour map; C Ground Verification Regional Geological Map.

5. Conclusion

FSVD uses SVD to decompose the original signals. Based on fractal theory, the singular value-energy measure curve after decomposition is analyzed. According to the principle of minimum residual square sum, the curve is fitted and the intersection points are obtained. The critical singular values can be calculated quantitatively, and the complicated geological structures can be distinguished. The partial singular values are reconstructed, and the anomaly signals caused by uranium ore in the survey area are successfully extracted. Through the experimental analysis of the airborne gamma-ray spectrum production data, it is found that the original data is decomposed by singular values, and different singular value sections represent different anomaly information: The high singular value section(>50) represents the background signal, the medium singular value (13-50) represents the ore-caused anomaly information, the low singular value part (<13) represents other noises; and the banding anomaly signal caused by flight error exists in the high singular value section. The method can suppress the background better and highlight the anomaly. The banding false anomalies caused by time domain batches are weakened, and the positive and negative false anomalies appearing in the southern part of the survey area disappear. Combined with the ground verification map, it is found that the anomaly range of the known mine points is reduced, and the anomaly shape is consistent with the actual mine site, which provides reliable data support for the ground anomaly verification work.

FSVD is used to decompose, filter and reconstruct the aerial gamma energy spectrum data, which can effectively extract the anomaly and provide an indicative role for the selection of the prospecting target areas. However, when interpolating the original data, the interpolation method will affect the result. In the next step, the interpolation method suitable for the aviation gamma spectroscopy data should be further analyzed to improve the accuracy of the anomaly information extraction.

Acknowledgement
The authors acknowledge the financial support by the National Key R&D Program of China (Project No.2017YFC0602105), the National Science Foundation of China (Project No. 41774147 and Project No. 41774190) and the Key R&D Projects of Sichuan Province (Project No. 2018GZ0004).

References
[1] Ge Liangquan,Xiong Shengqing,Zeng Guoqiang,et al.2016.Airborne Gamma-ray Spectrum Detection and Application [M]. Beijing:Science Press:286.
[2] Yu Hao,Li Jingsong,Zhang Yan,et al.2013.Research Institute of Petroleum Exploration & Development[J]. Oil Geophysical Prospecting. 48(6):954-959.
[3] Chen Shangquan,Li Xiangyang.2015.State Key Laboratory of Petroleum Resources & Prospecting[J]. Oil Geophysical Prospecting. 50(2):213-218.
[4] Pei Tao,Bao Zhengyu.1998. Research on Noise Erasing Methods With Geochemical Data[J]. Geology-Geochemistry (4): 86-90.
[5] Zhu Dan, Liu Tianyou, Li Hongwei.2018. Separation of Potential Field Based on Singular Spectrum Analysis[J]. Chinese Journal of Geophysics. 61(09):3800-3811.
[6] LI Xinfu, LI Xiaofan, Wu Ye.2007. Application of Fractal Theory in Seismology[J]. Progress in Geophysics. 22(2):411-417.
[7] Xie Shuyun,Bao Zhengyu.2002.Continuous Multifractal Model of Geochemical Fields [J].Geochimica.31(2):191-200.
[8] Afzal P, Adib A, Ebadati N.2018.Delineation of Seismic Zonation Using Fractal Modeling in West Yazd Province, Central Iran[J]. Journal of Seismology.1-17.
[9] Zhou Ding,Chen Yongqing,Zhao Binbin.2014. The Application of Singular-value Decomposition (SVD) and Geochemical Blocks Methods to the Prospecting in Southern Huanggang-Ganzhu' Ermiao Metallogenic Belt [J]. Geology in China.41(2):621-637.
[10] Cheng Qiuming.2001.Multifractal and Geostatistic Methodsfor Characterizing Local Structure and Singularity Properties of Exploration Geochemical Anomalies.Earth Science-Journal of China University of Geosciences.26(2):55-60.

Journal of Physics: Conference Series **1423** (2019) 012018 doi:10.1088/1742-6596/1423/1/012018

Data comparison of the effects of feeding ginkgo leaf fermentation and Chinese herbal medicine on meat ducks

Jie Wang [1], Shanmao Ren [2*], Xixiu Fang[3]

[1, 2, 3]Jiangsu Agri-animal Husbandry Vocational College, Jiangsu, China

*Corresponding author and e-mail: Shanmao Ren, yonghuazheng2014@126.com

Abstract. In order to study the effect of feeding Ginkgo leaf fermentation and Chinese herbal medicine on meat duck, we added Ginkgo leaf fermentation and Chinese herbal medicine additive to the daily feed of meat ducks, and set up a control group (basic feed), a western medicine group (basic feed+150mg/kg aureomycin), an experimental group 1 (basic feed + Chinese herbal medicine 0.1%) and an experimental group 2 (basic feed + 0.1% Chinese herbal medicine + 2.5% Ginkgo leaf ferment). We fed the ducks for 45 days, and then completed the measurement of production performance and blood index. Based on the measurement method of related data, we used Excel and spss25 software for data statistics and analysis. The results showed that the diet supplemented with 0.1% Chinese herbal medicine and 2.5% Ginkgo leaf fermentation could significantly increase daily growth weight of meat ducks during the whole growth period ($P<0.05$), at the same time, it could reduce feed conversion ratio, and the content of total protein and albumin in serum could be significantly increased ($P<0.05$). The value of alkaline phosphatase in the experimental group was lower than that of the control group. The amount of immunoglobulin and complement in the serum were all higher in the experimental group than in the control group. The overall effect of Ginkgo leaf fermentation and Chinese herbal medicine in basic diet was significantly better than that of western medicine group ($P<0.01$).

1. Introduction

Ginkgo is a unique economic plant resource in China. The results show that, after processing, the content of ginkgolic acid is relatively higher. Although ginkgolic acid has a wide range of biological activities [1], such as anti-tumor, anti-allergy, anti-inflammatory, etc., it also has considerable toxicity to liver, kidney andembryo [2-3]. Experiments and research have proved that, after the components in Ginkgo leaves are decomposed and metabolized by some microorganisms, not only the content of active components such as flavonoids and lactones will increase, but also the decomposition and transformation of ginkgolic acid will be promoted, achievinga purpose of increasing efficiency and reducing toxicity for its components[4].

Domestic medical research and practice show that Chinese herbal medicine also plays a very important role in promoting immunity and animal growth [5]. Studies have shown that using microbial fermentation can promote the separation of effective ingredients in Chinese herbal medicine, and reduce the toxicity and side effects of Chinese herbal medicine [6]. Using traditional Chinese medicine or traditional Chinese medicine dregs as fermentation substrate and solid fermentation method for fermentation, not only the cost is relatively low, but also the operation is more convenient [7]. Therefore, we consider using solid fermentation method to process Ginkgo biloba leaves and Chinese herbal medicines, and make feed additives to test whether it has a favorable impact on the growth performance and serum indicators of ducks.

2. Materials and methods

2.1. Main test animals and materials

The test animal materials: Suyou No.2 meat duck [9], which is from Jiangsu Fengda waterfowl breeding farm. Gaoyou duck and wild duck were used as parents. After crossbreeding, cross fixation and generation selection, a new breed of meat duck with good meat production performance was obtained.

The test materials: Ginkgo leaves collected from May to August. Chinese herbal medicines were mainly (Fleece-flower root: Astragalus: Rhizoma atractylodis=1:1:1). The kit was purchased fromJiangsu Nanjing Jiancheng Bioengineering Research Institute.

2.2. Test methods

Ferment the ginkgo leaves for 0, 24, 36 and 48 hours respectively. After fermentation, add it to basic feed according to 2.5% and add Chinese herbal medicine to basic feed according to 0.1%. Based on the basic feed as the control, determine the nutrient composition, and finally select the ginkgo leaf fermented for 24 hours as the additive. Set up western medicine group: add 150mg of chlortetracycline per kilogram of feed; test group 1: basic feed+Chinese herbal medicine 0.1%; test group 2: basic feed + 0.1% Chinese herbal medicine + 2.5% Ginkgo leaf fermentation; control group: only basic feed. Select seven-day old healthy ducks and randomly divide them into four groups, 20 in each group. Then we carried out a 45-day experiment and determine the initial relevant indicators.

2.2.1 Analysis of feed composition

We use conventional methods to determine water, crude ash, crude fat, crude fiber and crude protein. Determination of flavonoids: take 4g of each group of samples and then extract itby Soxhlet extraction by using 70% ethanol for 4h, and then evaporate to obtain Ginkgo biloba extract, anddissolve it with 70% ethanol, take 1ml each, and then add 0.3ml sodium nitrite solution first, mix well, and let itstand for 6min. Then add 0.3ml aluminum nitrate solution, mix well, and let itstand for 6min. Then add 4ml sodium hydroxide solution, mix well, and let it stand for 6min. Finally, use70% ethanol to determine the volume. After making rutin standard solution, use spectrophotometer to measure it and draw standard curve. Then, determine the sample extract and calculate the flavone content. Determination of amino acid [8]: weighasame amount of feed sample and dissolve it in 100mL0.1mol/L HCl, filter it with 0.22μm filter membrane, inject 10μL, and finally get the result.

2.2.2 Determination of production performance

Record the beginning weight and end weight and daily gain of ducks, and calculate the feed conversion ratio. Determine the content of crude protein and crude fiber in feces.

2.2.3 Determination of the content of crude protein and crude fiber in feces

Collection and preparation of fecal samples: fecal samples were taken three times a day at 8:30 in the morning, 2:30 in the afternoon and 8:30 in the evening. Take about 5-10g of fresh feces each time, add 2.5ml of 10% tartaric acid solution to every 10g of fresh feces and mix it well, and after mixing the three samples, dry them at 75 ℃. We took samples for 3 consecutive days, and we took the sample mixture as the fecal sample.

2.2.4 Determination method for immune index

Collect5 ml blood from the vein under the armpit of duck. The blood sample was centrifuged at 3000r/min for 15min. Take the supernatant, sub-pack it and label it, and store it at -20℃.

(1)Determination of total protein by biuret method

Sample: take 50μl serum for determination.

Calculation formula: $$\frac{Total\ protein\ content}{(g/L)} = \frac{measured\ OD\ value - blank\ OD\ value}{standard\ OD\ value - blank\ OD\ value} \times Concentration\ of\ total\ protein\ standard(g/L) \times Dilution\ ratio\ before\ sample\ determination$$

Note: the concentration of protein standard is52.4g/L. OD is optical density

(2)Determination of albumin by bromocresol green colorimetry

Sample: take 10μl serum for determination.

Calculation formula:

$$\frac{ALB\ content}{(g/L)} = \frac{measured\ OD - blank\ OD}{standard\ OD - blank\ OD} \times Concentration\ of\ standard(g/L) \times Dilution\ ratio\ before\ sample\ determination$$

Note: the concentration of the standard is 28.9g/L.

(3)Determination of alkaline phosphatase (AKP)

Sample: take 50μl serum for determination (2-time dilution)

Calculation formula: $$Alkaline\ phosphatase\ (king\ unit\ /\ 100ml)) = \frac{measured\ tube\ OD\ value - blank\ tube\ OD\ value}{standard\ tube\ OD\ value - blank\ tube\ OD\ value} \times Phenol\ content\ of\ standard\ tube(0.005mg) \times \frac{100ml}{0.05ml} \times Dilution\ ratio\ before\ sample\ determination$$

(4)Determination of uric acid

Sample: take 0.1ml serum for determination

Calculation formula:

$$Serum\ uric\ acid\ (mg/L) = \frac{measured\ tube\ absorbance\ value}{standardtube\ absorbance\ value} \times Standard\ tube\ content(50mg/L)$$

(5)Determination of urea

Sample: take 0.02mlserum for determination

Calculation formula:

$$Blood\ urea\ nitrogen\ content\ (mmol\ /L)) = \frac{measured\ tube\ OD\ value - blank\ tube\ OD\ value}{standard\ tube\ OD\ value - blank\ tube\ OD\ value} \times Concentration\ of\ standard(mmol\ /L)$$

(6)IgA、IgM、IgG、complementC3、C4、IL-2

Sample: take 50μl serum for determination

Kit: Duck Immunoglobulin A (IgA) Elisa; Duck Immunoglobulin G (IgG) Elisa; DuckImmunoglobulin M (IgM) Elisa; Duck Complement Component 3(C3) Elisa; Duck Complement Fragment 4(C4) Elisa; Duck Interleukin 2(IL-2) Elisa.

2.3. Data analysis

Use Excel and spss25 software to carry out data statistics and analysis, and use One-Way ANOVA and t-test (test levelα=0.05) for the comparison between groups.

3. Result analysis

3.1. Analysis results of feed nutrients

Table 1. Analysis results of feed composition (%)

Grouping of experiments	/ater content	Crude ash	Crude fat	Crude fiber	Crude protein	Total flavonoids%
Basic feed	12.60	5.26	3.42	4.49	15.31	/
Chinese herbal medicine 0.1%	10.96	19.97	0.84	15.40	5.40	/
2.5% Ginkgo leaves (fermented for 24 hours)	9.66	13.30	8.59	8.86	13.50	5.56
2.5% Ginkgo leaves (fermented for 36 hours)	9.42	13.60	7.35	11.87	13.70	3.33
2.5% Ginkgo leaves (fermented for 48 hours)	10.09	14.03	9.26	15.40	14.05	3.26

Table 2. Content of amino acids in feed (g / 100g)

	Basic feed	0.1% Chinese herbal medicine	2.5% Ginkgo leaves (notfermented)	2.5% Ginkgo leaves (fermented for 24 hours)	2.5% Ginkgo leaves (fermented for 36 hours)	2.5% Ginkgo leaves (fermented for 48 hours)
Aspartic acid	1.2569	0.0798	1.3379	1.4032	1.3534	1.3915
Glutamic acid	2.6226	0.0785	1.4342	1.627	1.5622	1.5809
Serine	0.0682	0.222	0.0769	0.0744	0.0757	0.0729
Histidine	0.0892	0.2822	0.122	0.1361	0.1512	0.132
Glycine	0.059	0.1678	0.0501	0.047	0.049	0.0463
Threonine	0.0657	0.174	0.0651	0.0647	0.0675	0.0651
Arginine	0.0455	0.0699	0.0527	0.0566	0.055	0.0545
Alanine	0.0414	0.1352	0.0488	0.0489	0.0512	0.0468
Tyrosine	0.0695	0.2041	0.0788	0.0899	0.0925	0.0904
Cystine	0.0181	0.0019	0.0044	0.0053	0.0043	0.0042
Valine	0.0499	0.178	0.0522	0.0476	0.048	0.0479
Methionine	0.2917	0.0403	0.0025	0.0018	0.3664	0.0018
Phenylalanine	0.0578	0.2581	0.0587	0.0538	0.0566	0.054
Isoleucine	0.0628	0.232	0.0649	0.0584	0.0618	0.0593
Leucine	1.2926	0.146	0.0396	1.015	0.0383	1.0027
Lysine	0.0547	0.2204	0.0577	0.0552	0.066	0.0542
Proline	0.0441	0.1436	0.0643	0.0577	0.0589	0.0612
Total amino acids	6.1898	2.634	3.6109	4.8428	4.158	4.7656

According to the analysis results of feed nutrients in Table 1 and Table 2, the Chinese herbal medicine group and ginkgo leaf fermentation group are more comprehensive in nutrients than the basic feed. The total content of flavonoids and amino acids are significantly higher than that of the non-fermented ginkgo leaves.

3.2. Test results of production performance

Table 3. Daily gain and feed conversion ratio of meat ducks

Group	Average initial weight / g	Average final weight / g	Average daily gain / g	Feed conversion ratio
Control group	176.43±5.42[a]	1489.47±59.79[a]	29.18±1.28[a]	5.93[a]
Western medicine group	197.50±8.73[b]	1802.63±8.21[a]	35.66±0.17[b]	5.61[b]
Experimental group 1	203.13±4.13[b]	2383.97±152.14[b]	48.46±3.38[c]	5.34[b]
Experimental group 2	196.25±5.00[c]	2470.15±115.45[c]	50.53±2.59[c]	5.18[c]

Note: the different letters in the upper right corner of the data in the same row in the table indicate that the difference is significant ($P<0.05$).

For the daily weight gain effect, test group 2 > test group 1 > western medicine group, that is to say, the feed with basic feed + 0.1% Chinese herbal medicine + 2.5% gingko leaf ferment as formula is better than the basic feed with single addition of 0.1% Chinese herbal medicine, which can effectively reduce the feed conversion ratio (Table 3).

3.3. The content of crude protein in duck feces

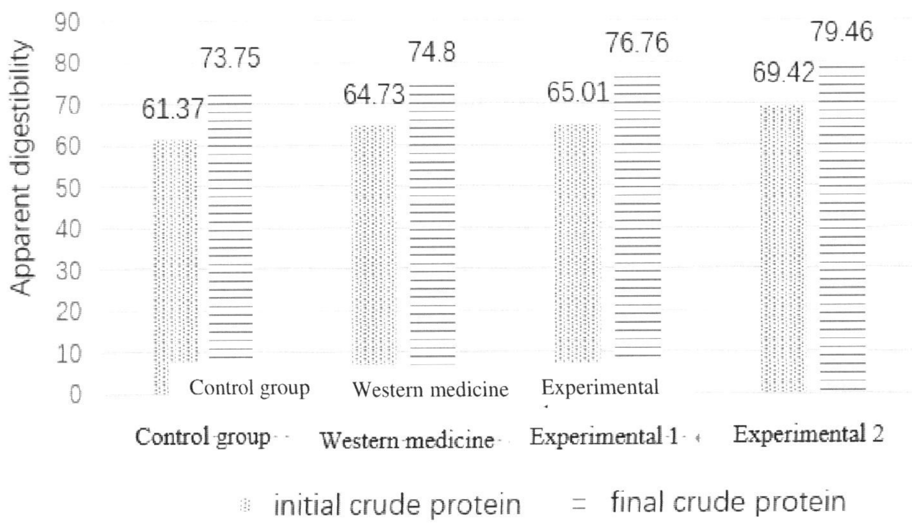

Figure 1. Apparent digestibility

From Figure 1, we can see that adding western medicine, Chinese herbal medicine and Ginkgo feed additives can promote the digestion and absorption ability of ducks, and the promotion effect of experimental group 2 > experimental group 1 > western medicine group, that is to say, whenthe duck is fed with the diet with basic feed + 0.1% Chinese herbal medicine + 2.5% Ginkgo leaf ferment as the formula, the digestion and absorption ability of meat ducks is the best.

3.4. Test results of immune indexes

Table 4. Index results of duck serum

Grouping	Control group	Western medicine group	Experimental group 1	Experimental group2
Total protein(g/L)	27.93±0.31^{a}	32.05±0.42^{b}	35.15±0.35^{bc}	38.49±0.36^{c}
Albumin (ALB) (g / L)	10.57±0.22^{a}	10.83±0.25^{a}	11.17±0.28^{b}	11.40±0.32^{b}
(King unit/100ml)				
Alkaline phosphatase(AKP)(King unit/100ml)	47.96±0.75^{d}	42.34±0.73^{c}	26.85±0.72^{a}	25.39±0.67^{a}
Uric acid(mg/L)	49.02±0.02^{a}	44.61±0.03^{b}	40.69±0.02^{b}	32.35±0.02^{c}
(mmol/L)				
Urea nitrogen(BUN)(mmol/L)	1.70±0.21^{a}	1.39±0.24^{b}	1.11±0.35^{c}	0.98±0.03^{d}
IgA(mg/ml)	0.8165±0.02^{a}	0.9552±0.11^{ab}	1.1107±0.13^{b}	1.2796±0.15^{c}
IgG(mg/ml)	4.6399±0.41^{a}	5.531±0.29^{b}	7.1393±0.19^{d}	7.6786±0.16^{de}
IgM(mg/ml)	0.5993±0.02^{a}	0.7498±0.01^{b}	0.859±0.01^{c}	0.9039±0.02^{d}
C3(ng/L)	196.8481±1.32^{a}	220.27±0.57^{b}	256.1904±1.29^{c}	408.8624±1.98^{e}
C4(ng/L)	102.7504±0.98^{a}	113.9385±0.86^{b}	147.285±0.69^{cd}	159.5384±0.39^{d}
IL-2(ng/L)	15.7699±0.03^{a}	21.0471±0.04^{b}	30.4405±0.04^{c}	33.6055±0.02^{d}

Note: the different letters in the upper right corner of the data in the same row in the table indicate that the difference is significant ($P<0.05$)

Compared with the control group, the test group has a increasing function in the total protein, albumin, immunoglobulin, complement and interleukin, and has a decreasing funktion in AKP, uric acid and urea nitrogen. The effect of the test group 2 > the test group 1 > the western medicine group (Table 4).

4. Discussion

4.1. The effect of fermentation on the nutritional components of Ginkgo leaves

Ginkgo leaves is a kind of abundant biological resources in China, and it has gradually produced scale in modern production. At present, fermentation is the most effective and practical method for the development of Ginkgo resources. The experimentresults show that microbial fermentation can increase the content of vitamin, enzyme and growth factor [10]. In this experiment, we also found that the total flavonoids of Ginkgo leaves treated by solid-state fermentation method, especially those fermented for 24 hours, were significantly better than those before fermentation. And the amino acid content of the fermented Ginkgo leaves was higher than that of the non-fermented ginkgo leaves.

4.2. Effects of feeding of fermented Ginkgo leaves and Chinese herbal medicine on growth performance of ducks

The results of this experiment show that feeding ducks with 0.1% Chinese herbal medicine and 2.5% Ginkgo leaf ferment can significantly increase the daily growth of ducks during the whole growth period, and at the same time reduce the feed conversion ratio, which is directly related to the rich active ingredients in the extracts of Ginkgo lead and Chinese herbal medicine, which can promote the growth and protein synthesis of ducks. This can be concluded from the digestibility of crude protein.

Because Ginkgo leaves are rich in flavonoid compounds, flavonoid of traditional Chinese medicine has always had highly-effective and low-toxic functions in actual animal production., and it has a broad development prospect [11]. At the same time, after 24 hours of fermentation, the flavonoids extracted were 90.08% higher than that of the unfermented ginkgo leaves. It is known that flavonoids can promote the growth of animals, and the content of extract increased after fermentation increases the promotion effect, which further proves that Ginkgo leaf fermentation products can be used as a new green and safe feed additive.

4.3. Effects of feeding of fermented Ginkgo leaves and Chinese herbal medicine on serum biochemical indexes of meat ducks

The results showed that the content of total protein and albumin in serum of ducks fed with basic feed supplemented with Ginkgo leaf ferment and Chinese herbal medicine could be significantly increased, and compared with the group only added with Chinese herbal medicine, the total protein increased nearly 9.5%. These two values can reflect the body's digestion and metabolism of protein, so it can be concluded that adding Ginkgo leaf fermentation and Chinese herbal medicine in daily feed can promote the growth of ducks. Alkaline phosphatase, which can reflect the body skeleton and hepatobiliary system disease, was lower in the test group than in the control group, and the indexes of uric acid and urea nitrogen, which can reflect the function of kidney, were also lower. It indicated that gingko leaf ferment and Chinese herbal medicine added to the feed had the function of health care for the duck, liver and kidney. At the same time, the amount of immunoglobulin and complement in serum were higher in the test group than in the control group, which significantly improved the antioxidant capacity and immune capacity of ducks. And through the comparison between the test group 1 and the western medicine group, the overall effect of the test group 1 is better than that of the western medicine group, but the difference is not significant, while the overall effect of the test group 2 and western medicine group is better than that of other groups, and the difference is significant ($p<0.05$). It can be concluded that Ginkgo leaf fermentation has a great influence on the growth performance and blood index of ducks.

5. Conclusion

The combination of Ginkgo leaf fermentation and traditional Chinese medicine can not only accelerate the digestion and absorption, but also accelerate the muscle growth, increase the antibacterial power of animals, and improve the immunity of the body. Because the traditional Chinese medicine is low toxic and has no side effects, it can make meat quality of meat ducks greener and healthier, and reduce people's worry about too much antibiotic intake. The combination feeding of Ginkgo leaf fermentation and traditional Chinese medicine compound can play a great role in the modern green cultivation. There is a great development space for the promotion of the overall growth performance of meat ducks. At the same time, it has significant effect on improving serum biochemical indicators, and can effectively improve immunity and regulate liver and kidney functions of the body.

Acknowledgment

This research was financially supported by Taizhou Science and Technology Support Plan (agriculture) Project in 2019(TN201918); supported by Jiangsu Agricultural Science and Technology Independent Innovation Project in 2018 (CX (18)1004); and supported by Jiangsu Modern Agriculture (Waterfowl) Industry Technology System Integration Innovation Center (JATS [2018]238).

Reference

[1] Sun Kai, Pan Xiaohai, Huang Shiying. The damage effect on the ginkgolic acid on mitochondria in rat liver [J]. Chinese pharmacological communication,2009, 26 (2) :56.
[2] HECKER H, JOHANNISSON R, KOCH E. In vitro evaluation of the cytotoxic potential of alkylphenols from Ginkgo biloba L.[J]. Toxicology, 2002, 177 (2-3) :167-177.
[3] Shen Qi, Li He, Lian Hong, etc. Research on the effects of ginkgolic acid on hepatotoxicity of

rats [J]. Chinese Journal of clinical pharmacology,2018,34(12):1457-1459+1462.
[4] Li Yinliang. Research on the detoxification effect of fermented ginkgo tea [D]. Xi'an: Shaanxi University of Science & Technology,2014.
[5] Liu Feng, Han Chunyang, Liu Cuiyan. Progress of the research on the microbial-fermentation chinese herb feed additives [J]. Amino acids and biological resources,2014,36(02):18-22.
[6] Chen Hua. Optimization of technological conditions of solid-fermentationAstragalus dregs [C]. Anhui Agricultural University,2016.
[7] Chen Zaixing, Zhu Xu, Wang Lin. Changes of total flavonoids, total flavonol glycosides and terpene lactones in gingko leaves in different seasons [J]. Chinese Journal of Hospital Pharmacy, 2010, 30 (12):1067-1069.
[8] Tao Yong, Ren shanmao, Yang Xiaozhi. Study on slaughtering performance and meat quality of Suyou No.2 duck [J]. Hunan Agricultural Science, 2014 (11):57-59.
[9] Du Zheng, Liu Wenjie, Mingpu. Study on the extraction of flavonoid compound from Ginkgo leaves [J]. Guangdong chemical industry, 2017, 44 (04):32-33.
[10] Zhang Jie, Bai Juan, Zhu qianyun. Determination of four amino acids in compound donkey-hide gelatin by HPLC method[J]. China Medicine and Pharmacy, 2017, 7 (01):60-63.

Research on knowledge mapping construction of big data

Xiulian Jiang[1], Peng Ju[1], Yining Song[2, 3], Lingcheng Mao[1] and Guifang Xie[1]

[1]Xuzhou University of Technology, Xuzhou, Jiangsu, 221008, China
[2]University of Chinese Academy of Sciences, Beijing, 100000, China
[3]Xinjiang Astronomical Observatory, CAS, Urumqi, Xinjiang, 830011, China

Corresponding author and e-mail: Jiang Xiulian, 447502622@qq.com

Abstract. The diversified development of network applications and the informatization of social life are triggering explosive growth of data. The core journal articles based on big data research in the 2011-2018 database, which are included in the CNKI database, are used as research objects, and CiteSpace, Netdraw, and Ucinet are used as tools for visualizing knowledge maps. Visualize the growth trend of big data literature, authors and institutional cooperation networks; use co-occurrence analysis and cluster analysis to mine research hotspots of big data; use mutation word detection and time zone view to study evolution path of big data development. The research hotspots of big data are shown in three directions of big data technology, application field and challenges brought by big data. The evolution path of the field of big data shows that in early stage, technology research related to big data was mainly conducted, and later, it developed into the research trend of technology and application.

1. Introduction

From a macro perspective, big data is a bridge connecting the physical world, the information world and human society. From the perspective of information industry, big data is also a powerful push for the new generation information technology industry. The definition given by the McKinsey Global Institute is: a collection of data that is large enough to capture, store, manage, and analyze much beyond the capabilities of traditional database software tools, with massive data scales, fast data flow, Diverse data types and low value density are four major features [1]. The knowledge map method is a process of visualizing the relationship of the object to be analyzed. That is to say, in scientometrics, the knowledge mapping method can be used to combine citation analysis with data and information visualization to form various patterns with multidimensional structural laws [2]. China's research on the construction of scientific knowledge maps began in 2004. The research team is best known for the WISE laboratory led by Professor Liu Zeyuan of Dalian University of Technology. At present, knowledge mapping applications are mainly concentrated in the disciplines of library and information, education, medicine, computer science, and management [3].

2. Basic knowledge map analysis of big data

2.1. Analysis of the law of document growth

The law of the growth of the literature reveals to some extent the growth law of the amount of scientific knowledge, which is closely related. Therefore, observing and analyzing the volume and variation of journals in the field of big data helps us to understand the research prospects and development trends in this field. The publication volume statistics of core journals related to big data from 2011 to 2018 are shown in Table 1.

Table 1. Statistics on the publication of big data research papers in 2011-2018

Publication year	Issue volume	Publication year	Issue volume
2011	2	2015	1067
2012	21	2016	1286
2013	263	2017	1337
2014	701	2018	1231

Among the 5,908 related articles collected, the research on big data in China started around 2011, and only 2 papers were published in that year, which was consistent with the time when the concept of big data was formally proposed. In the two years from 2011 to 2012, there were only 23 papers on big data research, accounting for 0.39% of the total length of the eight-year literature. However, since 2013, research papers have been developed in a blowout manner. By 2015, the number of documents has exceeded 1,000, and big data research is on the rise. In the three years from 2015 to 2017, the growth rate of the papers slowed down and showed a steady growth. By 2018, the relevant research papers had a slight downward trend, but in terms of volume, the amount of relevant research papers was still considerable. It can be seen that the research on big data is not reduced, and big data is promoted to national strategy in comprehensive time. Internet+, artificial intelligence and blockchain have become hot topics, and it is not difficult to predict the research prospects of big data.

2.2. Author's total knowledge map analysis

The scientologists Katz and Martin define scientific cooperation as: scientific cooperation is the joint work of research scholars to produce new scientific knowledge [4]. In the actual process, the forms of scientific cooperation are diverse. The scientific cooperation mentioned in this article refers to the emergence of different authors in a paper. Then, there is a cooperative relationship between the authors. Through the analysis of the co-occurrence knowledge of the co-authors, it helps us to understand the intensity of cooperation between scholars in the field of big data in 2011-2018, and whether there are obvious co-subgroups.

This paper uses SATI (Documents and Statistics Information Analysis Tool) to calculate the author frequency in the sample literature, automatically generate a 100*100 author co-occurrence matrix, and then import the matrix into Ucinet software to generate a matrix format that Netdraw can recognize. Finally, it is drawn into the co-occurrence knowledge map of the author. The map is shown in Figure 1. The square nodes in the figure represent the key elements in the cooperation map. The larger the node, the higher the degree of criticality. Whether there is a cooperative relationship between nodes mainly depends on whether there is a connection between nodes, and the thicker the connection, the more cooperation between the two scholars, the closer the cooperation relationship. The links between are loose. Overall, the research team of 2-3 people is the main one, indicating that the researchers are not very concerned about the cooperation between groups, but there are also obvious co-groups. Academic research groups such as Wang Wei, Yu Feng, Zhang Hao, Lan Yuexin, Gao Hong, Liu Bingyue and other scholars. These scholars are mainly from Nanjing University, Harbin Institute of Technology, Jilin University, Tianjin University and other units. They are the backbone of the big data research community and guarantee the publication of high-quality academic research results.

2.3. Institutional co-existence knowledge map analysis

Research institutions can be understood to some extent as synonymous with academic groups. Analysis of core institutions helps to understand the distribution of key research forces in the big data arena. At the same time, the cooperation between institutions helps to realize the optimal allocation of scientific research resources, which is of great help to the improvement of scientific research production capacity. Therefore, the co-existing analysis of scientific research cooperation is also very necessary. This paper uses Excel and CiteSpace output files to calculate the institutional distribution of the core journals in the 2011-2018 big data domain, which is a total of 5,778 domestic and foreign academic institutions. In order to better demonstrate the academic productivity of the big data field, the academic institutions of the statistics include first-level units and second-level units. This paper selects the top 10 academic institutions in the field of publications, which represent the core research and production institutions in the field of big data research. The distribution of core institutions is shown in Table 2.

Table 2. Distribution of core institutions

Serial number	institution name	Issue volume
1	Wuhan University School of Information Management	56
2	School of Journalism, Renmin University of China	53
3	Chongqing University of Technology School of Accounting	48
4	Chinese Academy of Sciences University	43
5	School of Information Management, Nanjing University	41
6	Wuhan University Information Resources Research Center	32
7	School of Information Resource Management, Renmin University of China	26
8	Research Center for Journalism and Social Development of Renmin University of China	19
9	China Telecom Corporation Guangdong Research Institute	18
10	Chongqing Institute of Technology Cloud Accounting Institute	18

The CiteSpace software was used to export the document information in the Reforks format as data preparation before visualization analysis. Before running the software, set the time zone to 2011-2018, the time slice is set to 1 year, the default “Threshold” is top 50, the node type is set to "institution", the relationship strength parameter selects the default value, click the "Go" button to debug Generate co-occurrence maps between agencies. There are 416 nodes in the map, 170 connections, and the network density is 0.002, indicating that there is less cooperation between organizations. This article selects the largest sub-network in the organization cooperation network, and its co-occurrence map is shown in Figure 2.

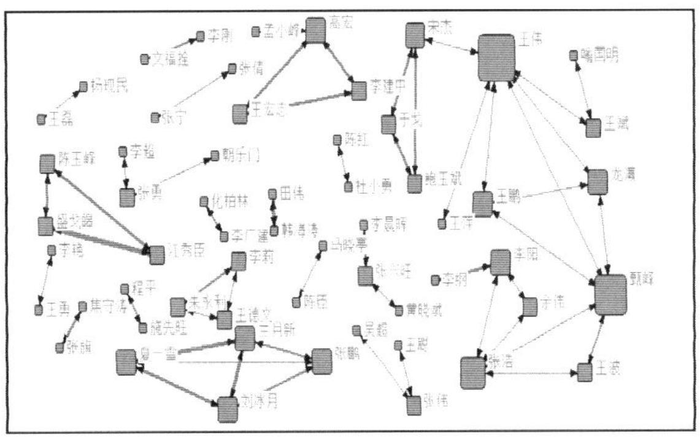

Figure 1. Author co-occurrence map

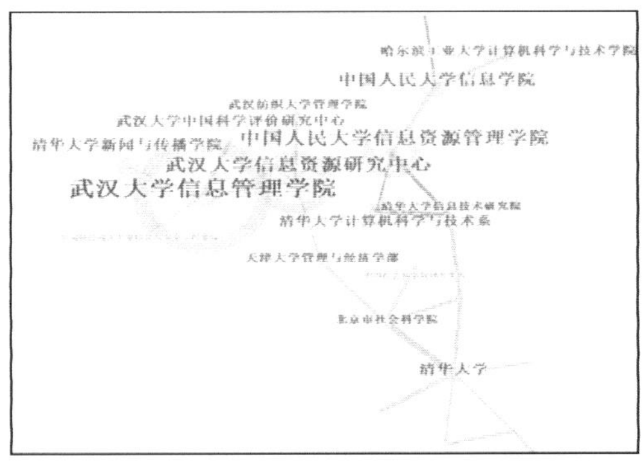

Figure 2. The largest sub-network map of issuing organizations

Through the largest sub-network map, it is not difficult to find that the group of the organization is mainly composed of the faculties and research institutes of well-known domestic universities such as Renmin University of China, Wuhan University, Tsinghua University, and Harbin Institute of Technology. The most prominent research institution is the School of Information Management of Wuhan University. From Table 2, we can also see that the cumulative number of publications reached 56, ranking first. Of course, Renmin University of China and Tsinghua University are also the backbone of research in the field of big data. Obviously, effective cooperation between academic groups is the key to ensuring scientific research capacity, but on the whole, there is no particularly fixed research group in the domestic big data field. This is a problem that needs to be addressed and resolved in the future development of big data in China.

Studying the geographical distribution of academic institutions can help us judge the main positions in the field of big data research, sort out the benchmark areas, and conduct targeted learning exchanges. The main positions in the field of research big data are concentrated in the eastern coastal and economically developed areas such as Beijing-Tianjin-Hebei, Yangtze River Delta and Pearl River Delta, while the research on big data in the central and western regions of China is relatively rare. It can be seen that the distribution of academic institutions in the field of big data in China is relatively concentrated.

3. Research on Big Data Hotspot Based on Keyword Co-occurrence and Clustering Map

3.1 Keyword co-occurrence map

The co-occurrence analysis of the keywords is to analyze the keywords provided by the authors in the dataset, and to reflect them in a co-occurrence pattern, which can well reflect the current research hotspots in a certain research field and the research hotspots that have been generated in the past[5]. The author imported the 5908 data collected into the CiteSpace software. The node type selects keyword, the time slicing is set to 1 year, and the default "Threshold" is top 50. Using the path finding algorithm, click the "Go" button to debug. Generate a keyword co-occurrence map. There are 212 nodes in the map, 238 connections, and the network density is 0.0106. The software generated keyword co-occurrence map is shown in Figure 3.

In Figure 3, the larger the node, the more times the keyword appears, such as cloud computing, data mining, artificial intelligence, etc, which are relatively large keywords. To a certain extent, they embody the research in the field of big data. The centrality of the keyword is another indicator to measure the research focus. The higher the centrality of a keyword, the greater the probability that the keyword appears in the same document as other keywords. The key to higher centrality, the word is probably the topic that scholars focus on for some time [6]. Therefore, the combination of centrality and frequency is the best way to measure the importance of keywords.

The centrality of high-frequency keywords such as big data, cloud computing, data mining, Internet of Things, Hadoop, MapReduce, visualization, innovation, and smart city is relatively high, and comprehensive influence is relatively large, which can better reflect the research hotspots in the field of big data. Special attention is paid to cloud accounting, although its frequency is relatively high, but its centrality is 0, indicating that its comprehensive influence is lower than other high-frequency keywords.

3.2 Keyword clustering map

After using CiteSpace to obtain the co-occurrence map of the keywords, the author obtained the keyword clustering map through the software automatic clustering function. The clustering map is shown in Figure 4. The measurement of clustering maps mainly depends on two indicators: module value (Q value) and average contour value (S value). In general, Q value>0.3, S value>0.5, indicating the clustering effect is reasonable. In the clustering map generated, the Q value is 0.5999 and the S value is 0.8626. The clustering effect is reasonable. The map shows that this clustering divides keywords into 13 categories. The larger the number in the figure, the larger the cluster size, which is precision marketing, database, financial integration, challenge, information services, hadoop, cloud computing, blockchain, media convergence, data news, smart city, deep learning, Internet+. These cluster names can reflect the focus of scholars to a certain extent.

4. Research on Big Data Evolution Path Based on Mutant Words and Time Zone Diagram

4.1 Mutant words detection

Mutant words refer to "keywords or phrases whose word frequency contribution changes significantly in different time windows and show a sudden rise or fall [7]." The mutant words in the keyword and found it in 2011-2018. There were 33 mutant words in the year. The keyword with the highest mutation intensity is information service, reaching 8.8279. The frequency of use increased sharply between 2013 and 2015. It can be seen that information service has become a focus of scholars in the field of big data during this time. At the same time, the keywords with high mutation intensity include digital publishing, library, data mining, and internet finance. Mainly focused on the application of big data. At the same time, the keywords of online public opinion, small data, applications, and privacy. They reflect the research hotspots in the field of big data. The words mutated from 2016 to 2018. There is a mutant tendency for further. From the macro analysis of the time span of mutant words, in the early days of big data, the research hotspots mainly focus on big data computing frameworks such as Mapreduce, Hadoop, cloud computing and related technology levels. There are also nouns, such as data news, big data technology, big data era and so on. After that, the research focus in the field of big data quickly expanded to application areas, such as libraries, enterprises, commercial banks, and finance. In recent years, the application of big data has continued to receive attention, but the focus has also begun to change, such as more emphasis on data security, privacy and so on.

4.2 Time zone view

In order to further study the law of the evolution of key words in the big data field over time, to explore the evolution path of the big data research topic, the author uses the time zone view provided by CiteSpace for path analysis. The time zone view collects the keywords in the same time in the same time zone, where the same time refers to the time when the keyword first appears, and the time series are arranged in order from far to near. The time zone view obtained by this study is shown in Figure 5.

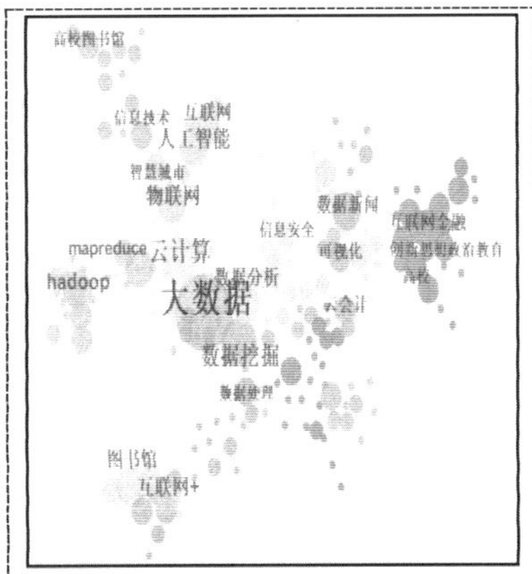

Figure 3. Keyword co-occurrence map

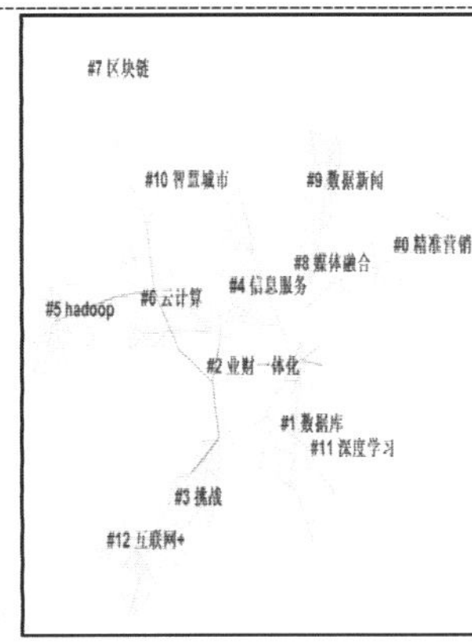

Figure 4. Keyword clustering map

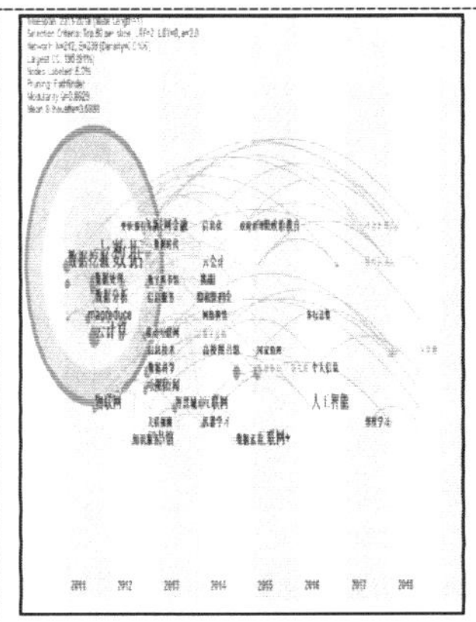

Figure 5. Keyword annual evolution map

From Figure 5, we can see that in the initial stage of big data research, the research content mainly includes data mining, data analysis, mapreduce, cloud computing and other technical foundations. During 2012-2013, the research theme has grown dramatically, except with the introduction of new technical terms such as visualization, association rules, mobile Internet, etc., we also found terminology at the technical application level represented by Internet finance, libraries, information services, digital libraries, etc. In 2014, the application scope of big data continued to expand, and application fields represented by cloud accounting, online public opinion, digital publishing, and university libraries appeared. At the same time, we can also see that the research theme also involves challenges, privacy protection and other aspects. The research scope of big data has been further broadened, and relevant scholars have realized the challenges and problems in the era of big data. In 2015, with the introduction of the national big data strategy and the Internet+, national governance appeared in the field of big data research, and scholars began to explore the integration of Internet+ and big data. At the same time, the research scope of big data has been extended to the ideological and political education in colleges and universities, and the innovative reform of ideological and political education in colleges and universities has begun to be explored. During 2016-2018, the frequency of new nodes began to decrease. In 2017, the integration of artificial intelligence and big data technology has been widely concerned by scholars, and the combination of big data and current affairs policies has become more and more obvious. At the same time, new technical level terms such as parallel computing and deep learning have also been proposed. From the evolution of the above topics, it is not difficult to see that the main content of big data research will continue to focus on the technical level and application level. The integration of big data and other technologies will also generate new kinetic energy for the economy and society.

5. Conclusions

Using Netdraw and CiteSpace as representative tools, the knowledge map construction of sample data is studied from perspectives of literature growth law, literature authors and research institutions. The conclusions are as follows. First, the volume of publications in the field of big data in 2011-2018 is generally on the rise, and the relationship between the volume of publications and time is linearly fitted. The fitting equation is F(t) =-4.51E5+2.24E2* t. Second, scholars in the field of big data in China are relatively loosely connected with the overall research team of 2-3 people. Third, domestic big data research groups are mainly concentrated in colleges and universities (including computer, information management and other departments) as well as university research centers (institutes, research institutes). The most prominent research institution is the School of Information Management of Wuhan University. There is less cooperation and the organization's distribution is relatively concentrated.

Acknowledgement

This research was supported by the Research on the Application of Big Data Technology in College Management under Grant 2017-R-59479.

References

[1] Xiao Ming, Kong Achievements. Big Data: Where to Go—Based on the Perspective of Bibliometrics[J].Journal of Library Science,2014,36(11):110-117.
[2] Wang Cenlan, You Jianxin. Review of Big Data Literature: Visualization Research Based on Software Citespace[J]. Science and Technology Management Research, 2017, 37(21): 180-189.
[3] Li Xin. Research on the construction of knowledge maps in management science and engineering disciplines [D]. Hubei University of Technology, 2016.
[4] Lei Shuiwang. Research Progress of Big Data at Home and Abroad Based on Knowledge Mapping[J]. Library and Information Research, 2017, 10(02): 62-69.
[5] Yang Ruixian. Bibliometric Analysis of Big Data Research[J]. Information Science, 2015, 33(08): 152-156.
[6] Hou Yuanyuan, Huang Yurong, Zhang Hong, Lu Huaqiao. Analysis of the Progress of Big Data Research in China Based on Bibliometrics[J]. Library and Information Service, 2014, 58(S2): 204-208.
[7] LIU Minjuan, ZHANG Xuefu, YAN Yun. Research on the subject evolution method based on core words, mutated words and new words[J].Journal of Information,2016,35(12):175-180.

The heterogeneity of village cadres and poverty reduction-Based on the survey data of 757 administrative villages in western minority areas

Yuke Luo[1, a], Degong Ma[2, b]

[1, 2]School of economics, Sichuan University, Chengdu, China

Corresponding author: Luo Yuke and e-mail: 405548241@qq.com

Abstract. By using the data of 7 provinces, 757 administrative villages and 31615 famers in the minority regions of the Midwest, this paper investigates the relationships between heterogeneity of village cadres and poverty reduction. And we find that the age of the village cadres, ethnic identity, investment promotion, income and the education background have a significant positive impact on farmer poverty reduction. Using village collective financial income as the instrumental variable, we find that village cadres of every 1% increase in wage income, the poverty rate will be reduced by 0.007%. Therefore, we table four proposals: the choice of the village and village officer should follow the principle of "young and old "; encouraging to choose village cadres of ethnic minorities; encouraging investment promotion and investment stay in the village; setting income of village cadres within a reasonable range. The conclusions above is of great political significance to solve the problem of poverty reduction of the heterogeneity of village cadres.

1. Introduction and principle

The Chinese party and government have been paying close attention to poverty alleviation work, and poverty alleviation has been at the core. To ensure that farmers are lifted out of poverty, the government should raise the spirits of the masses of cadres in the poverty-stricken areas and resolutely change the confidence and determination of poverty and backwardness. Do precisely lies in grassroots governance for poverty alleviation, and public officials such as village cadres and party members and cadre's responsibility are clear and know whether or not in place, whether solid style, methods properly, to a certain extent determines the effect of poverty reduction, affects the image of the government. Therefore, the impact of the heterogeneity of village cadres on poverty reduction is of great practical significance to the current 70 million poor people. The existing literature rarely deals with the influence of the heterogeneity of village cadres on poverty reduction. Therefore, from the theoretical perspective, studying the heterogeneity of village cadres also has theoretical significance for the impact of poverty reduction. The main contribution of this paper is to use the central university for nationalities in 2012 school of economics and the Chinese academy of social sciences institute of ethnology and anthropology research data, in view of the poor areas the Midwest 7 provinces, empirical research on the effects of heterogeneity research on poverty reduction effect of village cadres, and from the two aspects of the farmers' income and the income gap analysis on various aspects of village cadres quality influence on rural economic development.Arnim Scheidel (2014) believes that the poverty problem mainly comes from the shortage of land resources. Farmers need to make a diversified livelihood and make new economic activities to achieve poverty reduction. Swati Dutta (2015) tested that malnutrition in most places would affect the accumulation of farmers' assets, while the impact of illiteracy was uneven. In the study of India, Deaton&Dreze (2002) believed that factors such as health and education could have an impact on poverty. Cheng, Jin Yanhong and others (2014) also believe that the human capital represented by health and education will affect the income level of farmers and have an impact on rural poverty alleviation. Chen qi (2012) believes that human capital investment will have a greater impact on poverty reduction. While investigating the causes of poverty, Armando Barrientos (2010) found that the majority of respondents believed that poverty would be long-lasting and pessimistic about breaking out of poverty. MICHELLE s. MOOD (2005) believed that the economic goals and election of village cadres influenced the development of villages, and the ability of village cadres to be leaders led economic growth.

2. Materials and experiments

On the basis of the above literature, we carried out the following research with questions. Based on the definition of village cadres' heterogeneity and poverty reduction, we make several assumptions: Hypothesis 1: the time of the village cadres and the age of the village cadres are positively correlated with the farmers' poverty reduction.Hypothesis 2: the education of village cadres and the presence of village cadres are positively correlated with the reduction of poverty reduction.Hypothesis 3: village cadres are positively correlated with poverty reduction among ethnic minorities and farmers.Hypothesis 4: the income standard of village cadres is positively correlated with the poverty reduction. The datawas being investigated by central university for nationalities school of economics and the Chinese academy of social sciences institute of ethnology and anthropology. This study based on primary data collected from 757 village leader spreading over 757 villages belonging to 7 provinces. Such as the Inner Mongolia autonomous region, Ning Xia hui autonomous region, Qing Hai province, Xin Jiang uygur autonomous region, Guang Xi Zhuang autonomous region, Gui Zhou province, Hu Nan province.Though the study area is located in the poor rural areas, the results can be generalized across to the whole country. Rural survey gathers administrative villages population employment and economic, agriculture and village collective land, finance and management, public service aspects of the micro information, fully reflects the basic situation of the Midwest rural economy.

Model setting: probit regression model: The model is establishing in this section to prove the relationship between the heterogeneity of village leaders and the investment of farmland irrigation facilities. Specific as follows:

$$p_i = \text{irrigation facilities}_i = \beta_0 + \beta_1 x_1 + \beta_2 x_2 + \beta_3 x_3 + \beta_4 x_4 + \beta_5 x_5 + \sum_{z=1}^{21} B_i x_{zi} + \mu_i$$

In the model (1), the explanatory variable whether povertyi represents the first person to be out of poverty, 1 represents poverty reduction, and 0 representatives have not yet reduced poverty. The 15 control variables involved in the four aspects of farmers' individual characteristics, social environment, natural environment and public service are included. This paper makes a certain arrangement of data variables, and the descriptive statistics of all variables form table 1.

Table 1. Descriptive statistics of variables

Variable	Variable definitions	average value	standard deviation
income1	The farmers' 2011 net income	4336.633	2169.076
whether poverty	Whether farmers are out of poverty	0.1120112	0.3153854
Time	From which year began to act as village cadre	2004.292	9.376653
age1	The age of the village cadre in 2011	47.316475	7.717646
education background	1= primary school and below; 2= middle school, 3= high school, 4= high school, technical school, 5= secondary school, 6= college, 7= university degree and above;	2.844623	1.361544
nationality	The ethnic minority of village cadres = 1, han = 0;	0.6344412	0.4815954
Earning	The annual income of the village cadres in 2011	6298. 603	752.0976
invite investment	Whether the village cadre has the attractors project to the village.	0.0631938	0.2433156
age2	The age of the farmer	33.91623	19.79277
Gender	Men = 1, women = 0.	0.4776596	0.4995086
household Category	It refers to the state of the peasants. Individual business and cadre households = 3, common = 2,	2.090543	0.2907237
cooperative economic	The cooperative economic organizations that farmers voluntarily participate in. Participation = 1, not participating = 0;	0.0226744	0.1488655
Outside	Whether or not farmers are going out for employment is equal to 1, or not equal to 0;	0.1286727	0.3348487
medical treatment	The medical mutual aid system that the farmer volunteers to participate in, participation =1, not participating = 0;	0.987681	0.1103068
county level cadre	If there are any cadres at the county level or above in the party and government departments at the county level and above, yes=1, no=0;	0.5639988	0.4958952
old areas	The village is a revolutionary old area. Is equal to 1, or not equal to 0;	0.1714249	0.3768859

Table 1, cont			
ghettoize	Is the village a minority area? yes=1;No=0;	0.7146605	0.4515832
Endemic	Is this village an endemic area? yes = 1;No=0;	0.0290694	0.1680038
Distance	The village is about the distance from the nearest county	31.64783	25.09245
Terrain	Mountain = 3, hilly = 2, plain = 1;	2.380824	0.8300797
Disaster	Whether there is a natural disaster affecting the production and living of farmers in this village. Is = 1; N = 0;	0.5656919	0.4956737
concede land	It refers to the situation of reducing the area of cultivated land due to the implementation of the conversion of farmland to forestry.	0.6965101	0.4597725
public program	The fund originates from the superior or external level, affecting the economic life of the village project.	0.1793333	0.3836389

3. Discussion and conclusion

The measurement software used in this paper is stata13, and before verifying the results, the correlation between all variables is tested to prove that there is no correlation between all variables. The correlation coefficient between all variables is under 0.5, which satisfies the criteria of the regression.After a collinear test, it is proved that there are no multiple collinear problems between the models, so this paper argues that the regression results of the model are more accurate and reliable. Table 2 main report the heterogeneity of the village cadres regression results on poverty reduction, of which the first (1), (3) column is in poverty reduction/various poverty), village cadres as time (year), age (age1), education (education) and income (earning), investment promotion and capital introduction (invite) and national (nationality), not the introduction of control variable Probit and OLS regression results. The results show that the age and investment attraction of village cadres have a significant positive impact on farmers' poverty reduction and are significant at the 1% level. The income and ethnic groups of village cadres have a positive impact on the poverty reduction of farmers and are significant at the 1% level of the Probit model, which is significant at the 5% level of OLS model. The education of village cadres has a positive impact on poverty reduction, which is significant at the 10% level.The (2) and (4) columns are the analytical results after the regression of Probit and OLS with all four class control variables.The results showed that the age of the village cadres to villager's poverty reduction coefficient is 0.02, the influence of the symbol is positive, and the 1% significance level, prove that the age of village cadres has a positive promoting effect on the farmers' poverty reduction, prove the hypothesis of a set. The influence of the village cadres education on poverty reduction coefficient is 0.05, the symbol is positive, and the 10% significance level, and whether the village cadres to carry on the investment promotion and capital introduction on poverty reduction coefficient is 0.39, the influence of the symbol is positive, at 5% significant level, education and investment promotion and capital introduction of village cadres in the farmers' poverty reduction has a positive role in promoting, prove the hypothesis 2 set is correct. The influence of village cadres on poverty reduction is 0.3, the symbol is positive, and in the Porbit model, it is significant at the 1% level. In the OLS model, it is significant at 5% level. It shows that the minority attributes of village cadres can have a positive impact on poverty reduction and proves the inference of hypothesis 3. The influence coefficient of village cadres' income on poverty reduction is 0.0003 and the symbol is positive. It's significant at the 10% level. It shows that the income of village cadres has a positive impact on poverty reduction, which is consistent with hypothesis 4.

Table 2. The empirical results

Variables	Dependent Variable:whether poverty			
	(1)	(2)	(1)	(2)
Variables	Probit	Probit	OLS	OLS
earning	0.0005***	0.0003*	0.0357**	0.0389*
	(0.0000)	(0.0002)	(0.0003)	(0.0003)
age1	0.0238***	0.0213***	0.0075***	0.0066***
	(0.0018)	(0.0035)	(0.0084)	(0.0051)
education	0.0547*	0.0533*	0.2353*	0.2084*
	(0.0101)	(0.0276)	(0.0531)	(0.1244)
Year	0.023	0.0219	0.0012	0.0009

Table 2, cont				
	(0.0012)	(0.0035)	(0.0022)	(0.0829)
invite	0.351***	0.3854**	0.8792***	0.8425**
	(0.0542)	(0.1514)	(0.0982)	(0.0252)
nationality	0.3328***	0.3569***	0.5779**	0.6002**
	(0.0289)	(0.1055)	(0.0073)	(0.0071)

Note: ***, **, * respectively indicates the significance of parameters at the level of 1%, 5%, and 10%. The values in parentheses are the robust standard deviations. It remains the same in the following table.

According to the conclusions of this paper, the following policy recommendations for poverty reduction are as follows: (a) Village cadres should attach great importance to the culture of their own quality, and strive to improve the level of record of formal schooling education, become the high quality human capital, should also actively learn the experience of senior, knowledge and experience will have the biggest effect. (b)The government should encourage to break on the village of village cadres election ethnic discrimination, properly enlarge the proportion of ethnic minority cadres, not only to a certain extent, to get more policy support,combined with the feelings of the relationship among nationalities, ethnic minorities and honest will eliminate some of rent-seeking behavior of village cadres, to reduce incidents of infringement to the interests of farmers.(c) The village cadres shouldaccording to the strategic development plan of the village, actively draw up the plan and encourage the enterprises to invest in the factory. On the premise of not affecting the ecological environment of the village, the economic development of the villages and the employment proportion of the villages will be improved, so as to realize the harmonious coexistence between man and nature and economy. (d) At the economic level of its village, the income level of village cadres is reasonably set. It is necessary to avoid the problem that the low income of village cadres can cause the lack of enthusiasm of the village cadres, and also avoid the harm of the rent-seeking behavior of the village cadres to the economic income of the peasants.

References

[1] Arnim Scheidel, 2014. Land poverty and emerging ruralities in Cambodia: insights from Kampot province [J]. Environment, Development and Sustainability: 823-840p.
[2] Swati, 2015. Identifying Single or Multiple Poverty Trap: An Application to Indian Household Panel Data [J]. Social Indicators Research: 157-179p.
[3] Deaton&Dreze, 2002. Income and Expenditure Switching Among the Impoverished During Structural Adjustment in India. Economics of Planning: 221-252p.
[4] Cheng, 2014. Rural poverty reduction: it should pay more attention to education or health - empirical evidence based on income growth and gap narrowing. Economic research: 130-144P.
[5] Chen qi, 2012. Human capital and income poverty of rural households in contiguous areas. Jiangxi social science:231-235p.
[6] Armando Barrientos, 2011. Attitudes to Chronic Poverty in the Global village. Social Indicators Research:101-114p.
[7] MICHELLE S. MOOD, 2005. Opportunists, predators and rogues: the role of local state relations in shaping Chinese rural development. Journal of Agrarian Change: 217-250p.

Tourism spatial structure of urban agglomeration in the middle reaches of the Yangtze River based on social network analysis

Chun Liu[1, 2, a*], **Guihua Nie**[1]

[1]School of Economics, Wuhan University of Technology, Wuhan, 430070, China
[2]Center for Hubei Business Service Development Studies, Wuhan, 430070, China

Corresponding author and e-mail: Chun Liu: www.lc576@126.com

Abstract. Based on a revised gravity model, this paper establishes an evaluating indicator and model of the spatial structure of tourism economic linkages in urban agglomeration by using social network theory and method. Tourism spatial economic data of 56 urban in the urban agglomeration of the middle reaches of the Yangtze River were collected. With the help of Ucinet software, we discuss the tourism spatial structure features in the urban agglomeration of the middle reaches of the Yangtze River, and the tourism spatial structure of urban agglomerationis optimized by network density, centrality degree, structural holes, core-periphery structure and condensing subgroup model.

1. Introduction

In August 2012, *Opinions on Vigorously Implementing the Strategy of Promoting the Rise of the Central Region* enacted by the State Council mentioned that the three sub-urban agglomerationsin the middle reaches of the Yangtze River (Wuhan Metropolitan Area, Changsha-Zhuzhou-Xiangtan Urban Agglomeration and the Urban Agglomerations around the Poyang Lake)need to collaborate to promote the integrated development of the Urban Agglomerations in the middle reaches of the Yangtze River (State Council of the PRC, 2012). In April 2015, the State Council approved *the Development Plan of Urban Agglomerations in the Middle Reaches of the Yangtze River*, pointing out that the innovative modes and paths should be created and developed to build an International Urban Agglomeration which would be an important backbone of the Yangtze River Economic Zone and a new growth pole of China. (the State Council of the PRC, 2015).

Under the guidance of the national policy, the regional tourism cooperation of the urban agglomeration in the middle reaches of the Yangtze River has been carried out step by step and developed extremely well. . Since 2013, the provincial capitals of the urban agglomerations in the middle reaches of the Yangtze River have jointly signed the Wuhan Consensus, the Changsha Declaration, the Hefei Outline, the Nanchang Action and the Action Plan, and established the Urban tourism development cooperation organization and the Tourism Enterprise Alliance of the Urban Agglomeration in the middle reaches of the Yangtze River to build a barrier-free tourist area with multi-industry integration and low carbon. This area would be an important backbone of the Yangtze River Economic Zone and have impact on the world. The construction of the barrier-free tourist area based on the integrated development of the Urban Agglomeration in the middle reaches of the Yangtze River, so the optimization of tourism spatial structure of this area is an important measure to build a world-famous tourist destination.

2. Literature analysis on tourism spatial structure of urban agglomeration in the middle reaches of the Yangtze River

Everything is related to everything else but things nearby are more relevant than things far away (Tobler, W. R., 1979). According to this theory, no region is isolated, and each region is developing on the basis of its correlation with other regions. The evaluation index system of overall and individual tourist route network was estanblished to discuss the network and spatial structure characteristics of tourism destinations of Yunnan Province (SUN Y., et al., 2016). The tourism spatial cooperation pattern of the main node cities in China along the "The Belt and Road" is recommended on the basis of the analysis on the network spatial structure of tourism economic connection of those node cities. (ZOU Y.-G., 2017). WANG L.-J. , et al. (2019) criticized the spatial differences and spatial clustering of Chinese mainland tourism economy by adopting 2001-2014 panel data of 31 provinces and Tobit mode. WANG X. & GUAN W.-H. (2018) studied the spatial structures of 13 tourist cities with economic connection in Jiang Su Province and analyzes the dynamic mechanism of spatial structure formation.

Theoretical reflections about the transformation of urban spaces under the impact of tourist traffic can also be found in works by the Polish social geographer Liszewski,S. (1999).The impact of tourism on HSR demand is usually related to the difficulties of estimating HSR induced demand in the modelling stage (Guirao, B., & Campa, J.L., 2015).However, Albalate, D. (2016) comes to a statistically significant negative impact on tourism outcomes by adopting the method based on differential panel data and the strategy of double fixation. WANG D. G., & NIU Y.,et al. (2018) demonstrates that the High-speed rail has strengthened tourism economic relations among cities through studying the relationship among 338 cities, with using economic-relation model and spatial analysis method from ArcGIS, and a "corridor "effect of the spatial distribution of the change rates of tourism external economic relationships has presented.

As a tourist geography concept with a spatial attribute, tourist flow represents both the nerve center and the links of the tourism system (WU J.-F., &BAO H.-S., 2002). A spatial model of tourist flow mainly contains a hierarchical model of tourist flow (Marrocu, E.,&Paci, R., 2013), a city tourist flow model (ZHANG Y., XU J. & ZHUANG P., 2011), the core-edge space model of tourist flow (Asero, V., & Gozzo, S., et al., 2016) and the circle and layer structure model of tourist flow (PENG H., et al., 2016). JIN C., & CHENG J., et al. (2017) analyzed the relationship between the tourists' movement networks and the lengths of their trip, and later found the time heterogeneity in the movements.

Capone, F. & Boix, R. (2008) analyzed the spatial dynamics related to the sources of tourism competitiveness within a small scale. Weidenfeld, A., &Williams, A. M., et al. (2014) provided theoretical and empirical contributions to the understanding of space competition by examining the tourist attractions of two different business clusters (low and high) in Cornwall, England, and found that the competition between local tourist attractions is related to the differences among the levels of agglomeration, spatial proximity and thematic products when compared to the regional scale. João, R., & Nijkamp, P. (2017) analyses whether and how regional systems of innovation influence the competiveness of tourism destinations in Europe.HUANG T., &XI J.-C., et al. (2017) constructed the gravity model of tourism space competition under the interaction of multi-destination and multi-tourist. Furthermore, they used GIS and spatial econometric methods to observe the structure changes of tourism competition system of the Yangtze River Delta Urban Agglomeration under the situation of whether there is the Ningbo-Hangzhou high-speed railway.

Romão, J., &Saito, H. (2017) applied an exploratory spatial analysis and a spatial econometric model to identify the specific spatial characteristics prevailing in Japanese regional tourism dynamics, and determined the spatial heterogeneity and agglomeration processes. More recently, tourism researchers explore and investigate tourism economy and tourists' spatial behavior by applying techniques and methods that were not part of traditional tools for tourism research, such as GIS and GPS. (Zoltan, J., & McKercher, B., 2015). Similarly, there are also a lot of literatures using moderate centrality, intermediary centrality and close centrality of SNA to analyze the structure of tourist attractions. (Asero, V., & Gozzo, S., et al., 2016; Jin, C., & Cheng, J., et al., 2017.). WANG Q.-L., &TANU Y.-Y., et al. (2017) used the neighboring index, geographical concentration index, the gini coefficient, Lorentz curve and hotspot analysis to explore the evolution characteristics of spatial distribution, the degree of concentration and Equilibrium level of A-class tourist attractions in Anhui Province. SONG L., & LI Y., et al. (2016) analyzed the spatial structure of tourism in the Circum-Bohai-Sea theoretically by using the point-axis theory, the location entropy and the city flow. DUAN B. (2015) developed the data model of regional tourism center's scale and spatial structure by using fractal geometry method, and took the Yangtze River Delta as an example for empirical study.

At present, scholars have not formed a stable research paradigm for the study of urban tourism spatial structure, because the research results vary with different regions, scales, time series and methods. The research object of this paper will cover 16 urban of Wuhan urban agglomeration, 22 urban of Changsha-Zhuzhou-Xiangtan urban agglomeration, and 18 urban of Poyang Lake urban agglomeration in the middle reaches of the Yangtze Rive. This paper will apply the social network analysis method to measure and compare the network density, centrality, and network cohesion of the tourism spatial structure of urban agglomeration in the middle reaches of the Yangtze River to identify the characteristics of tourism spatial structure of Urban Agglomeration in the middle reaches of the Yangtze River, and finally propose the optimal layout of tourism.

3. Spatial structure analysis model of urban tourism economy in urban agglomeration in the middle reaches of the Yangtze River.

3.1. Urban tourism economic linkage model

Normally, "attribute data" is used in economic research, while social network analysis adopts "relational data", which is some quantitative value of the "relationship" between things. Referring to the classical gravity correction model in econometric geography, the "attribute data" can be transferred into "relational data". The formulas are as follows:

$$R_{ij} = k_{ij}\frac{\sqrt{P_i q_i}\sqrt{P_j q_j}}{d_{ij}^2} \qquad k_{ij} = \frac{q_i}{q_i + q_j} \tag{1}$$

In the formula (1), R_{ij} is the index of urban tourism economic linkages between city i and city j. k_{ij} is the contribution rate of city i to R_{ij}. P_i and P_j are respectively the total number of tourists in city i and city j (unit: 10,000 person-time). q_i and q_j are respectively the gross tourism revenue in city i and city j (unit: RMB 100 million). d_{ij} is the distance between city i and city j, with adopting the shortest length of highway between cities in urban agglomerations (unit: km).

3.2. Network density

Network density refers to the closeness degree of the connection between cities in the network, obtained by the ratio of the actual number of relationships between cities and the possible number of relationships in theory. The more connections between urban systems, the greater the density of the network, the greater the overall openness of the network and the ability to access resources. The calculation expression of urban agglomeration network density is:

$$D = \frac{\sum_{t=1}^{n} d_i(c_i)}{n(n-1)} \quad d_i(c_i) = \sum_{t=1}^{n} d_i(c_i, c_j) \tag{2}$$

In the formula (2), n stands for the urban scale of urban agglomerations in the middle reaches of the Yangtze River, namely the number of cities. If there is a correlation between city i and city j, $d_i(c_i, c_j)$ is equal to 1; if not, 0.

3.3. Centrality degree

Centrality is one of the focuses of the social network analysis methods, which can be divided into the network centralization and the node centrality degree. In general, the node centrality degree can be divided into two types: relative centrality degree and absolute centrality degree. In this study, relative centrality degree is adopted to measure the centrality degree of each node in the network.

3.3.1 Node centrality degree

It is used to measure the ability of a city to communicate with other cities in the network diagram. It indicates the association strength between one city and the others. The calculation formula is as follows:

$$C_{RD}(i) = \frac{C_{AD}(i)}{n-1} \tag{3}$$

In the expression (3), $C_{RD}(i)$ is the relative centrality degree of city i, which refers to the ratio of absolute centrality degree of city i to the maximum possible degree of any node in the topological graph of urban system connection network. $C_{AD}(i)$ is the absolute centralitydegreeof city i, indicating the number of the other nodes connected to city i in the network.

3.3.2 Closeness centrality

It can be used to measure the degree that one city in the network is not controlled by the others, representing the sum of the shortest traffic distance between one city and the others. The smaller the closeness centrality value, the stronger the closeness centrality degree, which suggests that the better the traffic accessibility, the tighter the connection between this city and the others, and the closer the city is to the network center. The calculation formula is as follows:

$$C_{RPi}^{-1} = \frac{C_{APi}^{-1}}{n-1} = \frac{\sum_{j=1} d_{ij}}{n-1} \tag{4}$$

In the expression (4), C_{RPi}^{-1} and C_{APi}^{-1} are respectively the relative closeness centrality and the absolute closeness centrality ofcityi. d_{ij} indicates the shortcut distance between city i and city j.

3.3.3 Betweeness centrality

Betweeness centrality measures the spatial intermediary position of one city between the other cities in the topological graph of urban system connection network, used as an index of "control ability" macroscopically measuring the control and dependence on the other cities in the urban system connection network. The greater the center centrality, the stronger the control power to the urban agglomeration, the more advantages in the network structure system, the higher the dependence of the other cities on this city. The calculation formula is as follows:

$$C_{RBi} = \frac{2C_{ABi}}{(n-1)(n-2)} = \frac{2\sum_{j}^{n}\sum_{k}^{n} b_{jk}(i)}{(n-1)(n-2)} = \frac{2\sum_{j}^{n}\sum_{k}^{n} g_{jk}(i)/g_{jk}}{(n-1)(n-2)} \tag{5}$$

In the expression (5), C_{RBi} is called the relative betweeness centrality of city i. C_{RBi} is called the absolute betweeness centrality of city i. $b_{jk}(i)$ is the ability of city i to control the communication between city j and k. g_{jk} is the number of shortcuts between city j and k. $g_{jk}(i)$ represents the number of shortcuts through city i existing between city j and k. here, $b_{jk}(i) = g_{jk}(i)/g_{jk}$

4. Overview of study area and sources of data

4.1. Overview of the study area

In this paper, the research object is urban agglomeration in the middle reach of Yangtze river, consisting of 56 cities in Wuhan Urban Agglomeration, Changsha-Zhuzhou-Xiangtan Urban Agglomeration and the Urban Agglomerations around the Poyang Lake. Specifically, Urban Agglomeration in the middle reach of Yangtze river covers 16 cities in Wuhan Urban Agglomeration, including Wuhan, E'zhou, Daye, Huangshi, Hanchuan, Xiaogan, Yingcheng, Xiantao, Xianning, Tianmen, Wuxue, Huanggang, Anlu, Qianjiang, Chibi and Macheng; 22 cities in Changsha-Zhuzhou-Xiangtan Urban Agglomeration including Changsha, Xiangtan, Zhuzhou, Yueyang, Hengyang, Yiyang, Liuyang, Leiyang, Changde, Loudi, Chenzhou, Yuanjiang, Xiangxiang, Miluo, Shaoshan, liling, Zixing, Lianyuan, Changning, Lengshuijiang, Linxiang and Jinshi; also including 18 cities-- Nanchang, Xinyu, Fengcheng, Gao'an, Zhangshu, Jiujiang, Jingdezhen, Guixi, Yingtan, Ji'an, Leping, Yichun, Gongqingcheng, Ruichang, Shangrao, Fuzhou, Dexing and Jinggangshan within the Urban Agglomerations around the Poyang Lake (Fig.1). the spatial pattern of tourism economy of urban agglomeration in the middle reach of Yangtze River will be analyzed by observing the tourism economic relationship among those 56 cities.

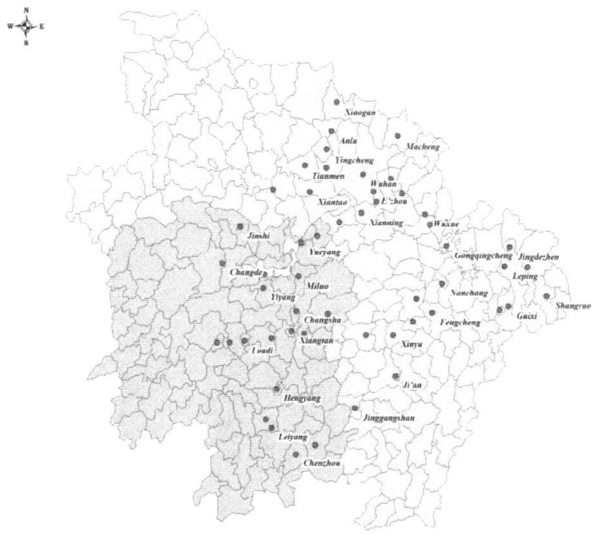

Figure 1. Urban distribution map of urban agglomeration in the middle reaches of the Yangtze River

4.2. Sources of date

In this study, the initial data including total number of urban tourists (unit: 10,000 person-times) and total tourism revenue (unit: RMB 100 million) are from Hubei Statistical Yearbook (2017), Hunan Statistical Yearbook (2017) and Jiangxi Statistical Yearbook (2017). In addition, referring to the Statistical Bulletins for national economic and social development of 56 cities the above mentioned in the urban agglomeration in the middle reach of Yangtze River in 2017. The shortest traffic distances (unit: km) between the cities in the urban agglomerations are collated according to the shortest highway distance between the cities in Baidu maps to reflect the latest tourism economic spatial structure of the urban system of the middle reaches of the Yangtze River.

5. Social network analysis of tourism spatial structure of urban agglomeration in the middle reaches of the Yangtze River

5.1. Analysis of spatial relationship of tourism economy in urban agglomeration of the middle reaches of the Yangtze River

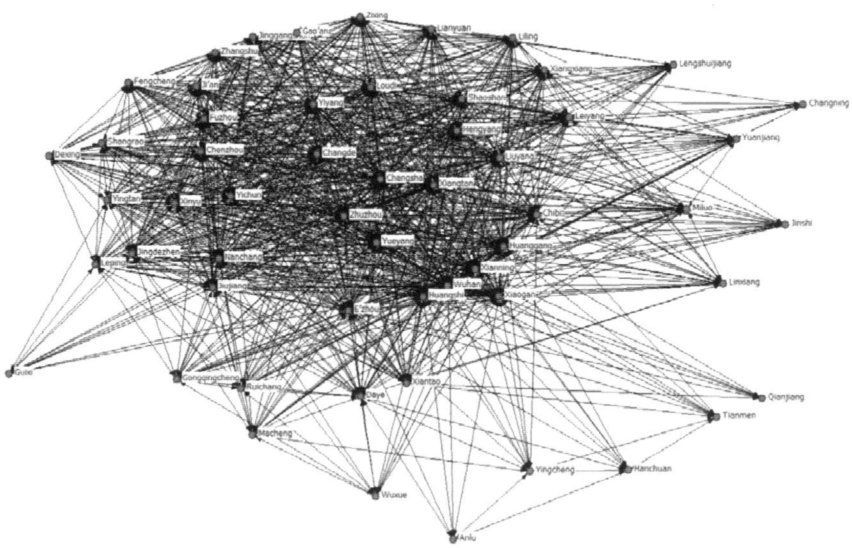

Figure 2. Network Structure basic pattern of economy connection in each city of urban agglomeration in the middle reaches of the Yangtze River

This study uses the tourism economic relationship model to calculate the tourism linkage degree among 56 cities in urban agglomeration in the middle reaches of the Yangtze River. Based on this, Ucinet-netdraw method is adopted to form the basic form of tourism economic connection network of urban agglomeration in the middle reaches of the Yangtze River (Fig.2). The arrow line in the figure represents the tourism economic connection from one city to another.

By analyzing the structure chart of tourism economic connection network of urban agglomeration in the middle reaches of the Yangtze River, we can more intuitively judge the connection strength, tourism radiation and cooperative behavior among cities. From Figure 2, it is obvious that the intensity of tourism connection nodes of Wuhan, Changsha, Zhuzhou, Xiangtan and Nanchang in urban agglomeration in the middle reaches of the Yangtze River is at the highest level, which shows that the clustering and radiation ability of these five cities are the strongest in tourism economics, followed by that of Yueyang, Jiujiang and Xianning. All these are inseparable from the city's overall economic strength and administrative division status, and the traffic location in the urban agglomeration in the middle reaches of the Yangtze River.

Ucinet-Density method is used to calculate the overall tourism economy spatial network density of the urban agglomeration in the middle reaches of the Yangtze River in 2016 (Tab.1). The overall density is 0.3344, which shows that the level of network density is low, indicates that the tourism economic links between cities are weak and in a loose state. The tourism economy spatial network density of Changsha-Zhuzhou-Tan urban agglomeration and the urban agglomeration around Poyang Lake is respectively 0.5758 and 0.5451. The network density is higher, which indicates that the tourism economic links between cities in the urban agglomeration are strong. Furthermore, urban agglomerations are in a strong state of tourism economy connection. The tourism economy network density is 0.4250 in Wuhan urban agglomeration, suggesting the network density is not high, indicates that the tourism economic links between cities and urban are not strong, and in a weak connection state.

Table 1. Characteristics of social networks in urban agglomeration in the middle reaches of the Yangtze River (2016)

	Density	Average distance	Distance-based cohesion
Wuhan Urban Agglomeration	0.4250	1.575	0.713
Changsha-Zhuzhou-Xiangtan Urban Agglomeration	0.5758	1.424	0.788
Urban Agglomerations around the Poyang Lake	0.5451	1.392	0.812
Urban Agglomerationsin the middle reaches of the Yangtze River	0.3344	1.681	0.665

Ucinet-Distance method is used to calculate the geodesic distance between any two cities in the Urban Agglomerationsin the middle reaches of the Yangtze River. By applying such method, the following two indexes of distance are achieved: The average of the cumulative sum of the geodesic distances of any two cities is 1.681in the Urban Agglomerationsin the middle reaches of the Yangtze River and the Tourism distance-based cohesion is 0.665. The above-mentioned two indices indicate that the overall tourism cohesion is weak in the Urban Agglomerations in the middle reaches of the Yangtze River, and the structure of city tourism economy presents a loose form. However, the average of the geodesic distances cumulative sum of Wuhan urban agglomeration, Changsha-Zhuzhou-Tan urban agglomeration and Urban Agglomerations around the Poyang Lake included Urban Agglomerationsin the middle reaches of the Yangtze River are 1.575, 1.424 and 1.392, respectively. Correspondingly, the respective tourism distance-based cohesion of the three urban agglomerations is 0.713, 0.788 and 0.812 (Tab. 1), which shows that the tourism cohesion of Urban Agglomerations around the Poyang Lake is strong and city tourism economy spatial structure is more compact. However, the tourism cohesion of Changsha-Zhuzhou-Tan urban agglomeration and Wuhan urban agglomeration is weak, which shows that the provincial administrative division is the obstacle factor of cross-regional economic activities.

With the urban agglomerations in the middle reaches of the Yangtze River stepping into the "two-hour tourism circle" and the integration acceleration of the urban agglomerations in the middle reaches of the Yangtze River, the inter-city tourism activities and cross-regional tourism activities are increasingly frequent, the spatial tourism economic links in the urban agglomeration in the middle reaches of the Yangtze River will increase year by year, and the administrative barriers between provinces will gradually increase.

Provincial administrative barriers will gradually be eliminated and the advantages of cluster and dispersion will emerge.

5.2. Centrality degree analysis of spatial relationship of tourism economy in urban agglomeration of the middle reaches of the Yangtze River

With the help of three indexes, namely degree centrality, closeness centrality and betweenness centrality, we can determine the location of the city in tourism economic connection spatial network of urban agglomerations in the middle reaches of the Yangtze River. By adopting Ucinet-Degree method, Ucinet-Closeness method and Ucinet-Betweenness method, the network centrality of tourism economic links among 56 cities in the middle reaches of the Yangtze River is calculated. The specific location and status of each city in the tourism economic connection of urban agglomerations are shown in table 2. We can draw the following conclusions:

Table 2. Network centrality table of urban tourism economy connection of cities in urban agglomeration in the middle reaches of the Yangtze River

Ranking	Degree centrality		Closeness centrality		Betweenness centrality	
	Sort	Value	Sort	Value	Sort	Value
1	Wuhan	54	Wuhan	54.5	Wuhan	404.3
2	Xiannin	43	Xiannin	49	Xiannin	95.4
3	Changsh	43	Changsh	49	Changsh	77.8
4	Zhuzhou	39	Zhuzhou	47	Yueyang	57.5
5	Xiangtan	39	Xiangtan	47	Zhuzhou	46
6	Yueyang	38	Yueyang	46.5	Xiangtan	46
7	Liuyang	34	Liuyang	44.5	Nanchan	45.1
8	Nanchan	32	Nanchan	43.5	Xiaogan	42.4
9	Jiujiang	32	Jiujiang	43.5	Huangsh	35.6
10	Hengyan	31	Hengyan	43	Yingtan	26.8
11	Yichun	29	Yichun	42	Liuyang	24.2
12	Jingdezh	29	Jingdezh	42	Huangga	24.2
13	Changde	28	Changde	41.5	Jiujiang	21.2
14	Yingtan	27	Yingtan	41	Hengyan	20.1
15	Ji'an	27	Ji'an	41	Changde	14.3
16	Fuzhou	27	Fuzhou	41	Jingdezh	12.4
17	Yiyang	26	Yiyang	40.5	Yichun	9.9
18	Xinyu	26	Xinyu	40.5	Yiyang	8.5
19	Chenzho	25	Chenzho	40	Chenzho	5.8
20	Loudi	24	Loudi	39.5	Fuzhou	5
21	Huangsh	24	Huangsh	39.5	Ji'an	4.7
22	Huangga	24	Huangga	39.5	Loudi	4.6
23	Xiaogan	22	Xiaogan	38.5	Xinyu	3.4
24	Jinggang	22	Jinggang	38.5	E'zhou	3.1
25	Shangra	21	Shangra	38	Shaosha	2.3
26	Zhangsh	19	Zhangsh	37	Shangra	1.4
27	Shaosha	17	Shaosha	35.8	Jinggang	1.4
28	Lianyua	15	Gao'an	35	Chibi	0.9
29	Gao'an	15	Fengche	35	Yingche	0.7
30	Fengche	15	Lianyua	34.8	Leiyang	0.7
31	Zixing	13	Leping	34	Lianyua	0.6
32	Leping	13	Dexing	34	Zhangsh	0.6
33	Dexing	13	Zixing	33.8	Zixing	0.4
34	Lengshu	12	Lengshu	33.3	Linxiang	0.4
35	Xiangxia	11	Xiangxia	32.8	Leping	0.2
36	Yuanjian	10	Yuanjian	32.3	Dexing	0.2
37	Leiyang	10	Leiyang	32.3	Lengshu	0.1
38	E'zhou	10	E'zhou	32.3	Gao'an	0.1
39	Ruichan	9	Ruichan	32	Fengche	0.1
40	Miluo	9	Miluo	31.8	Yuanjian	0
41	Chibi	8	Chibi	31.3	Xiangxia	0
42	Linxiang	7	Linxiang	30.8	Xiantao	0
43	Liling	7	Liling	30.8	Wuxue	0
44	Gongqin	6	Gongqin	30.5	Tianmen	0

45	Changni	6	Changni	30.3	Ruichan	0
46	Yingche	5	Yingche	29.8	Qianjian	0
47	Machen	5	Machen	29.8	Machen	0
48	Daye	5	Daye	29.8	Liling	0
49	Xiantao	4	Xiantao	29.3	Jinshi	0
50	Wuxue	4	Wuxue	29.3	Hanchua	0
51	Jinshi	4	Jinshi	29.3	Guixi	0
52	Hanchua	4	Hanchua	29.3	Miluo	0
53	Anlu	3	Anlu	28.8	Gongqin	0
54	Tianmen	2	Tianmen	28.3	Daye	0
55	Guixi	2	Qianjian	27.8	Changni	0
56	Qianjian	1	Guixi	24.7	Anlu	0

(1)Among the 56 cities of urban agglomerations in the middle reaches of the Yangtze River, Wuhan, Xianning, Changsha, Zhuzhou, Xiangtan, Yueyang, Liuyang, Nanchang, Jiujiang and Hengyang have much higher degree centrality than other cities, especially Wuhan. With their superior tourism location and natural tourism resources, these cities are in the central position of the tourism economic connection network of the urban agglomerations in the middle reaches of the Yangtze River, while Wuhan is the core city of the diffusion radiation of traffic and tourist flow among urban agglomerationin the middle reaches of the Yangtze River. Due to the lack of tourism resources and poor accessibility of tourism, the degree centrality ranking of urban tourism economic connection network is the lowest in Xiantao, Wuxue, Jinshi, Hanchuan, Anlu, Tianmen, Guixi and Qianjiang, and they are in the periphery position in the urban tourism economic connection network of urban agglomerations in the middle reaches of the Yangtze River. With the development of regional tourism integration and transportation network, central cities have an external effect on other cities in the same region through the development of tourism economy, and form a pattern of interconnected development of tourism economy in urban agglomerations.

(2)The closeness centrality of the urban tourism economic links of urban agglomerationsin the middle reaches of the Yangtze River shows that the closeness centrality of Wuhan, Xianning, Changsha, Zhuzhou, Xiangtan, Yueyang, Liuyang, Nanchang, Jiujiang and Hengyang is high, which indicates that the tourist accessibility between these cities and other cities in the urban agglomeration is better, the city connectivity is closer, also shows that these cities are in the center of the tourism economic network and less controlled by other cities. Yingcheng, Macheng, Daye, Xiantao, Wuxue, Jinshi, Hanchuan, Anlu, Tianmen, Qianjiang and Guixi have poor tourist accessibility, and loose relationship with other cities in urban agglomerations. However, in terms of the level of closeness centrality, the overall connectivity of the tourism economic network of the urban agglomeration in the middle reaches of the Yangtze River is better.

(3)According to Betweenness Centrality of urban tourism economy connection of cities in urban agglomeration in the middle reaches of the Yangtze River (Tab. 2). The betweenness centrality of Wuhan, Xianning, Changsha, Yueyang, Zhuzhou, Xiangtan and Nanchang are the highest, these cities have strong control over other cities in the urban agglomerations, and are at the key nodes of the urban agglomerations tourism economic network. By contrast, Yuanjiang, Xianxiang, Xiantao, Wuxue, Tianmen, Ruichang, Qianjiang, Macheng, Liling, Jin, Hanchuan, Guixi, Miluo, Communist Qingcheng, Daye, Changning and Anlu all have a betweenness centrality of only 0, which explains that these cities are at the edge of tourism economic connection in urban agglomeration in the middle reaches of the Yangtze River, and the spatial tourism economic interaction with other cities in the urban agglomeration are relatively weak.

In conclusion, the network centrality of each city in the middle reaches of the Yangtze River show that Wuhan has the highest score in the three indexes of network centrality, which indicates that Wuhan has the greatest influence in the entire tourism economic connection network, the closest relationship with the other cities, and the strongest control ability to the other cities. Changning, Yingcheng, Macheng, Daye, Xiantao, Wuxue, Jinshi, Hanchuan, Anlu, Tianmen, Guixi and Qianjiang have the lowest values of the three network centrality indices. It indicates that these 12 cities have the least impact in the whole tourism economic network, the less connections with other cities and the weaker control ability of other cities, and they are in a relatively isolated position in the network.

6. Conclusion

The urban system of the Yangtze River city group overall tourism cohesion is weak with loose structure of urban tourism economy; the tourism of Poyang Lake city group cohesion is strong; the Changsha Zhuzhou Xiangtan city group and Wuhan city circle tourism has weak cohesion.

In terms of the degree centrality, closeness centrality and betweenness centrality, Wuhan scores the maximum, indicating that Wuhan in the tourism economy network in Yangtze River city group is the most influential city. Changning, Macheng, Daye, Xiantao, Wuxue, Tianjin, Hanchuan, Anlu, Tianmen, Guixi and Qianjiang has the lowest score in the three center index, and are at a relatively isolated position in the network.

Acknowledgment
The authors acknowledge the research grants from the project *On Docking of the Wanli Tea Ceremony and the Belt and Road* (S2016013).

References

[1] Albalate, D. (2016). High speed rail and tourism: Empirical evidence from Spain's provinces and cities [J]. Transportation Research Part A, 85:174-185.
[2] Asero, V., Gozzo, S., & Tomaselli, V. (2016). Building tourism networks through tourist mobility [J]. Journal of Travel Research 55(6): 751–763.
[3] Capone, F. & Boix, R. (2008) Sources of growth and competitiveness of local tourist production systems: an application to Italy (1991–2001) [J]. Annals of Regional Science 42: 209–224.
[4] DUAN B. (2015). A Study on the Scale and Spatial Structure of Tourism Center in Yangtze River Delta Based on Fractal Theory [J]. Journal of Henan University (Natural Science) 45(2): 181-186.
[5] Guirao, B., & Campa, J. L. (2015). The effects of tourism on HSR: Spanish empirical evidence derived from a multi-criteria corridor selection methodology [J]. Journal of Transport Geography 47:37-46.
[6] HUANG T., XI J.-C., & GE Q.-S. (2017). Tourism Spatial Competition Pattern of Urban Agglomeration Under the Influence of High-Speed Railway Using Spatial Econometrics [J]. Economic Geography 37(8):182-191.
[7] JIN C., CHENG J., & XU J. (2018). Using User-Generated Content to Explore the Temporal Heterogeneity in Tourist Mobility[J]. Journal of Travel Research 57(6):779-791.
[8] João, R., & Nijkamp, P. (2017). Impacts of innovation, productivity and specialization on tourism competitiveness - a spatial econometric analysis on European regions[J]. Current Issues in Tourism 2:1-20.
[9] Liszewski, S. (1999). Przestrzeń turystyczna miasta (przykład Łodzi) [J]. Turyzm 9(1): 51-71.
[10] Marrocu, E., & Paci, R. (2013) Different tourists to different destinations - Evidence from spatial interactionmodels [J]. Tourism Management 39(39): 71-83
[11] National Development and Reform Commission of the PRC (2015). Development planning of urban agglomeration in the middle reaches of the Yangtze River [EB/OL]. http://www.sdpc.gov.cn/zcfb/zcfbtz/201504/t20150416_688229.html, 2015-04-13/2017-03-12.
[12] Romão, J., & Saito, H. (2017). A spatial analysis on the determinants of tourism performance in Japanese Prefectures [J]. Asia-Pacific Journal of Regional Science 1(1): 243-264.
[13] SONG L., LI Y., & XU K. (2016). Research on the tourism space structure of the Bohai sea ring area based on the pole-axis theory [J]. World Regional Studies 25(3): 99-105.
[14] State Council of the PRC (2012). Some opinions on vigorously implementing the strategy of promoting the rise of the central region [EB/OL]. http://www.gov.cn/zwgk/2012-08/31/content_2214579.htm, 2012-08-31/2017-03-12.
[15] SUN Y., SHI C.-Y., & TANG W.-W., et al. (2016). Research on the Spatial Network Characteristics of Travel Itinerary in Yunnan Province [J]. Human Geography 31(1): 147-153, 160.
[16] Tobler, W. R. (1979). Smooth pycnophylactic interpolation for geographical regions [J]. Journal of the American Statistical Association 74(367): 519–530.
[17] WANG D. G., NIU Y., & QIAN J., et al. (2018). Evolution and optimization of China's urban tourism spatial structure: A high speed rail perspective [J]. Tourism Management. 64: 218-232.
[18] WANG G.-M., LI X.-M., & HU Z.-D., et al (2012). Study on traffic networks of urban agglomeration of Chang-Zhu-Tan based on structural holes theory [J]. Computer Engineering and Applications 48(15):1-6.
[19] WANG G.-M., LI X.-M., & HU Z.-D., et al. (2012). Study on traffic networks of urban agglomeration of Chang-Zhu-Tan based on structural holes theory [J]. Computer Engineering and Applications 48(15):1-6.
[20] WANG L.-J. , ZENG G.-J., & BI D.-D. (2019). Spatial spillover effects of ICT on tourism industry growth [J]. ACTA GEOGRAPHICA SINICA 74(2):366-378.
[21] WANG Q.-L., TANU Y.-Y., & MAO N. (2017). Analysis of spatial pattern evolution of A-grade scenic spots in Anhui Province [J]. Engineering of Surveying and Mapping 26(10): 45-51, 58.

[22] WANG X., & GUAN W.-H. (2018). Research on the Spatial Structure and Mechanism of Tourism Economy in Jiangsu Province [J]. Modern Urban Research (10):45-51.
[23] Weidenfeld, A., Williams, A. M., & Butler, R. W. (2014). Spatial competition and agglomeration in the visitor attraction sector [J]. Service Industries Journal 34(3):175-195.
[24] WU J.-F., & BAO H.-S. (2002). Spatial structure mode of tourist system [J]. Scientia Geographica Sinica 22(1):96–101.(in Chinese)
[25] ZHANG Y., XU J. & ZHUANG P. (2011). The Spatial Relationship of Tourist Distribution in Chinese Cities [J]. Tourism Geographies 13(1): 75-90.
[26] Zoltan, J., & McKercher, B. (2015). Analysing intra-destination movements and activity participation of tourists through destination card consumption [J]. Tourism Geographies 17(1):19–35.
[27] ZOU Y.-G. (2017). Research on the Spatial Structure and Space Layout of Tourism Economic Relationship between Node Cities of China on the "The Belt and Road" [J]. Economic Management 39(5): 22-35.

Development of on-line monitoring system for insulation of high voltage electrical equipment

Hang Li[1], Yaoyao Luo[1, *], Yang He[1], Fangying Li[1], Ting Wu[1]
[1] Chengdu University of Technology, Chengdu 610059, China

[*] Corresponding author and e-mail: Yaoyao Luo, 280460115@qq.com

Abstract. The development of on-line monitoring technology for high-voltage electrical equipment insulation, the working principle of online monitoring device, and the problems in operation were comprehensively expounded. The on-site verification technology of high-voltage electrical equipment insulation online monitoring system was studied, and online monitoring was analyzed. Management and technical issues with the device were found. On this basis, the on-site calibration device for the on-line monitoring system of high-voltage electrical equipment insulation was developed. The results of on-site non-power-off verification of high-voltage electrical equipment insulation online monitoring system produced by several manufacturers were introduced. Analysis and elaboration of its key technologies.

1. Introduction

The aging and failure of high-voltage electrical equipment insulation is a slowly developing latent fault. The traditional preventive experiment has a long experimental period, and the equipment cannot be found to deteriorate in a short period of time. The experimental voltage is much lower than the normal operating voltage of the equipment, and it is susceptible to temperature and humidity. Many defects such as strong electromagnetic field interference and influence cannot be discovered. Moreover, the sudden development cannot be discovered in time, which has serious consequences for power supply and production. Online monitoring of power equipment insulation is an important technical means to ensure safe power generation and power supply, to detect hidden dangers in time, to reduce power outages and repairs, and to improve power supply reliability. On-line monitoring system for power equipment insulation can continuously monitor the leakage current, dielectric loss, equivalent capacitance, operating voltage, ambient temperature and humidity of high-voltage electrical insulation equipment, and collect data and automatic analysis by computer, make equipment final evaluation of normal work can improve the early prevention of insulation damage of high voltage electrical equipment. At present, the advanced and reliable equipment of domestic substation equipment is generally low, and the reserve factor of equipment is also low, which cannot guarantee the safe operation of power equipment. Therefore, it is of great practical significance to develop and promote the online monitoring technology of insulation.

2. Status of online monitoring of domestic high voltage electrical equipment insulation

The development and application of on-line monitoring technology for power equipment insulation in China has been carried out for decades. This work has played a positive role in improving the operation and maintenance level of power equipment, detecting hidden troubles in time, and reducing the occurrence of power outages. But there are still many problems. According to the survey results of 57 sets of centralized online insulation monitoring systems installed by the China Electric Power Research Institute in 1998 in some provinces and municipalities, it is only 30% of normal or relatively normal, and it is determined that it cannot operate normally or is in a state of paralysis. It accounts for 35% of the total. The main performance is that the insulation online monitoring system is not reliable, for the following reasons:

At present, some online monitoring systems adopt a centralized signal processing method, that is, after a signal is extracted by a sensor placed near the field device, the weak signal to be measured is directly introduced into a computer placed in the main control room through a large number of cables, and the host pair Field devices perform tour detection and data processing. Other detection systems use a centralized processing method of partitioning, which is, dividing the device under test into several regions according to the distribution of equipment in the substation, respectively performing aggregation and signal strobing, and passing the gated analog signal through a special multi-core shielded cable. It is sent to the host of the main control room, and the host performs tour detection and data processing. Although the second method reduces the amount of cable used, it still cannot solve the distortion problem caused by the analog signal after long-distance transmission, and both of these methods have the characteristics of large on-site installation workload and difficult maintenance.

The sensor's ability to resist environmental temperature, humidity changes and electromagnetic interference is relatively poor, resulting in sensor failure, damage, voltage signal distortion, causing media measurement is not accurate enough, stability, repeatability is poor, and measurement error is large.

In the traditional insulation online monitoring system, communication is carried out through analog signals, so the system communication anti-interference ability is poor, the signal is easily distorted, the data transmission and processing part is faulty, and data loss occurs.

3. Working principle of high-voltage electrical equipment insulation online monitoring system tester

High-voltage electrical equipment insulation online monitoring system High-voltage standard tester is an electrical device that can change the dielectric loss with power. The device under test can work under the system voltage. The change of dielectric loss is known and tested by standard. The high-voltage standard tester is mainly composed of a high-voltage non-destructive standard capacitor, a low-voltage adjustable non-inductive standard resistor, and a switching device. The high-voltage standard capacitor has a capacitance of 1 000 pF, and the low-voltage non-inductive standard resistor is combined with different values to form a high-voltage standard loss tester with a dielectric loss of 0.33 to 1.337 and a total of 7 stops. The high-voltage standard loss tester can be used in the device under test. Adjust the gear of the standard loss tester with the operating voltage always applied. The sensor of the insulated online monitoring system is connected to the low voltage side of the high voltage standard tester. The schematic diagram of the comparison test of the high-voltage standard tester for the insulation online monitoring system is shown in Fig. 1. The checker is mainly composed of AC filter power supply, 30 kVA voltage regulator (B1), 80 kV cascade test transformer (B2, B3), high voltage standard capacitor (C), 100 kV/100 V standard voltage mutual inductance. Device (B4) composed with discharge tube (P-350), standard non-inductive resistor and other components. High-voltage electrical equipment insulation online monitoring system standard tester generated by the current IN (current sampling sensor is also called I / U converter converted into voltage signal UIN) and standard voltage signal UN (taken from the voltage transformer secondary) input Data processing to the insulation online monitoring system. From this, the dielectric loss tan δ, capacitance C and leakage current IN can be measured.

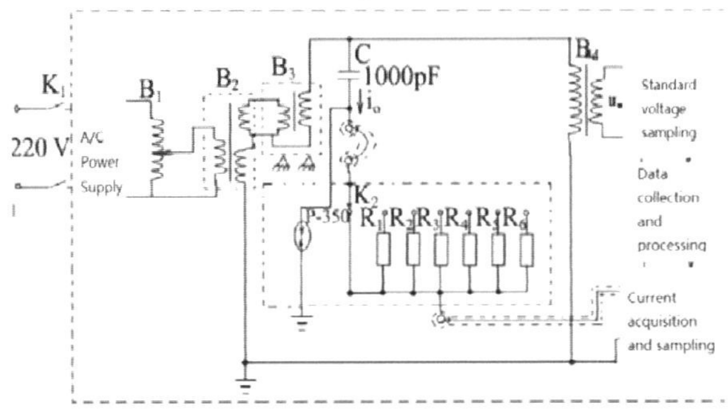

Figure 1. Electrical schematic diagram of high voltage standard tester for insulation online monitoring system

When the high-voltage standard tester performs the comparison test on the on-line monitoring system in the field, there are many devices used, and sometimes the detection is inconvenient. For this reason, the ZJB-I type low-voltage standard loss device with variable dielectric loss value is developed. The low-voltage standard tester that can change the dielectric loss value consists of a filter power supply (external), a voltage regulator, an isolation transformer, CN1, CN2, CN3 standard capacitors, standard non-inductive resistors R1 to R21, a voltmeter, an ammeter, and a switch. The working principle of the comparison test on the insulation online monitoring system is shown in Fig. 2.

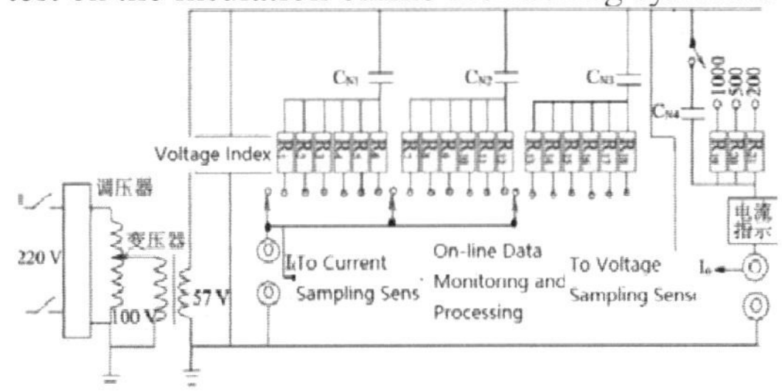

Figure 2. Insulation online monitoring system ZJB-I low voltage standard tester electrical schematic

According to the sampling sensor current of the insulation online monitoring system, select the loss selection 1 ~ 3. The low-voltage standard tester is also equipped with a full-current and resistive current test circuit for the zinc oxide surge arrester.

4. Edge online monitoring system hardware design

Insulation online monitoring system should be able to accurately and accurately reflect the insulation of high voltage equipment. The hardware circuit used to obtain the required signal must have high monitoring sensitivity and good anti-interference ability. Due to the required signals and various interferences in the power system, in order to carry out accurate on-line monitoring of insulation, the pre-hardware circuit must be sensitive to capture the required voltage and current signals, and within the precision range allowed by the project. The resulting signal is pre-processed to reduce or suppress interference, and is subjected to analog-to-digital conversion to a digital signal that can be processed by the computer. Since the tgδ value is small, the performance requirements for the hardware circuit are also relatively high. In order to improve the accuracy and accuracy of the measurement, the hardware circuit to be used has high sensitivity, low temperature drift, and high common mode rejection ratio. At the same time, this paper samples the software analysis method based on FFT. The hardware circuit is relatively simple. Some integrated circuit chips and circuits with better performance can be used to eliminate the interference and prevent the signal from being distorted. In the insulation online monitoring system studied in this paper, the front circuit part realizes the functions of signal conversion, amplification, filtering and sampling. This chapter will introduce the working principle of each part from the selection of sensors, the amplifying circuit, the design of the filter circuit, and the analog-to-digital conversion.

$$P = UI\cos\Phi = \frac{U^2\cos\Phi}{\sqrt{R_s^2 + \frac{1}{(\omega C_s)^2}}} = \frac{U^2\omega C_s\cos\Phi}{\sqrt{1+(\omega R_s C_s)^2}} = \frac{U^2\omega C_s\sin\delta}{\sqrt{1+(\omega R_s C_s)^2}}$$

When tgδ is very small:

$$\text{tg}\ \delta = \frac{U_R}{U_C} = \frac{IR_s}{I\frac{1}{\omega C_s}} = \omega R_s C_s$$

So:

$$P = \frac{U^2\omega C_s \text{tg}\ \delta}{\sqrt{1+\text{tg}^2\ \delta}}$$

$$P = Ui\cos\Phi = UI_R = UI_C\text{tg}\ \delta = U^2\omega C_P\text{tg}\ \delta$$

$$\text{tg}\ \delta = \frac{I_R}{I_C} = \frac{R_P}{U\omega C_P} = \frac{1}{\omega R_P C_P}$$

A medium or a device, whether represented by a series circuit or by a parallel circuit, flowing the same voltage.

The flow should be the same, i.e. the total impedance represented by the series loop and the total impedance represented by the parallel loop should be equal. At the same time, under the same conditions

The loss and tgδ should also be equal. Therefore:

$$C_p = \frac{C_s}{1+(\omega R_s C_s)^2} = \frac{C_s}{1+\text{tg}^2\ \delta}$$

$$R_P = R_s\left[1+\frac{1}{(\omega R_s C_s)^2}\right] = R_s\left(1+\frac{1}{\text{tg}^2\ \delta}\right)$$

Since the monitor uses sensors, and the accuracy of the measurement results largely depends on the sensor, in view of the importance of the sensor, an overview of the sensor principle, selection, etc. is made. The sensor is a device that converts the information of various physical quantities reflecting the state of the device into the same or other kinds of output according to a certain law, and is the primary link for realizing online monitoring, which directly affects the development of the monitoring technology. The accuracy of the monitoring system has a lot to do with the accuracy of the sensor. If the sensor's monitoring parameters cannot be accurately and reliably measured, no matter how accurate and reliable the hardware circuit design is, how accurate the software analysis is , the measured data is inaccurate or even useless, so it is necessary to accurately perform online monitoring and select The sensor is the primary condition. When the insulation monitors the tgδ value of the high-voltage equipment online, the value of the dielectric loss current is generally small, and the minimum level of micro-ampere is required to accurately obtain the signal. At the same time, the insulation online monitoring system and the monitored system are required to have no electrical connection to ensure the safety of the monitoring system and monitoring personnel will not affect the normal operation of the monitored system due to the failure of the monitoring system. This puts higher demands on the sensor used to acquire the signal. For the sensor used, it should be considered from several aspects: 1, In order to accurately measure the parameters, and take into account the convenience of monitoring system installation, commissioning and normal operation, the sensor installation position must be properly selected; 2, According to the power system requirements, technically based Different devices and different measurements are selected to select reasonable sensors, and their peripheral circuits are reasonably designed. 3, From an economic point of view, the sensor requires good performance, economical and durable, convenient for commissioning and maintenance. In view of the importance of the sensor, the following technical requirements are imposed on the sensor for the specific parameters to be monitored. 1, It can detect the feature quantity reflecting the state of the device, and has good static characteristics (including linearity, sensitivity, resolution, accuracy, stability and hysteresis) and dynamic characteristics (including step response, sinusoidal response and dynamic response).. 2, No impact or weakness on the device under test, the energy absorbed by the system to be tested is small, and the subsequent units can be well matched. 3,. Good reliability and long life.

Since the voltage and current signals are taken from the grounding line of the transformer's last screen, the signal is very weak after the sensor (generally mA level), which contains a large number of common mode interference such as power frequency, static electricity and coupling, and must be accurately amplified by a certain multiple. It can meet the signal requirements of the measurement system. The preamplifier circuit is designed to have a high input impedance to reduce the effects of on-resistance. Usually to improve the accuracy of the amplifier circuit, the integrated amplifier is generally used to design the preamplifier. The integrated op amp has a high open-loop amplification factor (generally up to 410) and a high degree of integration. When it forms a negative feedback amplifying circuit, its voltage amplification factor depends only on the ratio of the external resistor, independent of the op amp parameters. Therefore, the preamplifier consisting of the integrated op amp as the main component is characterized by high gain and common mode rejection ratio, while the offset voltage and offset current, temperature drift, and noise are small, which is a versatility and strong amplifier. Since the amplifier circuit composed of the integrated operational amplifier has a high common mode rejection ratio and input impedance, low noise and low offset voltage, the resistance in the circuit should be carefully selected so that the circuit satisfies the symmetry requirement, and the resistance value due to process variations, it is difficult to select. If the circuit can't reach symmetry, it will have a great impact on the performance of the circuit. This is not only difficult to debug, but also inconvenient to manufacture. To this end, this article directly uses an integrated amplifier to design the preamplifier system. Among the many integrated amplifiers, we chose OP07 high-precision amplifier with high input impedance, low offset voltage, low noise, and large open-loop amplification. This was according to the performance and usage of the amplifier.

5. Software design of insulation online monitoring system

The purpose of online monitoring is to obtain a true description of the operation of the equipment by analyzing and processing various raw data reflecting the condition of the operation of the electrical equipment. Therefore, online monitoring software design is particularly important. The hardware circuit part of the online monitoring is used to complete various functions such as voltage and current signal conversion, transmission amplification, filtering, and sampling/holding. After the original data is obtained, it has not been analyzed, calculated and displayed. It is not possible to visually describe the operating state of the power equipment. The software must be used to analyze and process the data in detail, and complete the functions of data storage, result analysis and fault diagnosis. System operators provide an intuitive, simple human-machine interface.

After power-on, the monitor begins to execute the application software that is cured at the EPROM address 0000H. Through the assignment statement command: (1) Set AT8051 internal memory 5FH-7FH is the stack area for calling subroutine and field protection when responding to interrupt. (2) Set the special function registers TMOD and TCON, and the state of the interrupt control enable register IE to determine the control and operation of the timer.

6. Anti-interference problem of insulated online monitoring system

Differential mode interference refers to the interference voltage that acts on the instrument after being superimposed in series with the effective signal, mainly from the space electromagnetic field generated by the high-voltage line and the power line laid parallel to the signal line. Therefore, differential mode interference is usually caused by electromagnetic coupling. The source of the interference is electromagnetically and electrostatically coupled to the signal line to an induced voltage of up to millivolts. In addition to the signal line, the inherent drift, ripple and noise of the signal source, as well as the poor shielding and voltage-stabilizing filtering effect of the power supply introduce differential mode interference. Common mode interference refers to the interference voltage common to the two inputs of the input channel. Because there is often a certain potential difference VC between the references ground point of the signal under test and the reference ground point of the instrument. The amplitude of this interference is generally high, depending on the environmental conditions in which the interference occurs in the field and the grounding of the instrument. Common mode interference is more common, and the main hazard is the insulation damage of the instrument and the safety of the operator. Common mode interference can also be converted to differential mode interference when the common mode voltages on the two signal lines are not equal, or when the resistance of the two signal lines to ground is unbalanced. Interference caused by capacitive coupling is generally common mode interference. Suppressing common mode interference is related to whether the device can be used in the field.

Table 1. The test result compartment

Range	Standard Value	Dielectric loss Test Value	Capacitance Test Value
0	0.313	0.53	979
1	0.426	0.59	973
2	0.541	0.73	973
3	0.770	0.97	972
4	0.887	1.10	972
5	1.111	1.33	973
6	1.384	1.62	967

There are many causes of interference, such as lightning, substation switch operation, corona, radio emissions, and so on. In short, all emitters of electromagnetic energy may become sources of interference, and those who receive this energy and affect their own performance become disturbed, such as monitoring devices placed at the substation site. Because the insulated in-line device is placed directly at the high-voltage equipment, and the electromagnetic environment there is extremely harsh, a simple analysis of the electromagnetic interference problem is made here. There are two types of ways in which electromagnetic interference propagates from the source to the affected equipment: radiation and conductance. Radiated interference is transmitted to the affected equipment through space; conducted interference is transmitted to the affected equipment through the circuit. When the distance S between the interference source and the monitoring device is large relative to the wavelength λ of the interference signal (i.e., $S>\lambda/(2\pi)$), the nature of the interference source appears as a source of radiation interference, and the nature of the field is expressed as radiation electromagnetic field. The radiated electromagnetic field belongs to the radiation field. This kind of interference is usually the interference with high frequency; when the distance S between the interference source and the monitoring device is small which was relative to the wavelength λ of the interference signal (i.e. $S<\lambda/(2\pi)$), The nature of the interference source is the source of conducted interference, and the nature of the field is mainly represented by the electric field or the main representation of the magnetic field. This interference frequency is generally low.

In the case where two circuits have a common wire (such as a power line, an input/output signal line, a ground line, etc.), the current flows of the two circuits pass through a common impedance, so that the current in one circuit is at the common impedance. The pressure drop constitutes interference with another loop. Zg in the figure below is the common impedance of loop 1 and loop 2.

There are always stray capacitances between different conductors in the medium. When two circuits are at different potentials, the charge of one circuit will be induced to the other. Generally, the circuit with large ground impedance is easily interfered by the electromagnetic field, so the ground resistance is reduced, the stray capacitance between the two circuits is reduced, and the interference caused by the capacitive coupling can be reduced; The higher the frequency of the interference source, the easier it is. Other loops create capacitive coupling. There is a mutual inductance between the two circuits, and a change in current in one circuit can affect the other circuit through the flux linkage. The induced potential is proportional to the angular frequency of the interferer. It can be seen that the higher the frequency of the interference source, the greater the mutual inductance between the two circuits, and the greater the interference caused by the inductive coupling.

To suppress the interference source is to reduce the du/dt, di/dt of the interference source as much as possible. This is the most important and most important principle in anti-jamming design in PCBs, and it often has a multiplier effect. Reducing the du/dt of the interference source is mainly achieved by connecting capacitors across the interference source. Reducing the di/dt of the interferer is achieved by placing the inductor or resistor in series with the source loop and adding a freewheeling diode. The main measure is to connect a high frequency capacitor of 0.01ìF~0.1ìF to each IC on the circuit board to reduce the impact of the IC on the power supply. Avoid 90-degree fold lines when wiring will reduce high-frequency noise emissions.

Pay attention to the crystal wiring. The crystal oscillator and the MCU pins are as close as possible, and the clock region is isolated by the ground wire. The crystal oscillator case is grounded and fixed. The board is reasonably partitioned, such as strong and weak signals, digital and analog signals. Keep interference sources (such as motors, relays) away from sensitive components (such as microcontrollers) as much as possible. Use a ground wire to isolate the digital zone from the analog zone and finally connect it to the power ground at one point.

Improving the anti-interference performance of sensitive devices refers to the method of minimizing the pickup of interference noise and recovering from an abnormal state as soon as possible from the sensitive device side. The measures to improve the anti-jamming performance of sensitive devices are as follows: Minimize the loop area during wiring to reduce the induced noise. When wiring, try to increase the power supply and ground wires. In addition to reducing the voltage drop, it is more important to reduce the coupling noise. For the idle I/O port of the MCU, it is grounded or connected to the power supply. On the idle side of other ICs, ground or power is connected without changing the system logic. The system uses the watchdog circuit MAX813L, which can greatly improve the anti-jamming performance of the entire circuit.

7. Conclusion

Through an on-line monitoring technology for parameters such as dielectric loss, leakage current, equivalent capacitance, and operating voltage of power equipment, an on-line monitoring system based on monitoring leakage current and dielectric loss angle of high-voltage equipment has been developed. Find hidden dangers in time, reduce power outages, and improve power supply reliability.

Acknowledgement

The authors acknowledge the financial support by the Sichuan Provincial Department of Education (Project No. 14ZB0074), the National Natural Science Foundation of China (Project No. 41604116) and the Pilot projects for innovation and entrepreneurship education reform of CDUT (Project No. YJ2017-JD002).

References

[1] Nie Peng, Zhao Xuezeng, Zhang Yan, et al. Development of On-line Monitoring System for High Voltage Electrical Equipment Insulation [J]. Journal of Northeast Electric Power University, 2000(1): 49-52.

[2] Zhan Xiaomeng, Zhao Rui, Li Qijun, et al. Application of on-site calibration technology in on-line monitoring system for high-voltage electrical equipment insulation [J]. Science and Technology Innovation, 2014(26): 106-106.

[3] Liao Xinzheng, Zhao Guojie, Zhang Yuanyuan, et al. Development of on-line monitoring system for high-voltage electrical equipment insulation based on DSP [J]. Automation, 2015(12):98-99+101.
[4] BAI Wei, ZENG Cheng-bi, WANG Tao. Research on On-line Monitoring of Insulation of High Voltage Electrical Equipment [J]. China Journal of Test and Measurement, 2006, 32(1): 64-66.
[5] Lü Zecheng, Deng Yurong. Distributed On-line Monitoring System for Distributed High Voltage Electrical Equipment Based on Fiber Synchronization and Transmission [J]. China Southern Power Grid Technology, 2012, 6(Z2): 2-5.
[6] Zhang Ying, Li Junwei. Development and Application of On-line Monitoring System for Insulation Status of SF6 Electrical Equipment [J]. Industrial Safety and Environmental Protection, 2017(5).

Journal of Physics: Conference Series **1423** (2019) 012023 doi:10.1088/1742-6596/1423/1/012023

Research on brand value improvement strategies of fresh products based on cluster analysis-taking cold fresh meat as an example

Yanghao Ye, Qing Wang*, Shengwen Liu, Yajuan Huang

Changsha University of Science & Technology, Changsha, China

Corresponding author and e-mail: Qing Wang, simon.wq@qq.com

Abstract. Aiming at the strategies to improve the brand value of fresh products, this paper uses the field research data, official data, public information data in the industry, etc. in the cold fresh meat market as the data basis of the research, and uses cluster analysis to divide the cold fresh meat consumer groups into potential group and main force group, conservative group and follow-suit group. And it uses SPSS and mathematical modeling to analyze the characteristic portraits of the four categories of people, and gives strategies to improve the brand value of fresh products through correlation analysis.

1. Introduction

China is the world's largest consumer of pork, most of which is hot fresh meat, while cold fresh meat is a minority. Traditional hot fresh meat is usually marketed immediately after slaughter. The meat temperature after slaughter is as high as 40° C, which is in the period of stiffness of the meat without undergoing the change of sufficient maturity. The meat is tough and relatively rough in taste. At the same time, the high temperature of hot fresh meat will breed lots of microorganisms, easy to deteriorate the meat and produce harmful substances. Cold fresh meat is quickly cooled and acid-treated after swine slaughtering. The pork falls to 0-7° C within 24 hours and undergoes the process from stiffness after slaughtering to the maturity of the meat, which improves the meat quality. In addition, the production, transportation, and storage of cold fresh meat are all under low temperature conditions, which can effectively inhibit the growth of microorganisms [1].

At present, in the fresh meat market in the northern region, hot fresh meat still accounts for more than 80% of the market share. Consumers have low awareness of cold fresh meat, and market expansion is difficult. However, the cold fresh meat in developed countries accounts for over 90% of the market share. Therefore, this paper will carry out investigation and research on the brand value of cold fresh meat in combination with the new media environment to provide a certain reference for the return of the cold fresh meat market to the standardized, stable and healthy development environment.

2. Research Principles and Materials

2.1. Overall design

The investigation and research in this paper takes the national pork purchasing decision-makers as a whole, and investigates the people who have had pork consumption behaviors in the past month. The consumer group is mainly people aged from 21 to 65.

The investigation was conducted in two stages. The first stage was the pre-investigation period, which lasted five days. Cold fresh meat wholesale markets, large supermarkets in the city center and rural communities were treated as the sampling population for random sampling, and a total of 188 questionnaires were collected. 149 valid questionnaires were selected, and SPSS and mathematical modeling were used to analyze and form the final results. The second stage was the formal investigation period that took 15 days. Main consumers of cold fresh meat from four provinces were chosen as the sampling population for random sampling, and a total of 913 questionnaires were selected. After selection, 848 valid questionnaires were obtained. Finally, we entered the information of 848 valid questionnaires into the software of SPSS for subsequent analysis.

2.2. Concept and method selection

Cluster analysis is to classify according to the characteristics of individuals, and it aims to make individuals within the same category have a higher similarity, while different categories have greater differences, so as to effectively analyze according to the characteristics of different categories.

According to the attribute of the title, we adopted the hierarchical cluster method, which treats the cases participating in the cluster as one category, and then gradually merges them according to the distance or similarity between the two categories until the final case is merged into one large category.

2.3. Data sources

2.3.1 First-hand research data

- Collect data on the awareness of urban residents and promotion and suggestions on cold fresh meat in various places through questionnaires.
- Interviews relevant experts, manufacturers, distributors, and business (sales) personnel in the industry to obtain the latest first-hand market information.

2.3.2 Second-hand research data

- Official statistics: Statistics and information of government departments and official institutions such as the National Bureau of Statistics, the National Development and Reform Commission, the Development Research Center of the State Council, industry associations, the General Administration of Customs, and industrial and commercial taxation bureaus.
- Public information in the industry: Quarterly reports and annual reports of key enterprises in the cold fresh meat industry and upstream and downstream enterprises; articles, comments, opinions, etc. of the mainstream media; and opinions delivered by senior experts in the industry.
- Literature retrieval: Literature of various Chinese and English journal databases, libraries, research institutes, and colleges and universities.

3. Cluster Analysis Experiments and Discussion

After sorting out and analyzing the data, the cluster analysis was carried out to 500 consumers, who were classified as groups with different characteristics. And four factors were used to describe and name different consumers to more intuitively see their types and characteristic portraits.

3.1 Hierarchical cluster

The 500 respondents were firstly treated as a category, then the distance between categories was specified. The pair with the smallest distance was merged into a new category, and the distance between the new category and other categories were calculated, and the two categories with the closest distance were merged, and each time one category was reduced, and all the variables were divided into four categories. Through frequency statistics, we found that the number of the first category is the largest, and followed by the second and the third categories, and the number of the fourth category was the smallest. Please refer to Table 1 for the specific frequency.

Table 1. Final cluster center

	Cluster			
	1	2	3	4
Province	3.20	1.56	2.57	2.61
Family information	1.30	.71	1.22	1.07
Age	3.5	3	3	3
Current family monthly income (yuan)	2	4	3	3
Education background	3	4	3	4
Frequency of family pork purchasing	4	3	3	3
Total score	1	3	1	3
After you buy the cold fresh meat, do you feel that the quality and taste are better than those of the traditional pork?	3	2	-3	2
The production of cold fresh meat needs to go through a very complicated process. Compared with traditional pork, how much higher the price of cold fresh meat can you accept	2	2	2	2
How many kilograms of pork you buy in your daily life (1 kg = 500 g)	11	14	2	2
After learning about the cold fresh meat, are you willing to buy it?	1	-3	1	-3
Cold fresh meat, hot fresh meat, and forzen meat, which one do you think more healthy?	1	-3	2	-3
Are you willing to promote cold fresh meat to the people around you	1	1	1	1
Do you think that cold fresh meat will replace the traditional pork in recent 5 years?	2	1	2	2

3.2 *Analysis of clustering results*

After clustering, the four categories of people are named and characterized. The formula of the consumption intensity is obtained by the principal component analysis method (please refer to Table 2), and the scores of factors are calculated according to the factor score calculation formula table, and the consumers are named according to the scores of the four types of factors.

Table 2. Factor score calculation formula table

Consumption attitude	a_1F_1
Consumption capacity	a_2F_2
Consumer preference	a_3F_3
Consumption level	a_4+F_4

The four categories of consumers are described and the people are named according to the factor score table of each category of people. Please refer to Table 3.

Table 3. Division of people

People division	Potential group	Main force group	Conservative group	Follow-suit group
Beijing	4%	28%	31%	31%
Shandong	13%	26%	17%	28%
Hunan	22%	23%	25%	21%
Gansu	60%	23%	27%	20%

3.2.1 The first category of people (potential group):

Table 4. Wilcoxon Signed Rank Test table for the potential group

Original hypothesis	Test	Significance	Decision
The difference of the median between the understanding way and the way to try to understand is 0	Wicoxon symbol test of relevant samples	0.0036	Refusal of the original hypothesis

The potential group consists of two parts: the general potential group and the competitors. The so-called potential group refers to the people who have the intention to buy but have not become members of any similar products or organizations, as well as people who once were the population of certain organizations but when they make purchase decisions, they have more casual recognition of the brand (namely, the organization); the so-called competitor group is relative to the group of the enterprise, that is, the group owned by the competitors; this group of people can be intermediate people (such as agents, wholesalers, retailers), and they can also be the final consumers, that is to say, they are what we call the group in the general sense.

In this paper, the consumption behavior of this category of people is relatively low with relatively small income, who are aged at 30 - 50 in developing cities. They know little about cold fresh meat, but they are optimistic about it. We named this category of people the potential group, that is, this group of people do not know much about the cold fresh meat, and they seldom buy products of cold fresh meat. However, they have an optimistic attitude to the future development of it. This group of people may have not consumed the cold fresh meat due to some factors, but there is a great development room among them.

3.2.2 The second category of people (main force group):

Table 5. Wilcoxon Signed Rank Test table for the main force group

Original hypothesis	Test	Significance	Decision
The difference of the median between the understanding way and the way to try to understand is 0	Wicoxon symbol test of relevant samples	0.000	Refusal of the original hypothesis

The consumption behavior of this category of people is the highest among the four categories of groups. They are at the age of 30 - 40, and they are usually office workers living in cities, or people who run their own businesses in cities, with greater demand and education. And they know well about the cold fresh meat, and they have an obvious preference for cold fresh meat. At the same time, this group of people are very optimistic about the future development of the cold fresh meat. We named this category of people the main force group, that is, based on their clear understanding of the cold fresh meat, a group of people who are willing to consume products of the cold fresh meat category due to their preference for the cold fresh meat or the attraction of the cold fresh meat to them, and who believe that the cold fresh meat will have a better development.

This category of people is very important for enterprises' achievement of their sales goals, and they account for or will account for a large proportion of sales revenue. If the cold fresh meat market loses this consumer group, the development of the market will be seriously affected, and it will be difficult for the sales performance to be recovered in a short term. At the same time, it is difficult to quickly establish other sales channels. That is to say, the cold fresh meat market has a certain dependency on this consumer group; there is a stable cooperation relationship between them, and they have a great potential for the future business of enterprises. Their developments are in line with the development goals of enterprises, and then the time is ripe, enterprises can carry out back-ward integrated strategy to form a strategic alliance relationship, and use the advantages of the main force group, which are beneficial to the growth of enterprises.

3.2.3 The third category of people (conservative group):

Table 6. Wilcoxon Signed Rank Test table for the conservative group

Original hypothesis	Test	Significance	Decision
The difference of the median between the understanding way and the way to try to understand is 0	Wicoxon symbol test of relevant samples	0.022	Refusal of the original hypothesis

The consumption behavior of this category of people is the lowest among the four categories, with medium income and lower education. They have not purchased cold fresh meat, and know little about cold fresh meat products. They have no obvious requirements and preferences for cold fresh meat, and they are not optimistic about the future development of it. We named this category of people the conservative group, namely, they are unwilling to know about new things, and they do not understand the products. And, limited by the traditional food concept, they do not think that the cold fresh meat has development room.

The consumption characteristics of this category of people are that they are unwilling to undertake any risks. They are sensitive to price and critical to products. However, they believe in the recommendation of experts and good friends. For this category of consumers, enterprises must provide mature products that are thoroughly tempered, and become recognized brands in the society. Conservative consumers refer to consumers with characteristics of pursuing reality, low price, convenience, savings, and popularity. Seeking reality: their main purpose is to pursue the use value of the commodity; seeking low price: they pay great attention to the price, and they pursue good products with low price under the circumstances the needs are basically met; seeking convenience: they aim at convenience; seek savings: this is caused by various psychologies, such as saving money for purchase, and endowment insurance; seeking popularity: in fact, this is a psychology of fear of losses.

3.2.4 The fourth category of people (follow-suit group):

Table 7. Wilcoxon Signed Rank Test table for the follow-suit group

Original hypothesis	Test	Significance	Decision
The difference of the median between the understanding way and the way to try to understand is 0	Wicoxon symbol test of relevant samples	0.048	Refusal of the original hypothesis

The consumption behavior of this category of people is the highest among the four categories of groups. However, they do not know much about the cold fresh meat, and their demand for pork is less. At the same time, they are not optimistic about the future development of the cold fresh meat. We named this category of people the follow-suit group. In fact, this category of people do not know about the cold fresh meat and they are not optimistic about its future. Although their consumption on cold fresh meat is high, this is not out of their love to it, but may be due to its popularity on the Internet or many recommendations, and they are just following the trend.

4. Conclusion

4.1 Marketing strategy for the potential group

4.1.1 Keeping real customers

This is the most basic way to compete for potential customers, and enterprises must pay great attention to it. To keep real customers, it is a must to make them satisfied and cultivate their loyalty. The loyalty of customers not only can make them resist the temptation of the promotion of other brands and purchase again or purchase products or services of the same enterprise in large amounts, but also can make them actively recommend products or services of the enterprise to relatives and friends and people around them. Cultivating customer loyalty is the highest pursuit of an enterprise [2] or organization's customer satisfaction quality strategy.

4.1.2 Fighting for general potential customers

The number of the people in the potential group that choose the publicity means of (mobile phone tweets) is the most. Therefore, this marketing method is currently effective. Through the above-mentioned Wicoxon symbol test of relevant samples, consumers believe that there are differences and improvements shall be made. Through the above picture analysis, the combination of (mobile phone tweets and government promotion and publicity) is the most effective for the potential group [6]. In daily life, it is not difficult for us to find that most of people in the potential group of the cold fresh meat are the middle aged. Although these people do not know much about the cold fresh meat, they pay much attention to mobile phone network and new policies and regulations and so on. Therefore, the adoption of mobile phone tweets can strengthen their understanding of the information of the cold fresh meat. And the publicity of the government can make the potential consumers trust it and want to have a try. In the case that the potential group only accounted for 4% in Beijing, we can focus on the use of mobile phone tweets supplemented by the government publicity. While for Gansu where the potential group accounted for 60%, the government publicity shall be mainly carried out to achieve the goal of widespread popularization in a short time.

4.1.3 Key links

- Establishing brands, creating brands, and winning the market with quality and reputation.
- Grasping services, shaping images, and pleasing customers with sincerity and integrity.
- Strengthening communication, expanding publicity, and timely and effectively disseminating enterprise information to every customer.

4.2 Marketing strategy for the main force group

The number of the people in the main force group that choose the publicity means of (indicate cold fresh meat at the brand trademarks of pork) is the most. Therefore, this marketing method is currently effective. Through the above-mentioned Wicoxon symbol test of relevant samples, consumers believe that there are differences and improvements shall be made. Through the above picture analysis, the combination of (brand trademarks of pork and newspaper and journal publicity) is the most effective for the main force group. In daily life, the main force group of the cold fresh meat has already had enough understanding of it. Therefore, in terms of promotion and publicity, we can appropriately weaken the role of the network. Under the premise of sufficient understanding of the cold fresh meat, what we need to do is to enable them to buy the genuine cold fresh meat at ease, and marking is no doubt the best way to make them to buy with confidence. The cold fresh meat pursues nutrition and health more. For the main force group of cold fresh meat, what they pursue is definitely a healthy and upward lifestyle. As a kind of paper media, newspapers and magazines will undoubtedly produce more positive psychological hints for these people.

4.3 Marketing strategy for the conservative group

The number of the people in the conservative group that choose the publicity means of (pork varieties and popularity of the sales person in the market) is the most. Therefore, this marketing method is effective. Through the above-mentioned Wicoxon symbol test of relevant samples, consumers believe that there are differences and improvements shall be made. Through the above picture analysis, the combination of (advertising on TV and marketing) is the most effective for the conservative group. The conservative group are mostly of the middle-aged and the aged. The age characteristics of this category of people determine their suspicious psychology of the virtual network. The popularity of the pork brand by the sales persons can not only make use of the more leisure time of this category of people who are willing to listen to you, but also can make it easier for them to accept. There is also the phenomenon that this category of people often watch lots of TV programs, and TV advertising is also an important way for them to obtain information, just like that most of the health care products for the elderly choose TV advertising.

4.4 Sales strategy for the follow-suit group

The number of the people in the follow-suit group that choose the publicity means of (mobile phone tweets) is the most. Therefore, this marketing method is effective. Through the above-mentioned Wicoxon symbol test of relevant samples, consumers believe that there are differences and improvements shall be made. Through the above picture analysis, the combination of (mobile phone tweets, marketing and government promotion and publicity) is the most effective for the follow-suit group. Although the follow-suit group do not know much about the cold fresh meat, they are the consumers of it. They just follow the footsteps of some people to consume, which indicates that they rely on Internet more. Popularization of pork varieties through mobile phone tweets, and sales persons in the market can make re-recognize the taste, nutrition, etc. of cold fresh meat under the circumstance that they have consumed it. This can be compared to a kind of "hindsight", achieving the effect of synchronization of consumers' feelings and information. For those who have not yet consumed, it is undoubtedly a driving force for their consumption, which makes them feel that buying the cold fresh meat is a trend. And the government publicity will have an effect on those follow-suit people.

Acknowledgement

Hunan Provincial Department of Education Scientific Research Project, "The Internet +" Background of Regional Agricultural Public Brand Building and New Media Promotion Research. (18C0245)

Reference

[1] Junhua HU. The popularity of cold fresh meat in China is less than one-third of that in Europe and America Shuanghui has laid out for 20 years to wait for the breakout [EB/O L]. http://news.hbtv.com.cn/p/156822.html,2016-05-29.
[2] Jianfei NAN, Zhijian XIONG. A Comparative Analysis of Potential Customers and Real Customers [J]. Journal of Xi'an Shiyou University: Social Sciences Version, 2002, 1:14-18.
[3] Zhenghai WU, Zhichao ZHOU. Micro-marketing Environment Analysis and Market Positioning of High-end Brand Cold Fresh Meat - - Taking Besun Cold Fresh Meat in Xi'an Market as an Example [J]. New West (Theoretical Version), 2014 (08): 21-22.
[4] Xiangquan XIE. Formation Mechanism of Agricultural Regional Brand and Its Value Enhancing Strategies [J]. Journal of Xinyang Agriculture and Forestry University, 2019 (03): 54-55+58.
[5] Linlin BAO. The Brand Communication Strategy in New Media Environment [J]. Media Forum, 2019, 2 (14): 168-169.

[6] Alan Dutka. How to change potential customers to real ones [J]. Innovation Science and Technology, 2006 (12): 44-45.

Journal of Physics: Conference Series **1423** (2019) 012024 doi:10.1088/1742-6596/1423/1/012024

Analysis and forecast for growth of lottery sales based on optimal control model

Zhang Liu[1,2], Canhua Li[1], Jian Wu[1]

[1] Computer and Information Engineering College, Jiangxi Agricultural University, Nanchang 330045, China
[2] School of Mathematics and Statistics, Wuhan University, Wuhan 430072, China

Corresponding author and e-mail: Zhang Liu, liuzhang1006@163.com.

Abstract. Based on the optimal control theory, this paper analyzes the relationship among the growth of lottery sales and the accumulation of capital and the rate of accumulation. By the introduction of Hamilton function, we established a mathematical model and calculate the satisfied equation. Using a simple model, we realized data fitting for the corresponding formula by Matlab through dealing with the sales and the funds of prize pool of the super lotto from 2014 to 2018. We made some forecasts according to the results.

1. Introduction

The issuance of lottery tickets is now a common form of fund-raising. It is usually based on legal funds and fairness, and redistributes social idle funds. Since the recovery of the lottery industry in China, it has been franchised by the China Welfare Lottery Distribution Center and the China Sports Lottery Distribution Center.。 According to the data of the Ministry of Finance, the cumulative sales of lottery tickets in China for the whole year of 2018 was 511.472 billion yuan, a year-on-year increase of 20%, and the lottery volume increased. Fucai and Sports Lottery have played a big role in the development of social welfare and sports, and have also made great contributions to China's philanthropy. With the continuous development of the lottery business, the mathematics problems contained in it have gradually attracted the interest of many experts and scholars. For example, in 2002, the National College Student Modeling Competition was based on the question of the probability of winning the lottery in the "mathematics in lottery". Meirui et al. (2010) used ruin probability to analyze the risk problem in lottery; Zou and Yin (2006) studied the probability of consecutive numbers in computer welfare lottery; Hu and Qian (2007) discussed the mathematical analysis and investment strategy of lottery tickets; Zhang and Zeng (2006) studied the optimal measurement and its application in the consumer behavior of lottery;Xue et al. (2006) used the "Xinjiang Style" computer welfare lottery to conduct an in-depth study on the mathematical analysis and application of lottery tickets; Li and Li (2002) investigated the composition and purchase behavior of Beijing sports lottery buyers, and studied the favorable and unfavorable factors for the purchaser and made recommendations through the analysis of the results.

In recent years, there have been more research issues in the field of lottery. For example, Jia and Xie (2009) applied multiple regression analysis to analyze the correlation between the welfare lottery sales in China's various provinces and the five major categories of economic life's predictors; Wei and

I CLC number: O211.5, O211.6 Document code: A.

Lei (2013) analyzed the sales of sports lottery and the spatial heterogeneity of its driving factors in China by using exploratory spatial data analysis (ESDA) method and geographic weighted regression (GWR) model through various index data that affect social development; Han (2015) used the panel data-based method to establish a dual logarithmic linear regression model and a general linear regression model to quantitatively analyze the factors affecting the overall lottery sales and Internet lottery sales in China. Zhang (2017) studied the factors affecting the sales of welfare lottery tickets in China under the capping system, and proposed some reasonable countermeasures; Liu (2017) explores the distribution strategy of Chinese sports lottery bonuses and how to help the lottery department design the optimal winning probability and winning prizes by constructing the general equilibrium model of the lottery market to promote the scientific development of the lottery industry; Li (2019) studied the interactive effects of China's lottery market and stock market from the perspective of behavioral finance, and combined with theoretical and empirical results, proposed policy recommendations for improving related fields. By now, the research focus on the lottery market is mainly focused on the sales plan of lottery tickets and the direct consumption behavior of lottery players. The analysis of the lottery market is a comprehensive multidisciplinary issue involving mathematics, economics, sociology and more. This paper starts from the current most popular and high-selling lottery ticket in the Chinese market----Super Lotto, and analyzes the relationship between the growth of lottery sales, accumulated funds and the cumulative rate. In addition, this paper also introduces the Hamilton function to establish a mathematical model, calculates the equation that it satisfies, and uses the simplified model to realize the data fitting of the equations by Matlab. Through the collection and consolidation of the sales volume and prize pool funds of the Super Lotto between 2014 and 2018, the future sales were analyzed and forecasted. Finally, based on the relevant results, Reasonable comments and suggestions was proposed for the healthy development of the Chinese lottery market.

2. General model

The sales amount of a certain lottery ticket at time t is recorded as $x(t)$, and the portion for accumulating funds (referring to the balance of the current prize pool funds) is recorded as $y(t)$, and the proportion of accumulated funds in the lottery sales is recorded as $u(t)=x(t)/y(t)$ (Accumulation rate), the optimization control model is established below. The purpose of modeling is to find the optimal accumulation rate $u(t)$, so that the total sales of the lottery market $x(t)$ will grow bigger and faster.

The growth rate $z(t)$ of the lottery sales revenue depends on the total income $x(t)$ and the cumulative rate $u(t)$ of the lottery sales at that time, so their relationship can be expressed as

$$z(t)=f\left(t,x(t),u(t)\right). \tag{1}$$

Considering that for a period of time T (such as one year or three years), to make $x(t)$ from the initial value $x(0)$ to the largest possible value $x(T)$, the initial condition is

$$x(0)=x_0, \tag{2}$$

$$\max x(t)=\{\max x(t):0\le t\le T\} \tag{3}$$

Then the problem of the study can be summarized as how to find the extreme value of (3) under the conditions of (1) and (2).

The above problem is equivalent to its dual problem: under the fixed endpoint condition $x(T)=x_1$, the time T used to make the total revenue of the lottery from x_0 to x_1 is the smallest, so

$$x(0)=x_0,\ x(T)=x_1, \tag{4}$$

$$G(u(t))=\int_0^T dt. \tag{5}$$

The method of finding the extremum problem by function can express the above problem as follows. To find the optimal cumulative rate under conditions (1) and (4) to make the minimum of G in (5).

The Lagrange multiplier method is used to transform conditional extreme value into the unconditional extreme value, and the Lagrange multiplier factor $\lambda(t)$ is introduced to obtain the function,

$$L(x(t),u(t))=\int_0^T\left[1+\lambda(t)\cdot\left(f(t,x(t),u(t))-z(t)\right)\right]dt. \tag{6}$$

Introducing the Hamilton function

$$H(t,x(t),u(t))=1+\lambda(t)\cdot f(t,x(t),u(t)), \tag{7}$$

so equation (6) can be written as

$$L(x(t),u(t))=\int_0^T\left(H-\lambda(t)\cdot z(t)\right)dt. \tag{8}$$

According to the necessary condition that the function (8) reaches the extreme value, the Euler equation is

$$\begin{cases}\dfrac{\partial(H-\lambda\cdot z)}{\partial x}-\dfrac{d}{dt}\left[\dfrac{\partial}{\partial z}(H-\lambda\cdot z)\right]=0,\\[2ex]\dfrac{\partial(H-\lambda\cdot z)}{\partial u}-\dfrac{d}{dt}\left[\dfrac{\partial}{\partial(u')}(H-\lambda\cdot z)\right]=0,\end{cases} \tag{9}$$

that is,

$$\begin{cases}\dfrac{\partial H}{\partial x}+\lambda'(t)=0,\\[2ex]\dfrac{\partial H}{\partial u}=0,\end{cases} \tag{10}$$

so the whole problem is equivalent to solving the following equivalent equations

$$\begin{cases}\lambda'(t)=-\dfrac{\partial H}{\partial x}=-\lambda(t)\cdot\dfrac{\partial f(t,x,u)}{\partial x},\\[2ex]\lambda(t)\cdot\dfrac{\partial f(t,x,u)}{\partial u}=0,\\[2ex]z(t)=x'(t)=f(t,x,u),\\x(0)=x_0,\ x(T)=x_1,\end{cases} \tag{11}$$

Solving the equations can find the optimal function $u(t)$ and the state curve $x(t)$.

3. Data and empirical analysis

3.1. Data collection and processing

This article takes the Super Lotto of China Sports Lottery as the research object. The sample taken by empirical modeling is the real-time sales data of Super Lotto from 2014 to 2018. The data comes from the Chinese lottery website. The lottery draw is for three issues a week (every Monday, Wednesday, and Saturday). Due to the different interval between each lottery, the single sales are affected by many factors. In order to accurately study the nature of the problem, we selected the one-month time as the smallest research unit from a macro perspective, and calculated the average value of multiple lottery sales in a month. After sorting and calculation, the data of the prize pool at the end of each month during the 2014-2018 period and the average sales of the monthly lottery are given. See Table 1 below. The trend chart of the change is shown in Figure 1-2.

Table 1. Average monthly sales of Super Lotto and the prize pool balance during the 2014-2018 period.

month	Prize pool fund balance(yuan)	Average sales (yuan)	month	Prize pool fund balance(yuan)	Average sales (yuan)
2014-01	324,041,671	127,895,072.46	2016-07	3,206,441,463	179,966,295.00
2014-02	325,669,431	128,793,210.11	2016-08	3,449,353,717	180,868,686.64
2014-03	293,099,517	132,254,991.71	2016-09	3,502,839,878	182,755,096.08
2014-04	427,750,617	129,852,745.38	2016-10	3,742,471,490	180,594,100.07
2014-05	576,508,734	156,968,476.69	2016-11	3,857,570,669	188,990,544.69
2014-06	729,191,341	175,685,479.15	2016-12	3,235,556,086	200,199,759.54
2014-07	929,910,085	184,106,520.77	2017-01	3,296,873,997	192,723,541.73
2014-08	595,465,040	175,753,372.62	2017-02	3,427,151,759	187,535,170.91
2014-09	759,165,112	165,263,532.46	2017-03	3,606,190,192	192,786,536.69
2014-10	980,248,197	184,053,199.69	2017-04	3,661,625,450	203,691,244.15
22014-11	1,251,566,141	214,223,099.92	2017-05	3,602,783,336	222,328,107.86
2014-12	1,135,404,076	209,299,295.64	2017-06	3,790,728,402	216,539,231.17
2015-01	922,796,421	194,949,233.77	2017-07	3,970,858,480	204,014,110.00
2015-02	1,052,353,522	182,887,707.56	2017-08	4,125,812,710	203,501,900.46
2015-03	1,230,022,010	158,525,391.85	2017-09	4,288,325,250	209,612,396.31
2015-04	1,201,607,522	162,773,880.92	2017-10	4,224,508,242	201,946,730.69
2015-05	1,275,356,607	180,416,114.77	2017-11	4,219,678,773	213,370,776.92
2015-06	1,369,791,223	173,538,394.23	2017-12	4,478,126,302	220,315,411.85
2015-07	1,470,621,295	155,740,439.00	2018-01	4,726,825,681	213,185,240.50
2015-08	1,449,994,950	162,825,559.50	2018-02	5,016,451,627	213,498,688.56
2015-09	1,512,320,825	163,098,424.62	2018-03	5,297,754,793	217,227,654.31
2015-10	1,707,307,787	165,095,955.31	2018-04	5,368,233,194	244,673,841.92
2015-11	1,964,244,295	171,323,855.00	2018-05	5,703,821,164	248,867,827.31
2015-12	2,174,681,996	176,055,021.38	2018-06	5,958,857,007	220,105,063.77
2016-01	2,308,385,677	181,598,760.08	2018-07	6,175,967,288	207,896,708.77
2016-02	2,423,340,376	179,281,063.40	2018-08	6,308,725,026	207,638,184.31
2016-03	2,586,646,951	180,529,878.46	2018-09	6,552,450,316	208,220,686.54
2016-04	2,815,398,495	203,681,497.23	2018-10	6,594,664,860	215,470,012.50
2016-05	3,132,059,059	208,380,693.69	2018-11	6,827,035,600	221,977,626.00
2016-06	3,255,540,118	183,587,770.31	2018-12	7,118,917,925	213,793,699.21

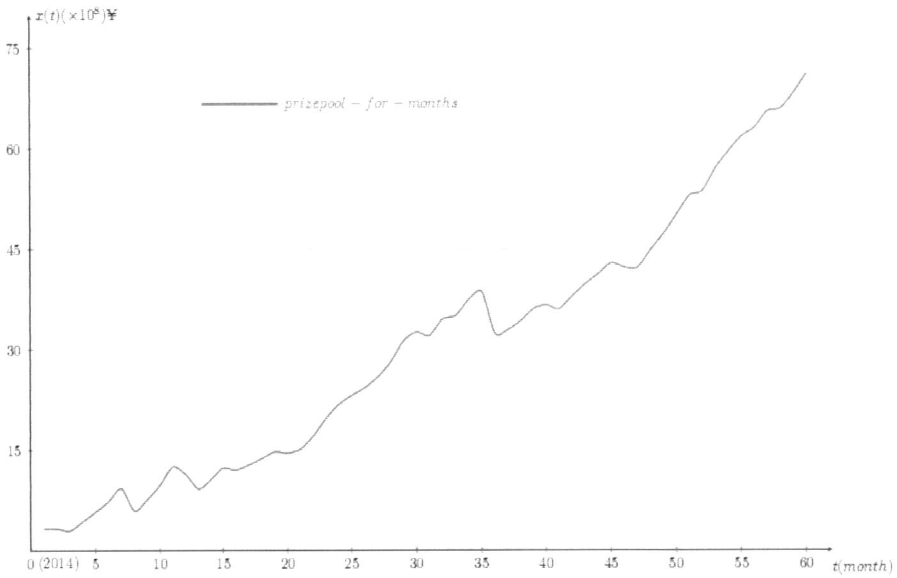

Figure 1. The balance of the lottery prizes at the end of each month from 2014 to 2018 (the horizontal axis is the month and the vertical axis is the month-end balance).

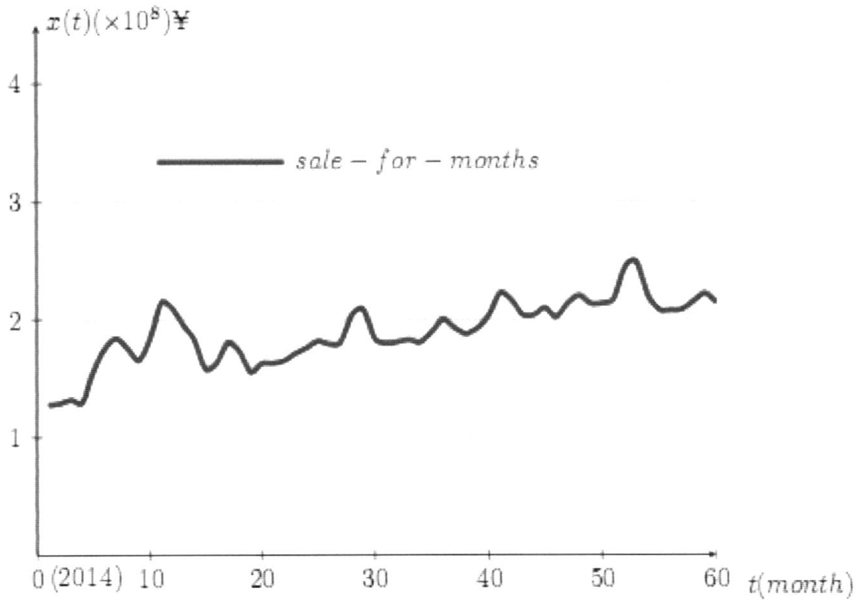

Figure 2. The average monthly sales volume from 2014 to 2018 (the horizontal axis is the month and the vertical axis is the monthly average sales).

3.2. Empirical simplified model

The following questions are discussed by a simplified model. In order to make the function $f\left(t, x(t), u(t)\right)$ discussed above have a concrete and concise expression, it is possible to set the sales growth of the lottery ticket to have such a rule: when the accumulation rate u is small, the relative growth rate z/x of sales increases with the increase of u. This is because the accumulation of funds in the prize pool promotes the consumption of the lottery. However, as u

becomes larger, the growth rate of z/x is slower and slower. When u increases to a certain extent, z/x becomes smaller. This is because the total amount of lottery consumption is limited, and the increase in the amount of input will make the accumulated prize pool less (the number of first prizes will increase), and this result is consistent with the actual average monthly sales (Figure 2). Therefore, the above model can be simplified to

$$z/x = u\left(r - su\right), \tag{12}$$

where r, s is a constant and x is the total monthly sales. z is the growth rate and u is the cumulative rate. The sales data of 2014-2018 is counted, and the values of r and s can be calculated by statistical methods.

The next step is to further simplify the equation to $y = -sx^2 + rx$. Using the known data, we can find the values of 60 sets of y and x, and then use Matlab to achieve least squares fitting. Then the running result is obtained: r=0.0163, s=-0.0149; combining (1), (4) can obtain $f = u\left(s - ru\right)x$, thus we obtain the equations

$$\begin{cases} \lambda'(t) = -\lambda(t)\cdot u\left(r - su\right), \\ \lambda(t)\cdot\left(r - 2su\right)x = 0, \\ z = u\left(r - su\right)x, \\ x(0) = x_0,\ x(T) = x_1. \end{cases} \tag{13}$$

Solve it we can get

$$\begin{cases} u(t) = r/2s, \\ x(t) = x_0 \exp\{r^2 t/4s\}, \\ T = 4s\left(\ln x_1 - \ln x_0\right)/r^2. \end{cases} \tag{14}$$

That is, when the accumulation rate is a fixed value $r/2s$, then the lottery sales revenue increases in the form of an approximate exponential function, and the shortest time is T.

3.3. Empirical Analysis: Prediction and Analogy

According to the conclusions in (12-14), combined with the data in Table 1, we can predict the sales of Super Lotto in the future. Using Matlab programming, we can predict the relevant value of monthly average sales for the first nine months of 2019, see Table 2. Note that the actual sales data for the first nine months of 2019 are also given in Table 2, we compare the two and present the relative errors in Table 2.

Table 2. Comparison of the monthly average sales forecast of the Super Lotto from January to September in 2019 and the true value.

month	Prize pool fund balance(yuan)	Forecast of average sales per month(yuan)	The true value of the average monthly sales per month (yuan)	Relative error
2019-01	7261573924	214865506.06	215615698.38	0.3%
2019-02	7260411038	216714259.02	231554865.11	6.4%
2019-03	6990724180	220735472.18	266684995.15	17.2%
2019-04	5385687606	231912314.46	272816429.85	14.9%
2019-05	4808403292	246667997.12	281250266.08	12.2%
2019-06	4372011525	263076639.05	283034572.00	7%
2019-07	3755102164	275946565.77	275733593.79	0.07%
2019-08	2965514601	278208715.43	278215424.67	0.02%
2019-09	2618860320	279558464.14	289671755.15	3.4%

It can be seen from the forecast results that the sales in the first three months of 2019 are basically in line with our forecast, especially the relative error of the forecast results in the first month and the next few months is basically within 3%. However, with the difference in prediction time and the influence of various random factors in reality, the relative error of prediction will be significantly larger in some months. Therefore, after the model is established, it is necessary to continuously replenish new data and update the values of r and s in the model to realize dynamic prediction of the model, thereby maintaining high prediction accuracy. On the other hand, we also forecast the sales of the Super Lotto for a long time in the future. We forecast the sales data of the Super Lotto in October-December 2019 and December of the next three years, see Table 3 below. In Figure 3, we present a comparison of projected sales and real sales for the Super Lotto during 2019. As can be seen from the figure, the sales forecast of the Super Lotto is increasing as a whole, and the increasing trend is not completely exponential growth, but according to the law of the expression (12-14); and the real monthly average sales are affected by a variety of random factors, which makes analysis errors between the predicted and actual values. However, in the long run, the forecast data of this method is basically consistent with the real situation, indicating that the forecasting effect of this method is relatively good, which can provide a reasonable decision basis for the benign operation and supervision of China's lottery business.

Table 3. Forecasts sales for October-December 2019 and the next three years.

Time	2019-10	2019-11	2019-12	2020-12	2021-12	2012-12
Forecasts	281663836.40	285321152.50	290718042.18	323843696.56	348386681.91	366567393.29

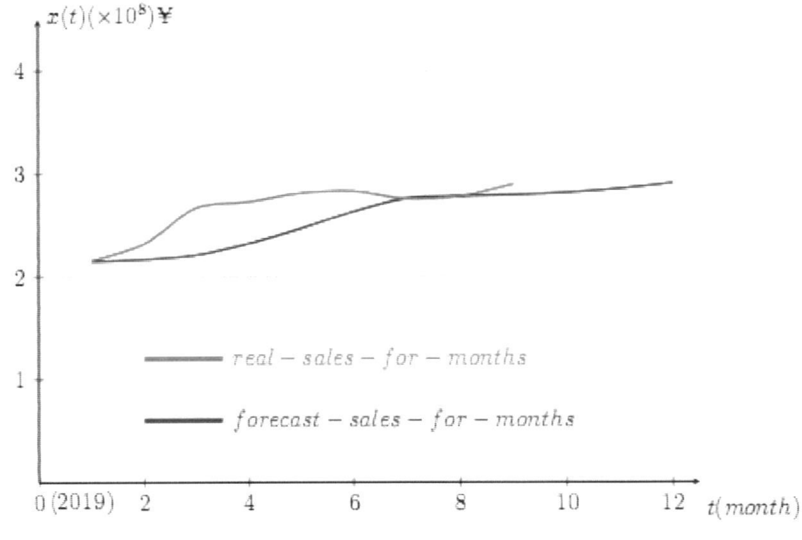

Figure 3. Comparison of projected sales and real sales for the Super Lotto during 2019.

4. Related conclusions analysis and recommendations

Recently, lottery has become the sixth largest industry in the world. The progress of the lottery business is not only related to the development of China's economy, but also closely related to sports, politics, environmental protection, charity, social welfare, etc. From the perspective of statistical mathematics, analyzing and forecasting the trend of China's lottery market is also a realistic issue that we are concerned about. Based on the above analysis, we put forward some opinions and suggestions on the sound development of the Chinese lottery market.

Reasonably adjust the amount of bonuses and the proportion of awards. It can be seen from the above that the growth rate of lottery sales (z/x) is closely related to the cumulative rate (u) of the lottery prize pool. That is, how much funds are left in the prize pool during each lottery draw, as the prize pool funds enter the next lottery fund flow, and how much funds are returned to the lottery pockets through the award. It is necessary to properly stimulate the consumption of lottery players and reserve reserves for future capital turnover. This paper calculates that when the cumulative rate is a fixed value $r/2s$, that is, u=-57338, the growth rate of sales is the largest.

Appropriate increase in bonuses to stimulate consumption. Super Lotto has added an extra bonus every year from 2014 to 2018, respectively

From the period of 14052 to 14066 (2014-5-5 to 2014-6-9), the additional bonus amount is 150 million yuan.

From the period of 15051 to 15072 (2015-5-4 to 2015-6-24), the additional bonus amount is 360 million yuan.

From the period of 16042 to 16061 (2016-4-11 to 2016-5-28), the additional bonus amount is 500 million yuan.

From the period of 17044 to 17063 (2017-4-17 to 2017-6-3, the additional bonus amount is 500 million yuan.

From the period of 18041 to 18060 (2018-4-9 to 2018-5-26), the additional bonus amount is 600 million yuan.

Comparing the time interval of the above additional bonuses with the corresponding sales, it can be clearly seen that the sales during the extra bonus period is significantly higher than the usual sales, and the sales tend to increase slowly. In addition, the enthusiasm of the lottery has gradually increased, so the appropriate extra bonus is beneficial to the increase in lottery sales and the healthy development of the lottery market.

Contact the actual and adjust the strategy. Sorting the lottery sales data in recent years, we can find that the sales of lottery tickets are not only affected by the prize pool funds and the amount of lottery, but also interfere with many realities, such as newly issued laws and regulations, political patterns, festivals and holidays. Therefore, the sales strategy of lottery tickets cannot be uniform, but it is necessary to adjust strategies according to the actual situation to achieve greater growth in sales and better serve social welfare undertakings.

Acknowledgement
This work is supported by Jiangxi Science and Technology Plan Fund Project (GJJ20180201).

References

[1] MEI R., ZHANG C., ZHANG W.H. Analysis of Risk Issues in Lottery Using Ruin Probability. Journal of Hebei North University (Natural Science Edition). 2010,(1).
[2] Zou Z.M., Yin T.K. The Probability of Serial Numbers in Computer Welfare Lottery. Mathematical Statistics and Management. 2006, (4).
[3] Hu R.S., Qian G.H. Mathematical Analysis and Betting Strategies of Lottery Tickets. Journal of Changshu Institute of Technology. 2007, (10).
[4] Zhang D.M., Zeng Z.L. Optimization Measurement and Its Application in Comparative Study of Consumer Behavior in Lottery. Statistics and Decision Making. 2006, (15).
[5] Xue C.G, Dong X.Y., Zhang G.Y. Mathematical analysis of lottery tickets and its application statistics and decision making. 2006, (1).
[6] Li Zhi, Li Yong. Analysis and research on the current situation of computer sports lottery buyers in Beijing. Journal of Beijing Sport University 2002, (2).
[7] Jia C., Xie Z.J. Analysis of Factors Affecting China's Welfare Lottery Sales and Forecast Based on Residual Principal Component Analysis. Mathematical Statistics and Management, 2009, 2: 191-203.
[8] Wei D.Y., Lei W. Research on Spatial Heterogeneity of China Sports Lottery Sales and Its Driving Factors Based on ESDA-GWR. Journal of Tianjin Physical Education Institute, 2013, 28(4): 304-309.

[9] Han J.J. Analysis of the Factors Affecting the Sales of Lottery Tickets and the Strategy of the Development of Lottery Industry in China. China University of Mining and Technology, 2015.
[10] Zhang C.G. Research on the Factors Affecting the Lottery Sales in China under the Capping System. Southwest Jiaotong University. 2017.
[11] Liu S.W. Research on Distribution Strategy of Chinese Sports Lottery Bonus under the Background of Uncertainty. Shandong University. 2017.
[12] Li Y.Z. Research on the Interaction Effect between China's Stock Market and Lottery Market Based on Behavioral Finance. Shanghai Normal University. 2019.
[13] Jiang Q.Y. Mathematical Model. Higher Education Press. 1993
[14] Shen J., Zhou T.H. Mathematical Prediction Methods in Lottery Tickets. Journal of Liaocheng Teachers College (Natural Science Edition), 2002, (1).
[15] Yu Q., Zhou F.Y., Zhang Y.S. Mathematical Model of Lottery Analysis. Journal of Qiannan Normal College for Nationalities. 2002, (6).

Journal of Physics: Conference Series **1423** (2019) 012025 doi:10.1088/1742-6596/1423/1/012025

Feature extraction of non-linearity and low noise-signal ratio signals and its application to engine fault diagnosis

Fengli Wang [1] and Yannian Cai [2]

[1] College of Marine Engineering, Dalian Maritime University, Dalian 116026, P. R. China

[2] College of Marine Engineering, Dalian Maritime University, Dalian 116026, P. R. China

Corresponding author and e-mail: Fengli Wang, wangflsky997@sina.com

Abstract. Since the vibration signals of engine cylinder head come from different interference sources, it is difficult to detect weak but useful signals in the background of noise. Aiming at nonlinear and low noise characteristics of engine vibration signals, a method based on integrated empirical mode decomposition (EMD) and fractal dimension is proposed. Firstly, the integrated EMD method is used to decompose the nonlinear low noise signal into a set of intrinsic mode functions from high to low. Then the morphological fractal dimension is applied to the intrinsic mode function (IMFs). The fractal dimension of the characteristic IMF is calculated. The vibration signals of a diesel engine exhaust valve under normal and leakage conditions are analysed. The results show that this method can extract fault features of diesel engine exhaust valve.

1. Introduction

Traditional vibration signal processing techniques, including time-domain and frequency-domain analysis, are based on the assumption that the signal generation process is linear and stable. In view of the non-stationary and non-linear of vibration signal, several advanced signal analysis technologies are introduced and applied to the fault diagnosis of machinery, such as engine [1]. Because engine has many and complex moving parts, many excitation sources and wide frequency range, all kinds of excitation are reflected in its surface vibration through corresponding transmission and coupling, combined with the integration of noise, the engine vibration signal is extremely complex, and when engine is in different states, due to the influence of nonlinear factors such as clearance, dynamic load, contact force, stiffness, etc. The unified society presents complex nonlinear behavior. It is difficult to analyse the geometry of the vibration signal of the engine by using the time-frequency analysis method.

Fractal geometry provides a geometric structure analysis method for complex signals, and the development of fractal geometry theory provides a new idea for diesel engine fault diagnosis [2]. Fractal dimension is an important parameter used to quantitatively describe the behavior of nonlinear system in state space. At present, many fractal dimensions have been proposed, such as Hausdorff dimension, correlation dimension and box dimension. The fractal dimension is based on the basic idea of measuring the fractal set in different scales, that is, the set of irregular geometry to describe the complexity of the signal, which is consistent with the idea of measuring the geometric shape of the analysed object in different scales by multi-scale morphology. In order to study the method of estimating signal fractal dimension by using multi-scale morphological operator, Maragos. P proposed to estimate signal fractal dimension by using morphological filter operator [3]. The fractal dimension algorithm based on morphology has the advantages of low computational complexity, good stability of calculation results, sensitivity to the change of signal geometry, and is not affected by signal amplitude range and other factors [3-5].

Since most of the vibration signals detected in practical engineering contain noise, the noise component has a great influence on the calculation of fractal dimension. In order to get the fractal dimension of shape, it is necessary to denoise the vibration signal. Integrated empirical mode decomposition (EMD) can adaptively decompose the nonlinear and non-stationary signals into the sum of the finite basic mode components (IMF) according to the local time characteristic scale of the signal, without preselecting the basis function [6]. In this paper, multi-scale fractal dimension and integratedEMD are introduced into vibration signal analysis.

2. New method

2.1. Morphological fractal dimension

Morphology is a mathematical method developed on the basis of set theory, integral geometry and topology. Erosion and swelling are two basic operations of mathematical morphology. This coverage can be obtained by using functional structural elements of one-dimensional corrosion and swelling, defined as,

$$f\Theta g(n) = \min_{m \in G}\{f(n+m) - g(m)\} \tag{1}$$

$$f \oplus g(n) = \min_{m \in G}\{f(n-m) + g(m)\} \tag{2}$$

Hence, for one-dimensional discretetime signal $f(n)$, $n = 0, 1, \cdots, N$, the structuring element is defined at scales ε [3],

$$\varepsilon g(n) = g \oplus g \oplus \cdots \oplus g \ \ \varepsilon\text{TIMES} \tag{3}$$

Taking $\varepsilon = 1, 2, \cdots, \varepsilon_{\max}$ as the range of analysis scale, $\varepsilon_{\max} \leq N/2$, then the results of erosions and dilations of the signal $f(n)$ at different scales ε are $f\Theta\varepsilon g(n)$ and $f \oplus \varepsilon g(n)$ respectively.

The cover area of the signal at the scale ε is defined,

$$A_g(\varepsilon) = \sum_{n=1}^{N}(f \oplus \varepsilon g(n) - f\Theta\varepsilon g(n)) \tag{4}$$

According to [3], $A_g(\varepsilon)$ satisfies the following conditions,

$$\log\frac{A_g(\varepsilon)}{\varepsilon^2} = D_M \log\frac{1}{\varepsilon} + c \tag{5}$$

where $\varepsilon = 1, 2, \cdots, \varepsilon_{\max}$.

The Minkowski-Bouligand dimension D_M of the signal is defined as,

$$D_M = \lim_{\varepsilon \to 0} \frac{\log[A_g(\varepsilon)/\varepsilon^2]}{\log(\frac{1}{\varepsilon})} \quad (6)$$

According to the analysis of literature [4], we use the flat structure unit with length of 3 as the unit structure unit. Using flat structure element, the estimation result of fractal dimension is not affected by the range of signal amplitude, and the amount of calculation is reduced. According to the method of determining the maximum grid scale of periodic signal, combined with the working cycle characteristics of l engine vibration signal, the maximum scale is 60.

2.2. Integrated empirical mode decomposition

Empirical mode decomposition (EMD) can decompose any time-varying signals into finite intrinsic mode components (IMFs). The IMFs represent the natural oscillatory mode embedded in the signal and work as the basis functions. However, one of the major drawbacks of EMD is the mode mixing problem. To overcome the problem, integrated EMD is proposed, which is a noise-assisted data analysis method and defines the true IMF components as the mean of an integrated trials [7]. The procedures can be given as follows.

(1) The time series $N(t)$ of white noise is added to the signal $y(t)$.

$$S(t) = y(t) + N(t) \quad (7)$$

(2) The data series $S(t)$ is decomposed into finite IMFs.

$$S(t) = \sum_{j=1}^{n} c_j(t) + r_n(t) \quad (8)$$

(3) Repeat steps 1 and 2, but use different white noise series each step, $i = 1 \sim m$,

$$S_i(t) = y(t) + N_i(t) \quad (9)$$

$$S_i(t) = \sum_{j=1}^{n} c_{ij}(t) + r_{in}(t) \quad (10)$$

(4) The average value of IMF is obtained and considered as the final decomposition result.

$$c_j(t) = \frac{1}{m}\sum_{i=1}^{m} c_{ij}(t) \quad (11)$$

3. Simulation analysis

In order to verify the effectiveness of integrated EMD decomposition, EMD and integrated EMD are used to decompose the simulation signal, which is composed of low-frequency sine and small impact components. Figure 1 shows the time-domain waveform of the simulation signal. Figure 2 shows the decomposition results of EMD. Due to the abnormal interference of noise, EMD produces mode mixing, which leads to the pseudo IMFs, and cannot meet the requirements of feature extraction. The sinusoidal signal and impulse signal are decomposed into IMF component C1, and then the sinusoidal signal is decomposed into IMF component C1 and C2. The decomposition result has serious distortion, and the EMD generation mode mixing, so the pseudo IMF cannot meet the requirements of feature extraction.

The simulation signal is decomposed by integrated EMD, the average decomposition times is 100, and the noise amplitude is 0.01 times of the signal standard deviation. As shown in Figure 3, C1 component corresponds to impulse signal component, C2 component corresponds to sinusoidal signal, and small impulse component and sinusoidal signal are decomposed accurately.

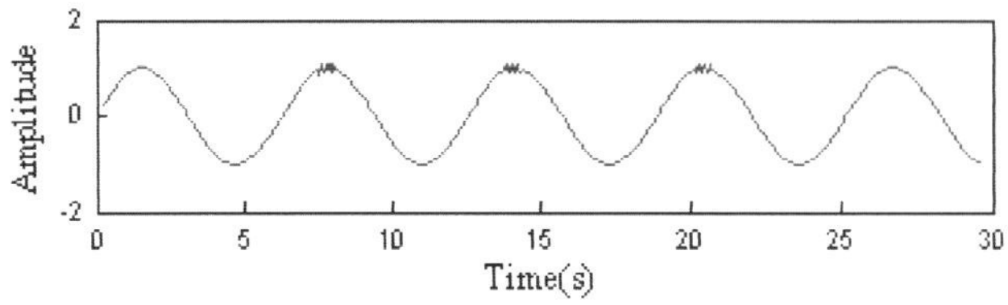

Figure1. Simulation signal

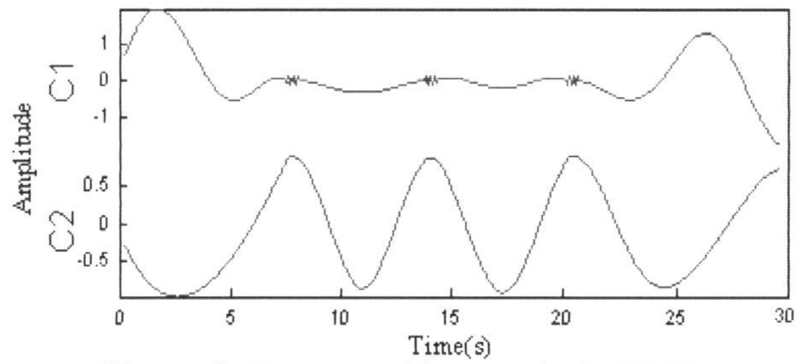

Figure 2. Decomposition results by EMD

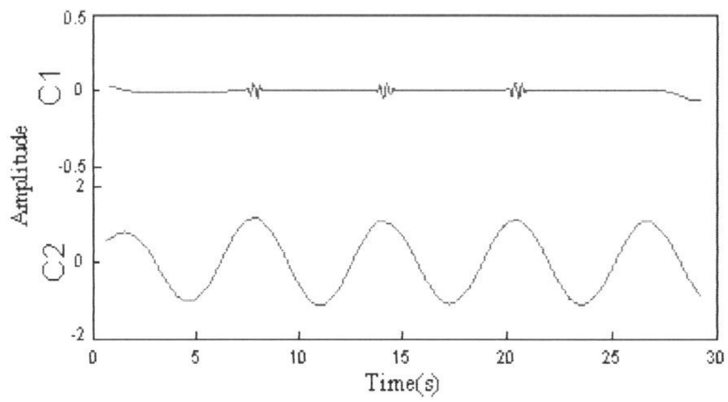

Figure 3. Decomposition results by integrated EMD

4. Experimental analysis

A four cylinder four stroke diesel engine is under the condition of normal working condition and different degree of exhaust valve leakage. The vibration response signals of diesel engine on the same cylinder surface are measured, which represent that the engine is in normal, slight and serious leakage state respectively. The sampling frequency is 25.6 kHz. The crankshaft speed of diesel engine is 1100r/min, and the vibration signal of diesel engine exhaust valve under three working conditions is shown in Figure 4. The fractal dimension of vibration signal is calculated directly by the fractal dimension estimation algorithm based on morphology, as shown in Figure 5. The fractal dimension is very close, so it is difficult to distinguish the state of exhaust valve.

By integrated EMD, the vibration response of diesel engine can be divided into several IMF, which represent the high frequency component to the low frequency component of the original vibration signal. In the leakage fault of exhaust valve, the inherent vibration component contains fault information, which belongs to high frequency component. Therefore, by calculating the fractal dimension of imf1, the working state of exhaust valve can be quantified. As shown in Figure 6, the fractal dimension can be clearly separated from the three regional conditions of diesel engine exhaust valve leakage.

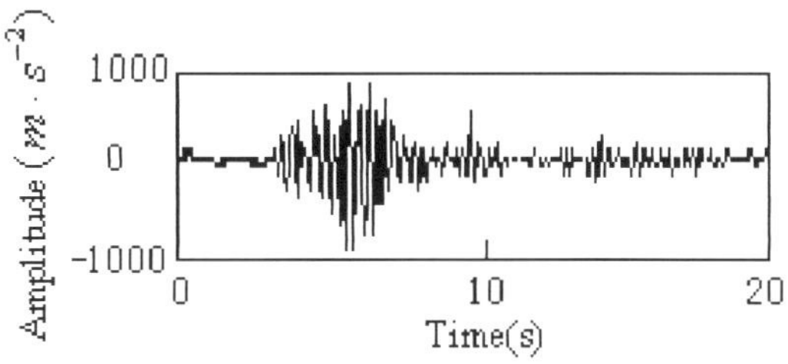

(a) Normal state

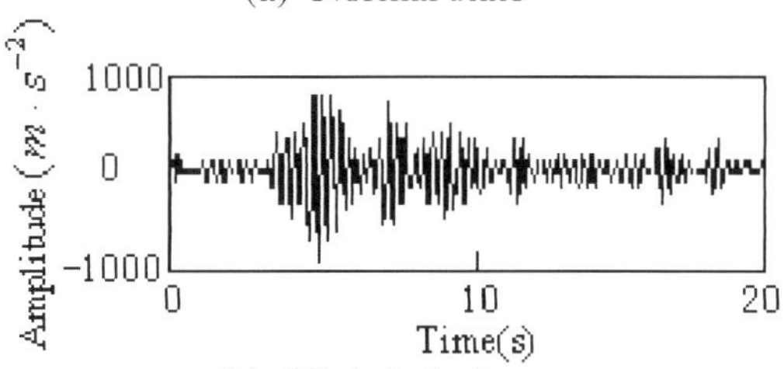

(b) Slightly leakage

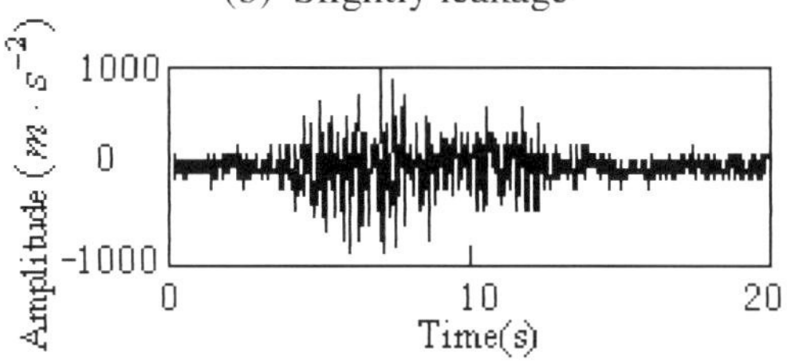

(c) Serious leakage

Figure 4. Vibration signals of diesel engine in three conditions

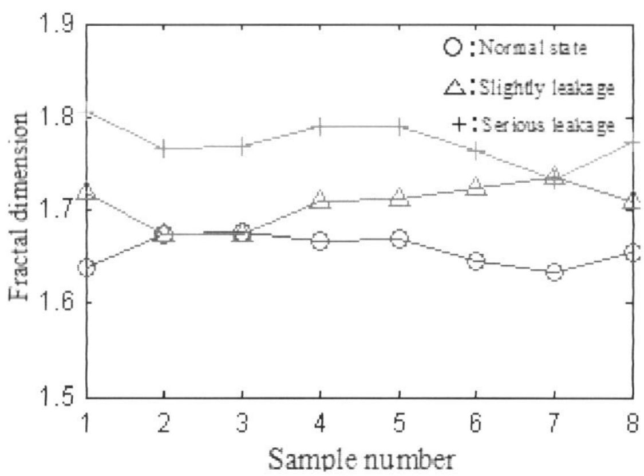

Figure 5. Morphological fractal dimensions of vibration signal in three conditions

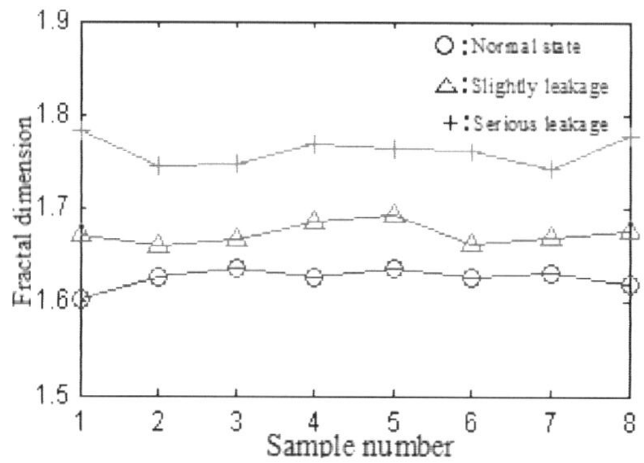

Figure 6. Morphological fractal dimensions of characteristic IMF in three conditions

5. Conclusion

Through analysis of fractal dimension of measured vibration signals of diesel engine, it is verified that the noise component will greatly affect the calculation result of the fractal dimension of the measured vibration signal. Therefore, it is not reliable to directly calculate the fractal dimension of the measured vibration signal of diesel engine to describe the working state of diesel engine, so it is very necessary to denoise the vibration test signal. The integrated EMD method can effectively separate the characteristic components of the measured non-linear non-stationary vibration signal from the noise and other interference components, thus improving the signal-to-noise ratio of the signal fractal analysis. The IMF component with fault features is analyzed by morphological fractal dimension, which can quantitatively describe the geometric characteristics of the vibration response signals of diesel engine, and the different working states of diesel engine are clearly distinguished. It is feasible to use morphological fractal dimension of the IMF component containing fault feature as the feature to judge the working state and fault identification of air valve leakage.

Acknowledgement

This research was supported by the Fundamental Research Funds for the Central Universities (3132019330).

References

[1] Z. Geng, J. Chen, J. B. Hull 2003*International Journal of MechanicalSciences*.vol 45 p 1391
[2] B. B. Mandelbrot 1982*The fractal geomety of nature*, (New York: Freeman).
[3] P. Maragos, FK Sun 1993*IEEE Transactions On Signal Processing*. vol41 p 108.
[4] P Maragos. E 1989 *Transactions on Pattern Analysis and Machine Intelligence*.vol11 p 701
[5] P Maragos 1987 *IEEE International Conference on ICASSP*, Dallas, p. 241.
[6] Z H Wu, NEHuang 2009 *Advances in Adaptive Data* Analysis vol 11 p 1 .

Design of fast digital spectrum stabilization method for UAV radiometry system

Xiaofeng Yang[1], Liangquan Ge[1], Guoqiang Zeng[1], Zongyi Yao[1], Lei Yan[1], Yang Hou[1], Mingtao Luo[1], Peng Yu[1]

[1]Applied Nuclear Techniques in Geosciences Key Laboratory, Chengdu University of Technology, Chengdu 610059, China

Corresponding author and e-mail: Guoqiang Zeng, 24829500@qq.com

Abstract. The drone radiation measurement system has a sharp increase in demand for environmental radiation monitoring due to its easy manoeuvrability, low aerial survey cost, high patrol efficiency and no personnel safety concerns. In order to adapt to the fast, flexible and efficient characteristics of the drone, the detector for the Unmanned Aerial Vehicle (UAV) radiation measurement generally consists of a plurality of small-sized detectors to form a detection array. Due to the low single crystal count rate and the influence of electronic system temperature and packaging, many detectors of the UAV radiation measurement system will have "spectral drifts", which affects the quality of the synthesized spectrum and affects the accuracy of the whole system. Aiming at the characteristics of UAV spectrum measurement system, a three feature point recognition model is established, and the K-40 feature peak (1.46MeV) is approximated by reverse piecewise summation. Thus, the backscattering peaks and characteristic peaks K-40 of the measured spectrum can be obtained to achieve fast spectrum stabilization. Finally, through Gaussian Mixture Model (GMM) peak estimation and experimental validation, the method can obtain the characteristic peak position accurately, and the time between instrument start-up and automatic spectrum stabilization is less than 2 minutes.

1. Introduction

UAV is an unmanned aerial vehicle operated by radio remote control equipment or its own program-controlled device. Because of its wide use, low cost, no casualty risk and good mobility, quite a few countries, such as the United Kingdom, Canada, the United States and so on, have developed UAV geophysical detection equipment. And it plays an important role in modern society. After the Fukushima nuclear accident in Japan, RMAX-G1 was also used to carry out aerial surveys to detect radiation levels around the nuclear power plant [1-3]. In China, with the development of UAV technology and the needs of environmental radiation measurement, quite a few domestic units have developed UAV radiometric measurement system for environmental monitoring, such as Nuclear Industry Remote Sensing Center, Nanjing University of Aeronautics and Astronautics, Chengdu University of Technology and so on.

In order to adapt to the fast, flexible and efficient characteristics of the drone, the detector for the UAV radiation measurement generally consists of a plurality of small-sized detectors to form a detection array [4]. Due to the low single crystal count rate and the influence of electronic system temperature and packaging, many detectors of the UAV radiation measurement system will have "spectral drifts", which affects the quality of the synthesized spectrum and affects the accuracy of the whole system [5-6]. Therefore, the problem of spectrum stabilization must be solved.

Recently, the main methods of spectrum stabilization are hardware spectrum stabilization, reference source spectrum stabilization, characteristic peak spectrum stabilization, etc. They have their own pros and cons [7-11], which are not discussed here. This paper develops a fast dual reference peak spectrum stabilization method based on the characteristic peak of spectra-stabilization and verified in the field environment.

2. Material and methods

In this paper, the dual reference peaks spectrum stabilization method of backscatter peak and K-40 characteristic peak () are studied. Taking the UAV detection system developed by our research team as an example. The system consists of four 1.5 inch LaBr3: Ce scintillation detectors. Fig. 1 shows the spectrum obtained from the environmental background of two detectors, which has the following characteristics:

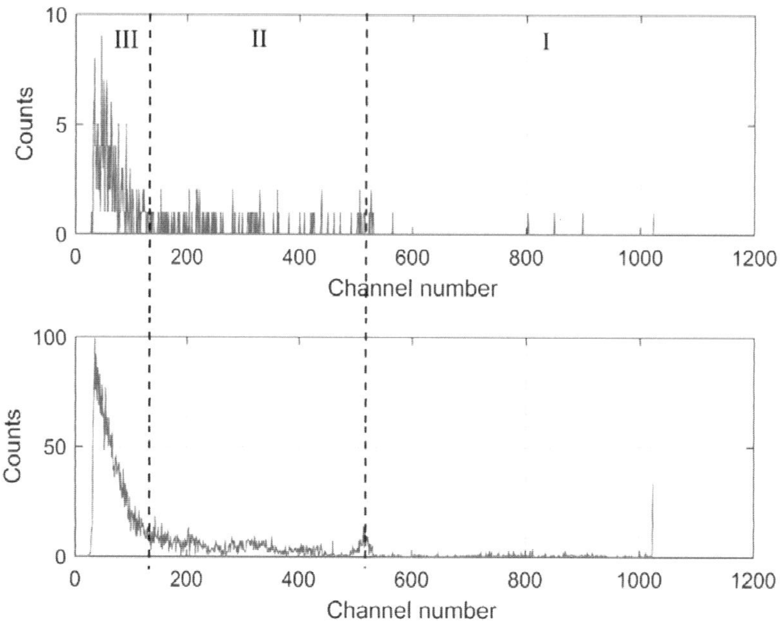

Figure 1. LaBr3: Ce spectra of Initial Spectrum of Environmental Background at 5s and 60s.

1) The spectral data have three distinct zones: I, II and III. And at the boundary of the partition, the relative counting rate changes abruptly.
2) The counting rate of small size crystals is low. Furthermore, the cumulative counting is still very low in a transitory time. After smoothing the spectrum, the shape is complex and can't be used for direct peaks searching.
3) The peak of backscattering is the most obvious in the region III and has a significant trend of change. Although the K-40 peak appears the highest probability of the junction of I and II, the peak is not obvious. In this case, automatic spectra-stabilization can be achieved unless accumulated over a long period of time.

Through the above analysis, we need to use the backscatter peak and the K-40 peak to achieve the correction of the spectrum. The key is to identify the K-40 peak position quickly. Fig.1 shows that the probability of K-40 peaks is the highest at the junction of I and II. Therefore, this method uses the reverse piecewise summation method to approach K-40 peaks one by one and extracts two characteristic points in the backscattering region III to represent the backscattering peaks. In this way, a bimodal recognition model is established to obtain the amplitude and peak position of the eigenvalue. Then the spectrum correction is carried out by the ratio of the characteristic peak position on the target peak position.

2.1. "B&K" bimodal recognition model

The identification model of "B&K" is based on the characteristics of three partitions of spectral data. And the backscattering peaks and K-40 characteristic peaks at abrupt ratio is taken as the research objects. The recognition principle is shown in Figure 2, which mainly includes three key steps: data preprocessing, segmented peak parameter calculation and peak seeking. Firstly, the measurement spectrum data is segmented from high energy segment to low energy segment, and the change rate between adjacent segments and the peak to total ratio of each segment are calculated. Secondly, set the threshold value of peak to total ratio to eliminate false peak. Thirdly, cluster the coefficient of change rate, find out the abnormal value and get the segmentation of characteristic peak value. Finally, the eigenvalues (peak value, channel address) of the characteristic peak are obtained in the peak segmentation, and the spectrum is stabilized by software. Therefore, the specific implementation of the above process is as follows:

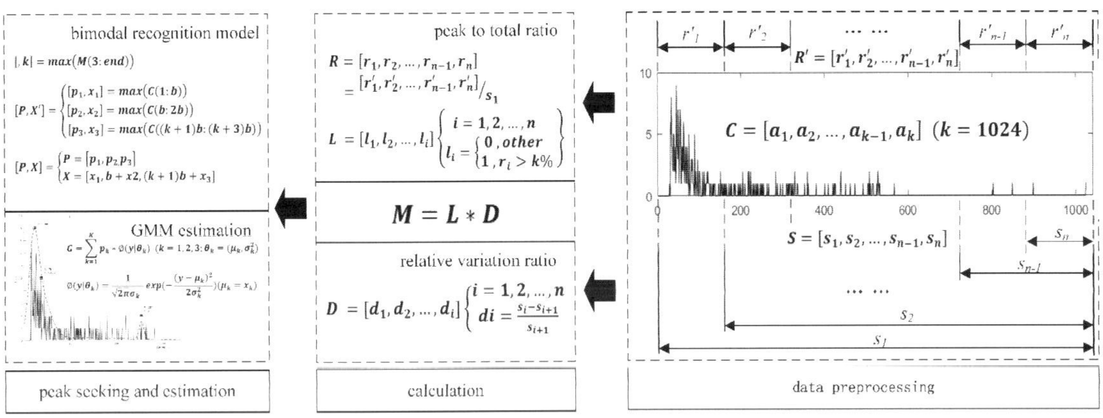

Figure 2. "B&K" bimodal recognition process.

1) Spectral data preprocessing

Let the total channel address of spectral data be j and each channel count was represented by a_j. If the reverse step distance is b, the total number of segments are R' and the total number of segments is $n = j/b$. Therefore, the spectral data C, the counts R' of each segment and the reverse summation S of the segments are expressed as follows:

$$\left.\begin{aligned} C &= [a_1, a_2, \ldots, a_{j-1}, a_j] \\ R' &= [r'_1, r'_2, \ldots, r'_{n-1}, r'_n] \\ S &= [s_1, s_2, \ldots, s_{n-1}, s_n] \end{aligned}\right\} \quad (1)$$

2) Computation of peak to total and relative variation ratio

The two key parameters of the model are the peak to total and relative variation ratio. After the spectral passes through the reverse segment, there is a peak at each segment, and the false peak is eliminated by the condition of peak formation. Through experiments and knowledge of probability theory, it is found that the radiation measurement itself is a statistical event, and the objective condition for the formation of spectral peaks are that the counts of peak area regions have a larger growth probability than the non-peak region. This feature can be measured by the ratio R of segment peaks to total. And it has the possibility of local peaks when the ratio is greater than k%. So, the piecewise peak ratio and the binarization are as follows:

$$\left.\begin{array}{l} R = [r_1, r_2, \dots, r_{n-1}, r_n] = [r'_1, r'_2, \dots, r'_{n-1}, r'_n]/s_1 \\ L = [l_1, l_2, \dots, l_i]\left\{\begin{array}{c} i=1,2,\dots,n \\ l_i=\begin{cases}0, other \\ 1, r_i > k\%\end{cases}\end{array}\right\} \end{array}\right\} \quad (2)$$

The peak threshold k% in upper formula needs to be paid more attention. Taking the UAV radiation detector developed by our team as an example, for the counting rate of 100 cps at 5-second, the peak forming condition is that the peak to total ratio is more than 1%.

The relative variation ratio D of subsection counts is expressed by the quotient of the difference between the counts of segment *i+1* and *i* divided by the counts of segment *i+1*. The calculation formula is shown in 3.

$$D = [d_1, d_2, \dots, d_i]\begin{cases} i = 1,2,\dots,n \\ di = \frac{s_i - s_{i+1}}{s_{i+1}} \end{cases} \quad (3)$$

3) Peak seeking

Peak-seeking algorithm is actually a calculation method based on the characteristics of three-part spectral data. No matter what the spectral is, the peak of the backscattering is obvious, and there are two adjacent relatively large change rates after segmentation. Therefore, we first find the backscattering peak, which is expressed by two points. Secondly, we find the maximum value in the remaining relative change rate, that is, the K-40 peak position. The specific implementation algorithm is as follows:

$$\left.\begin{array}{ll} M & = L * D \\ [,k] & = max\big(M(3{:}end)\big) \\ [P, X'] & = \begin{cases} [p_1, x_1] = max\big(C(1{:}b)\big) \\ [p_2, x_2] = max\big(C(b{:}2*b)\big) \\ [p_3, x_3] = max\Big(C\big((k+1)*b{:}(k+3)*b\big)\Big) \\ [P, X] = \begin{cases} P = [p_1, p_2, p_3] \\ X = [x_1, b + x2, (k+1)*b + x_3] \end{cases} \end{cases} \end{array}\right\} \quad (4)$$

2.2. GMM peak estimation

When spectral data are processed by reverse segmentation, the interval satisfying the conditions of *L* and *D* is the peak maximum probability interval. After the above treatment, it is apparent that the backscattering peaks have two relatively large change rates of d_1 and d_2. Therefore, the position of the other maximum between d_3 to d_n is K-40 peak. Next, we take the results of the peak search as the characteristic parameter and estimate the peak pattern by using the Gaussian mixture model. The formula is as follows:

$$\left.\begin{array}{l} \boldsymbol{G = \sum_{k=1}^{3} p_k * \emptyset(y|\theta_k)\ (k = 1,2,3; \theta_k = (\mu_k, \sigma_k^2))} \\ \boldsymbol{\emptyset(y|\theta_k) = \frac{1}{\sqrt{2\pi}\sigma_k} exp\left(-\frac{(y-\mu_k)^2}{2\sigma_k^2}\right)(\mu_k = x_k)} \end{array}\right\} \quad (5)$$

In formula 5, p_k and x_k are the results of formula 4. And p_k is used as the amplitude parameter of GMM model and x_k as the center of the distribution [12-13]. At the same time, in order to better match the peak, the algorithm inserts a point between the backscattering peak and K-40 peak, which constitutes a 4-point GMM peak shape estimation.

2.3. Automatic spectra-stabilization

The software automatic spectrum stabilization process is shown in Figure 3. First, the K-40 target peak is set to x0 (for example, x0=512@1024). Then, the third point peak eigenvalue x_3 is obtained through the "B&K" bimodal recognition model, which is the position of real-time K-40 peak recorded as x_i . And the software gain G_i in that state is read. Finally, the ratio formula is used to calculate the next software gain adjustment [14].

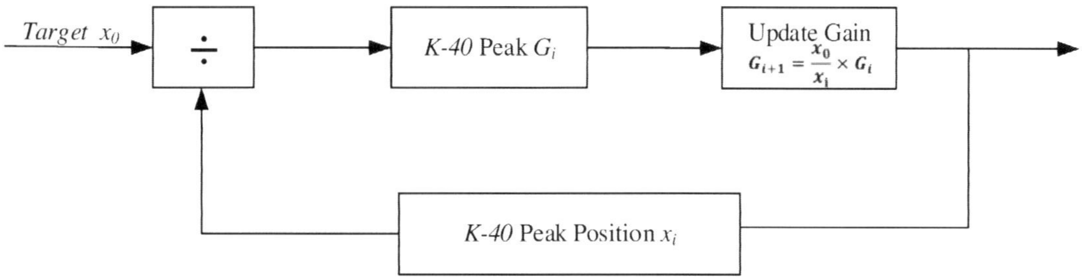

Figure 3. Principle of automatic spectra-stabilization.

So repeat the process of feature peak recognition and software gain calculation. And then write software gain $G_{(i+1)}$ to detector continuously until $|x_i - x_0| \leq 1$, that is, the accuracy of spectrum stabilization is ±1 channel.

3. Simulation and results

The method of spectrum stabilization designed in this paper has been validated in the UAV energy spectrum measurement developed by our team. As shown in Figure 4, it is the main process of automatic spectrum stabilization.

Fig. 4 (a) is the 5-second spectrum measured by the LaBr3 detector. The total count of the whole spectrum is about 484. The backscatter peak is located in 50 channels, but the K-40 peak position is not obvious. Fig. 4 (d) - (f) is the "B&K" bimodal recognition process. Among them, figure 4 (d) is the spectrum after 16 segments. It can be seen that the first two segments count obviously and the third 16 segments have statistical fluctuation. Therefore, it can not be directly used to search peaks. Fig. 4 (e) shows the relative variation ratio of each segment. And the red line represents the peak threshold, which equals 1 when the total ratio of segments are greater than 1%. Finally, the backscatter peak and the K-40 peak are located in the first and the ninth segments respectively, as shown in Figure 4 (f). Therefore, the backscatter peak and K-40 peak are located in 50 and 507 channels respectively. After automatic spectrum stabilization, the 60-second cumulative spectrum of the same detector is processed the same way. So the backscatter peak and the K-40 peak are located in 42 and 517 channels respectively. Moreover, the GMM peak estimation shows that the matching degree is more than 90%, as shown in Figure 4 (c).

At the same time, we validated another set of data. Fig. 5 (a) is a 5-second spectrum with a particularly low count of 475. Using this algorithm, the peak position of K-40 is 473 channels, which is 34 channels different from the actual position. The reason for this deviation is that the maximum values of the spectral lines between 450 and 550 channels are all 1. In this case, the maximum peak estimation accuracy is only related to the number of segments. The more segments, the more accurate the estimation result, which will make the calculation iterative process complex. Therefore, we can increase the cumulative time of spectral data to improve the accuracy. As shown in Figure 5 (b), when the spectrum accumulates to 10 seconds, the model can correctly find the peaks.

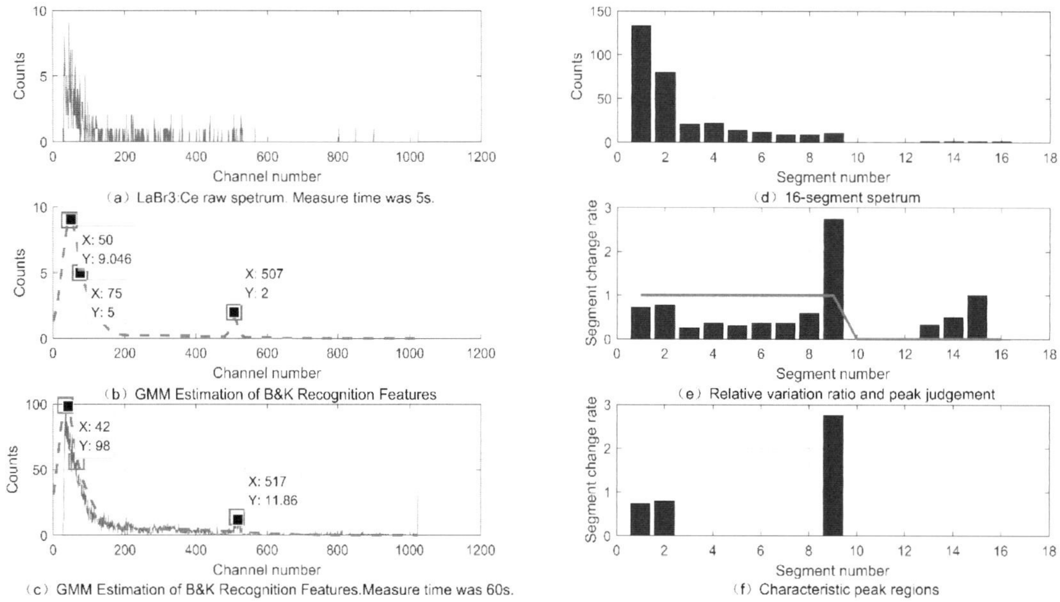

Figure 4. Main Processes of Automatic Spectrum Stabilization.

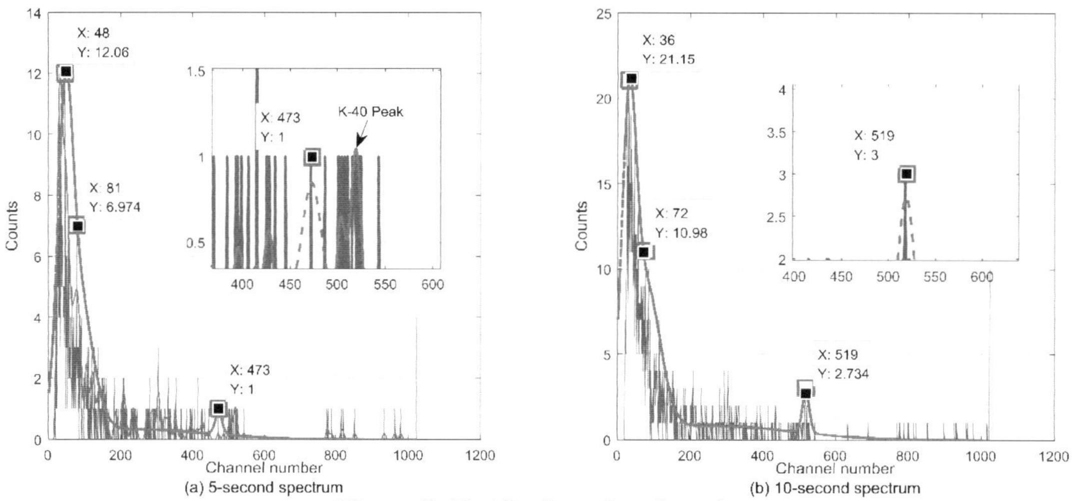

Figure 5. Verification of weak peak.

The above data validation shows that the method can quickly carry out spectrum stabilization. The recognition accuracy of backscatter peak and K-40 characteristic peak can be improved by adjusting the peak forming threshold and accumulation time. At the same time, the number of data segments is also an important parameter that affects the accuracy of recognition.

4. Conclusions

In this paper, the radiation measurement and the spectrum characteristics of UAV are analyzed and a fast peak identification method of "B&K" is proposed. The back scattering peak and K-40 characteristic peak can be obtained quickly by using the segmented peak to total ratio, relative counting rate and GMM estimation. This method does not need additional reference sources, does not need the smooth calculation of multiple spectra, and can automatically process in real time. Fig.6 is a

4-channel multi rotor UAV radiation measurement system developed by our team, which consists of four 1.5-inch lanthanum bromide detectors. We have verified the automatic spectrum stabilization of the system in the field environment, and the spectrum stabilization time is less than 2 minutes, and the accuracy is as high as 95%.

Figure 6. Radiation measurement system for multi-rotor UAV. It consists of four 1.5 inch LaBr3: Ce probes. The time from start-up to spectrum stabilization is less than 2 minutes.

On the other hand, this method has certain constraints. In the field environment, the instrument requires the background to be "clean", that is, the K-40 peak in the background of the environment will not be disturbed. Meanwhile, the model involves many parameters, such as relative variation ratio, peak-to-total ratio, peak forming thresholds, number of segments, standard deviation, etc. Therefore, the following research will carry out the identification accuracy under the dynamic optimization of multi parameters, and carry out the test in the actual environment.

Acknowledgement
This work was supported by the National Key R&D Program under Grant 2017YFC0602100.

References

[1] Ondřej Šáleka, Matolín, Milan, & Gryc, Lubomír. (2018). Mapping of radiation anomalies using uav mini-airborne gamma-ray spectrometry. Journal of Environmental Radioactivity, 182, 101-107.
[2] Sanada, Y. , Kondo, A. , Sugita, T. , Nishizawa, Y. , & Torii, T. . (2014). Radiation monitoring using an unmanned helicopter in the evacuation zone around the fukushima daiichi nuclear power plant. Exploration Geophysics, 45(1), 3-7.
[3] Xiao-Yu Ma & De Meng. (2014). Application of UAV Nuclear Radiation Aerial Survey Technology in Wide Area Radiation Monitoring in Japan. Nuclear Electronics & Detection Technology (3), 409-413.
[4] Miller, A. , Machrafi, R. , & Mohany, A. . (2014). Development of a semi-autonomous directional and spectroscopic radiation detection mobile platform. Radiation Measurements, 72, 53-59.
[5] Olmos, P. , Diaz, J. C. , Perez, J. M. , Aguayo, P. , Gomez, P. , & Rodellar, V. . (1994). Drift problems in the automatic analysis of gamma-ray spectra using associative memory algorithms. *IEEE Transactions on Nuclear Science, 41*(3), 637-641.
[6] Qin, Z. J. , Ge, L. Q. , Wu, Q. F. , & Cheng, J. P. . (2011). A new algorithm for spectral drift calculation in a gamma-ray spectrometer. *Advanced Materials Research, 346*, 705-710.
[7] Kosina, Z. . (1983). Real time spectrum stabilization by software. *Nuclear Instruments and*

Methods in Physics Research, 213(2-3), 551-552.

[8] Guo-Qiang Zeng, Jian Yang, Shi-Long Wei Liang-Quan Ge & Lei Yan(2016). Development of Large Volume Digital Vehicle-borne Gamma Spectrometer. Atomic Energy Science and Technology, 50(11).

[9] AOQi, WEIYi-xiang, & QUJian-shi. (2009). Design of Digital Spectrum Stabilization Method for NaI Spectrometer. Nuclear Electronics & Detection Technology, 29(1), 19-22.

[10] Stromswold, D. C. , & Meisner, J. E. . (1979). Gamma-ray spectrum stabilization in a borehole probe using a light emitting diode. *IEEE Transactions on Nuclear Science, 26*(1), 395-397.

[11] Zeng, G. , Tan, C. , Ge, L. , Zhang, Q. , & Gu, Y. . (2014). Frequency spectrum analysis for spectrum stabilization in airborne gamma-ray spectrometer. *Applied Radiation and Isotopes, 85*, 70-76.

[12] Hang Li. (2012). Statistical Learning Method.

[13] Dong-Wen Ying, Yong-hong Yan, Qiang Fu, & Yan-Meng Guo (2010). Noise power spectrum estimation based on constrained GMM. Papers of the 2010 Annual Conference on Communication Theory and Signal Processing.

[14] Guo-Qiang Zeng, Liang-Quan Ge, Sheng-Qing Xiong, Wei-Chong Ni, Zheng-Guo Fan, & Wan-ChangLi(2010). Application of Kalman Filter in Automatic Spectrum Stabilization of Aerial Gamma Spectrum Exploration System. Nuclear Electronics & Detection Technology, 30(5), 698-702.

Journal of Physics: Conference Series **1423** (2019) 012027 doi:10.1088/1742-6596/1423/1/012027

Application research of a data stream clustering algorithm in network security defense

Canshi Zhu[1], Xiaoyang Wang[2], Lin Zhu[*3]

[1]Information Engineering College, Xijing University, Xi'an 710123, China
[2]Xi'an Satellite Control Center, Xi'an 710043, China
[3]Equipment Management and Safety Engineering College, Air Force Engineering University, Xi'an 710051, China

*Corresponding author and e-mail: Lin Zhu, 3399241158@qq.com

Abstract. The traditional intrusion detection system feature model is based on static data mining. Its mining algorithm relies on too many assumptions, which makes it difficult for intrusion detection systems to adapt to dynamic and real-time system detection requirements. Using attenuated sliding window technology, data stream mining technology and fusion technology with intrusion detection system, a data flow clustering algorithm based on attenuated sliding window is designed to improve and optimize the feature pattern extraction method of intrusion detection system to solve the dynamics of intrusion detection system. Through algorithm design, algorithm application and intrusion detection system simulation verification, the feasibility and accuracy of the algorithm and the optimized intrusion detection system are proved.

1. Introduction

Intrusion detection system is an effective network security defense strategy [1]. The method is to detect data packets that want to flow into the network under test, andreview security threats to the defensed system at the front end of the defensed system [2]. It can block the access operations of the threat system to protect the protected network system resources from attack.

Intrusion detection system is simple and convenient to implement. It does not need to cross multiple physical network segments, but only monitors on ports [3]. It can passively and silently collect the information flowing into the network at the entrance of the protected network system, analyze and detect the behavior mode of entering the network. It allows access to the network for normal and authorized user operations, and prevents and counteracts user access for abnormal and unauthorized behavior, so as to achieve the purpose of network security defense [4]. The key link of intrusion detection system is the extraction of intrusion feature patterns or rules, and the accuracy of feature patterns or rules extraction. Timeliness and blocking ability are the key technologies of intrusion detection system [5].

Traditional intrusion feature extraction is based on static data analysis and mining [6]. This method needs periodic analysis and frequent updating of feature pattern library. Therefore, the traditional intrusion detection is less effective and dynamic, and the security defense effect is limited. The traditional intrusion detection or anomaly detection schematic is shown in Figure 1.

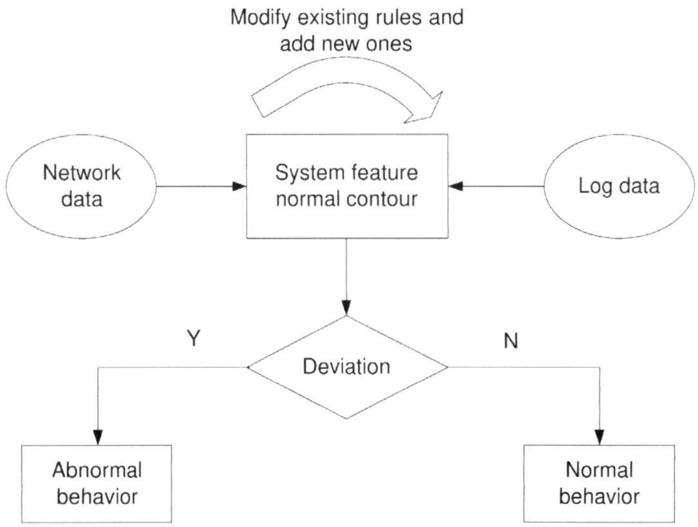

Figure 1. Schematic of traditional intrusion detection or anomaly detection

Commonly used intrusion detection is classified according to the extraction technology rules, such as intrusion detection technology based on expert system, intrusion detection technology based on neural network, intrusion detection technology based on immune algorithm and data mining-based intrusion detection technology [7-8].

The intrusion detection technology based on the expert system encodes the intrusion or attack behavior to form an expert system rule base. Each of these rules is a code for an intrusion scenario that detects intrusion behavior by matching the audit record with the rules.

The intrusion detection technology based on neural network is mainly neural network algorithm. The algorithm allows the neural network to simulate user phase behavior and adjust to new changes. The network receives the input event data and compares it with the historical behavior of the reference to determine the similarity or degree of deviation between the two.

The intrusion detection technology based on immune algorithm is to achieve the purpose of eliminating the dissident and maintaining the balance of the system through the affinity of the antibody and the antigen. The most basic ability of the immune system is to be able to identify tissues that belong to the normal body, and those that are not part of the normal body are abnormal.

Data mining-based intrusion detection technology applies data mining technology to intrusion detection. This method can comprehensively and effectively audit data and obtain characteristic patterns of attack behavior, so as to accurately distinguish and capture actual intrusion and normal behavior patterns.

With the rapid development of network technology, network attack methods and technologies are constantly developing and improving [9]. Coupled with the increase of network bandwidth, the increase of network speed and the rapid increase of network traffic, these pose new challenges to network security defense. Traditional intrusion detection technology has been difficult to cope with this challenge. The main reason is that system feature extraction is implemented in a static data environment. The timeliness of feature patterns lags, and real-time processing capabilities cannot meet system requirements [10]. Therefore, how to dynamically extract feature patterns in massive data streams not only solves the accuracy of feature patterns, but also improves the timeliness of pattern extraction, which is the key to improving the performance of intrusion detection systems.

2. Analysis of Data Stream Clustering Mining Algorithm

The traditional clustering mining algorithm is based on static data, its data will be reused and updated in small amount in real time. It relies on too many assumptions, which makes it difficult for intrusion detection systems to adapt to dynamic and real-time system detection requirements.

Data streams are characterized by continuous fast, short-lived, unrecoverable, and unpredictable. Therefore, designing data stream mining algorithms should solve the following technical problems.

Firstly, the algorithm requires solving the problem of unlimited data volume mining with limited memory resources. The traditional data is limited, the storage space can be satisfied, but the data stream has an unlimited amount of data, and the memory capacity cannot be satisfied.

Secondly, the algorithm has high requirements for time complexity. Traditional data mining is carried out in static data environment, and multiple iterations can be performed using traditional data mining algorithms. However, data stream mining is performed in a dynamic data network environment. The data stream is irreversible and the data can only be processed once.

Thirdly, the data stream is a specific type, requiring the data stream mining algorithm to adapt to changes. For example, as time progresses, the knowledge or models that are mined should also change constantly. This can be called concept drift or evolution

As far as the data stream mining algorithm is concerned, it can be divided into clustering and classification. This paper is based solely on clustering algorithm analysis. The clustering algorithm generally divides the mining objects into N groups. The similarity of the objects in the group is as high as possible, and the similarity of the objects outside the group is as low as possible. Typical data stream clustering algorithms are k-means, CluStream, and DBSCAN algorithms.

The k-means algorithm first determines the k-value, and divides the data object into k groups. Each group of data objects is not empty. Each object belongs to a group if and only belongs to one group. Each group forms a cluster, and the algorithm loops to find the optimal cluster center.

The CluStream algorithm is an evolution of the data stream clustering algorithm, which mainly deals with the clustering of data streams in different time domains. There are two stages in the process of dealing with data stream which may be changing constantly. One is the online maintenance of candidate micro-clusters, the other is the offline generation of clustering results.

The DBSCAN algorithm is a density clustering algorithm, which uses data space D as a data class and D as a high-density object region. If we want to study a point p of the space D, and use p as the core, query the neighborhood of the D area, and all the points in the neighborhood and p belong to the same class. These points will be the next object of investigation. In this iterative way, the class is constantly expanded until a complete class is found.

The research content of this project comes from the part of the team of Professor Zhu Canshi of Xijing College in 2019, supported by the Shaanxi Provincial Science and Technology Department, China, "Big Data Security and Application Research in the Age of Sharing". Aiming at the shortcomings of data stream clustering algorithm, this paper proposes and designs a clustering mining algorithm based on attenuated sliding window technology, which improves the clustering quality and efficiency of traditional clustering algorithm.

3. A Data Stream Clustering Algorithm Based on Attenuated Sliding Window (ASWD Stream)

Data stream mining is a new technology for dynamically extracting feature patterns in data streams. The core technology is data stream mining algorithm, which is based on the full analysis of classical clustering algorithms.

The processing idea of the algorithm is to analyze the critical candidate micro-cluster and determine whether it will be converted into a candidate micro-cluster or an out-of-group micro-cluster. Clustering is performed in a sliding window. The sliding window is divided into N sub-windows. Firstly, clustering operations are performed on each sub-window to obtain N candidate micro-cluster sets, and then candidate micro-clusters are used instead of specific records for processing. Thereby reducing the impact of the order of data inflow on the clustering results.

The algorithm sequentially reads the network data into each sliding sub-window, and performs clustering in each sliding sub-window to obtain a plurality of candidate micro-clusters and a plurality of critical candidate micro-clusters. The candidate micro clusters are saved into the corresponding set, and the critical candidate micro clusters are saved to another corresponding set. After all the sub-windows in the sliding window have been processed, the saved critical candidate micro clusters and candidate micro clusters are processed. After obtaining the micro-cluster in the sliding window, the algorithm merges the micro-clusters into the pyramid structure that has been established in advance by using an offline method, and saves the current cluster cluster in a certain snapshot form. It uses a clustering result based on the preservation history of the attenuation model. After the sliding sub-window processing is completed, the candidate micro clusters and the critical candidate micro clusters in the set are processed. The processing flow of the data stream algorithm is shown in Figure 2.

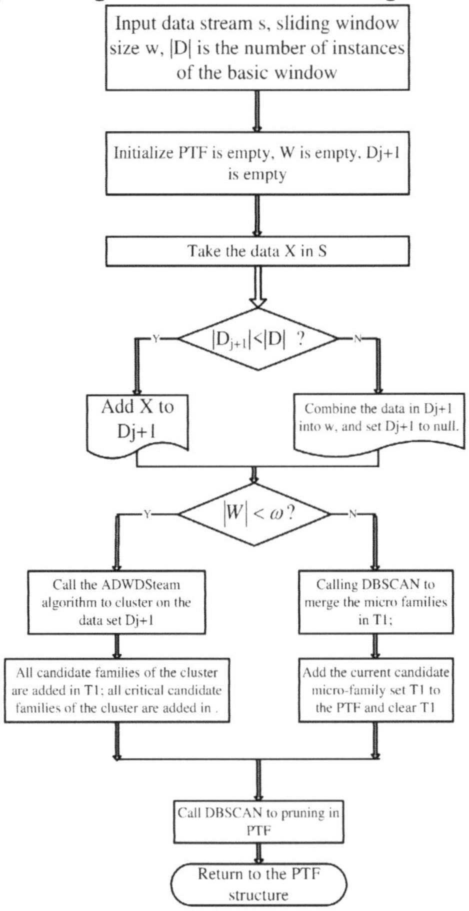

Figure 2. Processing flow of data stream algorithm

4. Simulation Verification of Data Stream Clustering Algorithm (ASWD Stream)

The validation data set is based on the data provided by the 3rd international knowledge discovery and data mining competition (KDD CUP 1999). The KDD CUP 1999 data is highly authoritative in assessing the performance of intrusion detection systems.

When verifying, the following parameter settings are used. The initial data points are N=1000, the data flow velocity is $v = 1000$, the attenuation factor is $\lambda = 0.25$, $\varepsilon_1 = \varepsilon_2 = 13$, $\mu = 10$,, and the outlier threshold is $\beta = 0.2$.

The data stream clustering algorithm (ASWDStream) and the CluStream algorithm use the same parameters as above. The clustering process of the data stream clustering algorithm (ASWDStream) after MATLAB simulation is shown in Figure 3.

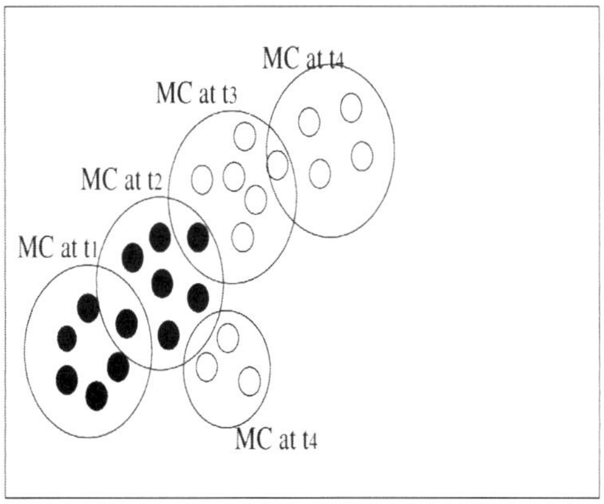

Figure 3. Micro-cluster formation process of ASWD Stream algorithm

The two algorithms are compared in terms of time complexity. The simulation results of the two algorithms are compared as shown in Figure 4.

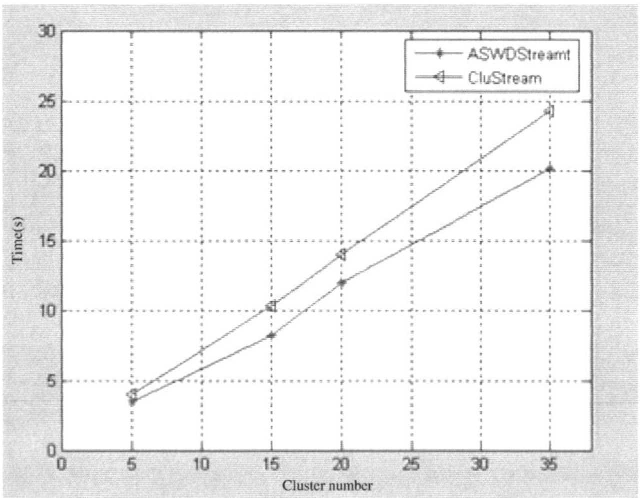

Figure 4. Running time comparison of the two algorithms

Through the above verification analysis, it can be seen that the ASWDStream algorithm is superior to the similar CluStream algorithm in terms of processing speed and clustering quality.

5. Application of ASWD Stream Clustering Algorithm in Intrusion Detection System

The intrusion detection system based on ASWDStream Clustering Algorithm is much more complicated than the traditional intrusion detection system. This is because it implements real-time mining, analysis of intrusion feature patterns, and real-time dynamic detection. After the inflow of external network information, the data stream clustering mining algorithm is first used to mine and analyze. For the suspicious feature pattern mined by the algorithm, the intrusion detection system will block the attack.Thereby realizing the security of the intranet system.

The design principle and structure of intrusion detection system based on data stream clustering mining algorithm are shown in Figure 5.

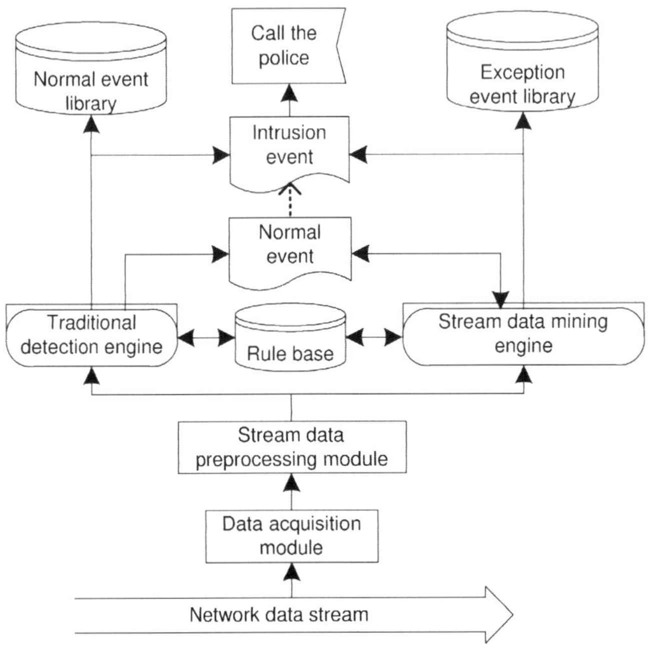

Figure 5. Intrusion detection system model based on data stream clustering mining

Firstly, the system performs sample mining on the normal data stream using the designed clustering algorithm (ASWDStream), and the obtained feature pattern is used as the normal model.

Then, the intrusion detection system is laid out, and the real-time data flow into the network system is mined and analyzed by using the clustering algorithm (ASWDStream). The mining result is passed through the system and matched in real time with the normal mode. If the matching detection result is normal, the data stream can enter the network system, otherwise the data stream is blocked from entering the system.

Of course, the intrusion detection system must analyze the suspicious mode and continuously update and optimize the normal feature pattern library.

6. Intrusion Detection System Simulation Based on Data Stream Clustering Mining Algorithms

There are 7 000 training samples, including 7 types of invasion. They are neptune, pod, ftp_write, spy, perl, buff_overflow, ipsweep, respectively.

There are 3000 test samples, including 9 types of invasion. They are neptune, pod, ftp_write, spy, perl, buff_overflow, eject, ipsweep, nmap, respectively.

The Distribution of test data types is shown in Table 1.

Table 1. Distribution of test data types

Data type	Number of training samples	Percentage of training samples	Number of test samples	Percentage of test samples
Normal	4743	67.76%	1867	62.23%
DoS	786	11.23%	345	11.5%
R2L	653	9.32%	289	9.63%
U2R	456	6.51%	168	5.60%
probing	362	5.17%	331	11.03%
Total	7000	100%	3000	100%

The Test results and Statistical results are shown in Tables 2 and 3.

Table 2. Test results

Detected sample set	Training sample set	Test sample set
Accuracy	99.87%	98.23%
Misinformation rate	0	0.57%
Misreporting rate	0.186 %	1.13 %

Table 3. Statistical results

Intrusion type		Number of training samples	Number of test samples	Detection rate
Dos	pod	754	231	30.64%
	neptune	32	114	
R2L	ftp_write	432	143	33.10%
	spy	221	146	
U2R	perl	320	65	20.31%
	buff_overflow	136	32	
probing	ipsweep	362	178	49.17%
	nmap	0	153	

From the simulation analysis, after applying ASWDStream algorithm to the intrusion detection system based on data stream mining, the intrusion detection system shows good detection performance, such as high detection rate, low misinformation rate and misreporting rate. It has better recognition ability for specific intrusion types, especially for R2L and probing intrusion types, which have high detection rate. In addition, it has better detection ability for new intrusion types.

7. Conclusion

In this paper, in view of the fact that the feature extraction of traditional intrusion detection system can not adapt to the dynamic and real-time system detection requirements, based on the analysis of traditional intrusion detection system and data stream mining technology, a data stream clustering algorithm based on attenuated sliding window (ASWD Stream) is designed. The algorithm is integrated with intrusion detection system to design an intrusion detection system based on data stream mining, and the system is simulated and verified. The simulation results show the feasibility of data stream clustering algorithm based on attenuated sliding window. The intrusion detection system based on data stream mining constructed by this algorithm solves the problem that the feature pattern extraction of traditional intrusion detection system can not adapt to dynamic and real-time system detection.

Acknowledgment

This work is supported by the Key Project of Soft Science Research Program of Department of science and technology of Shaanxi Province, China, No. 2019KRZ014.

References

[1] Zhu Lin, Zhu Canshi. Data stream sliding window clustering algorithm applied in IDS [J]. Computer Engineering and Applications, China. 2014, 50(1):87-90.

[2] Sun Ruxia,Zhang Sun,Yin Chunyong,Wang Jin,Min Seungwook.Strategies for data stream mining method applied in anomaly detection [J]. Cluster Computing. 2018.

[3] Phan H, Le B. A Novel Parallel Algorithm for Frequent Itemsets Mining in Large Transactional Databases [M]// Advances in Data Mining. Applications and Theoretical Aspects. Springer, Cham, 2018.

[4] Zymbler M. Parallel Algorithm for Frequent Itemset Mining on Intel Many-core

Systems [J]. 2018.

[5] V. R. B. Rohini, Dr. G. P. Saradhi Varma. FIDOOP-DP: Implementation of Data Partitioning in Frequent Itemset on Bigdata using Hadoop Pseudo Distributed Environment [J]. International Journal of Scientific Research in Science, Engineering and Technology. 2017, 8(3): 663-668.

[6] Liu Y, Ding W. A KNNS based anomaly detection method applied for UAV flight data stream[C]// 2015 Prognostics and System Health Management Conference (PHM). IEEE, 2015.

[7] Wang B, Liu D, Wang W, et al. A hybrid approach for UAV flight data estimation and prediction based on flight mode recognition [J]. Microelectronics Reliability, 2018, 84:253-262.

[8] Shabtai A. Anomaly Detection Using the Knowledge-based Temporal Abstraction Method [J]. 2016.

[9] Bangalore P, Letzgus S, Karlsson D, et al. An artificial neural network-based condition monitoring method for wind turbines, with application to the monitoring of the gearbox [J]. Wind Energy, 2017.

[10] Ahmed R, Srinivasa Pai P, Sriram N S, et al. Comparison of wavelet based denoising schemes for gear condition monitoring: An Artificial Neural Network based Approach [J]. IOP Conference Series: Materials Science and Engineering, 2018, 310:012010.

Journal of Physics: Conference Series **1423** (2019) 012028 doi:10.1088/1742-6596/1423/1/012028

Research on image matching algorithm based on improved SIFT UAV

Rong Guo[1], Shixin Li[2], Ruyi Cai[3] and Xiali Sun[4]

Tianjin University of Technology and Education, Tianjin, China

Corresponding author and e-mail: Shixin Li, li_shixin@sina.com

Abstract. Aiming at the low accuracy of UAV image registration algorithm, and the complex calculation of SIFT algorithm feature descriptor and the low time efficiency, an image matching algorithm based on improved SIFT is proposed. Firstly, the image of UAV is preprocessed, and the feature point operator is detected by Shi-Tomasi corner, then is extracted by SIFT feature vector, and the dimension of feature vector is reduced by PCA. Then the nearest neighbor method is used for feature matching, and the BBF algorithm is used to search the feature values to improve the matching speed. Finally, the RANSAC algorithm is used to purify the feature points for accurate matching. Experiments show that compared with the traditional image matching method, the proposed algorithm improves the efficiency and accuracy of the UAV image registration technology.

1. Introduction

In 1999, David Lowe first proposed the classic Scale-Invariant Feature Transform (SIFT) algorithm, which was improved in 2004. The algorithm combines the scale space pyramid with the Gaussian kernel to improve the detection rate of feature points, which is very invariant to scale scaling, translation rotation, and illumination variation. However, the algorithm has problems such as complex description and long calculation time. Yan Ke et al. reduced the feature description sub-dimension and reduced the amount of computation by using the PCA-SIFT algorithm of principal component analysis. By improving the Random Sample Consensus (RANSAC) algorithm, Wu et al. proposed to use the fast iterative strategy to filter the feature point pairs to obtain more accurate transformation parameters, and achieve accurate matching of remote sensing images. The image matching algorithm based on improved SIFT proposed in this paper can achieve automatic image registration and improve the efficiency and accuracy of UAV image registration technology.

2. SIFT algorithm

The basic idea of the SIFT algorithm is that the execution of the SIFT method is divided into two parts: first, the feature points are identified for the image, and then the feature points are described. The algorithm detects a stable feature point in an image with different degrees of blur, and then constructs a multi-dimensional descriptor by using some information retained in the detection process to express the feature point. Specific steps are as follows:

(1) Scale space construction: Search for images in all scale spaces, and identify potential points of interest and rotation invariant through Gaussian differential functions. The construction of the scale space is shown in the following formula:

$$L(x, y, \sigma) = G(x, y, \sigma) * I(x, y) \tag{1}$$

$$G(x, y, \sigma) = \frac{1}{2\pi\sigma^2} e^{-(x^2+y^2)/2\sigma^2} \tag{2}$$

The $L(x, y, \sigma)$ of equation (2) is a scale space image at a certain scale, which can be obtained by convolving the original image $I(x, y)$ with a Gaussian kernel $G(x, y, \sigma)$.

(2) Construction of Gaussian pyramid: DOG (difference of guassians) pyramid is defined for the difference of pixel points in statistical images at different scales:

$$\begin{aligned} D(x, y, \sigma) &= (G(x, y, k\sigma) - G(x, y, \sigma)) \times I(x, y) \\ &= L(x, y, k\sigma) - L(x, y, \sigma) \end{aligned} \tag{3}$$

(3) Detection of spatial extremum: The key point is the local extremum point with direction information detected under images of different scale spaces. These points will not disappear due to changes in lighting conditions, such as corner points, edge points, bright spots of dark areas, and dark spots of bright areas. Since the two images have the same scene, they respectively extract their respective stable points. There will be a matching point between them.

(4) Construction of SIFT descriptor: After the above steps, the feature points existing at different scales have been found. In order to achieve image rotation invariance, the direction of the feature points needs to be assigned. The gradient distribution characteristics of the neighborhood pixels of the feature points are used to determine the direction parameters, and then the gradient histogram of the image is used to obtain the stable direction of the local structure of the key points.

3. Improved SIFT UAV image matching algorithm

The accuracy of the UAV image registration algorithm is low, and the SIFT algorithm feature descriptor is computationally complex and the time efficiency is low. This paper uses the Shi-Tomasi corner to detect the feature point operator and uses the SIFT feature vector to extract the feature points, then uses PCA to reduce the dimension of the feature vector. The specific algorithm flow chart is shown in Figure 1. The algorithm improves the time efficiency while ensuring high accuracy and good stability.

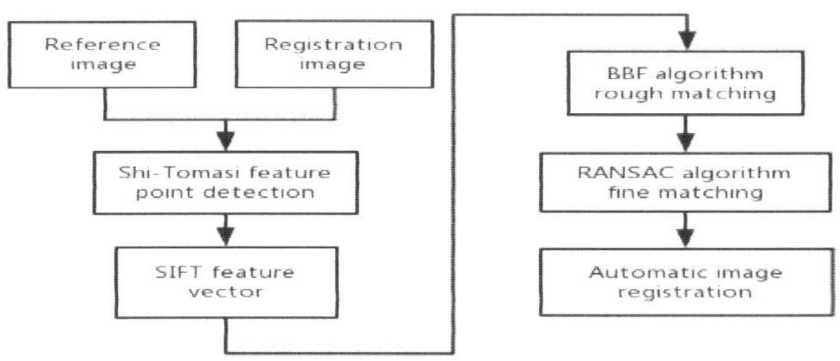

Figure 1. Algorithm flow chart

3.1. Shi-Tomasi detection feature points

The Shi-Tomasi algorithm was proposed by J. Shi and J. Tomasi in 1994. This method is an improvement to the Harris algorithm. In the Harris algorithm, it is determined whether or not a corner point is based on a combination of two eigenvalues of the covariance matrix M. In the Shi-Tomasi algorithm, it is determined whether the corner point is based on whether the smaller eigenvalue is greater than the threshold. This judgment is based on the fact that a smaller eigenvalue indicates that the variance in the direction of the eigenvalue is smaller, and if the smaller variance can be greater than the threshold, the change in this direction satisfies the judgment requirement of the corner point.

3.2. Feature point extraction

For the extraction of feature points, this paper uses the improved SIFT algorithm to experiment $D(x, y, \sigma)$ to improve the Gaussian pyramid difference function, and perform three binary Taylor expansion to achieve sub-pixel precision.

$$D(X) = D + \frac{\partial D^T}{\partial X} X + \frac{1}{2} X^T \frac{\partial^2 D}{\partial X^2} X \tag{4}$$

In equation (4) $X = (x, y, \sigma)$, in order to enhance the stability of the key points, the key points with an extreme value less than 0.03 are filtered out.

Since the Gaussian difference function has a strong edge response, the Gaussian difference function has a large curvature along the edge, and the vertical edge has a small curvature. The eigenvalue of the matrix is proportional to the curvature of the Gaussian difference function, then dropping the unstable key points near the edge.

In order to ensure the rotation invariance of the feature vector, the coordinate axis is rotated by θ angle in the nearby neighborhood, that is, the coordinate axis is rotated into the main direction of the feature point, and the new coordinates of the pixels in the neighborhood after the rotation are:

$$\begin{bmatrix} x' \\ y' \end{bmatrix} = \begin{bmatrix} \cos\theta & -\sin\theta \\ \sin\theta & \cos\theta \end{bmatrix} \begin{bmatrix} x \\ y \end{bmatrix} \tag{5}$$

3.3. Feature matching

After extracting the feature points according to the above analysis, the BBF algorithm can be used to first perform coarse matching on the feature points, and then the RANSAC algorithm is used to eliminate the mismatched points generated in the rough matching process to achieve accurate matching of the images.

BBF (Best Bin First) search algorithm: The BBF algorithm is a search algorithm developed on the basis of the K-D tree structure, and performs better in high-dimensional features than the K-D tree search algorithm. The BBF algorithm is an improved method of the K-D tree. It can be seen from the above-mentioned standard K-D tree query process that the "backtracking" in the search process is determined by the "query path", and does not consider some of the data points themselves on the query path. A simple improvement is to sort the nodes on the "query path". The BBF algorithm is an approximation algorithm proposed for high-dimensional data. This algorithm can ensure the priority of retrieving the space with the most likely neighbors. Therefore, we choose to use K-D tree to organize the feature points and use the BBF algorithm to search for feature points.

The RANSAC algorithm adopts the sampling and verification strategy. The main idea is to solve the eight parameters of the homography matrix H that can be satisfied by one sample point. This is an important part of removing the wrong matching point, (x, y) and (x', y') are set to be matched Point coordinates. While the homography matrix H describes the transformation relationship between the corresponding two sets of coordinates, that is, the geometric transformation relationship between images:

$$\begin{bmatrix} x' \\ y' \\ z' \end{bmatrix} = \begin{bmatrix} h_0 & h_1 & h_2 \\ h_3 & h_4 & h_5 \\ h_6 & h_7 & 1 \end{bmatrix} \begin{bmatrix} x \\ y \\ 1 \end{bmatrix} \tag{6}$$

4. Experiment and analysis of results

In order to verify the effectiveness of the algorithm, this paper uses the aerial image of the crater in Bergen, a city in southwestern Norway, to experiment and compare the experimental results with the results of the SIFT algorithm. The experimental equipment parameters are: processor Intel Core 2 Duo T5750 2.00GHZ, memory 2GB) (DDR2667MHZ). The software platform used in the experiment is:

Microsoft Visual C++ 2017 Express Edition, OpenCV 3.4.7, cmake-3.11.1-win64-x64. This algorithm is implemented using C++ language programming.

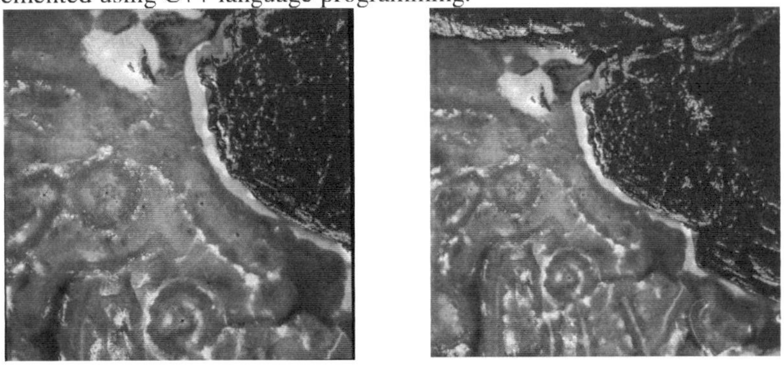

Figure 2. Raw image data

Figure 2 shows the original image data. Figure 3 shows the comparison of the feature point extraction results. The experimental process is to extract the key points of the two images first; then extract the SIFT descriptors from the four gradient patterns of each image; the gradient values of each key point and its surrounding points are sampled as descriptor vector elements, and the dimension of the feature vector is reduced by PCA. Figure 3 is a graph of the results obtained after the key points are extracted. The key points are marked by small circles.

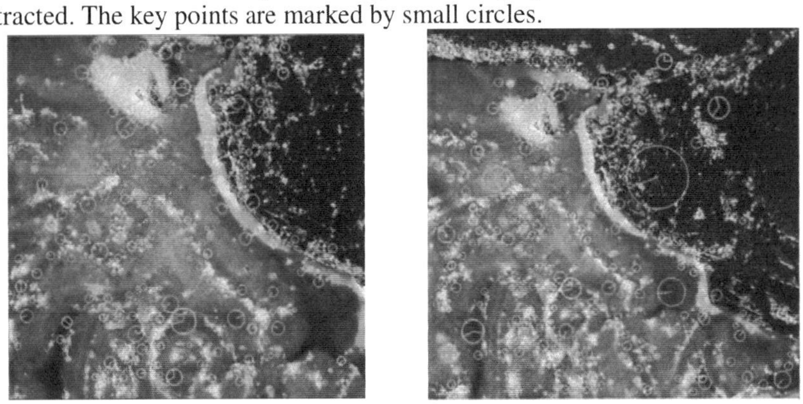

Figure 3. Comparison of feature point extraction results

Figure 4. SIFT algorithm (left) and the algorithm matching and running results

Figure 4 shows the SIFT algorithm (left) matching and running results of the algorithm. We compare the two algorithms and calculate the matching time, the detected matching point pairs, the correct matching point pairs, and the matching rate.

It can be seen from Table 1 that the algorithm uses the Shi-Tomasi algorithm to calculate the feature corners only, and the detection time consumption is short. The improved SIFT algorithm proposed in this paper extracts more feature points than the traditional SIFT algorithm. The number is larger than the number of matching of the traditional SIFT algorithm, which shows the efficiency of the proposed method in extracting feature points. The matching rate of the above experiment shows that the matching accuracy of the algorithm is relatively high.

Table 1. Comparison of test results of two algorithms

	Match time	Detected match point pair	Correct match point pairs	Match rate
SIFT algorithm	9.56 s	216	156	56.3%
This paper algorithm	3.12 s	828	549	69.6%

5. Conclusion

In this paper, different strategies are adopted in different stages of the matching algorithm, and it shortens the running time. Using Shi-Tomasi corner detection and SIFT algorithm, RANSAC algorithm ensures the calculation accuracy. The algorithm makes full use of the gray information and corner position of the image. Experiments show that the proposed algorithm can achieve accurate and fast matching for images with large chromatic aberration and deformation, and its accuracy and speed are better than traditional matching algorithms, and increases the robustness of the algorithm.

Acknowledgement
Fund Project: Tianjin Science and Technology Commissioner Project (19JCTPJC54800); Tianjin University of Technology and Education Graduate Innovation Fund (YC19-30).

References

[1] ZHANG YIMING, FU HUI. A method of splicing aerialimage sequences of UAV forest land based on gap degree [J]. Journal of Beijing Forestry University, 2017, 39(6):107-115. (In Chinese)
[2] JIA YINJIANG, XU ZHENAN, SU ZHONGBIN, et al. Remote sensing crop image mosaicing based on opti-mized SIFT algorithm [J]. Journal of Agricultural Engi-neering, 2017, 33(10):123-129. (In Chinese)
[3] QIN XUJIA,WANG QI,WANG HUILING,et al. Sequen-tial remote sensing image mosaic algorithm based on opti-mal sutures[J]. Computer Science, 2015, 42 (10):306-310. (In Chinese)
[4] DENG TAO, HOU DONGPING, LE QINLAN, et al. Im-age mosaic method of small drone combined with SIFT and Harris algorithm [J]. Journal of Information Engineering University, 2015, 16(3):321-326. (In Chinese)
[5] Li Yan. Based on the improved SIFT algorithm of block model, the image matching method of UAV is studied [D]. Guiyang: Guizhou Normal University, 2016.(In Chinese)
[6] Feng Wenbin, Liu Baohua. Improved SIFT algorithm image matching research [J]. Computer Engineering and Applications, 2018, 54(3): 200-205. (In Chinese)
[7] Han Y, Choi J, Byun Y, et al. Parameter optimization for the extraction of matching points between high-resolutio multisensory images in urban areas [J]. IEEE Transactions on Geoscience and Remote Sensing, 2014, 52(9):5612－5621.
[8] Cai G R, Jodoin P M, Li S Z, et al. Perspective-SIFT: An efficient tool for low-altitude remote sensing image registration [J]. Signal Processing, 2013, 93(11):3088-3110.
[9] BEKELE D, TEUTSCH M, SCHUCHERT T. Evaluationof binary keypoint descriptors[C]//IEEE International Conference on Image Processing. Melbourne, Australia: IEEE, 2013:3652-3656.
[10] Yan Ke, Sukthankar R. PCA-SIFT: A more distinctive representation for local image descriptors [J]. 2004, 2(2): 506－513.

Journal of Physics: Conference Series **1423** (2019) 012029 doi:10.1088/1742-6596/1423/1/012029

Research on the competency of teachers of ideological and political theory courses in undergraduate colleges

Guang Li

School of Education Baoshan University, Baoshan, China 678000

Corresponding author and e-mail: Guang Li, 781216468@qq.com

Abstract. Teachers of ideological and political theory courses in colleges should have the consciousness of ideological and political education, be able to guide the practice of education and teaching, and cultivate students to adhere to and apply the basic principles of Marxism to solve practical problems. Using literature method, questionnaire method and mathematical statistics method, the article puts forward that the competence of teachers in ideological and political theory courses in colleges and universities in China includes loyalty, political literacy, initiative, self-confidence, training others, learning ability, knowledge and skills. The mathematical statistics method was used to explore and verify it. It was found that the second-order 4-factor first-order 7-factor competency model has a good fitness.

1. Introduction

General Secretary Xi Jinping clearly pointed out at the National Symposium on Ideological and Political Theory Teachers: "The ideological and political theory course is the key course to implement the fundamental tasks of Lide Shuren." "The key to running the ideological and political theory course is the teacher, the teacher's enthusiasm and initiative. Sex and creativity, building a team of ideological and political theory teachers who are credible, respectable, reliable, ambitious, courageous and promising is an important guarantee for the ideological and political theory. At present, the ability of teachers in the ideological and political theory courses in colleges and universities is insufficient, the work system is not fully integrated, and the pattern of "three-round education" has not yet been formed, especially the teachers of the ideological and political theory courses in the colleges and universities of the prefectures are not up to standard, and some universities Nearly 30% of the gap has brought difficulties to ideological and political work.

Strengthening the study of the competence of teachers in ideological and political theory courses and improving their professional ability are important ways to promote ideological and political work. Through literature review, there is very little research on the competency structure of teachers in ideological and political theory courses. The situation of ideological and political education in colleges and universities in China is very serious. Exploring the competency model of ideological and political theory teachers is to promote the professionalism and professional level of the ideological and political theory course teachers in China as a whole, and promote the professionalization of the ideological and political theory course teachers in colleges and universities. Professional construction is of great significance.

2. The construction of teachers and teachers in the ideological and political theory courses of undergraduate colleges in the prefecture

2.1. Ideological and Political Theory Course Teacher Competency Dictionary Construction

Through the network carrier such as WeChat, QQ, email, etc., the questionnaires for the teachers and teachers of the ideological and political theory courses of undergraduate colleges and universities are sent to the questionnaires, and the respondents are asked to list 10 entries that can summarize the competencies of ideological and political theory teachers. After the vocabulary combing, merging and categorizing, 80 items of professional expectations, ability factors, self-motivation and occupational conservation were collected. Inviting a master's degree or associate professor's title to a total of 3 people to college and university ideological and political theory course teachers competency terms (80) back-to-back merger classification, select at least three or more approved entries, and finally get 26 entries.

Table 1. College Ideological and Political Theory Course Teacher Competency Dictionary Entry

Dimension	entry
Professional expectations	(including expertise and skills, comprehensive knowledge and skills)
Competence factors	Interpersonal communication, relationship building, self-control, teamwork, team leadership, awareness, innovation, learning, communication, execution
Self-motivated	loyal, initiative, self-confidence, self-awareness, optimism, pleasantness, openness, self-motivation
Occupational conservation	political literacy, responsibility, training others, caring for others, organizational identity, empathy, professionalism

According to the above table, the competency dictionary of some colleges and universities is used to construct the competency coding dictionary of the ideological and political theory course teachers in the undergraduate colleges. Each term consists of a name, a definition, a rating, and a description of the behavior. The level indicates the intensity or complexity of the performance. Different levels of behavioral descriptions are primarily representative of behavioral performance in student work.

2.2. Exploring the competence of teachers in ideological and political theory courses in colleges and universities

2.2.1 Participant

Through the IQ group of ideological and political theory teachers and the qualified ideological and political theory teachers around the members of the research group, they agreed to interview. Because some teachers of ideological and political theory take into account the long interview time and have an influence on them, they propose to provide work-related case texts for research use.

2.2.2 Research materials and tools

The self-edited "Interview Outline of Key Events for Teachers and Teachers of Ideological and Political Theory Courses in the Undergraduate Colleges and Universities" mainly includes research introduction, confidential description, basic information, and interview questions for the interview host.

According to the Dictionary of Competency of Teachers and Teachers of Ideological and Political Theory Courses in the Undergraduate Colleges and Universities, the paper analyzes the key events collected in the interviews. After coding, analysis and assignment, spss15.0 software is used for correlation analysis and factor analysis to explore the competency model of ideological and political theory teachers.

2.2.3 Competency factor coding

Two members of the research team were invited to carry out the formal coding work after training, and all the key event text materials were familiar before coding. According to the Dictionary of Competency of Teachers and Teachers of Ideological and Political Theory Courses in the Undergraduate Colleges and Universities, each coder independently analyzes and identifies the rank, frequency and average score of the competency indicators appearing in each event text. Finally, collate and merge the results of the competency coding, and discuss and determine the differences in the coding.

2.2.4 Data analysis

The competency coding data is collated, and the highest score, average score and frequency of each competency factor are counted. After entering the spss19.0 system, statistical analysis was performed.

2.2.5 Constructing a competency model for teachers of ideological and political theory courses

According to the statistical analysis of the highest score, average score and frequency index of competency factors, the elements with significant differences in each index are selected to construct the competency model of the teachers of ideological and political theory courses in undergraduate colleges.

2.3. Analysis of results

Spencer believes that when constructing a competency model, it can be explored based on the frequency, average score, or highest score of the competency factor. In the study, the statistics of the three indicators were comprehensively analyzed. In order to improve the efficiency, accuracy and objectivity of the analysis, in the analysis process, the indicators with too many missing values after the coding of the competency indicators are excluded. In the analysis results of 26 competency indicators, statistical analysis was performed after removing the indicators with more than 14 missing values. Through the analysis of the highest scores, average scores and frequency indicators of competency and other indicators, the statistical results are organized as follows.

2.3.1 Statistics on the highest scores

Table 2. Statistics on the highest scores of competency of teachers in ideological and political theory courses in colleges and universities

Loyalty						
Initiative	0.032					
Political literacy	0.013	0.005				
Self-confidence	0.021	0.404	0.032			
Cultivating others	0.019	0.081	0.001	0.140		
Learning ability	0.414	0.007	0.046	0.289	0.005	
Knowledge and skills	0.016	0.004	0.082	0.011	0.013	0.019

From the relevant statistics of the highest score indicators, the competency indicators that remove too many missing values, and the indicators that finally enter the analysis are statistically compiled. The results show that the indicators have a significant difference in the $p<0.05$ level. Including loyalty, initiative, political literacy, self-confidence, training others, learning ability, knowledge and skills.

2.3.2 Relevant statistics of the average sub-indicators

Table 3. Statistics on the average scores of teachers' competency in ideological and political theory courses in colleges and universities

Loyalty						
Initiative	0.440					
Political literacy	0.013	0.021				
Self-confidence	0.032	0.014	0.423			
Cultivating others	0.069	0.031	0.011	0.012		
Perception power	0.585	0.011	0.055	0.012	0.015	
Knowledge and skills	0.266	0.008	0.042	0.007	0.007	0.000

From the relevant statistics of the average sub-indicators, the competency indicators that remove too many missing values, and finally enter the analysis of the indicators, the results show that the indicators have a significant difference in the $p<0.05$ level of competency indicators. Including loyalty, initiative, political literacy, self-confidence, training others, awareness, knowledge and skills.

2.3.3 Statistics on frequency indicators

Table 4. Statistics on the Competency Frequency Index of Teachers in Ideological and Political Theory Courses in Colleges and Universities

Loyalty						
Empathy	0.002					
Political literacy	0.021	0.001				
Self-confidence	0.134	0.009	0.001			
Cultivating others	0.001	0.331	0.006	0.263		
Learning ability	0.013	0.005	0.013	0.549	0.031	
Knowledge and skills	0.033	0.000	0.032	0.044	0.001	0.000

From the statistics of the frequency index, the competency indicators with too many missing values are eliminated, and the indicators that finally enter the analysis are statistically analyzed. The results show that the indicators of the indicators have significant differences at the $p<0.05$ level, including the loyalty indicators. Empathy, political literacy, self-confidence, training others, learning ability, knowledge and skills.

2.3.4 The competency model of teachers and teachers of ideological and political theory courses in prefectural colleges and universities

From the statistics of the highest score, the average score and the frequency index, the competency indicators that do not significantly distinguish the indicators are not shown in the table. There are at least two indicators on the three indicators that have significant competencies. After finishing, the competency models of the ideological and political theory teachers include loyalty, initiative, political literacy, self-confidence, training others, learning ability, knowledge and skills.

2.3.5 Reliability analysis

The consistency result of the coder's independent coding is a key indicator to measure the reliability of the construction of the competency model, mainly from the consistency of the classification. Interpretation is made by the percentage of scorers who are consistent with the same text material as a percentage of the total number of codes. The calculation formula is: $CA = (2 \times T1 \cap T2) / (T1 \cup T2)$. T1 denotes the number of codes of the coder A, T2 denotes the number of codes of the coder B, $T1 \cap T2$ denotes the number of codes of the coders, and $T1 \cup T2$ denotes the sum of the code numbers of the coders. In the study, T1=1608, T2=1589, $T1 \cap T2$=1466, CA=91.7%.

2.3.6 Validity analysis
Seven teachers of ideological and political theory were invited to analyze the seven competency indicators. After the feedback from the ideological and political theory teachers, the consistency reached 99%. The seven indicators of teachers and teachers of the ideological and political theory courses in the undergraduate colleges and universities have good Theoretical validity.

3. The competency model verification of teachers and teachers of ideological and political theory courses in colleges and universities of prefecture

3.1 Methods and processes

3.1.1 Research tool
According to the seven dimensions of the competence of teachers in the ideological and political theory course, the questionnaires for teachers and teachers in the ideological and political theory courses of colleges and universities are compiled. The questionnaire includes demographic variables and competency dimensions. The questionnaire was compiled and discussed by three research team members. Three students from the ideological, philosophical and literary backgrounds were invited to participate in the revision, and three experts were invited to give final guidance and improvement on the questionnaire.

3.1.2 Participants
In accordance with the principle of random sampling, the test of the competency questionnaire for teachers of ideological and political theory was completed in Yunnan Province. After finishing, a total of 428 valid questionnaires were collected, all of which were sorted and entered into the system software, and SPSS 19.0 statistical analysis was used.

3.2 Results and analysis

Demographic variable analysis
The demographic variables for the sample data are statistically as follows.

Table 5. Statistical analysis of respondents

Basic information		Frequency	Percent
sex	Male	150	35.0
	Female	278	65.0
Political outlook	Member	3	0.7
	Member of the Communist Party of China	310	72.4
	Democratic Party	25	5.8
	Other	90	21.0
Age	30 years old and below	38	8.9
	31-40 years old	145	33.8
	41-50 years old	171	40.0
	51 years old and above	74	17.3
Education	High School and below	0	0
	Specialist	10	2.3
	Undergraduate	216	50.5
	Graduate and above	202	47.2
Technical position	Primary	42	9.8
	Intermediate	168	39.3
	Advanced	214	50.0
	No title	4	0.9
Engaged in ideological and political class time	2 years and below	84	19.6
	3-5 years	128	29.9
	6-8 years	107	25.0
	9 years and above	109	25.5

In the demographic statistics of the questionnaire, the female sample was high, accounting for 65.0%. Party members are high, accounting for 72.4%. Young and middle-aged teachers accounted for 82.7% of the total, and the income of student staff was relatively low. The degree of education is mainly undergraduate and graduate students, accounting for 97.7%. Professional and technical positions are mainly senior, accounting for 89.3%. Most of the time spent working in ideological and political education is more than 2 years, accounting for 80.4%.

Exploratory factor analysis
Bartlett spherical test and KMO value analysis were performed, and the p value was close to 0. Under the condition that the significance level α was 0.05, the P value was smaller than α, and there was a significant difference. The KMO value is 0.919 (close to 1) and is suitable for factor analysis. The factor analysis result statistic and the total variability interpretation rate, as well as the eigenvalues and contribution rates of each factor are shown in the table.

Table 6. Factor analysis statistics and indicator total variation interpretation rate

Factor	eigenvalue	percentage change (%)	cumulative percentage of variation (%)
1	7.118	28.621	28.621
2	5.755	20.438	49.059
3	2.837	9.152	58.211
4	1.566	5.103	63.314
5	1.493	4.852	68.166
6	1.276	3.942	72.108
7	1.181	3.076	75.184

After exploration and analysis, when the cumulative rate of variation tends to be stable, the competency model of the teachers of ideological and political theory courses in 7 dimensions is obtained, and the cumulative variance interpretation rate is 75.184%. Therefore, seven factors can be used to describe the competence of teachers in ideological and political theory courses in colleges and universities. Considering the rationality and interpretability of the measured variables, the seven factors are named: loyalty, initiative, political literacy, self-confidence, training others, learning ability, knowledge and skills.

Reliability and structural validity analysis
On the basis of exploratory factor analysis, in order to further verify the consistency and stability of the ideological and political theory course teacher competency model, the reliability and structural validity are analyzed.

Table 7. Correlation between factors and the total score of competency

	Loyalty	initiative	political literacy	self-confidence	cultivating others	learning ability	knowledge and skills	competence
Loyalty								0.683
initiative	0.449							0.724
political literacy	0.463	0.528						0.773
self-confidence	0.484	0.498	0.478					0.742
cultivating others	0.519	0.575	0.569	0.509				0.799
learning ability	0.314	0.588	0.589	0.546	0.599			0.772
knowledge and skills	0.421	0.478	0.543	0.552	0.602	0.591		0.781

The correlation coefficient between the seven factors is between 0.314 and 0.599, indicating that each factor can contribute to the total score of competence, and at the same time has a certain relative independence. The correlation coefficient between each factor and the total score of competence is between 0.683 and 0.799, which is significantly higher than the correlation coefficient between the factors, indicating that the structure validity of the questionnaire is good.

Factor internal consistency reliability and split-half reliability

The internal consistency of the seven factors in the questionnaire was analyzed, and the Cronbach-coefficient and the split-half reliability coefficient were calculated. The results are shown in the table:

Table 8. Internal consistency and split-half reliability of each factor

	Loyalty	initiative	political literacy	self-confidence	cultivating others	learning ability	knowledge and skills	competence
α coefficient	0.866	0.867	0.861	0.880	0.852	0.859	0.858	0.819
Semi-trustworthiness	0.861	0.872	0.798	0.843	0.853	0.849	0.876	0.884

The sum of each factor is summed to obtain the total score of competence. The Cronbach-coefficient is 0.819. The correlation between the factors is higher, and the internal consistency between the factors is higher. Because the competency of each dimension of the questionnaire is randomly arranged. Further, the correlation between the first half of the questionnaire and the second half of the questionnaire is calculated. The first part is 0.809, the second part is 0.844, and the sub-trust reliability of the total score of the questionnaire is 0.884, indicating that the internal reliability of each factor is relatively high ideal.

Verification and Analysis of the Competency Model of Teachers in Ideological and Political Theory Courses in Colleges and Universities

From the relationship between the seven factors and the total score of competency, there is a certain correlation between the competency factors of teachers in ideological and political theory courses in colleges and universities, and it is obviously lower than the correlation between each factor and the total score of competency. Do further verification analysis. The confirmatory analysis was performed using Amos7.0 software, and the fitting results of the 7-factor related structural equation model are shown in the table.

Table 9. Statistics of fitted indicators of the measurement model

χ^2	df	χ^2/df	RMSEA	CFI	NFI	TLI
1606	612	2.252	0.057	0.938	0.926	0.945

It can be seen that the fitting index $\chi 2$/df=2.252 is between 2.0 and 5.0, RMSEA=0.057 is lower than 0.08, and the CFI, NFI and TLI values are all greater than 0.9, indicating that the model fit is good, from the fitting index. The model has a high degree of fit and stability.

Since there is correlation between the 7 factors, model correction is also needed to extract higher order factors. According to the analysis, the results are: loyalty, initiative, self-confidence, extracting high-order factors named self-driven; training others to extract higher-order factors named occupational literacy; learning ability extracts high-order factors named ability factors, political literacy Knowledge and skills are extracted from high-level factors named professional literacy. The results of the fitting index of the above second-order 4-factor first-order 7-factor model are shown in the table.

Table 10. Statistics of second-order 4-factor first-order 7-factor model fitting index

χ^2	df	χ^2/df	RMSEA	CFI	NFI	TLI
1586	612	2.399	0.068	0.955	0.946	0.958

It can be seen from the table that the fitting index $\chi2/df=2.399$ is still between 2.0 and 5.0, RMSEA=0.068 is lower than 0.08, and the CFI, NFI and TLI values are all greater than 0.9, indicating that the model has good fit and high Stability, the model structure is more reasonable.

4. Discussion

4.1 The construction method of the competency model of teachers in ideological and political theory courses in colleges and universities

In constructing the competency model of ideological and political theory teachers in colleges and universities, using the key event interview methods recognized in the current research, combined with the factor analysis and structural equation model in mathematical modeling, explored and verified the competence of teachers in ideological and political theory courses in colleges and universities. The second-order 4-factor first-order 7-factor structure of the force model. In the construction of the competency dictionary for teachers of ideological and political theory courses in colleges and universities, the author has done a full gestation of the collection, sorting, screening and supplementation of the terms, combined with the analysis of the work of ideological and political theory teachers, and invited the experts to revise and determine the competency dictionary. perfect. In the lexicon construction and questionnaire sample selection, the test is based on random sampling and convenient sampling. The sample covers a wide range and is representative.

The research involves the analysis of the work of teachers in the ideological and political theory class, which plays an important role in the construction of the competency model of teachers in ideological and political theory courses in colleges and universities. Niu Duan, Zhang Minqiang (2012), Wu Hongyan, Li Wendong, Xie Yizhong et al. (2006) used the path of work analysis in the study of competency model, which proved to be feasible and effective. Therefore, the follow-up study on the competency model of ideological and political theory teachers can also try to combine the methods of job analysis.

4.2 Analysis of the competency structure of teachers in ideological and political theory courses in colleges and universities

After research and analysis, the competency model of teachers in ideological and political theory courses in colleges and universities is a second-order 4-factor first-order 7-factor structure: loyalty, initiative, and self-confidence are self-motivating factors; cultivating others as professional literacy factors; learning ability as ability factors, Political literacy, knowledge and skills are professional literacy factors. The model verification shows that the structural model is more reasonable.

In general, the capabilities required for a particular job described in a competency model consist of 7-9 dimensions (Shippman, Ash, Battista, Carr, Eyde, Hesketh, et al., 2000). The model in the study is a first-order 7-factor structure, consistent with the conclusions.

4.3 The use of competency of teachers in ideological and political theory courses in colleges and universities

As the backbone of the ideological and political education work, ideological and political theory teachers have put forward higher requirements for their professional ability under the background of the big political policy, and the competency model of the ideological and political theory teachers not only provides the individual's professional ability improvement. The specific direction can also provide a reference for the management of the ideological and political theory course teachers in the personnel department of colleges and universities.

4.4 Research gaps and prospects

Although the research constructs and verifies the competency model of ideological and political theory teachers, and puts forward the structural dimension of teachers' ideological and political theory teachers' competence, there are still many things to be further explored. In the research, the demographic variables of the competency structure of ideological and political theory teachers have not been analyzed, which may have an impact on the use of the competency of teachers in ideological and political theory courses under different variables. To this end, future research can be further refined, commonality and difference analysis for different variables, exploring regular content, and proposing specific practical measures for different variables, the effect is more ideal.

Acknowledgement

Funds: Baoshan University Teaching Quality and Teaching Reform Project Construction Project "Exploration on Optimization of Teaching Staff Management in Transforming Colleges and Universities" (subject number: ZHP201806) fund;

Baoshan University Teaching Quality and Teaching Reform Project Construction Project "Frontier Basic Education Professional Group Construction" (subject number: ZXZ201805) fund.

References

[1] Wang Wei, Wu Qiyang. Study on the Competency Structure Model of College Ideological and Political Theory Courses-Based on the Grounded Theory Perspective [J].Educational academic monthly, 2019.9

[2] Laura, S., Serap, K., Mary, W. J., Joumana, B., Brenda, C., Matyl-da, F., et al. (2013). Culture and competencies: A multi -country examination of reference service competencies. Interna-tional Journal of Libraies & Information Services, 63, 33-46.

[3] Spencer L. M, Spencer S. M. Competence at Work: Models for Superior Performance [M] . New York: John Wiley & Sons, Inc.1993.86.

[4] Shippman, J. S., Ash, R. A., Battista, M., Carr, L., Eyde, L. D., Hesketh, B., et al. (2000). The practice of competency modeling. Personnel Psychology, 53, 703-740.

[5] Me Celland D C.Testing for Competence Rather Than for "Intelligence" [J].American Psychologist, January 1973.

[6] Spencer L.M., Spencer S.M.Competence at Work: Models for Superior Performance. New York: John Wiley&Sons Inc, 1993:199-220.

[7] Danielson C. Enhancing professional practice: Aframe work for teaching [M]. Alexandria: As-sociation for Supervision and Curriculum Development, 1996:311.

Review on big data application of medical system based on fog computing and IoT technology

Baoling Qin[a], Huiying Tang[b], Hongtao Chen[c], Liangyun Cui[d], Jianjie Liu[e], Xueke Yu[f]

College of Electronic Information Engineering, Foshan University, Guangdong Foshan, China
[a]2978879663@qq.com,[b]1017833417@qq.com,[c]247538141@qq.com,[d]1243384137@qq.com, [e]982083608@qq.com, [f]1095132588@qq.com

Corresponding author and e-mail: Baoling Qin, 2978879663@qq.com

Abstract. From the current general situation of big data in medical systems, most of the medical system architectures of comprehensive related literature are based on cloud computing. With the deepening of big data application in the medical system, the drawbacks of medical network systems are apparent, for example, long network time delay, cloud computing overload, information leakage. A new intelligent medical system based on fog computing and IoT technology is proposed to effectively solve the problem of big data application through the comprehensive analysis of many works of literature. Compared with the traditional cloud computing model, the system which adopts a distributed fog computing model based on edge computing has the advantages of reducing network delay and plays its benefits of distributed, edge and decentralization. That the system takes advantage of IoT's features of multi-node, low-energy consumption and big data technology is a new type of medical system that integrates intelligence and big data application. The system is effective through the application and practice of many medical institutions. In short, this system has several characteristics. For example, it can effectively improve medical intelligence and big data processing ability, significantly reduce network delay, effectively improve data security and effectiveness, largely improve the management level and service quality of medical institutions, play the important role of artificial intelligence and big data in the medical field, and greatly improve disease screening rate and reduce misdiagnosis rate. Ultimately it leads to reducing the contradiction between doctors and patients, providing patients with more convenience, faster and more efficient service quality, and alleviating the current situation of difficult and time-consuming medical treatment.

1. Introduction

In recent years, with the continuous development of IoT technology and integration into various fields of society, different networked carriers have also been continuously increased, resulting in more and more data. Therefore, the processing requirements for data are also increasing. Thus, the demand for data processing is also steadily increasing. However, the problems of long-time data processing, easy data loss, insufficient bandwidth resources, and network congestion, have begun to limit the application of cloud computing in some areas, where high-speed processing of data and quick feedback are needed. Especially in the medical system, whether to deal with the long queuing up time for drug retrieval or to reflect the patient's real-time physical condition by wearing a networked instrument for the patient, it is necessary to feedback the data obtained by the sensing device in a very short time. After processing and returning to the devices in the hands of people, these different types of data are transmitted from various sensors. If the system is still based on cloud computing, the system will be overburdened and even paralyzed. Aiming at the serious problems in the current big data application, an intelligent medical system based on fog computing and IoT technology is proposed. The purpose is to solve the problem of big data processing and application of medical service by using the characteristics of fog computing and IoT technology. [1] At the same time, using fog computing to uniformly migrate related data processing and applications to network edge devices, process and save them, and then transfer valuable residual data to the cloud server for storage or further processing. The mechanism effectively solves the pressure of system cloud computing, ensures the limited resources of the system to be applied to key tasks, and improves the real-time data processing capability of the medical system. Based on these characteristics, we will make full use of new big data security technologies to adequately protect the security of big medical data and prevent information leakage. This research has practical significance in the current medical system big data application.

2. Fog computing and architecture model of IoT technology

The fusion of fog computing and IoT technology is the hotspot of current network architecture model research. How to make use of their respective advantages to make the medical system more stable and advantageous service is the subject of our research.

2.1. Basic concepts and characteristics

Fog computing is an extended concept of cloud computing, and data communication processing is closer to the end-user. It is similar to but different from edge computing, which is to release some of the cloud's capabilities to the LAN of the IoT device such as the edge of the network, gateway or router to reduce the network and computing load of the cloud data center. However, fog computing cannot replace cloud computing. [2]It is mainly to expand and supplement the current cloud computing, and it is an enhanced version. In general, it is composed of a variety of weak performance, widely distributed functional computing devices. It is a distributed computing that extends the computing paradigm of the network center to the edge of the network. This kind of distributed computing can meet the demands of some areas where real-time data feedback is faster, such as intelligent transportation, intelligent industrial production, data processing of intelligent medical systems, etc. These features effectively reduce network latency and improve system computing power.

2.2. Architectural model

The architecture model of fog computing is based on the cloud computing model. A layer of fog is added between the mobile terminal and the cloud layer. The fog layer includes networking devices such as gateways and routers, which can process some data and use them. The data is uploaded to the cloud, and the edge layer is an essential part of IoT. It merges with the fog layer to form a new architecture. Its architecture is shown in Fig.1.

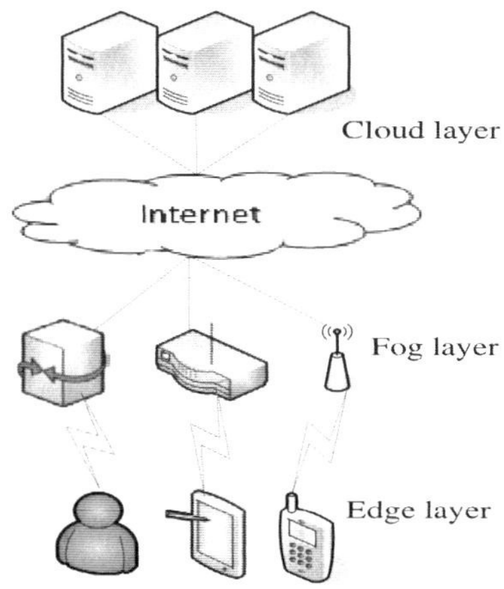

Figure 1. Fog computing architecture model

3. Analysis of big data application of medical system based on fog computing and IoT technology

Medical system big data includes different structural data such as pictures, images, data, etc., and puts a large amount of data to be processed into cloud computing. In addition to the burden, some data may not be processed, resulting in system stagnation or inability to work. Based on these, we use fog computing and IoT technology.

3.1. Medical system model

The widely used medical system is designed and built in the B/S structure. The system can be divided into three layers, namely the browser layer, the server layer and the database layer [3], as shown in Fig.2.

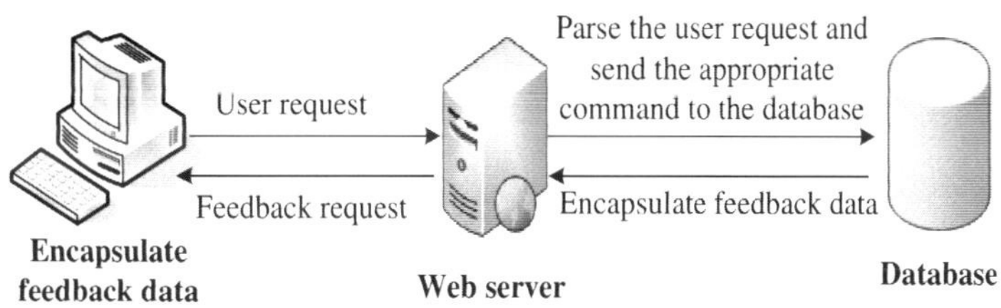

Figure 2. Medical system structure model

The website visualization interface browsed by the system users such as doctors and patients are deployed in the browser layer. The user can access the system by inputting the correct server address through the browser and realize the corresponding request service and output of the required result. The server layer is actually the middle layer. It is the medium for interaction between the browser layer and the database layer. It has the functions of analyzing, processing, and transmitting data. [4]When the layer receives the request transmitted by the browser layer, it can parse, authenticate, etc. according to the established rules, and the corresponding data request is sent to the database. After receiving the request information, the database layer sends the corresponding data to the server layer, and the server layer processes the data collected from the database layer and sends the data to the system for display. It is not difficult to find out that the server layer is indispensable, and its function is good or bad, which can affect the performance of the entire system.

3.2. Fog computing and IoT technology fusion service framework

Fog computing is an extension of cloud computing. The use of cloud computing requires a lot of bandwidth. But the current wireless network bandwidth is still limited. In comparison, the bandwidth requirements of fog computing are much lower, and a considerable part of the data can be realized by fog computing. [5]Localized computing for valuable data can still be transmitted through the cloud platform, but because a part of the data has been localized, it can significantly reduce the burden of the cloud while also saving costs. Integrating fog computing into the Internet of Things not only preserves the powerful features of cloud computing but also increases the speed of computing and reduces the computational cost.

The fog computing is closer to the end of the data acquisition than the cloud calculation, and the data transmission distance is closer, so the data transmission rate is fast. And a fog computing layer can be extended between the cloud centre and the device. The fog computing layer is deployed in the sensing layer of the Internet of Things. [6]The fog access point with computing and storage functions acts as the sensing node of the Internet of Things, enabling data collection, partial data, local processing and storage functions. These points can directly communicate directly with the IoT terminal in the area [7], and its converged service architecture is shown in Fig.3.

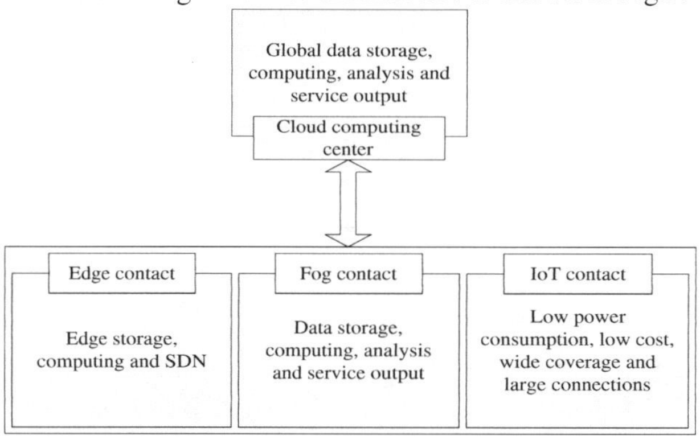

Figure 3. Fog computing and IoT integration services framework

3.3. Application Analysis of Artificial Intelligence and Big Data in Medical System

3.3.1 Application of AI in medical systems

At present, the fusion application of AI and big data is mainly based on fog computing (edge computing) and IoT technology which promotes its application and development in the medical field. Here are three examples. Firstly, the integration of hospital management and AI can not only effectively optimize resource allocation, improve hospital management efficiency and service quality, but also rationally optimize the information and time allocation of personnel, finance, materials and medical institutions and make up the loopholes of hospital management to the maximum extent. Secondly, the integration of medical image and AI, especially cardiovascular and tumor imaging. AI cardiovascular angiography can improve early screening and prevention of circulatory diseases in humans. [8]AI tumor imaging is particularly useful in enhancing the understanding and analysis of treatment and related technology solutions for human tumor diseases. At last, it can apply to disease diagnosis and prediction and AI fusion. Through artificial intelligence technology for timely and effective quality control for image capture, automatic analysis and diagnosis, unnecessary time-consuming can be avoided. Image screening can be effectively realized, and artificial intelligence technology is used to judge and predict the disease from the results of human behavior, biochemistry and imaging.

AI effectively promotes medical research and the standardization of medical data. For instance, it can use machine learning and natural language processing techniques to capture patient medical records and clinical variables automatically. It can also integrate multi-source heterogeneous medical data, structured Medical records and related literature to generate a standardized medical database to effectively achieve "secondary" utilization, reduce costs, and improve the quality and efficiency of medical services.

3.3.2 Application of big data in medical systems

Medical system data includes basic patient information, electronic medical records, medical imaging data, medical management, financial data, device parameter information, and meter data. These data are obtained by collecting various services from various subsystems of the medical system. They have basic data characteristics such as large scale, complex structure, fast growth, and huge potential value. In particular, due to this Class data collection comes from different medical edge devices, and inevitably, it will also have time, redundancy and privacy. Big data application analysis of medical data in the medical system can provide a more reliable basis for drug use, cause analysis, disease prevention and wearable device optimization [9].

For the future development of medical systems, as the scale of medical data will gradually become larger. By mining medical data, analysis and processing, we can further improve the accuracy and timeliness of clinical diagnosis. Compared with the traditional working mode where diagnosing depends on the doctor's expertise and diagnostic experience, the big data medical system can accurately calculate and deliver more decision information to provide reliable data support for clinical diagnosis and effectively improve the accuracy and timeliness of diagnosis. For patients, they could reduce artificial misdiagnosis and time delays. There is no doubt that big data used in the medical field will gradually become more common because it can be used for diagnostic analysis and decision making. [10]But it is also necessary to pay attention to protecting data security and considering how to protect the balance of data privacy. Therefore, it still needs to be added access control strategy with adequate security and an effective data encryption method. Still, the premise is that it cannot affect the overall efficiency, minimize the contradiction between doctors and patients and improve the comprehensive quality level of medical technology and services.

4. Key Technology Analysis of Big Data in Medical System

The key technologies of big data in medical systems mainly include techniques such as data fusion mechanism, data migration, data replacement and security.

4.1. Big data fusion mechanism

The medical system has a lot of data, such as patient name, doctor number, device model, electrocardiogram, sound wave, medical record, image picture, etc. There must be an efficient data fusion mechanism to extract useful information from these massive heterogeneous data which are different specifications for using structured and unstructured data. We use a hybrid structured big data fusion mechanism that is a combination of a structured data center and an unstructured data management platform [11].

A structured data center is a collection of multiple structured data sets that are constrained by database schemas to ensure data consistency, discipline, accuracy and integrity. [12]The unstructured data management platform is an application that uses heuristics to guide data fusion and knowledge acquisition. This fusion mechanism has different processing methods for structured data and unstructured data. The structured data is extracted and washed, and then written into the database directly, and the unstructured data is divided, washed and counted first. What's more, the unstructured data is divided into two categories, a type that can be converted into structured data, another type of unstructured data that cannot be converted or dynamic. The first type is converted into structured data by various application techniques and then written into the database, and for the second type, the data center is managed separately using a special database [13]. Data fusion of a large number of heterogeneous data that has been collected and stored in the current big data situation: there are three ways of data fusion: data combination, data integration and data aggregation, from combination to integration to aggregation, and gradually complete various data. The deep interaction between the two, to achieve data fusion [14], as shown in Fig.4.

4.1.1 Data combination

The data combination is a data table formed by the characteristic information of individual data, which is structured data. Data can be directly written into the database, and the nature of the data attribute does not change.

4.1.2 Data Integration

Data integration gathers multiple data groups with the same attributes to form a data mart, and multiple data group marts form a structured data center.

4.1.3 Data aggregation

Data aggregation further aggregates multiple data marts into one subject data, and multiple topics are collaboratively operated in the data management platform to realize information sharing and fusion, thereby further forming a heuristic algorithm of the system, and continuously improving the algorithm. Then, this heuristic algorithm is used to guide the fusion of big data of the system to form a ring of system self-improvement. This is a feature of the hybrid structured big data fusion mechanism, as shown in Fig.5.

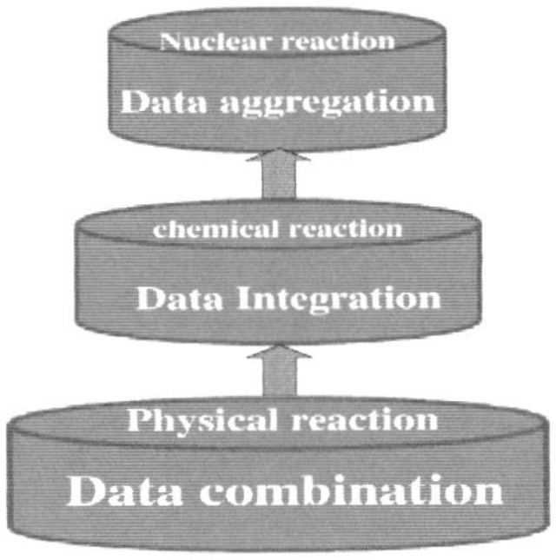

Figure 4. Big data fusion model

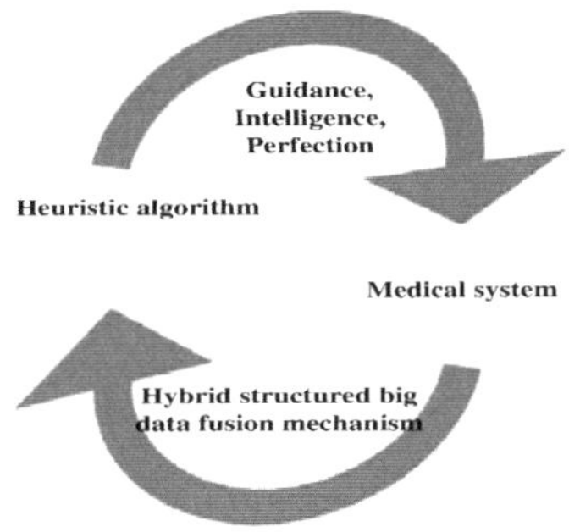

Figure 5. Big data fusion heuristic algorithm model

This big data fusion mechanism can structure and standardize the core data besides. It can flexibly process both custom and unstructured data. This not only ensures the feasibility and availability of data after the integration of big data, but also improves the efficiency of data usage, greatly improves the utilization rate of resources and reduces the occupancy rate of resources, and is a medical system that is scarce for the original resources. It is a very good mitigation solution. [15]

4.2. Big data migration technology

In today's Internet-based healthcare systems, the number of users of information systems is increasing, and the resulting data is exploding. As information technology and big data applications continue to evolve, there is a need to improve and update information systems continuously. In this process, a large amount of data needs to be transferred to the new database, so data migration is particularly important.

Big data migration technology is a kind of technology that migrates a large number of data sets obtained through certain methods and time to a specific database or information system and implements hierarchical storage.

4.2.1 Big data migration technology features: extraction, conversion and loading.

- **Data extraction**

The extraction process determines the extraction method for each data source, and data having a utilization value to the target host or system is extracted from the previously obtained data set. The extracted data is then cleaned, and data that is ambiguous, ambiguous, repetitive, incorrect, and incomplete is discarded.

- **Data conversion**

It is to convert the cleaned data into a specific data set according to the requirements of the new host system. The mapping relationship between databases in the old and new information systems is an important basis for extraction and transformation.

- **Data loading**

Data loading is to use the load tool or self-written program to save the converted data set to a specified target database.

4.2.2 Two big data migration technologies commonly used in medical systems.

- **Logical volume migration technology**

Connecting the host to an original volume and a new volume, and perform data synchronization mirroring between the original volume and the new volume. After the mirroring is completed, the original volume is detached for migration.

- **Data storage migration technology**

Data storage migration technology can be divided into homogeneous storage and heterogeneous. It can complete the data migration work in a relatively short time. Therefore, it can be used in the case of long transmission distance, massive data collection and long interruption time. Therefore, it is suitable for the case, including the long-distance transmission, the big data collecting, and the not-too-long interrupt service time.

4.3. Big data replacement technology

In the medical system, the early accessed data may not be replaced with the most recently accessed data due to a large amount of data, the limited number and storage capacity of cache nodes and the low speed of processing data.

Big data replacement technology is a technology that establishes a data channel between two or more data terminal devices and then takes an optimal scheduling method to solve the replacement problem between a large number by determining the scheduling frequency of certain data sets. The principle is that nodes in the network can implement data exchange between devices connected to the other end according to a certain transmission mode. This technology - an extension of data exchange technology in the field of big data - enables us too efficiently and accurately coordinate the transmission of a huge number of data traffic between various devices and at the same time ensure the security of data. Today, there are quite a few big data exchange service applications, such as FTP/sFTP, Web Service, Dubbo, Oracle, Kafka [16], and so on.

The big data replacement technology of the medical system not only has the essential characteristics of the big data replacement technology but also has the function of data verification and data filtering of the big data replacement technology. Therefore, for smooth transmission of data, there are requirements for hardware configuration. The big data application of medical system based on fog calculation allows the processing and result of data to be placed in the bottom edge terminal, which can significantly reduce the pressure on the server and effectively improve the processing efficiency and transmission rate of the data. This application can meet the requirements of the medical system for big data replacement. Through distributed storage, much important information can be backup well. The sharing of big data is also facilitated. The application of fog computing and big data replacement technology in medical systems, which can significantly alleviate the current situation of excessive medical information, tight storage resources and slow data processing, is a trend in the future medical system.

4.4. Big data security new technology

Since the beginning of applying data to the system, the data security issues have become popular. For example, more and more data leakage, fraud, extortion and cyber-attacks have occurred. It is the same in medical systems; the core issues are patients' data leakage and medical fraud, and so on, which bring great trouble to them. A large number of cyber-attacks also appear in enterprise systems, causing instability factors for society. With the extensive use of big data and Internet of Things technologies, cyberattack capabilities are increasingly powerful, so, it is necessary to improve the requirements of big data security technology. In addition to enhancing security technology functions, it is necessary to strengthen prevention and to use new security technologies to fight against advanced attacks on the Internet of Things. We also need to enhance digital business transformation and new computing technology applications, such as fog computing, edge computing, mobile computing and IoT technology.

A new big data security technology currently in use in medical systems.

- **Cloud platform load protection technology**

This technology allows a single console to manage different types of workloads, infrastructure, and locations, including physical/virtual machines and containers, except for private/public clouds, enabling common security policies to be deployed in all locations.

- **Security proxy technology of cloud access**

Multiple cloud services and applications are used within the enterprise, all of which are monitored by the cloud access security agent. The organization effectively implements security policies to deal with the risks of cloud services and ensure the compliance of all cloud services.

- **Deception technology**

Deception technology can provide insight to detect malicious attacks and take proactive security measures to defeat the attackers. The solutions of deception technology cover multiple layers of the enterprise network, including the data, applications and endpoints of the entire network.

- **Endpoint detection and response technology**

In the Internet of things, these security solutions monitor all endpoints of any abnormal / malicious behavior. The technology of endpoint detection and response focuses on detecting abnormal activities, and then investigates abnormal activities and repairs to mitigate the threat level.

- **Partitioning technology**

It separates the application and work of virtual data center, and the security model of virtualization software assigns fine-tuning security policies to each partition or even each work.

- **Software peripheral technology**

In the IoT system, different systems and enterprises have the same network connection. Software peripheral technology can enable companies to define the content of sensitive information they need to know, and can also effectively delete this information. It helps reduce the attack surface by checking the identity and status of the device before allowing it to access the application infrastructure.

5. Conclusion

The computational structural of distribution mode, aiming at the deficient application of big data of cloud computing platform in the medical service system, uses new IoT structure design with decentration, making the utmost of fog computing and Internet of Things technology. It combines with the technical feature of artificial intelligence and the applied technology of big data to build a new medical service system. This mode reduces the network delay in data transmission in the medical system and the bandwidth requirements greatly and ensures high efficiency, accuracy and security of data transmission. On the one hand, it can provide users with intelligent information functions which can improve the comprehensive management level and efficient service quality of medical institutions, such as hospital information, queuing appointments, medical diagnosis and image analysis, disease screening analysis, medical results printing, medical checkout, etc. On the other hand, it can provide users with a more convenient and competent medical service to solve medical problems such as “difficulties in registration, long queues, network congestion, information leakage”. What’s more, it can improve the efficiency of medical diagnosis, reduce the rate of misdiagnosis, reduce contradiction between doctors and patients and provide the brand-new big data technology application services for the medical establishment.

References

[1] Bonomi F, Milito R, Zhu J,et al.Fog Computing and its Role in the Internet of Things.Edition of the Mcc Workshop on Mobile Cloud Computing. ACM, 2012:13-16.
[2] Shi Yingjuan. Software Architecture Analysis and Application Research of Fog Computing Platform [D]. Zhejiang Normal University, 2016.
[3] Liu Xin. Analysis and Design of Application Function of Hospital HIS System [J]. Electronic Technology and Software Engineering, No.12, 2019 153-153.
[4] 2016 Big Data Development Trend_The Age of Fog Computing Analysis Opens [J]. Journal of Distance Education, 2016(1): 60-60.
[5] Guo Yuxing. Fog computing technology and its application of Internet of Things [J]. Telecommunications Science, 2018 (Z1): 90-97.
[6] Bonomi F, Milito R,zhu J,et al.Fog Computing and its Role in the Internet of Things.Edition of the Workshop on Mobile Cloud Computing.ACM,2012:13-16.
[7] Yu Fudong, Yang Limin, Yang Lin. Research on Internet of Things Application Based on Fog Computing [J]. Information and Communication Technology and Policy. No.5, 2019, 59-62.
[8] Lehrig S, Sanders R, Brataas G, et al. CloudStore - towards scalability, elasticity, andefficiency benchmarking and analysis in Cloud computing[J]. Future Generation Computer Systems-the International Journal of Escience, 2018, 78: 115-126.
[9] Zhao Xin. The application of big data and cloud computing in medical systems [J]. Science and Technology and Innovation. 2017, No. 22, 121-122.
[10] Song Jie, Shi Huijun. Prospects for the future development of human-car information interaction system [J]. Science and Technology Innovation Guide, 2014, 11 (32): 99.
[11] Zhai Baolin, Xiang Yufan, Li Yi. Structured Data Center and Unstructured Data Management Platform [J]. China Informatization 2019, No. 10.
[12] Li Zuozhao, Liu Jinxu.Application of Mobile Edge Computing in Internet of Vehicles [J]. Modern Telecommunications Science and Technology, 2017,47(03):37-41.
[13] Peng Liu. Research on Unstructured Big Data Fusion Mechanism of University Archives [J]. Inside and Outside Lantai, No. 5, 2019.
[14] https://www.cnblogs.com/itboys/p/9724250.html.
[15] Song Jie, Shi Huijun. Prospects for the future development of human-car information interaction system [J]. Science and Technology Innovation Guide, 2014, 11 (32): 99.
[16] https://www.jianshu.com/p/9763d4d5198f.

A method of fiducial-value normalizing to leveling the global average atomic number of lunar surface calculated by orbital Gamma spectroscopy data

Jiankun Zhao[1, 2, 3], Chao Xiong[1, 3], Wenxun Yan[1], Liangquan Ge [2*]

[1]School of Nuclear Science and Engineering, East China University of Technology, Nanchang 330013, China
[2]College of Applied Nuclear Technology and Automation Engineering, Chengdu University of Technology, Chengdu 610059, China
[3]Fundamental Science on Radioactive Geology and Exploration Technology Laboratory, East China Institute of Technology, Nanchang, Jiangxi, 330013, China

Corresponding author and e-mail: Liangquan Ge, glq@cdut.edu.cn

Abstract. As "banded structure" is a vexed question in orbital Gamma spectroscopy data inversion. In this paper, a new method to level the lunar global data of average atomic number by the fiducial-value in the stable composition circle is proposed. In order to verify the accuracy, the existing moon-sampling rock data as the "standard sample" and fast neutron data are also presented. The slope of the positive fit curve is 1.875 and the correlation coefficient is 0.72. And some geological characteristics, such as the structural boundaries and internal features of the Antarctic Aiken Basin, can identify obviously.

1. Introduction

The Chang'e-1 gamma ray system (CE1-GRS) has an on-orbit time of up to 1 year. During the flight detection process, it is affected by various factors such as solar activity period, spatial radiation variation, flight attitude change, and may generate various invalid data unavoidably (Ma Tao, 2010). A total of 2,467,169 spectral lines were collected and subjected to "abnormal screening →data correction →energy spectrum cumulation →smooth noise reduction" (Yang Jia, 2013). The spectral line pre-processing work concludes that dividing the gamma-ray spectrum detection line data of the whole moon into a grid of 5°×5° (150km×150km), and performing spectral-by-region spectral analysis on the gridded data (Yang Jia, 2010). While the result of elements distribution, such as ^{40}K, Th, U and others, still shows stripe distribution (Yang Jia,2017; Li Xiaoli, 2015), so as the average atomic number of lunar surface (Ge Liangquan, 2018).

Data leveling is be widely used in inversion, such as cutting-line level adjustment (Huang H P, 1999), measuring-line level (Green, 2003; Huang, 2008; White, Beamish, 2015), median-filtering level (Mauring et al., 2002; Mauring and Kihle, 2006), singular value decomposition (SVD) (Yang Jia, 2015; Smith et al., 2004), and so on. It is various that the orbital gamma data has not an actual measurement benchmark. In addition, CE1-GRS is a polar-orbiting flight vehicle, and the acquired gamma data has time-characteristics (significant differences in different longitude lines). So some methods mentioned above are not very effective in leveling the global average atomic number of lunar surface.

2. Fiducial-value normalizing level

2.1. The average atomic number of lunar surface

The average atomic number of lunar surface (Z_A) calculated by equation (1):

$$Z_A = a\frac{N_{0.511}}{\varepsilon_{0.511}\cdot\sum_{i=1}^{n}\mu_{E_i}\ln E_i\cdot(N_{E_i}^0/\varepsilon_{E_i})} \tag{1}$$

Where, $N_{0.511}$ and $N_{E_i}^0$ are the net peak areas with 0.511-MeV and E_i, ε_{E_i} is the peak efficiency of the gamma-ray spectrometer at energy E_i, μ_{E_i} is the linear attenuation coefficient of the lunar rock and soil for gamma rays with energy E_i, $\boldsymbol{a}$ is a coefficient. The deriving process of this formula can be found in (Ge Liangquan, 2018).

Calculate the distribution of the average atomic number of the lunar surface, and the spatial resolution of the result is 5°×5° (150km×150km). Table 6-2 shows the results of a measurement line crossing the "Mare Imbrium" basin in the range of 30°N and 55°W~15°E. According to formula (1), the effective atomic number Z_A and the ratio of positron annihilation radiant flux with the ratio of gamma ray flux at which electron pairing can occur have a positive correlation. The positive proportional coefficient is a=1.1, and preliminary calculation results of effective atomic number distribution of surface media over the whole moon are shown in Figure 1. Obviously, it is serious that banded structure cannot be removed.

Table 1. Calculation data of average atomic number in the 30°N region on lunar surface

Longitude	55°W	50°W	45°W	40°W	35°W	30°W	25°W	20°W
				count /N_i				
Z_A	11.55	11.78	12.20	12.36	12.86	12.39	12.43	12.66
Longitude				count /N_i				
	15°W	10°W	5°W	0	5°E	10°E	15°E	
Z_A	12.24	12.27	12.26	12.28	12.32	11.94	11.03	

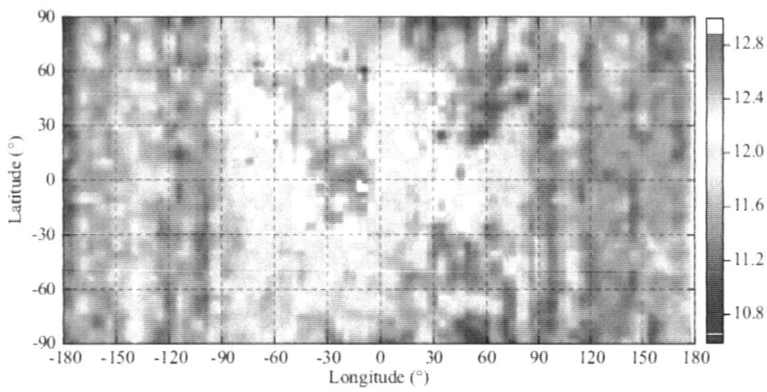

Figure 1. Average atomic number distribution of the lunar surface without data leveling

2.2. Fiducial-value normalizing data leveling

The main performance is as follows: the band profile of the elemental distribution shows obvious distribution along the meridian (Yang Jia, 2013). In addition, due to the obvious regional distribution characteristics of the lunar surface geological body, as shown in Figure 2 ("Chang'e-1 Full Moon Terrain Atlas" Committee, 2013), this makes the lunar gamma ray spectrum show regional differences at different latitudes. (Li Xiaoli, 2016).

Journal of Physics: Conference Series **1423** (2019) 012031 doi:10.1088/1742-6596/1423/1/012031

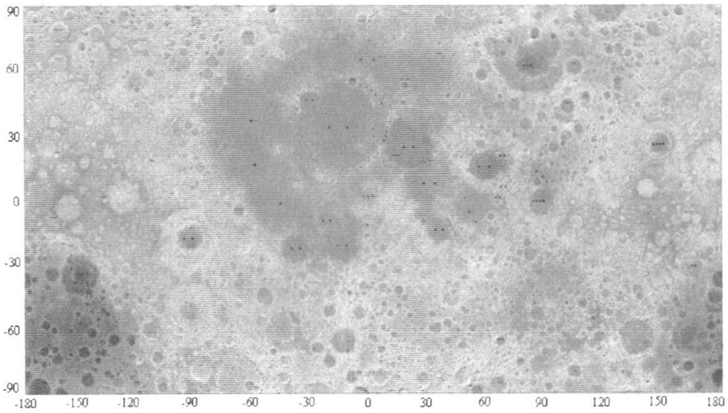

Figure 2. The whole moon topographic map collected by CE-1

Influenced by the above two objective factors, the deviation of gridding gamma spectroscopy data, and even some pseudo-abnormalities of data appears randomly. Therefore, the following methods for data screening and leveling are proposed:

(1) Setting a threshold to eliminate obvious pseudo abnormal data;

(2) Selecting the lunar surface stable element distribution area (latitude circle) as the baseline, and correcting the grid data to eliminate the strip effect of the data strip, which is "substance stability circle reference correction";

(3) For some of the remaining sporadic outlier data, it can be corrected according to the background characteristics of the geological body and the adjacent grid data.

The "strip" correction is performed according to the "material stability circle" reference value: extract the latitude data of 75°N~80°N in the average mass distribution map of the lunar surface medium (Figure 1). The uncorrected effective atomic number $Z_{A(i,75\sim80)}$ data distribution is shown in Figure 3.

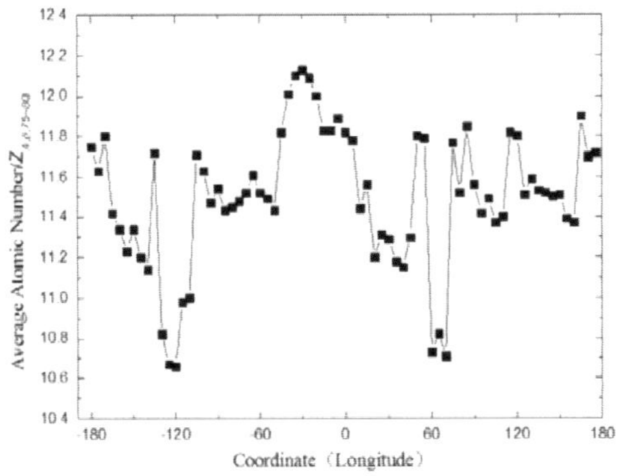

Figure 3. Effective atomic number distribution in the material stability circle

The data fluctuation ranges from 10.6 to 12.2. The standard deviation of this set of data is 0.34, the arithmetic mean is 11.50, and the relative standard deviation is 3%. In particular, the measurement data of the Mare Imbrium concentration area (-45°~0), on the front of the moon, generally shows a large positive deviation, while on the two meridian measurement belts centered on -120° and 65°. The measurement data is generally lower than the average of the whole moon, and this feature is consistent with the banded distribution of the data presented in Figure 1. Therefore, *"i" and "j" is set* as the longitude and latitude respectively. And extract the root mean square (rms) of the average atomic number value in the "material stability circle", as in formula (2):

$$Z^{rms}_{A,(total,75\sim80)} = \sqrt{\frac{\sum Z^2_{A,(i,75\sim80)}}{72}}, i \in (-180,-175,...170,175) \quad (2)$$

According to the above formula, the rms value of the "material stability circle" is 11.49. The calculation result of the "material stability circle" can be used as the reference value of the data of different longitude lines (different flight tracks) globally.

The method for correcting the data consistency between different lines using the result as a reference value is as follows: Firstly, calculate the system calibration parameter $\xi_{(i,\ 75\sim80)}$ of different line data under the condition of the reference value. The specific calculation method is as follows:

$$\xi_{(i,75\sim80)}=\frac{Z_{A,(i,75\sim80)}}{Z^{rms}_{A,(total,75\sim80)}},i\in(-180,-175,...,170,175) \tag{3}$$

The calculation results are shown in Table 3:

Table 2. System correction parameters for different line data $\xi_{(i,75\sim80)}$

longitude	parameter	longitude	parameter	longitude	parameter
-180	1.02	-120	0.93	-60	1.00
-175	1.01	-115	0.95	-55	1.00
-170	1.03	-110	0.96	-50	0.99
-165	0.99	-105	1.02	-45	1.03
-160	0.99	-100	1.01	-40	1.04
-155	0.98	-95	1.00	-35	1.05
-150	0.99	-90	1.00	-30	1.05
-145	0.97	-85	0.99	-25	1.05
-140	0.97	-80	1.00	-20	1.04
-135	1.02	-75	1.00	-15	1.03
-130	0.94	-70	1.00	-10	1.03
-125	0.93	-65	1.01	-5	1.03
longitude	parameter	longitude	parameter	longitude	parameter
0	1.03	60	0.93	120	1.03
5	1.02	65	0.94	125	1.00
10	0.99	70	0.93	130	1.01
15	1.01	75	1.02	135	1.00
20	0.97	80	1.00	140	1.00
25	0.98	85	1.03	145	1.00
30	0.98	90	1.01	150	1.00
35	0.97	95	0.99	155	0.99
40	0.97	100	1.00	160	0.99
45	0.98	105	0.99	165	1.03
50	1.03	110	0.99	170	1.02
55	1.03	115	1.03	175	1.02

Secondly, based on the acquisition of different orbital correction parameters, the system calibration parameters of the latitude circle of 90°S~90°N can be obtained, $\xi_{(i,j)}=\xi_{(i,75\sim80)}$. This parameter can be used to eliminate the system deviation between different tracks, and adjust the data of different tracks to the same level, then the corrected calculation $Z^{cal}_{A,(i,j)}$ is based on the following formula:

$$Z^{cal}_{A,(i,j)}=\xi_{(i,j)}\cdot Z_{A,(i,j)}$$
$$\begin{cases} i\in(-180,-175,...,170,175) \\ j\in(-90,-85,...,85,90) \end{cases}, \tag{4}$$

2.3. Result verification

According to the data leveling method in above section (1) to (3), the average atomic number distribution of the lunar surface is obtained, as shown in Figure 4. The red contour line in the figure is the main area geological structure boundary which is drawn according to the whole moon topographic map (Figure 4).

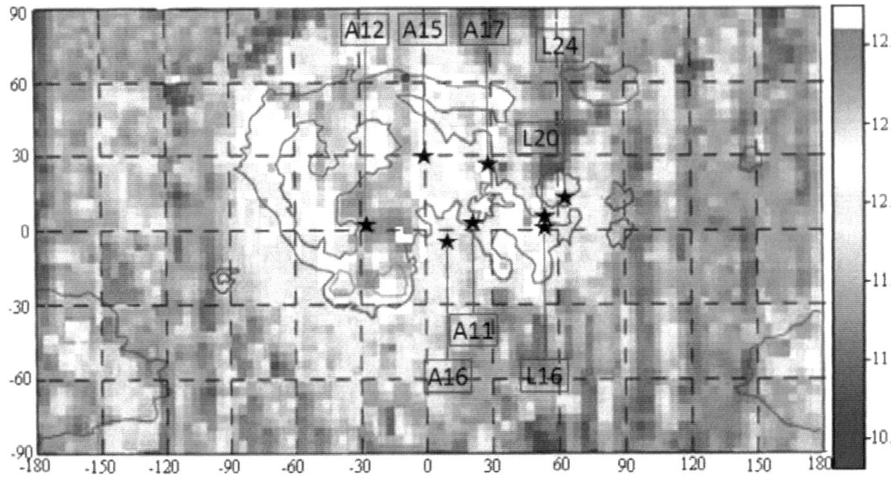

Figure 4. Average atomic number distribution after leveling

In order to verify the accuracy of the average atomic number distribution, the average atomic mass of the sampled rocks (Apollo 11, Apollo 12, Apollo 15~17, Lunar 16, Lunar 20, and Lunar 24 landing points (Gasnault, 2001)) are selected. The linearly fit the two sets of data of the average atomic number of the rock and the average atomic mass by the least squares method. The result is shown in Figure 5:

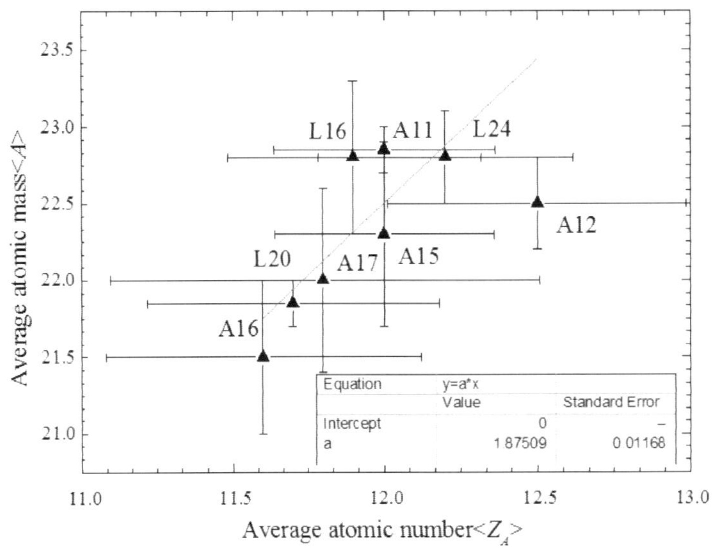

Figure 5. Average atomic number distribution accuracy check result

Compare inversion of atomic mass distribution by fast neutron flux characteristics with inversion of effective atomic number distribution by positron annihilation gamma radiant flux characteristics, as shown in Figure 6. The high and low characteristic range of the effective atomic number distribution is consistent with the basic geological structure distribution of the lunar surface, especially the medium with high average atomic number concentrated in the mare Imbrium area. In addition, the distribution feature can also identify the structural boundaries and internal features of the Antarctic Aiken Basin.

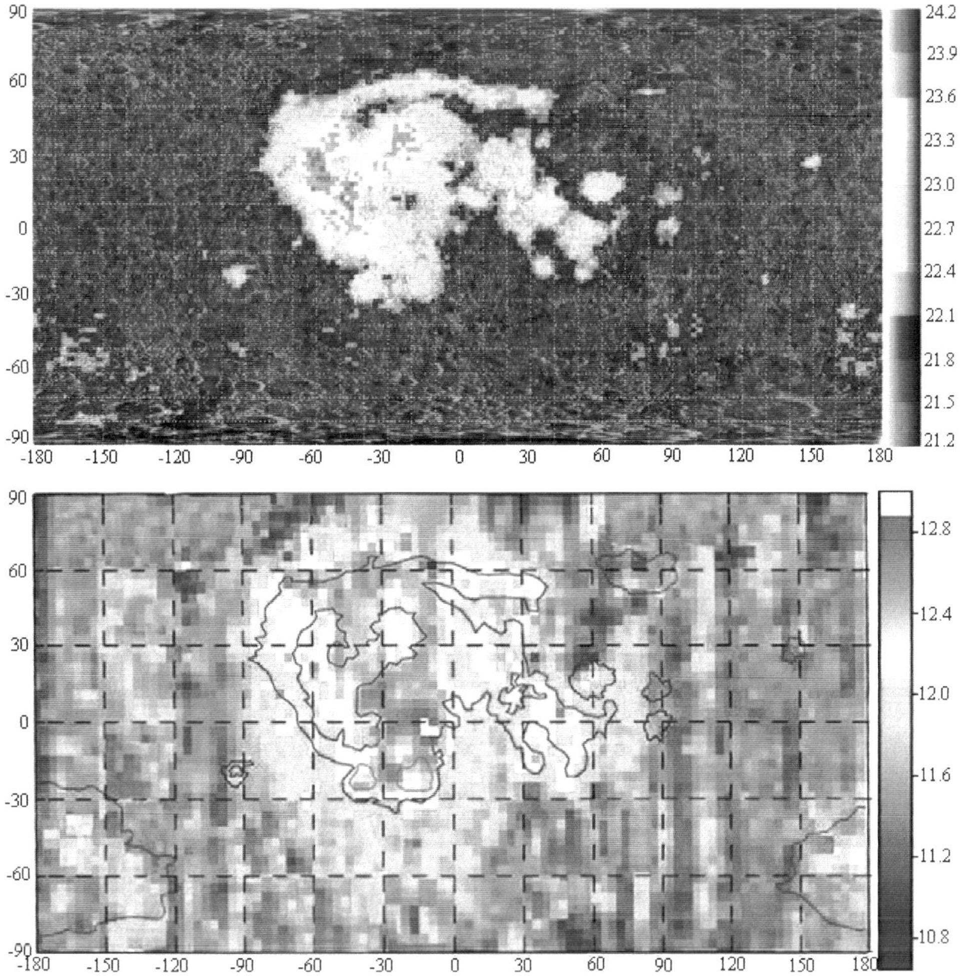

Figure 6. Comparison of inversion of atomic mass distribution by fast neutron flux characteristics and inversion of effective atomic number distribution by positron annihilation gamma radiant flux characteristics

3. Conclusions

After leveling, some new discoveries can be recognized, such as a certain number of high average atomic number points are scattered in typical impact crater areas on the surface around the moon, Schwarzschild crater (70.1N 121.2E). The impact crater is an impact structure formed after the impact of meteorites on the moon. In this structure, the overburden mantle material accumulated, gushed out or invaded the upper part of the lunar crust at the bottom of the impact crater. Resulting in a sharp rise in the content of basalt, which presents a high effective atomic number. This phenomenon is highly consistent with the mass tumor phenomenon involved in gravity. The fiducial-value data leveling is benefit for removing the stripings on average atomic number image mostly. It is meaningful to expand the application of lunar orbital gamma ray data.

Acknowledgement

This work is supported by the National Key Research and Development Project (No. 2017YFF0106503), Scientific Research Project of Jiangxi Education Department (GJJ170451) and RGET1912.

References

[1] Gasnault, O., Feldman, W. C., Maurice, S., et al. 2001 Composition from fast neutrons: application to the moon. Geophys. Res. Lett., 28(19), 3797–3800.

[2] Ge L Q, Zhao J K, Zhang Q X, et al, 2018 Mapping of the lunar surface by average atomic number based on positron annihilation radiation from Chang'e-1, Earth and Planetary Physics, 2, 238-246.

[3] Green A. 2003 Correcting drift errors in HEM data. ASEG Special Publications, (2): 1.

[4] Huang H P, Fraser D C. 1999 airborne resistivity data leveling. Geophysics, 64(2): 378-385.

[5] Huang H P. 2008.Airborne geophysical data leveling based on line-to line correlations. Geophysics, 73(3): 83-89.

[6] Li X L, Ge L Q, Yang J et al. 2015 Distribution of the Thorium Counting Rate on the Lunar Surface from CE2-GRS Based on Cluster NASVD. Nuclear Physics Review, 32(04):453-459.

[7] Ma, T., Chang, J., Zhang, N., Cai, M. S., et al. 2008 Gamma-ray detector on board lunar mission Chang'E-1. Adv. Space Res., 42(2), 347–349.

[8] Mauring E, Beard L P, Kihle O. 2002A comparison of aeromagnetic levelling techniqu es with an introduction to median levelling. Geophysical Prospecting, 50(1): 43-54.

[9] Mauring E, Kihle O. 2006Leveling aerogeophysical data using a moving differential m edian filter. Geophysics, 71(1): 5-11.

[10] Smith, R. S., M. D. O'Connell, and L. H. Poulsen. 2004 Using airborne electromagnetic surveys to map the hydrogeology of an area near Nyborg, Denmark. Near Surface Geophysics, 2:123-130.

[11] White J C, Beamish D. 2015.Levelling aeromagnetic survey data without the need for ti e-lines. Geophysical Prospecting, 63(2): 451-460.

[12] Yang J. 2013 Global Distribution Analysis of Potassium on Lunar Surface Using CE1-GRS Data. Atomic Energy Science and Technology, 47(08):1417-1421.

[13] Yang, J., Ge, L. Q., Xiong, S. Q., et al. 2013 Global distribution analysis of potassium on lunar surface using CE1-GRS data. At. Energy Sci. Technol., 47(8), 1417–1421.

[14] Yang J, Li X L, and Ge L Q. 2017 Analysis of the Lunar Gamma-ray Spectra from Chang'E-1 and Chang'E-2. Nuclear Electronics & Detection Technology, 37(02):207-210.

Journal of Physics: Conference Series **1423** (2019) 012032 doi:10.1088/1742-6596/1423/1/012032

Investigation about the practical teaching effect of financial transformation colleges and the research on the construction of evaluation model

Xuehong Zhao, Yao Xu

School of Management, Harbin Finance University, Harbin, China

Corresponding author and e-mail: Xuehong Zhao, 23212949@qq.com

Abstract. Practical teaching is an effective approach to consolidating theoretical knowledge and deepening theoretical cognition, and also the important means to cultivate students to master scientific methods and improve their operational capacity. From the perspective of cultivating students' occupational capacity, on the basis of reading lots of relevant references, the questionnaire is designed from several aspects, including the teaching methods, teaching mode, course evaluation, the enhancement of occupational capacity and the evaluation on the setting of practical teaching and designed to know about the feelings of students about practical teaching as well as to evaluate on relevant practical teaching conditions. Besides, it has also carried out factor analysis and regression analysis on investigation data respectively, and extracted four factors of occupational adaptability, occupational action capacity, practical teaching design and practical teaching assistance from the enhancement of occupational capacity and the setting of practical teaching. In combination with the questionnaire result and the four extracted factors, corresponding improvement thoughts are put forward based on the objective of building the practical teaching system focused on enhancing occupational capacity, establishing and perfecting practical evaluation system and teaching assurance system.

1. Introduction

Practical teaching is the generic term for all teaching activities, in respect of theoretical teaching. It is an effective approach to deepen theoretical cognition, and consolidate theoretical knowledge, so that students can know about theoretical knowledge, cultivate students to master scientific method, form the scientific attitude that combines theories with practice, and improve students' comprehensive practical capacity under certain actual scenes. *Marketing Practical Training Course* is a course set to further improve students' occupational capacity. In the course, the basic objective is to cultivate students' practical skills and help students understand and master theoretical knowledge, supplemented by exercising students' organizing capacity and cultivating students' occupational quality. Besides, students can make decisions independently, participate in competition, and give full play to the subjective initiative to study in the entire practical teaching process.

Practical training can help students get familiar with the theoretical system of marketing and the marketing environment, establish the marketing concept and innovation awareness in the modern market, know about how to carry out market investigation and make marketing plan, initially master all marketing strategies and the comprehensive utilization, control the comprehensive application of marketing dynamically, cultivate students' capacity to design investigation plan, carry out reference research, field observation research, questionnaire design, collect investigation data, write plans and the teamwork capacity as well as other skills required for marketing.

2. Research Program Design

Students attending Marketing Practical Training course are selected in the investigation as the tested objects, to fill in the questionnaire. Lots of domestic and overseas relevant references were checked, to design questionnaire based on the current questionnaire and model. The questionnaire consists of the title, description letter, background and the main part of the questionnaire.

The main part of the questionnaire is designed with closed questions related to the teaching method, teaching contents, teaching mode, course evaluation and other aspects, and is also set the suggestions of practical teaching as open questions. In respect of the enhancement of occupational capacity and the evaluation on the setting of practical teaching, Liker five-point scale is applied in the investigation, of which occupational capacity can be divided into 8 aspects, including thoughts creation capacity, and teamwork capacity, and then investigation is carried out, in respect of students' evaluation on the enhancement of occupational capacity as well as the evaluation on the setting of practical teaching in 8 aspects, including hardware equipment, and teaching environment. Adopt five-point measurement method for scoring; record 1 point, 3 points and 5 points respectively to express dissatisfied, general and satisfied. In order to improve the scientific validity of the questionnaire, and check the quality of the questionnaire, we can obtain suggestions from experts, teachers and students in the form of paper questionnaires and electronic questionnaires, and modify the questionnaire based on feedback opinions. Test partial students and classrooms in a small scale, revise all titles repeatedly, and form the final questionnaire.

3. Analysis on the Scale Investigation Result

This investigation involves two scales, including the enhancement of occupational capacity and the evaluation on the setting of practical teaching

Reliability is the measurement of reliability for the data obtained from questionnaires, and refers to the reliability and stability of measuring results as well as the consistency of measurement. Consistency coefficient, Cronbach's Alpha is adopted in the paper for reliability analysis. The value range of Cronbach's Alpha is 0-1. If the value is bigger than 0.7, this indicates that the questionnaire has a higher reliability. SPSS22.0 statistic software is adopted to carry out reliability analysis for two scale. We can obtain Cronbach's Alpha value for two scales related to the enhancement of occupational capacity and the setting of practical teaching, which are respectively 0.907 and 0.85, indicating that the questionnaire has higher reliability in the design of contents. KMO value is respectively 0.881 and 0.805. The significance of Bartlett sphericity test is 0.000, which has met the level of significance, and indicated that the relevant matrix of the scale has general factors, and correlation, so it is suitable for factor analysis.

3.1. Factor analysis

3.1.1. Evaluation scale for the promotion of occupational capacity

As can be seen from the characteristic value, the characteristic value for the first factor is 4.904, which occupies about 38.257% of the variance. Select characteristic value bigger than 1, extract the former two factors in Factor process, and the characteristic value totally occupies 74.902% of the variance, which indicates that the former two factors have provided most of the information related to the raw data. In respect of factor selection, it requires to find out the index with bigger factor loading capacity after rotation. According to the method adopted by Michael Tracey et al., the requirement for the index is to control the load capacity in one factor at more than 0.5, and the load capacity in other factors at less than 0.4. Otherwise, it shall be removed, and indexes shall be adjusted to 7 from 8. According to the characteristic value of the extracted factors and the gravel figure display, extract 2 factors for naming.

Table 1. Evaluation Factor Naming Related to the Enhancement of Occupational Capacity

Factor	Contained information	Factor naming
X1	Marketing and planning capacity, teamwork capacity, understanding and innovative capacity as well as interpersonal communication capacity	Occupational adaptability
X2	Business negotiation capacity, autonomous learning capacity, and thinking creation capacity	Occupational action capacity

Table 2. Load Matrix for the Evaluation Factor Related To the Enhancement of Occupational Capacity

	Components	
	1	2
Marketing and planning capacity	. 866	
Teamwork capacity	. 804	
Understanding and innovative capacity	. 787	
Interpersonal communication capacity	. 736	
Business negotiation capacity		. 862
Autonomous learning capacity		. 836
Ideological creation capacity		. 810
Market investigation capacity	. 505	. 681

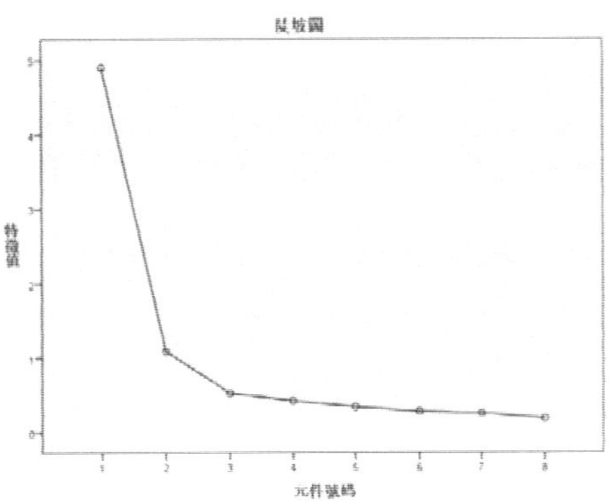

Figure 1. Gravel Figure for the Evaluation of Occupational Capacity Enhancement

3.1.2. Evaluation scale for the setting of practical teaching

As can be seen from the characteristic value, the characteristic value for the first factor is 3.962, which occupies about 36.072% of the variance. Select characteristic value bigger than 1, extract the former two factors in Factor process, and the characteristic value totally occupies 63.902% of the variance, which indicates that the former two factors have provided most of the information related to the raw data.

Table 3. Evaluation Factor Naming for the Setting of Practical Teaching

Factor	Contained information	Factor naming
X3	Evaluation mode, post setting, the interaction of teachers and students, teacher team, and course contents	Practical teaching design
X4	Hardware equipment, teaching environment, schedule arrangement	Practical auxiliary conditions

Table 4. Load Matrix for the Evaluation Factor Related to the Setting of Practical Teaching

	Components	
	1	2
Examination mode	. 804	
Post setting	. 791	
Interaction of teachers and students	. 786	
Teaching staff	. 646	
Course contents	. 632	
Hardware equipment		. 886
Teaching environment		. 803
Schedule arrangement		. 696

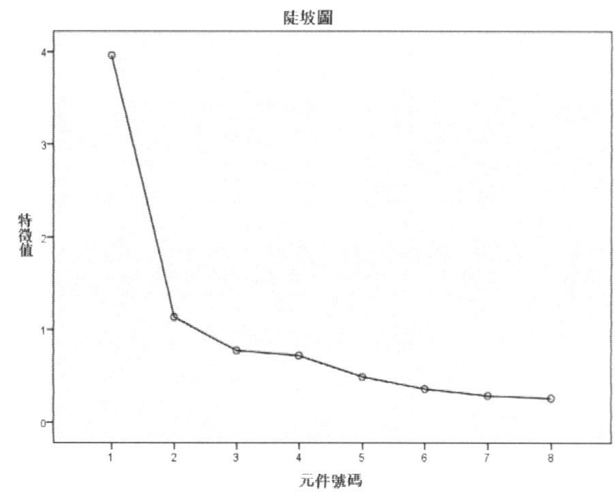

Figure 2. Gravel Figure for the Evaluation of Occupational Capacity Enhancement

3.2. Regression analysis

In respect of overall evaluation, the questionnaire is designed with two questions, i.e., the enhancement of occupational capacity is realized as expected; and course study can contribute to the combination of theory and practice. SPSS22 is applied to carry out regression analysis on factors influencing satisfaction and the overall evaluation result, and the result is as shown in the following table:

Table 5. Regression Coefficient Table Indicating the Enhancement of Occupational Capacity is Realized as Expected

Model		Non-standardization coefficient		Standardization coefficient	T	Significance
		B	Standard error	Beta		
1	(Constant)	4.258	. 050		84.441	. 000
	Occupational adaptability	. 134	. 051	. 187	2.641	. 009
	Occupational action capacity	. 119	. 051	. 166	2.344	. 020

As can be seen from the Regression Coefficient Table Indicating the Enhancement of Occupational Capacity is realized as Expected, Sig value for the two factors is less than 0.05, and can exert obvious influence on realizing the enhancement of occupational capacity as expected. According to the judgment based on the standard coefficient, the occupational adaptability is 0.187, which is 0.166 bigger than the practice auxiliary conditions. Besides, it has a more obvious influence on marketing and planning capacity, teamwork capacity, understanding and innovative capacity as well as interpersonal communication capacity in occupational adaptability.

Table 6. Regression Coefficient Table Indicating the Enhancement of Occupational Capacity is Realized as Expected

Model		Non-standardization coefficient		Standardization coefficient	T	Significance
		B	Standard error	Beta		
1	(Constant)	4.258	. 049		86.018	. 000
	Practical teaching design	. 164	. 050	. 229	3.297	. 001
	Practice auxiliary conditions	. 150	. 050	. 210	3.019	. 003

As can be seen from the Regression Coefficient Table Indicating the Enhancement of Occupational Capacity is Realized as Expected, Sig value for the two factors is less than 0.05, and can exert obvious influence on realizing the enhancement of occupational capacity as expected. According to the judgment based on the standard coefficient, the practical teaching design is 0.229, which is 0.21 bigger than the practice auxiliary conditions. Besides, it has a more obvious influence on evaluation mode, post setting, the interaction of teachers and students, teaching staff, and course contents in the practical teaching design

Table 7. Regression Coefficient Table Indicating that Course Study Can Contribute to the Combination of Theory and Practice

Model		Non-standardization coefficient		Standardization coefficient		
		B	Standard error	Beta	T	Significance
1	(Constant)	4.205	. 072		58.121	. 000
	Occupational adaptability	. 161	. 073	. 158	2.220	. 028
	Occupational action capacity	. 180	. 073	. 176	2.475	. 014

According to the Regression Coefficient Table Indicating that Course Study Can Contribute to the Combination of Theory and Practice, Sig value for the two factors is less than 0.05, and can exert obvious influence on contributing to the combination of theory and practice by course study. According to the judgment based on the standard coefficient, the occupational action capacity is 0.176, which is 0.158 bigger than the occupational adaptability, and effect is more satisfactory, in respect of the evaluation method, post position, the interaction of teachers and students, teacher team, course contents in occupational action capacity.

4. Thoughts to Improve Practical Teaching

According to the investigation result, Marketing Practical Training course is held as important by students, and 61.9% students think that the practical teaching course of Marketing Practical Training is extremely important, while 31.43% students think that the course is relatively important, and 6.67% students think that it is not that kind of important. The practical teaching course of Marketing Practical Training can improve professional knowledge and skills, but some students also express that practical teaching still has many problems. It has an unsound practical teaching system, and its practical teaching mode cannot link with social demands seamlessly. Besides, it is also unsound in cultivation measures and systems, which can occupy 57.14% and 56.19% respectively. Secondly, it is held that problems in practical teaching lacks of advanced teaching concept, and cannot motivate students' interest to study and the teaching methods are old-fashioned, and the two options that can easily cause learning fatigue respectively occupy 28.57% and 23.81%. Besides, factors restricting the development of practical teaching, including the insufficient enthusiasm of students, insufficient emphasis of the management level, unreasonable teaching design, insufficient expenditures and unsound evaluation system should also be solved urgently. Relevant suggestions are as follows:

4.1. Adhere to people-first, and build the practical teaching system focused on improving the occupational capacity

According to the questionnaire result, the most suitable mode to optimize practical teaching as far as students can think of is to perfect practical training facilities in the school at the selection rate of 78.1%, and perfect practical teaching evaluation system, innovate practical training method and the contents of practical training. The selection rate for these three options is 66.67%, 50.48% and 68.57% respectively. By combining with the factor analysis result, occupational adaptability includes marketing and planning capacity, teamwork capacity, understanding and innovative capacity, and interpersonal communication capacity. The occupational action capacity includes business negotiation capacity, autonomous learning capacity and thoughts creation capacity. Occupational adaptability can cultivate students' persistence to overcome difficulties in occupational practice, encourage students to take an active part in the course of explosive study, and closely connect with professional development trend. As thinking orientations are guided in class, and open homework is arranged after class, it encourages students to think actively, exploit students' thoughts, actively create spaces for students to think and explore, solve problems by personal endeavors, and reinforce their confidence in study. Occupational action capacity cultivates the students' capacity to adapt to work demands, obtain new knowledge and new skills and make improvement based on basic professional knowledge and skills. Students' occupational action capacity is mostly cultivated through the comprehensive training of professional capacity, method capacity and social capacity and overall study. It emphasizes on integrating into actual scenes, and regarding the investigation result about the teaching methods (optional) held by students as effective, with the highest selection rate reaching up to 83.81% as the scene for simulation. The selection rate of role play, case analysis, and students' independent discussion is above 60%. The selection rate of multimedia teaching, special investigation and field guidance is about 30%, and students' selection rate of traditional teaching method is only 12.38%. As can be seen from the data, students prefer the teaching mode of scenario simulation, role play, case analysis, etc. As the saying goes, experience is the best teacher. By the mutual iteration of theoretical teaching and practical teaching, it is aimed to make students truly feel about the enterprise environment, and improve professional capacity while experiencing the enterprise system requirements and enterprise culture.

4.2. Establish and perfect practical evaluation system

Achievement evaluation is a systematic, complicated and careful work. On one hand, it is requested to normalize the method to evaluate the tutor, who shall not only has the capacity to guide students in business operation, but also the capacity to make detailed records about students' performance, ensure the objectiveness and fairness of achievement evaluation. On the other hand, it is to formulate the method to evaluate students' performance, issue in the form of teaching documents, and raise students' attention, to ensure the practical teaching effect. As for the selection of evaluation contents, both knowledge and capacity should be focused in evaluation, especially students' comprehensive quality and practical innovation capacity. In respect of the investigation about the evaluation mode, 58% students think that the evaluation of course training process can decide their performance, and 19% students select test paper grading as the form of course evaluation. Investigation result also indicates that it's necessary to exert effective restriction and supervision in the course learning process of Marketing Practical Training.

In respect of the investigation about whether the current course evaluation mode can reflect the actual level", the ratio of people who think that the current course evaluation mode can partially reflect the actual level of students occupies the most, i.e., 62.86%, 20% people think that it can practically reflect the actual level; 14.29% people think that it can only reflect the general level, while 2.86% people think that it cannot reflect the actual level at all. In respect of the evaluation mode of practical teaching, adopt closed written examination, open written examination, oral test, simulated operation, case analysis report, plan design and other diversified examination and evaluation modes based on course characteristics; In respect of judging students' academic achievements, carry out the reform of the traditional academic evaluation system that focuses on the final examination result, and add various grading methods, such as unit test, class discussion, and the evaluation of operating skills. Promote students to develop comprehensively in both personality and capacity by the reform of examination and evaluation method.

The evaluation subject and evaluation contents of practical teaching should be diversified, and teachers should mainly examine whether students can complete the practice in each phase as well as the basic operation of practical training projects on schedule, whether they can truly understand relevant theoretical knowledge and solve actual problems with theoretical knowledge; Students' evaluation includes students' self-evaluation, students' mutual evaluation and group member evaluation. In respect of students' self-evaluation, it requires students to evaluate personally and objectively considering the performance, participation intention, the improvement of relevant occupational capacity in the practice process. As for the mutual evaluation of students, it is applicable to specify objects for students to specify randomly before implementing the practice, and students should focus on themselves and others simultaneously in the practice process.

4.3. Perfect practical teaching assurance system

The assurance system is the support to build integrated practical teaching system. It is also requested to focus on the design of practical teaching based on the questionnaire result, make students experience the actual task environment and reinforce adaptability. Reinforce the construction of teacher team, integrate the teaching resources at school, cultivate teachers who can be engaged in theoretical teaching and professional guidance and equipped with the occupational practice skills that can be synchronized to social economics and technical updating, and provide basic assurance for the occupational capacity of students. Focus on building teacher team, provide it with loose and long-term trust and assurance, improve treatment and reinforce motivation.

5. Conclusions

To sum up, schools should be fully aware of the importance and urgency of occupational practice teaching, build a reasonable and efficient practical teaching system that is oriented to employment and based on capacity, cultivate students' comprehensive professional skills and occupational quality, reinforce the adaptability of students to working posts and working environment by enhancing occupational practice education, realize "zero-distance" docking of learning contents and future work as well as "zero-distance" transmission from graduation to employment, and improve students' competitiveness in employment market.

Acknowledgement

Fund project: Major Entrusted Project for the Research on the Reform of Higher Education and Teaching in Heilongjiang Province in 2017 (Task No.: SJGZ20170028).

References

[1] Jing Shi et al. Exploration for the Route to Improve the Vocational Quality of College Students under New Media Background[J]. School Party Construction and Ideological Education, 2017, 03.
[2] Wei Shi. Discussion about the Practical Teaching System in Colleges[J]. Journal of Higher Education, 2013, (7).
[3] Peng Wang. Research on the Practical Teaching System of Applied Colleges under College Transformation Background[D]. Master's Thesis of Xi'an University of Architecture and Technology, 2017.
[4] Jia Wei, Jie Ni, and Yuanzheng Wu. Research on the Creation of Practical Teaching System in Applied Undergraduate Colleges Based on the Cultivation of Occupational Capacity[J]. Experimental Technology and Management, 2015 (03), 207-210.

Journal of Physics: Conference Series **1423** (2019) 012033 doi:10.1088/1742-6596/1423/1/012033

Monitoring method and test of wheat harvester loss rate based on PVDF sensor

D L Tan, D Y Geng, G L Su, J K Zhu, X R Yu and G D Zhang

School of agricultural engineering and food science, Shandong University of Technology, Zibo 255000, China

Corresponding author and e-mail: D Y Geng, dygxt@sdut.edu.cn

Abstract. Aiming at the problem that the cleaning loss of wheat harvester affects the harvesting quality, we developed a loss detection system based on PVDF (polyvinylidene fluoride) piezoelectric film cleaning loss sensor. After the study of the relationship between the impact speed and the piezoelectric effect of the scavenging screen, the filtering method for discriminating the grain signal is found. The effect of fan speed on grain quality distribution behind the clearance screen of longitudinal axial flow wheat harvester was studied. The results showed that the effect of fan speed 1100~1200r/min on grain quality distribution was not obvious. The optimum monitoring area of the sensor was found by studying the distribution law of grain quality of the clear separation loss, and the mathematical model of the monitoring of the cleaning loss was established. The result of the model test showed that the deviation of the model was less than 1.169, and the accuracy of the model was higher. This paper presents a method for calculating the cleaning loss rate based on the feed rate of the combined harvester. The influence of the feed rate on the monitoring accuracy of the cleaning loss rate was studied by bench test. The relative error of the monitoring results is between 1.84% and 3.15%, which indicates that the combined harvester cleaning loss monitoring system developed in this paper has a high monitoring accuracy.

1. Introduction

Cleaning loss refers to the number of grains mixed in the screenings of combine harvester (Jie, 2009; Zhao et al., 2011), which is generally measured by the rate of cleaning loss. It is an important index to evaluate the quality of harvester operation. Excessive cleaning loss will inevitably increase the loss of wheat harvest and reduce the economic benefits of farmers. In order to prevent the emergence of this problem, many domestic experts and scholars have conducted a lot of research on its monitoring technology and method. For example, Chen (2011) from China Agricultural University studied the law of the influence of feeding amount on the loss rate of combine harvester and established a mathematical model, the method indirectly monitors the cleaning loss by monitoring the feeding amount, but the actual factors affecting the harvest loss include not only the feeding amount, but also the rotational speed of the drum, threshing clearance, air flow size and other factors, so its promotion and use are greatly restricted. Zhou et al. (2010) of the Chinese Academy of Agricultural Mechanization developed a PVDF array cleaning loss detection sensor, by adjusting the sensor signal, the pulse signal number of grain impact plate was obtained to determine the grain loss; Li et al. (2012) from Jiangsu University developed an online monitoring and alarming system for cleaning loss by using piezoelectric ceramic sensor as sensing element and secondary instrument as display; Liang et al. (2015) developed a grain loss monitoring system with high precision, a monitoring model of rice cleaning loss was established. In conclusion, most of the studies on sorting loss are based on collecting the vibration frequency of grain impact plate, filtering it to get the signal of grain impact and counting, and ultimately determining the amount of grain loss. But the feeding quantity of harvester fluctuates, which makes it impossible to accurately describe the amount of loss and the quality of harvester operation. Therefore, this paper carried out research on sensor monitoring technology and method of wheat harvester loss rate.

2. Study on Detection Method of Cleaning Loss

In order to real-time monitor the grain loss caused by the cleaning process of combine harvester and display it in real-time in the form of loss rate, this paper developed a monitoring system of grain loss based on impact principle, the monitoring principle is shown in figure 1. Identification of hybrid grains in effluent by calibrating threshold difference between strike speed and sensor output voltage. The physical model between the seed impact signal and the amount of loss was established by studying the distribution law of lost grains behind cleaning screen, which provided a basis for the online monitoring of the amount of loss. In order to ensure that the selected monitoring points are representative, the model errors of each point are further analyzed and compared to determine the optimal detection location of the detection sensor. The relationship between the total amount of lost grains and the amount of feed was analyzed, and the Grain cleaning loss was converted into the rate of loss grains.

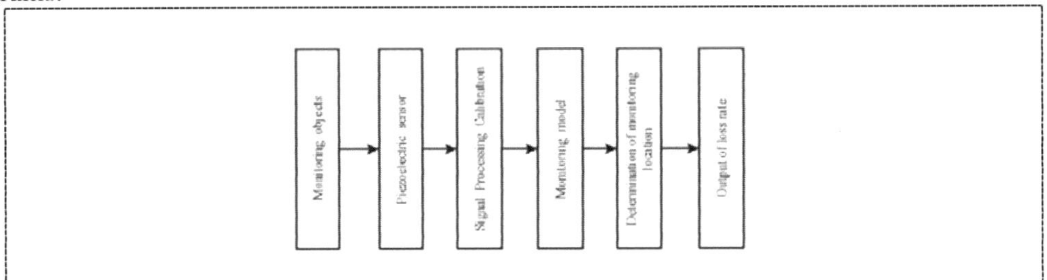

Figure 1. Schematic diagram of Monitoring system

2.1. Sensor selection

The commonly used piezoelectric impact sensors are piezoelectric ceramic type and piezoelectric thin film type (Ju et al., 2004). Because the piezoelectric constant of PVDF piezoelectric film is larger than that of piezoelectric ceramic sensor, it has better voltage output characteristics, and the piezoelectric film has good machinability, the size is determined according to the need, and can withstand large bending deformation. Therefore, the DT series PVDF piezoelectric film sensor of American MEAS brand was selected in this paper to monitor the cleaning loss. The sensor is 52um thick and soft, so this study selected 3mm thick steel plate as the mounting bracket, and the sensor is pasted on the bracket. In order to avoid the monitoring error caused by the secondary ejection between the grains and the mounting bracket, EVA foam double-sided rubber vibration isolation method is adopted between bracket and sensor to reduce noise, so as to ensure the monitoring accuracy of sensor, as shown in Figure 2.

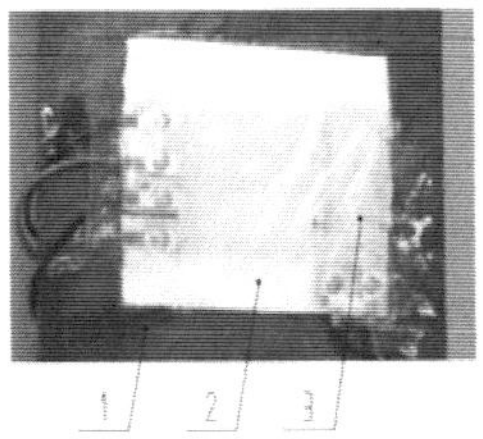

Figure 2. Sensor for cleaning loss

Figure 3. Amplifier for module charge

In order to control the detection noise caused by machine vibration, the rubber sheet is used to isolate vibration at the joint of the impact plate and the harvester frame to further improve the detection accuracy of the sensor. Considering that the small mass of the discharged material from the cleaning screen may lead to the problem that the impact signal is too weak to distinguish between the grain and the miscellaneous signal, a signal amplification module as shown in FIG. 3 is added in the system to amplify the output voltage of the piezoelectric thin film. The working voltage of this module is 5~9V, and the dynamic magnification is adjustable from 0 to 1000 times, which can effectively improve the resolution accuracy of the system.

3. Research on the monitoring model of cleaning loss

3.1. Calibration of seed impact signal

It is necessary to calibrate the impact signals of different impurities on the monitoring system in order to distinguish the impact signals of seeds from the impact signals of many sundries. According to the four components of grain, glume shell, shorter stalk and longer stalk, the effects of different impurity impact strength on the output signal were tested in laboratory in advance, which provided a basis for the calibration of seed impact signal. In this experiment, Jimai 22, the largest planting area in Shandong province, was selected as the experimental object to study the sensor output voltage of grain, glume shell, short stalk (< 10cm) and long stalk (> 10cm) under different impact velocity conditions. The test results are shown in FIG. 4 below. The glume shell is ignored because its floating speed is too small and it has too little impact on the sensor and the output voltage generated by its impact is too small. When the impact velocity of grain is 1.7-2.5m /s, the output voltage of grain is 2-3.3V, which is obviously different from the long and short stalk impact signal voltage of 0.1-1.2v. Therefore, the software filtering threshold is determined accordingly.

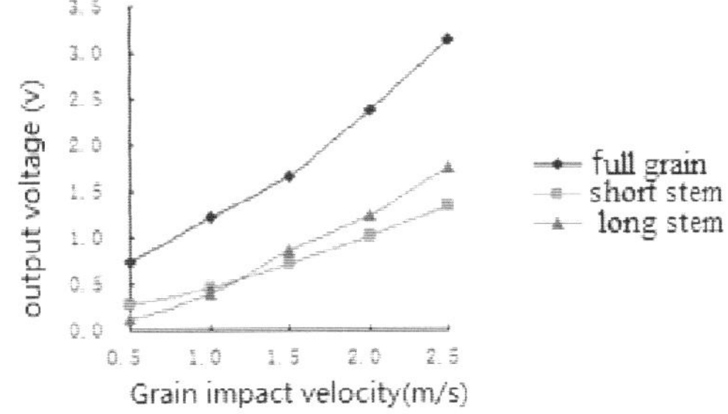

Figure 4. Related curve between striking speed and output

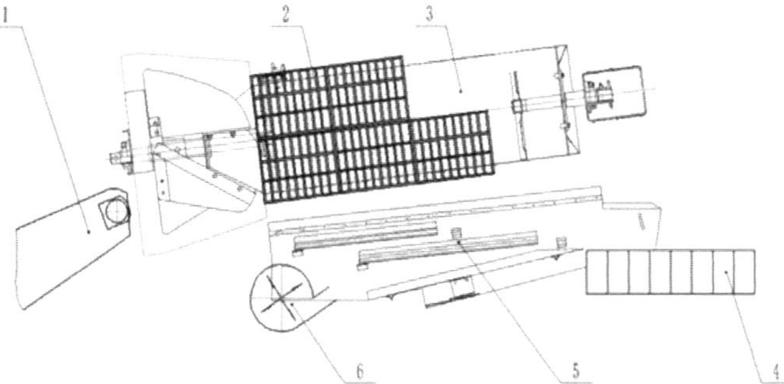

1 conveying bridge 2 concave plate sieve 3 threshing drum 4 connecting box
5 cleaning screen 6 cleaning fan

Figure 5. Schematic diagram of threshing and cleaning system

3.2. Research on mathematical model of grain mass distribution probability

The cleaning loss of the wheat harvester is distributed in a large area after Cleaning screen, so it is difficult to conduct comprehensive and accurate monitoring. Therefore, this study adopts the discrete point method for monitoring. Considering the inhomogeneity of the loss grain distribution in the longitudinal and transverse directions, the loss grain distribution law of cleaning was studied and a monitoring mathematical model was established. The threshing and cleaning test bench used in the experiment is based on the 4LZ-10 wheat harvester developed by Shifeng Group. It is mainly composed of threshing drum, concave plate screen, cleaning screen and fan, and its structure is shown in figure 5.

When working, wheat by conveying the bridge into the longitudinal axial flow threshing and separating drum for threshing, the wheat is separated from the stalk by the impact and rubbing of the concave plate and the ripple bar. The stalk moves backwards with the rotation of the threshing drum to complete the cyclic threshing of the grains without threshing and the separation of the grains from the stalks, which take off grain and some miscellaneous through separating concave plate into the cleaning system, while long stalk discharge outside. Cleaning system is mainly composed of cleaning screen and fan, most of the grains entering the cleaning screen are dropped from the cleaning screen hole, collected and sent to the granary. A small part of grains are discharged from the culm along with other impurities under the action of fan airflow and cleaning screen. This part of loss, together with the unseparated grains in the culm, forms the cleaning loss described in this study.

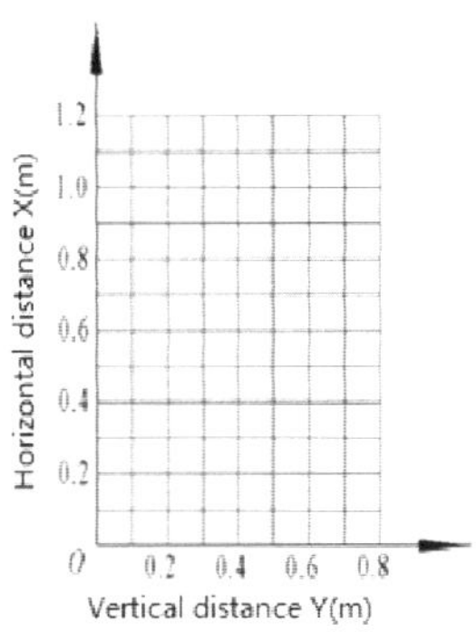

Figure 6. Schematic diagram of connecting box coordinate

Before the experiment, the cutting table of harvester and the culm shredding and throwing device were unloaded, and a receiving box was installed at the tail of the cleaning screen. The position of the receiving box was equal the tail of the cleaning screen. The distribution of the lost grains was determined by weighing the grains in the receiving box. Set the axis direction of the longitudinal flow threshing drum as y axis and its vertical direction as x axis, and the lower left corner of the receiving box as the origin of coordinates O to establish the coordinate axis, as shown in FIG. 6. The receiving box is divided into 12 horizontal grids and 8 vertical grids, with each grid size of 100×100×200mm.It can be seen from the literature (Li, 2013) that the fan speed is the significant factor affecting the loss grain distribution under the specific threshing and cleaning structure. Therefore, the fan speed was set as 1100r/min, 1200r/min, and 1300r/min in the test. The test conditions were 8kg/s feeding, 800r/min drum speed, 250r/min screen cleaning crankshaft speed, and the average value is obtained three times in each group.

During the experiment, wheat harvested from the field was evenly placed on the conveyor belt. The test bench was started and when it was running smoothly, the conveyor belt was opened to adjust the feeding amount of 8kg/s of the simulated harvester. At the end of each experiment, the grains in the box were cleaned and the clean grains were obtained and called their mass $m\ (x_i, y_j)$. The total quality of the lost grains was obtained by adding up the quality of the grains in all receiving boxes: $M = \sum_{i=0}^{8}\sum_{j=0}^{12} m(x_i, y_j)$. The proportion of grain quality in each square: $n\ (x_i, y_j) = \frac{m(x_i, y_i)}{M} \times 100\%$. The percentage of grain quality in each grid was calculated, as shown in table 1, table 2 and table 3 below.

Table 1. Distribution of the clear grains in the receiving box in 1100 r/min

Y the serial number	The x axis serial number											
	1	2	3	4	5	6	7	8	9	10	11	12
1	2.712	2.437	2.175	2.077	1.920	1.891	1.851	1.867	1.885	1.974	2.032	2.241
2	2.347	2.073	1.798	1.700	1.544	1.514	1.475	1.490	1.509	1.597	1.655	1.864
3	1.936	1.662	1.387	1.289	1.132	1.103	1.063	1.079	1.097	1.186	1.244	1.453
4	1.728	1.454	1.179	1.081	0.925	0.895	0.856	0.871	0.890	0.978	1.036	1.245
5	1.487	1.212	0.938	0.840	0.683	0.654	0.614	0.629	0.648	0.736	0.794	1.004
6	1.281	1.007	0.732	0.634	0.477	0.448	0.408	0.424	0.442	0.531	0.589	0.798
7	1.147	0.872	0.598	0.500	0.343	0.314	0.274	0.289	0.308	0.396	0.455	0.664
8	0.767	0.493	0.218	0.120	0.056	0.026	0.000	0.009	0.020	0.017	0.075	0.284

Table 2. Distribution of the clear grains in the receiving box in 1200 r/min

Y the serial number	The x axis serial number											
	1	2	3	4	5	6	7	8	9	10	11	12
1	2.668	2.394	2.131	2.033	1.877	1.847	1.808	1.823	1.842	1.930	1.988	2.197
2	2.314	2.039	1.765	1.667	1.510	1.481	1.441	1.457	1.475	1.563	1.622	1.831
3	1.910	1.635	1.361	1.263	1.106	1.077	1.037	1.052	1.071	1.159	1.218	1.427
4	1.718	1.444	1.169	1.071	0.915	0.885	0.846	0.861	0.880	0.968	1.026	1.235
5	1.492	1.218	0.943	0.845	0.688	0.659	0.619	0.635	0.653	0.742	0.800	1.009
6	1.292	1.018	0.743	0.645	0.489	0.459	0.420	0.435	0.454	0.542	0.600	0.809
7	1.167	0.892	0.618	0.520	0.363	0.334	0.294	0.309	0.328	0.416	0.474	0.684
8	0.869	0.594	0.320	0.222	0.065	0.036	0.004	0.019	0.030	0.118	0.177	0.386

Table 3. Distribution of the clear grains in the receiving box in 1300 r/min

Y the serial number	The x axis serial number											
	1	2	3	4	5	6	7	8	9	10	11	12
1	2.637	2.362	2.100	2.002	1.845	1.816	1.776	1.792	1.810	1.899	1.957	2.166
2	2.293	2.019	1.744	1.646	1.490	1.460	1.421	1.436	1.454	1.543	1.601	1.810
3	1.896	1.622	1.347	1.249	1.093	1.063	1.024	1.039	1.057	1.146	1.204	1.413
4	1.709	1.435	1.160	1.062	0.906	0.876	0.837	0.852	0.870	0.959	1.017	1.226
5	1.498	1.224	0.949	0.851	0.695	0.665	0.626	0.641	0.660	0.748	0.806	1.015
6	1.303	1.028	0.753	0.656	0.499	0.469	0.430	0.445	0.464	0.552	0.610	0.819
7	1.193	0.919	0.644	0.546	0.390	0.360	0.321	0.336	0.354	0.443	0.501	0.710
8	0.899	0.624	0.350	0.252	0.095	0.066	0.034	0.049	0.060	0.148	0.206	0.416

The percentage of grain quality distribution under each wind speed was cumulated vertically and horizontally to obtain the probability of grain distribution along the horizontal and vertical lines, such as table 4 and 5. The significance test showed that the fan speed had little influence on the grain distribution law of wheat when it was in the range of 1100~1300r/min, and the mathematical model under this condition was established.

Table 4. Transverse Distribution of Grain for different Fan rotational speeds

The x axis serial number	Fan speed (r/min)		
	1100	1200	1300
1	13.7794	13.4306	13.0303
2	11.4326	11.2338	10.9335
3	9.3679	9.0491	8.7488
4	8.3655	8.2667	7.9664
5	7.3116	7.0131	6.8128
6	6.9759	6.7771	6.5768
7	6.2675	6.4687	6.6684
8	6.3892	6.5911	6.7908
9	6.4311	6.7323	7.032
10	7.3375	7.4387	7.7384
11	7.5031	7.9043	8.104
12	9.2751	9.5771	9.9768

Table 5. Longitudinal Distribution of Grain under different Fan rotational speed

Y the serial number	Fan speed (r/min)		
	1100	1200	1300
1	23.5371	24.5383	25.783
2	19.1633	20.1645	21.1624
3	12.7148	15.3153	15.951
4	12.4161	13.0173	13.871
5	9.3017	10.3029	11.3062
6	6.6041	7.9053	9.150
7	5.3969	6.3981	7.7987
8	1.6385	2.8397	3.8349

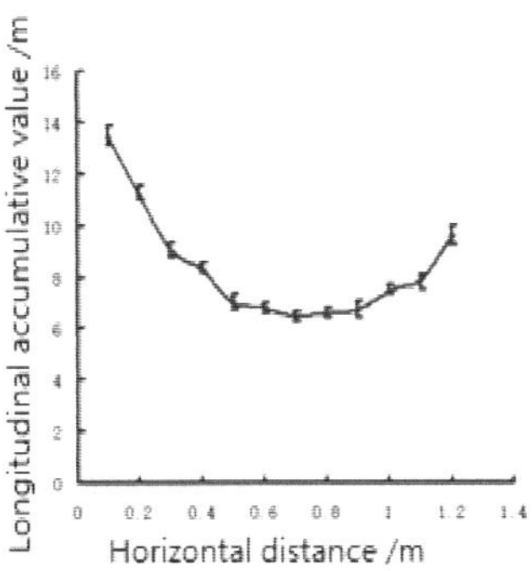

Figure 7. Grain distribution probability longitudinal fitting curve

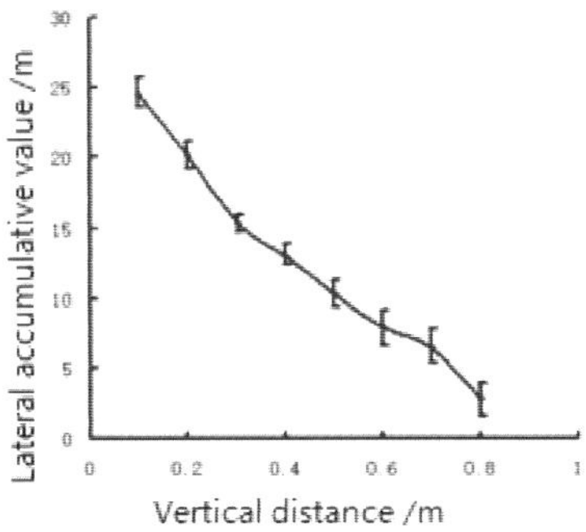

Figure 8. Transversal fitting curve of grain distribution probability

MATLAB was used to carry out polynomial fitting on the data in the above two tables, and the mathematical model (1) of grain distribution probability along the horizontal and the mathematical model (2) of grain distribution along the vertical were obtained respectively. The fitting curves were shown in FIG. 7 and FIG. 8 respectively. By fitting curve can be seen in the x axis between 0.6 ~ 0.8 m Δmi is less than 0.4009 (i.e. horizontal 7, 8 lines, Δmi for the ith row material box in a maximum of three kinds of wind speed under the condition of grain quality error), 0.3m to 0.4 m of y axis Δmj is less than 0.4559 (that is, vertical column 4, Δmj is the maximum value of grain quality error under three wind speed conditions of the junction box in row j) the minimum error limit of grain quantity distribution under different wind speed of the fan. Thus, it is determined that the more accurate monitoring value can be obtained by installing the sensor until (4, 7).

$$f(x) = 13.685x^4 - 40.091x^3 + 55.555x^2 - 38.728x + 16.875 \tag{1}$$

$$f(y) = -50.194y^3 + 97.587y^2 - 76.33y + 31.354 \tag{2}$$

Horizontal and vertical integrals were performed on the sensor monitoring area to obtain the percentage of lost grains in the monitoring area U:

$$U = \int_{x_0}^{x_0+b} f(x)dx \int_{y_0}^{(y_0+a)\sin\alpha} f(y)dy \tag{3}$$

where

x_0 = the transverse distance between the sensor and the tail of the cleaning screen, m;

y_0 = the longitudinal distance between the sensor and the tail of the cleaning screen, m;

a = the length of sensor, m;

b = the sensor width, m;

α = the installation angle between the sensor and the horizontal plane.

3.3. Inspection and modification of cleaning loss monitoring model

By formula (3), the percentage of loss in the area where the receiving box (4, 7) (i.e. transverse grid 7, longitudinal grid 4) and the receiving box (4, 8) are respectively calculated, and the deviation between them and the actual value is compared: $e = \frac{|U(x_i, y_j) - n(x_i, y_j)|}{U(x_i, y_j)} \times 100\%$. As shown in table 6, it can be seen that the deviation between the calculation result of the model and the actual value is 0.754%≤e ≤1.169%, the established model is relatively accurate and does not need to be revised.

Table 6. Deviation under different Fan turns into an exciting line

Squares,	The theoretical value	Deviation /% at each wind speed		
		1100 (r/min)	1200 (r/min)	1300 (r/min)
(4, 7)	0.908	1.065	1.169	1.126
(4, 8)	0.892	0.754	0.860	0.955

3.4. Research on calculation method of cleaning loss rate

The value measured by the sensor is the actual grain loss in the monitoring area, and the ratio of it to formula (3) can obtain the total grain loss per unit time H:

$$H = S / U \tag{4}$$

where

S = the monitoring value within the sensor monitoring time t;

The total quality of grain loss W is obtained by multiplying the total grain loss with the grain weight:

$$W = H \times \frac{p}{10^6 (1 - \frac{k}{100})} = \frac{H \times p \times 10^{-6}}{1 - 0.01k} \tag{5}$$

where

p = the thousand-grain weight of seeds, which is measured by the test, g;

k = the grain moisture content, %.

The ratio of the grain loss quality W to the total grains harvested by the harvester can obtain the grain cleaning loss rate L:

$$L = \frac{W}{Qvdt} \times 100\% = \frac{pS \times 10^{-6} \times 100\%}{Qvdt \times (1 - 0.01k)\ U} \tag{6}$$

where

Q = the yield per unit area of wheat, kg/m^2;

v = the operation speed of the harvester, m/s;

d = the operating width of the harvester, m;

t = the design time interval for the system, s.

4. Experiment

The cleaning loss monitoring system developed in this paper is installed on the above threshing and cleaning test bench, as shown in FIG. 9.

Figure 9. Experiment table for threshing and cleaning

The transverse distance x0 between the sensor and the tail of the cleaning screen was 0.7m, and the longitudinal distance y0 from the tail of the screening screen was 0.3m, the length of the sensor a was 0.03 m, the width of the sensor b was 0.05m, the angle α between the sensor and the horizontal plane was 30 degrees. Before the test, the 1000-grain weight p was 38.4g, the grain moisture content k was

17.2%, and the system time t was set as 2s (the system refreshed the loss rate once with the frequency of 2s). The above parameters were substituted into equation (6) to obtain:

$$L = \frac{W}{Qt} \times 100\%$$
$$= \frac{S \times 38.4 \times 10^{-6}}{2Q \times 0.808 \times \int_{0.7}^{0.72} f(x) \int_{0.3}^{0.325} f(y)} \times 100\%$$
$$= \frac{23.456S \times 10^{-3}}{Q} \%$$

Table 7. Results and errors of experiments

Feeding amount/ (kg/s)	Grain cleaning loss rate %		
	Monitor loss rate	Actual loss rate	The relative error
7.0	1.211	1.185	2.15
7.5	1.427	1.393	2.38
8.0	2.012	1.975	1.84
8.5	2.331	2.272	2.53
9.0	2.825	2.736	3.15

The formula of L is written into the program of cleaning loss monitoring system. The test method and wheat parameters used are the same as the above research on cleaning loss model. The relative error between the sensor monitoring value and the actual loss value under different feeding amounts is studied. The monitoring loss rate is measured by sensors and the actual loss rate is obtained by manual statistics. After the end of each group of tests, the grains harvested and the grains lost after cleaning and screening were weighed, and the actual loss rate was calculated. The sensor monitoring value was compared with the actual value, as shown in table 7.

The experimental results show that with the increase of system feed, the cleaning loss rate increases gradually, the relative error of sensor monitoring also increases, and the sensor monitoring value is larger than the actual value.

5. Conclusion

(1) A cleaning loss monitoring system based on PVDF piezoelectric film and a grain threshing and cleaning test bench were developed to realize the online monitoring of the cleaning loss test bench of the wheat harvester. The test results showed that the relative error range of the monitoring system was 1.84% to 3.15%, fully meeting the loss monitoring of the wheat harvester.

(2) The distribution law of cleaning loss caused by fan speed was studied, and the corresponding cleaning loss monitoring model was established, and that provided the decision-making basis for determining the optimal installation location of the sensor, for example, the optimal installation location of the test bench was 0.3m in longitudinal spacing with the cleaning screen and 0.7m in transverse spacing.

Acknowledgement

The work was supported by Shandong agricultural machinery equipment research and development innovation program (2016YF025).

References

[1] Jie, Z. (2009). Prospect of China's grain random loss rate test. *Agricultural mechanization research,* 31(07), 5-9. (In Chinese)

[2] Zhao, Z., Li, Y., Chen, L., Xu, J. (2011). Grain separation loss monitoring system in combine harvester. *Computers and Electronics in Agriculture.*

[3] Ni, J., Mao, H., Li, P. (2010). Design of grain cleaning loss monitor for array piezoelectric crystal sensor. *Journal of agricultural machinery*, 41(08), 175-177. (In Chinese)

[4] Chen, D., Wang, S., Kang, F., et al. (2011). Combined harvester feeding and harvesting process loss model. *Journal of agricultural engineering*, 27(09), 18-21. (In Chinese)
[5] Zhou, L., Zhang, X., Liu, Y., et al. (2010). Design and test of PVDF array sensor for grain loss measurement in combine harvester. *Journal of agricultural machinery*, 41(06), 167-171. (In Chinese)
[6] Li, Y., Liang, Z., Zhao, Z., et al. (2011). Research on real-time monitoring system of grain loss of combine harvester. *Journal of agricultural machinery*, 42(S1), 99-102. (In Chinese)
[7] Liang, Z., Li, Y., Zhao, Z., et al. (2015). Mathematical model for monitoring grain cleaning loss of vertical axial flow combine harvester. *Journal of agricultural machinery*, 46(01), 106-111. (In Chinese)
[8] Zhou, X. (2010). Research on the cleaning loss monitoring system of grain combine harvester. MS thesis. Shaanxi province: Northwest agricultural and forestry university. (In Chinese)
[9] Ju, D., Zhou, Z., Ou, J. (2004). Strain sensing characteristics of PVDF piezoelectric film. *Functional materials*, 04(04), 450-452+456. (In Chinese)
[10] Li, Y., Chen, Y., Zhao, Z., et al. (2013). Monitoring method and device for cleaning loss of combine harvester. *Journal of agricultural machinery*, 44(S2), 7-11. (In Chinese)

Journal of Physics: Conference Series **1423** (2019) 012034 doi:10.1088/1742-6596/1423/1/012034

Research on the optimization model of cold chain logistics distribution path under the constraint of carbon emission

Zhigao Liao[1], Yuncheng Quan[2], Kerong Jian[3], Yi Guo[4]

[1] Graduate Department, Guangxi University of science and technology, 545006
[2] Graduate Department, Guangxi University of science and technology, 545006
[3] Office of Financial Affairs, Guangxi University of science and technology, 545006
[4] School of Economics and Management, Guangxi University of science and technology, 545006

Corresponding author and e-mail: Zhigao Liao, liaozhigao@126.com

Abstract. The optimization of cold chain logistics distribution path under the constraint of carbon emission is considered. The restriction of hard time window and carbon emission is added to the model design. The carbon emission is converted into the final objective function of economic cost, transportation cost and goods loss cost in the way of carbon trading price. The problem of energy consumption is considered from the perspective of Economics. The cost of carbon emission is added to the mileage saving method, and the improved algorithm is used to increase the applicability of this kind of algorithm.

1. Introduction

With the rapid development of China's economy, the overall development process of cold chain logistics industry is relatively slow, which can not meet the needs of Chinese people for cold chain logistics. To change the existing cold chain logistics development mode and optimize the overall configuration is the primary task to improve the efficiency of logistics services. There is a certain amount of carbon emissions in the distribution process. If carbon emissions are included in the distribution cost as a cost, it will have an impact on the selection of the original distribution path. How to calculate the carbon emission is a difficult problem. Woensel. Creten[1] studied the carbon emissions generated by different fuel consumption, roughly calculated the carbon emissions generated by various types of fuel, established multiple quantitative optimization models, and finally calculated the fuel consumption in transportation activities, which was proved feasible by empirical analysis; Cai Hao and Xie Shaodong[2] comprehensively measured various factors affecting the carbon emission coefficient of transportation vehicles, and successively constructed Lu Pin[3] considers the distance and weight of goods moving in the process of goods transportation as the influencing factors and the amount of carbon dioxide emissions. Xie Shian [4] studies the effects of carbon emission costs on vehicle distribution. Cost impact, using the inertia weight method, improved the traditional particle swarm optimization algorithm.

Through the analysis of literature, it is found that the carbon emission of vehicles is related to driving distance, vehicle load and other factors. In this paper, carbon emissions as an economic cost is added to the saving algorithm, and the optimal objective function is constructed together with the cost of goods damage and transportation cost. According to the improved C-W saving algorithm, the optimal path is obtained.

2. Modelling

Cold chain logistics is a system engineering involving many links. The whole process of goods must be kept at a low temperature, so the overall investment and operation costs are high. Considering the characteristics and distribution mode of cold chain logistics, it is necessary to compare the common VRP problems from many aspects, such as how to reduce carbon emissions and save the total cost during the cold chain transportation. Therefore, the types of cold chain logistics distribution path optimization problems studied in this paper are: single distribution center, multiple regional distribution outlets, single type of products required, known demand of products, known geographical location of regional distribution outlets, distribution vehicles to and from the distribution center and outlets, each regional distribution network is served by one vehicle. The objective function of distribution is to minimize the total cost of distribution, which includes the total cost of transportation, the cost of goods loss and the cost of carbon emission.

2.1. parameter description

The parameter symbols in the carbon constrained cold chain distribution path optimization model studied in this paper are as follows. m: total number of vehicles in distribution centre of logistics base; n: total number of regional distribution outlets; C_0: Transportation cost per unit distance of vehicle; d_{xy}: Distance from regional distribution point x to regional distribution point y; t_{xy}: Time from regional distribution point x to regional distribution point y; d_y: Regional distribution network y goods demand; X_{xy}^k: it is a 0,1 variable. If the k-th vehicle passes the road section (v_x, v_y), X_{xy}^k=1, otherwise it is zero. X_y^k: a variable of 0,1. If the regional distribution point is served by vehicle K, thenX_y^k=1, otherwise it is zero. P_0: Unit price of carbon; F: carbon emission coefficient of fuel combustion; Q_{xy}^k: Carbon emission per unit distance; q_{xy}: refers to the vehicle load when the distribution vehicle travels from the regional distribution point x to the regional distribution point y; Q: maximum vehicle capacity; α: cargo loss coefficient during cargo distribution; β: cargo damage coefficient of loading and unloading; P: unit price of fruits and vegetables; t_y: The time point when the vehicle arrives at the regional distribution network; ET_y: Upper limit of time for regional distribution network y; LT_y: Time limit of regional distribution network y.

2.2. Distribution path optimization model

2.2.1. Carbon emission cost.

The carbon emission generated in the distribution mainly comes from the combustion of vehicle fuel and the operation of refrigeration equipment. This paper mainly studies the carbon emission of mechanical refrigeration vehicle in the distribution. For the convenience of research, it mainly considers the carbon emission generated by the driving distance of transport vehicles and the load of freight cars. The calculation formula (1) is as follows: $c_1 = \sum_{x=0}^{n}\sum_{y=0}^{n}\sum_{k=1}^{m} Q_{xy}^k\, d_{xy}\, P_0 F$ (1)

The traditional VRP Problem generally aims to minimize the vehicle transportation cost. In fact, the transportation cost is positively related to the vehicle driving distance, and the carbon emission cost is also related to the vehicle driving distance. In this paper, the calculation formula of carbon emission cost considers the carbon emission of distribution vehicles under different load capacities. A mechanical refrigerator car with a load capacity of 4 tons has the following relationship between its carbon emission per unit distance and its load capacity: $Q_{xy}^k = 1 + 0.25q_{xy}$ (2)

C_1 represents the cost of carbon emission during driving. In this paper, we choose to use carbon emission suggestion mechanism to cost carbon.

2.2.2. Transportation cost.

Because the refrigerated vehicles in this paper have been purchased, only the transportation cost of distribution vehicles is considered, which is proportional to the driving distance. The calculation formula (3) is as follows $c_2 = \sum_{x=0}^{n}\sum_{y=0}^{n}\sum_{k=1}^{m} c_0\, d_{xy}X_{xy}^k$ (3)

WhereX_{xy}^k is a 0, 1 variable. If the k-th vehicle passes the road section (v_x, v_y), then X_{xy}^k=1, otherwise it is zero.

2.2.3. Cost of goods damage.

The quality of fresh fruits and vegetables will change due to the temperature change and the time-consuming of loading and unloading, which leads to a certain proportion of goods loss, which is the cost of goods loss. In order to facilitate the study of this paper, only two factors are considered, one is the cumulative impact of transportation time on the quality of goods, the other is the impact of opening the door on the quality of goods when loading and unloading goods. The calculation formula (4) is as follows:

$$c_3 = P\sum_{x=0}^{n}\sum_{y=0}^{n}\sum_{k=1}^{m} X_y^k\left(\alpha t_{xy}q_{xy} + \beta d_y\right) \quad (4)$$

The objective function of the calculation model of fruit and vegetable product distribution cost is obtained from the above formula. The objective function model and its constraints are as follows:

$$\min c = P_0F\sum_{x=0}^{n}\sum_{y=0}^{n}\sum_{k=1}^{m} Q_{xy}^k\, d_{xy} + \sum_{x=0}^{n}\sum_{y=0}^{n}\sum_{k=1}^{m} c_0\, d_{xy}X_{xy}^k + P\sum_{x=0}^{n}\sum_{y=0}^{n}\sum_{k=1}^{m} X_y^k\left(\alpha t_{xy}q_{xy} + \beta d_y\right) \quad (5)$$

Subject to

$$\sum_{y=0}^{n} d_yX_y^k \le Q \quad (k = 1,2 \ldots m) \quad (6)$$

$$\sum_{k=1}^{m} X_y^k = 1 \quad (k = 1,2 \ldots m) \quad (7)$$

$$\sum_{x=0}^{n} X_{xy}^k = X_y^k \quad (y = 0,1 \ldots n; k = 0,1 \ldots m) \quad (8)$$

$$\sum_{x=0}^{n}\sum_{k=1}^{m} X_{0x}^k = \sum_{y=0}^{n}\sum_{k=1}^{m} X_{0y}^k \quad (9)$$

$$ET_y \le t_y \le LT_y \quad (y = 1,2 \ldots . n) \quad (10)$$

Formula (6): the vehicle load is greater than the total number of regional distribution outlets requiring distribution services; formula (7): each regional distribution outlet is only distributed by one vehicle; formula (8): only one vehicle arrives and leaves each regional distribution outlet; formula (9): each vehicle only goes to and from the logistics distribution centre and regional distribution outlets; formula (10): distribution within the time window specified by the regional network.

3. Model solution

Generally, C-W saving algorithm is used to solve the problem. This paper studies the path optimization problem of minimizing the total distribution cost, which is constrained by vehicle load and time. It is different from the traditional algorithm in algorithm, so it needs to add time window and vehicle load constraints to improve the algorithm to meet the research needs of this paper.

3.1. Optimal target improvement

The optimal objective function of the problem studied in this paper is the sum of carbon emission cost, transportation cost and goods damage cost. Therefore, when calculating the distance saving value (ΔL), it is necessary to convert the distance to the cost saving value (expressed in s(x, y)). S(x, y) =$C_{0x} + C_{0y}$+C_{xy} (x≠ y, 0 is the distribution centre), where C represents the sum of three costs.

(1) After connecting regional distribution network x and regional distribution network y, the carbon emission cost saved is $s_1(\mathrm{x,y})$,

$$s_1(\mathrm{x,y}) = \left(C_{0x}^1 + C_{x0}^1 + C_{0y}^1 + C_{y0}^1\right) - \left(C_{0x}^1 + C_{xy}^1 + C_{y0}^1\right) = C_{x0}^1 + C_{0y}^1 - C_{xy}^1$$

The carbon emission cost is calculated by the vehicle real-time load capacity and driving distance, so$C_{0x}^1 \neq C_{0y}^1$, similarly$C_{0y}^1 \neq C_{y0}^1$.

(2) After connecting regional distribution network X and regional distribution network y, the transportation cost saved is $s_2(\mathrm{x,y})$.

$$s_2(\mathrm{x,y}) = \left(C_{0x}^2 + C_{x0}^2 + C_{0y}^2 + C_{y0}^2\right) - \left(C_{0x}^2 + C_{xy}^2 + C_{y0}^2\right) = C_{x0}^2 + C_{0y}^2 - C_{xy}^2 = C_{0x}^2 + C_{0y}^2 - C_{xy}^2$$

Because the transportation cost and carbon emission cost are calculated separately in this paper, the transportation cost is only related to the driving distance, so$C_{0x}^2 = C_{x0}^2$, similarly,$C_{0y}^2 = C_{y0}^2$.

(3) After connecting regional distribution network x and regional distribution network y, the cost of goods loss saved is$s_3(\mathrm{x,y})$:

$$s_3(\mathrm{x,y}) = \left(C_{0x}^3 + C_{x0}^3 + C_{0y}^3 + C_{y0}^3\right) - \left(C_{0x}^3 + C_{xy}^3 + C_{y0}^3\right) = C_{y0}^3 + C_{0y}^3 - C_{xy}^3 = C_{0x}^3 + C_{0y}^3 - C_{xy}^3$$

The cost of goods damage discussed in this paper is affected by two factors, one is the loss caused in the course of distribution, the other is the loss caused by loading and unloading operations. The combination of lines will not change the cargo damage rate during loading and unloading operation, so the reduction of cargo damage rate is related to the reduction of driving distance saving, so the principle of saving is the same as the saving of transportation cost, so the principle of saving is the same as that of saving transportation cost, so the principle of saving is the same as that of saving transportation cost. Therefore, $C_{0x}^3 = C_{x0}^3$, similarly$C_{0y}^3 = C_{y0}^3$.

4) After connecting regional distribution network X and regional distribution network y, the total distribution cost saved is s (x, y): $s(x, y) = s_1(x, y) + s_2(x, y) + s_3(x, y)$

Finally, the time window constraint is added to the traditional saving algorithm, and the optimization target is changed to the minimum total cost, that is, the optimized algorithm is obtained.

3.2. Improve calculation steps

Step 1: select the base point. Based on logistics base, the distribution centre is set as P0. N regional distribution networks are connected with logistics base to form n initial solutions.

Step 2: calculate the distance saving value Δ L (x, y) after the connection of regional distribution point X and regional distribution point y, and then define them as arrays;

Step 3: if the values of Δ L (x, y) are all 0, the calculation will be stopped, otherwise, the saving values will be arranged in descending order from large to small, forming the saving distance value matrix;

Step 4: select the regional distribution network with the lowest total distribution cost as the first network of the first line. Connect the distribution centre with n outlets respectively, and calculate the total cost of vehicles from the distribution centre to each outlet, which is recorded as G_{C1}... G_{cn}. Find out the outlet with the lowest total cost of vehicle distribution by observation method. If it meets the requirements of$ET_k < LT_k$, andq_k<Q, it is recorded asP_k.

Step 5: calculate the second node after the first node. Connect the remaining regional distribution outlets with p, k respectively, and calculate the total cost savings of each outlet after connection compared with the total cost savings of the original direct connection with the distribution centre, , s (k, y) = $C_{0k} + C_{0y} - C_{ky}$ (k ≠y, 0 is the distribution centre), C represents the sum of the three costs, and find out the outlets with the largest total cost savings in the total cost savings obtained table;

Step 6: test the s (k, y) obtained from step 5. When it meets (1), (2) or (1), (3), skip to step 7. Otherwise, skip to step 5 and find out the outlets with the largest total cost saving among the remaining outlets.

(1) $q_k + q_y < Q$;(2) $E_{ky} = a_k + t_k + t_{ky} - a_y > 0$ and $\Delta y^+ = LT_y - a_y > E_{ky}$;

(3) $E_{ky} = a_k + t_k + t_{ky} - a_y < 0 \quad and \ \Delta y^- = a_y - ET_y > |E_{ky}|$;

Step 7: determine the second regional distribution network of the path as p_k2;

Step 8: repeat step 5-step 7. If the time and load force constraints cannot be met after the distribution network in any area is added, the combination of the lines ends;

Step 9: repeat step 4-step 8 to merge the new distribution path. When all regional distribution outlets are planned into the distribution path, the calculation ends.

4. Example

4.1. distribution data

For simple calculation, Suning logistics base is set as p_0, and 12 regional distribution outlets in Nanning are numbered as P1,P2,..., P12. The distance and demand between the distribution centre and the regional distribution network are shown in table 1, and the service time window and service time of each regional distribution network are shown in table 2. In order to simplify the calculation, the shortest driving distance from each regional distribution network to the logistics base on the map is used for the distribution distance data.

Table 1. Distribution distance and demand matrix

d/km	P0	P1	P2	P3	P4	P5	P6	P7	P8	P9	P10	P11	P12	d/t
P0	0													—
P1	12	0												0.88
P2	48	50	0											0.84
P3	60	38	72	0										0.55
P4	92	80	127	98	0									0.56
P5	50	62	131	110	60	0								0.55
P6	92	100	150	130	98	38	0							0.62
P7	120	124	168	159	156	96	58	0						0.43
P8	110	107	144	147	187	172	160	117	0					0.47
P9	22	28	77	67	110	100	157	117	76	0				0.85
P10	42	36	49	65	130	164	193	154	98	36	0			0.73
P11	42	35	96	95	140	72	117	77	84	42	67	0		0.65
P12	24	34	94	64	94	50	69	140	118	48	64	60	0	0.77

Table 2. Service time and time window of each outlet (after quantification)

Distribution point	service time window	service time /h	Distribution point	service time window	service time /h
P0	[0,7]	0	P7	[3,5]	0.2
P1	[1,3]	0.4	P8	[2.5,5]	0.2
P2	[1,3]	0.4	P9	[1,3]	0.4
P3	[2.5,5]	0.25	P10	[1,3]	0.35
P4	[2,4]	0.25	P11	[2,4]	0.3
P5	[2.5,5]	0.25	P12	[1,3]	0.35
P6	[3,6]	0.3	P7	[3,5]	0.2

4.2. model parameters

The average price of fruit and vegetable products in the distribution of Suning logistics base is p = 30000 yuan / ton (from the average price of the investigated products). The vehicles are mechanical refrigerated vehicles, with the maximum load Q = 4 tons, the average transportation costC_0= 1 yuan / km, the average vehicle speed v = 60km / h, the ratio coefficient of goods loss α is 0.02% (1 / h), the ratio coefficient of goods loss β is 0.01%, and the carbon dioxide emission coefficient f is 2.82 (kg / L), the unit cost of carbon emission is 0.05 yuan / kg (select the average price in the carbon trading market from October 2016 to April 2019).

4.3. Distribution path solution

According to the improved saving algorithm, the steps to solve the cold chain distribution path of Nanning Suning are as follows:

(1) Calculate that the time taken to drive from Suning logistics base to each regional distribution network is t_{0i}, in which the time window of regional distribution network i is $[ET_i\ , LT_i]$, and table 3 is obtained by calculating t_{0i}.

Table 3. Travel time of logistics base to all regional distribution outlets

t01	t02	t03	t04	t05	t06	t07	t08	t09	t10	t11	t12
0. 20	0. 80	1. 00	1. 53	0. 83	1. 53	2. 00	1. 83	0. 37	0. 70	0. 70	0. 40

a_i(i=1,2,3…12) is the time point from Suning logistics base to 12 regional distribution outlets. If $t_{0i} \leq ET_i$, thena_i=ET_i; if $ET_i \leq t_{0i} \leq LT_i$, then$a_i = t_{0i}$, so the initial solution ofa_i is shown in table 4.

Table 4. Time point of arrival of logistics base at all regional distribution outlets

a1	a2	a3	a4	a5	a6	a7	a8	a9	a10	a11	a12
1	1	2.5	2	2.5	3	3	2.5	1	1	2	1

(2) Calculate the distance saving value Δ L after connecting regional distribution network i and regional distribution network j. Since the data in the distribution distance matrix in table 4-1 is taken from the actual distance between two points in the map, not the straight-line distance, there will be negative values in the saving value matrix, and negative values will not affect the calculation and solution in this paper.

(3) Select P0 base point of logistics base, and connect P0 with P1, P2... P12 respectively, and 12 lines are obtained. GC1 and Gc12 are calculated respectively, and the lowest total cost $\boldsymbol{P_k}$ is selected, which meets the requirements of$\boldsymbol{ET_k} < \boldsymbol{LT_y}$ and $\boldsymbol{q_k} < \boldsymbol{Q}$. Get the cost of distribution.

It can be seen from that the total distribution cost from the logistics base to P1 is the minimum, and it is required to check whether P1 meets the time window and load constraints. a1=1$\leq LT_1$ and q1=0.88<Q=4, so P1 meets the constraint, and P1 is determined as the first regional distribution network of the first path.

The second, third, fourth and fifth distribution outlets can be determined respectively. After calculation, the time for vehicles to arrive at other regional distribution outlets after the completion of regional distribution outlets P5 does not meet the time window constraints, so the optimization of the route is over.

According to this method, the first path is p0-p1-p3-p4-p5-p0. Repeat the above steps to get all the path planning. Three vehicles are needed. The arrangement is as follows:

The first distribution route: p0-p1-p3-p4-p5-p0, 258km distribution distance, 2.54t vehicle load; The second distribution route: p0-p9-p8-p7-p6-p0, the distribution distance is 365km, and the vehicle load is 2.37t;The third distribution route: p0-p12-p11-p10-p2-p0, distribution distance: 235km, vehicle load: 2.96t;The C-W method is also used to solve the distribution path planning when Suning enterprises do not consider carbon emissions, and to find the path arrangement.

According to the solution results, three vehicles are needed, arranged as follows: the first distribution route: p0-p1-p2-p10-p9-p0, the distribution route is 169km, and the vehicle load is 3.3t; the second distribution route: p0-p12-p4-p5-p3-p0, the distribution distance is 348km, and the vehicle load is 2.43t;the third distribution route: p0-p11-p8-p7-p6-p0, 392km distribution distance, 2.96t vehicle load.

5. Conclusion

When considering the carbon constraints, three vehicles are needed to complete the distribution task. The total distance travelled by the distribution task vehicles is 858km, the carbon emission cost is 65.58 yuan, the cost of goods damage is 45.04 yuan, and the total cost is 969.02 yuan. However, without considering the carbon constraints, three vehicles are needed to complete the distribution task. The total distance travelled by the whole distribution task is 879km, the carbon emission cost is 115.42 yuan, the cost of goods loss is 65.54 yuan, and the total cost is 1059.96 yuan.

The results show that the optimal path model with carbon constraints is lower in carbon emission cost, goods loss cost and total cost than the optimal path model without carbon constraints. According to the analysis, if transport cost and goods damage cost are taken as the objective, the actual optimal solution may not be obtained, and if carbon emission constraints are added to the distribution optimization objective, the operation cost may not be increased.

Acknowledgement

This work is supported by Western project of NSFC (19XGL025), Guangxi Philosophy and Social Sciences Project (17FJY007) and Innovation Project of Guangxi Graduate Education (YCSZ2014203).

References

[1] Woensel TV, Creten R, Vandaele N. Managing the Environmental Externalities of Traffic Logistics: The Issue of Emissions[J]. Production Operations Management, 2001(10): 207-223.
[2] Cai Hao, Xie Shaodong. Determination of vehicle emission factors of different emission standards in China[J]. Journal of Peking University, 2010 (46): 319-326.
[3] LV pin. Study on optimization model of logistics network considering carbon emission[J]. Computer application research, 2013, 30.
[4] Xu conger, Qiu rongzu. Distribution path optimization considering carbon emission and transportation cost [J]. Mathematical practice and understanding, 2016, 46 (21): 89-94.
[5] Xie Shian. Research on the optimization method of delivery and delivery integrated distribution path considering carbon emission and time window [D]. Beijing Jiaotong University, 2017.
[6] Li Jin, Zhang Jianghua. Study on the influence of carbon trading mechanism on the optimization decision of logistics distribution path [J]. System engineering theory and practice, 2014, 34 (7): 1779-1787.

Watermarking in halftone image with maximal child node dependence coefficients in wavelet domain

Hangyu Gong[1], Zifen He[1*]

[1]Faculty of Mechanical and Electrical Engineering, Kunming University of Science and Technology, Kunming, China, 650500

Corresponding author and e-mail: Zifen He, zyhhzf1998@163.com

Abstract. To achieve high image quality, this paper proposes a novel approach but efficient algorithm for halftoning watermarking. Wavelet coefficients of edge details have strong relations, while the dependence of coefficients of texture is more weak. On the other hand, large wavelet coefficients contain the information of image brink. Considering the relationship of intra-scale and maximal coefficients, the boundary error measure is built. While the gray image turns to halftone image using proposed algorithm, binary watermark is embedded in the edge position. In order to enhance the ability to resist cropping and smearing attack, the watermark image is pre-treated with Arnold transformation processing. The proposed method is effective, because the halftone image is embedded as a watermark sequence during the halftone process. Experiment results show that this algorithm does not cause significant distortion of the image and is good resistance to cropping, daub, JPEG compression and noise attacks.

1. Introduction

Halftone is known as gray-scale and a technical indicator which reflects the image brightness level and the black and white contrast changes. Binary halftone image would try to compute a pattern of binary dots to achieve the illusion of a multi-bit image.

The halftoning algorithms can be further classified through various properties and features, including point based methods, such as screening [1], neighborhood based methods, such as error diffusion [2-5], iteration-based scheme such as Direct Binary Search (DBS) [6-8] and lookup based scheme such as Look Up Table (LUT) halftoning [9]. Iterative optimization methods use a human visual system model to minimize the perceived error [10]. Although all halftoning methods rely on some understanding of the properties of human vision and the display device, the goal of model-based halftoning techniques is to exploit explicit models of the display device and the human visual system (HVS) to maximize the quality of the displayed images [11].

However, the best halftone reproductions are obtained via iterative techniques that minimize the error between the continuous tone image and halftone image. They include one-dimensional Viterbi algorithm which to obtain the globally optimal solution [12], blocking with branch and bound minimization [13], diffusion-reaction model [14]. The LSMB halftoning approach attempts to produce an optimal halftone reproduction [15].

Digital watermarking has many applications, such as protecting the ownership rights of an image, tracking unauthorized uses of a work as in fingerprinting, or authenticating the originality of an image, that it has not been tampered with. Since the halftoning is widely used, halftone-based watermarking methods have been studied in recent years. The methods normally require higher computational complexity and produce higher image quality. These methods include using the concept of vector quantization to embed watermark into the most or least significant bit of an error-diffused image [16]. In [17], different dither cells were exploited to create a threshold pattern in the halftoning process, in which each dither cell represents the corresponding information bit of the watermark. In [18], the modified data-hiding error diffusion method was employed to embed data into an error-diffused image. In [19], the DBS was employed to achieve halftoning and watermarking simultaneously.

2. The halftoning of scales correlation coefficients fusion

For an input represented by a list of 2^n numbers, the Haar wavelet transform may be considered to simply pair up input values, storing the difference and passing the sum [16]. This process is repeated recursively, pairing up the sums to provide the next scale, finally resulting in 2^n-1 differences and one final sum.

Let ϕ^{LL} denote scale function, $\{\varphi^{LH},\varphi^{HL},\varphi^{HH}\}$ denote wavelet function. The scale function and wavelet function form $L^2(R^2)$ space orthogonal basis. For a given image $f(x,y)$ is decomposed in the space domain, which the size of $N \times N$. It can be expressed as

$$f(x,y)=\sum_{k,i=0}^{N_J-1} u_{J,k,i}\phi^{LL}_{J,k,i}(x,y)+\sum_{b\in B}\sum_{j=1}^{J}\sum_{k,i=0}^{N_{j-1}} w^b_{j,k,i}\varphi^b_{j,k,i}(x,y) \tag{1}$$

Where $b \in B$, $B=\{LH,HL,HH\}$ is the wavelet coefficients in three directions sub-band, $Nj=N/2$. The scale coefficients $u_{j,k,i}$ and wavelet coefficients $w_{j,k,i}$ in j scale sub-band B direction. The multiscale information are show as Figure1 (b).

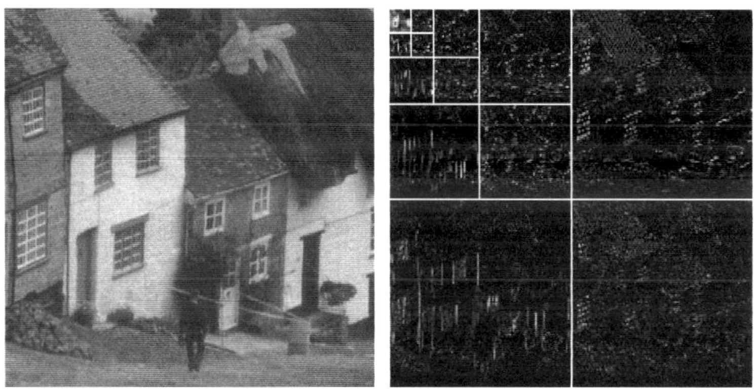

(a) (b)

Figure 1. The Original Image and the DWT (a) Original Image (b) Four Levels DWT

Upon the assumption that the wavelet coefficients are mutually independent, they are normalized. The boundary information of the image target is fused by the wavelet coefficients of the correlation between wavelet transform layers, which to increase the pixel resolution scale. We apply the inter-scale fusion method to gain fusion coefficient of the fine-scale, which take into account the detail of the image and approximate information. The fusion information of wavelet coefficients inter-scale is shown in Figure 2.

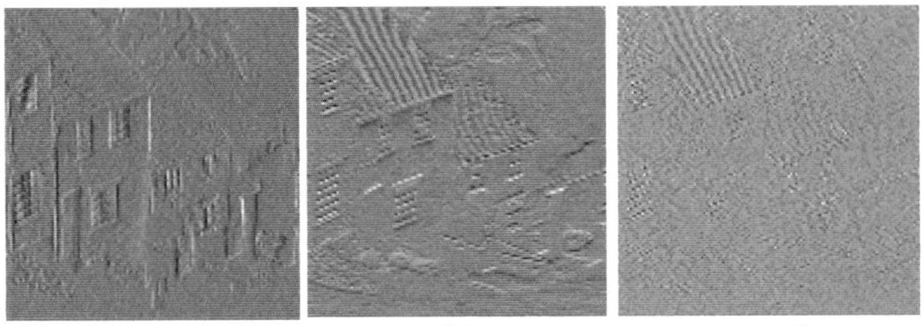

Figure 2. The Fusion Information of Wavelet Coefficients Inter-scale for HL, LH and HH Sub-band (a) HL (b) LH (c) HH

We will use the term contrast sensitivity here, since we have used this terminology throughout Mannos and Sakrison [18] proposed a model of the human contrast sensitivity function, which found the following filter frequency response to be good for predicting the subjective quality of coded images.

$$H_r(f_r) = 2.6(0.0192 + 0.114 f_r)\exp\left\{-(0.114 f_r)^{1.1}\right\} \tag{2}$$

fr in equation (2) is the spatial frequency of the visual stimuli given in cycles/degree. The function has a peak of value 1 approximately at *fr* =8.0 cycles/degree, and is meaningless for frequencies above 60 cycles/degree.

Least-squares model-based(LSMB) halftoning attempts to produce an optimal halftone reproduction[15]. The gray level of each pixel varies from 0 equal to white to 1 equal to black.

Assuming that the image has been sampled so there is one pixel per dot to be generated. Thus the gray-scale image array $[x_{i,j}]$ and the binary image array $[b_{i,j}]$ have the same dimensions. We are also given a printer model with the sliding-window form and an eye model of the form with a memory-less nonlinearity $n(.)$ followed by an FIR filter with impulse response $[h_{i,j}]$. In the LSMB approach we seek the halftone image that minimizes the squared error.

Weighted least squares model-based regression is useful for estimating the values of model parameters when the response values have differing degrees of variability over the combinations of the predictor values[19]. Optimal results that minimize the uncertainty in the parameter estimators are obtained when the weights used to estimate the values of the unknown parameters are inversely proportional to the variances at each combination of predictor variable values. $\lambda_i \propto 1/\sigma_i^2$.

In the proposed approach, we seek the halftone image that minimizes the weighted squared error. The related descriptions are formulated as below

$$E_{i,j} = \left\{ \sum_{l=1}^{4} (LH, HL, HH) + \sum_{k=1}^{4} \lambda_k \right\} \sum_{i,j} (z_{i,j} - w_{i,j})^2 \tag{3}$$

Note that we have allowed different impulse responses for the eye filters corresponding to the halftone and continuous-tone images. The halftoning image is show as Figure 3.

Figure 3. The Halftone Image

3. Watermark embedding and detection scheme

3.1. Arnold transformation processing

Image scrambling may transfer the image into a unorganized, which can't gain intuitive information. Therefore, the image is encrypted by scrambling transformation that enhanced the security of information. At the same time, scrambling transformation has a strong resistance to cropping and paint attack. Images matrix are transformation using Arnold transformation. In this paper, it used Arnold transformation to encrypt the watermark image, then inserted the watermark image into the important coefficients from different orientations of the wavelet transformation.

The watermark image is seen as a dual function on the flat area $G = F(x,y)$, $(x,y) \in R$. For arbitrary point (x, y) in R, then $F(x, y)$ is on behalf of image information, (such as the gray value, etc.). $G = F(x, y)$ is equivalent to a two-dimensional discrete lattice when the watermark image is digital. For digital image pixel coordinates $(x, y) \in \{1, 2,..., N\}$. So, Arnold transform is:

$$\begin{bmatrix} x' \\ y' \end{bmatrix} = \begin{bmatrix} 1 & 1 \\ 1 & 2 \end{bmatrix} \begin{bmatrix} x \\ y \end{bmatrix} (\mathrm{mod}\, N) \tag{4}$$

The result of iterative process is expressed as:

$$P_{xy}^{n+1} = AP_{xy}^{n} (\mathrm{mod}\, N), P_{xy}^{n} = (x, y)^T \quad n = 0,1,2,\cdots$$

where, n is the number of iterations, the iteration process is cyclical.

3.2. Arnold transformation processingWatermark embedding scheme

In order to enhance the security and robustness of watermark information, the scrambling image is gained after Arnold transform of binary watermark image. Arnold scrambling times n is a "secret parameter". This means that receiver will not be able to generate original information if it does not know this parameter. Watermark and scrambling image are shown in Figure 4.

(a) (b)

Figure 4. Arnold Transformation

Figure 5. Embedded Watermark Halftone Image

(a)32×32 Original Image

(b) n=10 Arnold Transformation

The watermark is embedded the edge of halftone image during the halftone processing, which are the optimal positions for watermark embedding. They are referred to as the key to the receiver. Ultimately embed watermark halftone image is shown as Figure 5.

3.3. Watermark extraction scheme

Firstly, our watermark detection scheme uses watermark key to determine watermark embedding position and extract the watermark value. Secondly, The extract watermark image is Arnold transform using watermark key and Arnold periodic table. Lastly, the embedded watermark in the halftone image become multi-tone image after the JPEG compression and add Gaussian noise. The embattled image is must binarization (choose threshold is 128) before extracting the watermark.

4. Experimental results and discussions

We test the proposed watermarking scheme on the popular test image 256×256 bit goldhill, as shown in Figure 1(a). The corresponding halftone images are generated by the fusion of scales correlation coefficients, as shown in Figure 3. We embed the watermark in the perceptually highly edge region, so the embedded watermark is less visible. Besides, the MSEv(mean square error value) and PSNR(peak signal-to-noise ratio) are used to measure the visual quality of the watermark halftone image. The visual quality valuation is show as Table 1.

Table 1. MSEv and PSNR between watermarked halftone image and original halftone image

Test image	MSEv	PSNR(dB)
goldhill	0.14	40.98

Here is the list of the different attacks, including *Noise addition*, *JPEG compression*, *Cropping*, *Daub* and *Rotations*, which are shown in Figure 6,7,8,9 and 10, respectively.

(a) (b) (c)

Figure 6. The Simulation Results for Noise Addition Obtained by Using the Proposed Scheme(a)Noise Addition(b)Watermark Extraction(c)Arnold Inverse Transformation

(a) (b) (c)

Figure 7. The Simulation Results for JPEG Compression Obtained by Using the Proposed Scheme (a)Noise Addition(b)Watermark Extraction(c)Arnold Inverse Transformation

(a) (b) (c)

Figure 8. The Simulation Results for Cropping Obtained by Using the Proposed Scheme (a) Noise Addition(b)Watermark Extraction (c)Arnold Inverse Transformation

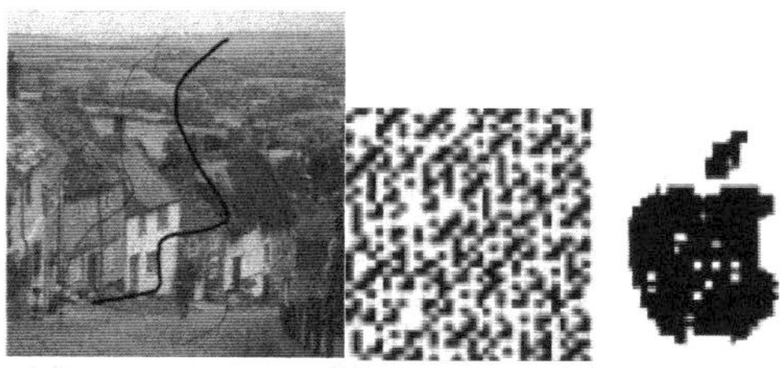

(a) (b) (c)

Figure 9. The Simulation Results for Daub Obtained by Using the Proposed Scheme (a)Noise Addition(b)Watermark Extraction (c)Arnold Inverse Transformation

(a) (b) (c)

Figure 10. The Simulation Results for Rotations Obtained by Using the Proposed Scheme(a)Noise Addition(b)Watermark Extraction(c)Arnold inverse Transformation

Table 2 summarize the detection results against several attacks. In this paper, we use the term "normalized cross-correlation,NC" to denote it. The NC value are 0.768, 0.991, 0.891, 0.984 and 0.935 between the extraction and original watermark, respectively.

Table 2. NC of detected watermark under several attacks

Attacks	**NC**
noise addition	0.768
JPEG compression	0.991
cropping	0.891
daub	0.984
rotations	0.935

The experimental results show that the image of error is spread out using Arnold transform. After the watermark pre-processing by Arnold, watermark can be able to resist clipping and daubing attack, the robustness of watermark is better. In the case of halftone image quality significantly decreased, extraction watermark can identify. Use proposed algorithm for edge enhance, invisibility of watermark is better in the edge position of the embedded watermark.

5. Conclusions

In many applications, it is desirable to embed data in halftone images, but the existing halftone image watermarking has the shortcomings of being not able to guard against attacks caused by cropping, daubing, noise and rotation, etc. In this paper, we proposed a digital method with Maximal Father-Child Node Dependence Coefficients in Wavelet Domain for halftone image with good visual quality. Experimental results show that the proposed scheme is robust against several attacks. At the edge of the halftone image, position watermark has good invisibility.

Acknowledgement

This work was supported by the National Science Foundation of China (NSFC) under Grants 61761024 and 61461022.

References

[1] B.E. Bayer, An Optimum Method for Two-Level Rendition of Continuous-Tone Pictures, IEEE 1973 International Con$ on Communications, (**1976**), Vol. 26, pp.11-15.
[2] R.W.Floyd and L. Steinberg, An adaptive algorithm for spatial grayscale, in Proc. Soc. Inf. Display (**1975**), pp. 36–37.
[3] J.F.Jarvis, C.N.Judice and W. H. Ninke, A survey of techniques for the display of continuous-tone pictures on bilevel displays, Compute. Graph. Image Process, (**1976**), Vol. 1, pp.13–40.
[4] P. Stucki, MECCA-A multiple-error correcting computation algorithm for bilevel image hardcopy reproduction, Res. Rep. RZ1060, IBM Res. Lab., Zurich, The Switzerland, (**1981**).
[5] J. N. Shiau and Z. Fan, A set of easily implementable coefficients in error diffusion with reduced worm artifacts, Proc. SPIE,(**1996**), Vol. 2658, pp.222–225.

[6] J. P. Allebach and Q. Lin, FM screen design using DBS algorithm, in Proc. IEEE Int. Conf. Image Process, (**1996**),Vol.1,pp.549–552.
[7] D. J. Lieberman and J. P. Allebach, A dual interpretation for direct binary search and its implications for tone reproduction and texture quality, IEEE Trans. Image Process,(**2000**),Vol.11,pp.1950–1963.
[8] S. H. Kim and J. P.Allebach,. Impact of HVS models on model-based halftoning, IEEE Trans. Image Process,(**2002**), Vol.3,pp.258–269.
[9] P. Li and J. Allebach, Look-up-table based halftoning algorithm, In Proc. 1998 IEEE Int. Conf. Image Processing,(**1998**), pp.559-569.
[10] M. Analoui and J. Allebach, Model-based halftoning using direct binary search. Electronic Imaging Science Technology,(**1992**), Vol. 1666, pp.96–108.
[11] Sang Ho Kim and Jan P. Allebach, Impact of HVS Models on Model-Based Halftoning, IEEE Trans. Image Process,(**2002**), Vol.3, pp.258-269.
[12] D. L. Neuhoff, T. N. Pappas and N. Seshadri, One-dimensional least-squares model based halftoning, in Proceedings of ICASSP, (San Francisco, CA)(**1992**),Vol. 3,pp.189-192 .
[13] A. Zakhor, S. Lin and F. Eskafi, A new class of B/W halftoning algorithms, IEEE Trans. Image Processing, (**1993**),Vol.10, pp.499-509.
[14] A. Sherstinsky and R. W. Picard, M-lattice: A novel non-linear dynamical system and its application to halftoning, in Proceedings of ICASSP, (Adelaide, Australia)(**1994**),Vol.4,pp. (11-565)-(11-568).
[15] N. Pappas and David L, Neuhoff. Least-Squares Model-Based Halftoning, IEEE Trans. Image Process, (**1999**), Vol. 8, pp. 1102-1116.
[16] J. R. Goldschneider, E. A. Riskin and P. W. Wong, Embedded color error diffusion, in Proc. IEEE Int. Conf. Image Process, (**1996**) Vol. 1, pp. 565–568.
[17] Z. Baharav and D. Shaked, Watermarking of dither halftone images, in Proc. Int. Conf. Security Watermark. Multimedia Content, (**1999**),pp. 307–316.
[18] M. S. Fu and O. C. Au, Hiding data in halftone image using modified data hiding error diffusion, Proc. SPIE Vis. Commun. Image Process, (**2000**),Vol. 4067, pp. 1671–1680.
[19] D. Kacker and J. P. Allebach, Joint halftoning and watermarking, IEEE Trans. Signal Process., (**2003**),Vol. 51, pp. 1054–1068.

Evaluation of the experimental project performance assessment mode with the application of descriptive statistical analysis method

Renbao Liu

School of Public Administration, Shandong Technology and Business University, Yantai, China

Corresponding author and e-mail: Renbao Liu, 6903759@163.com

Abstract: In order to solve the problem that currently, the performance assessment mode of most experimental projects only adopts the individual assessment while neglecting teamwork, puts forward the experimental project performance assessment mode based on team communication, which takes all the students participating in the project as a team, and divides the team into several groups according to project contents and tasks. Team members complete the project tasks jointly through team communication and group and individual quantitative scoring will be carried out according to the completion of tasks. After conducting the descriptive statistical analysis of an experimental project assessment performance of 223 students from 8 classes of 4 undergraduate majors in a university through Stata15.1, the results show that the results of the project assessment performance data all show a normal distribution by major, class, group and individual, the assessment results are ideal, and the assessment mode is feasible. The innovation is to use the Stata, and use the descriptive statistical analysis method to make an in-depth and careful statistical analysis of the experimental project performance assessment mode based on the team communication proposed. And the normality test is carried out to make an objective evaluation to the assessment results.

1. Introduction

The experimental project is an important part of the university curriculum, which combines the theory and practice of college students, improves their practical ability, and plays an important role in training their innovative spirit and innovative capabilities. [1] The experimental project performance assessment shall be fair and reasonable, and truly reflect the students' experimental situation· At present, most of the experimental project performance assessment modes only use individual assessment methods, including the assessment in terms of attendance, project completion, answer of questions, quality of experimental report writing, etc. There are some shortcomings in this assessment mode. Firstly, it emphasizes the individual performance of the students while neglecting the teamwork. Secondly, the assessment cannot be accurate to everyone since there are so many students participating in the project, and the assessment is not objective and accurate enough. Thirdly, it pays attention to the assessment of results while neglecting the assessment of the project process. Fourthly, it mixes the assessment of the project itself with the auxiliary assessments relevant to the project, and weakens the role of the project. [2-5] Based on this, this paper proposes an experimental project performance assessment mode based on team communication, which separates students' attendance, answer of questions and evaluation of experimental reports as a separate assessment of individual students, while taking all the students participating in the project as a team, and dividing the team into several groups according to the project contents and tasks, and taking the group assessment performance as the common performance of each group members to achieve the team communication quantitative assessment based on groups to mobilize the enthusiasm and initiative of each student to participate in the project, enhance their team communication awareness, and improve their team cooperation capabilities. This paper takes the actual assessment results of an experimental course in a university as the sample, and uses the Stata, and uses the descriptive statistical analysis method to conduct detailed descriptive statistics and normality test to evaluate the feasibility of this mode. [6]

2. Experimental Principle and Subject

2.1. Experiment principle

2.1.1 Assessment mode.

The core of the experimental project performance assessment mode based on team communication is the quantitative assessment of team communication that takes the group as a unit, and takes all the students participating in the project as a team, and divides the team into several groups and a group leader is determined for each group. During the project process, group members jointly complete the project tasks through team communication. And quantitative scoring is carried out according to the completion of tasks, and group scoring is treated as the common score of the group members. The average score is taken as the final assessment performance of the individual experimental project (the group leader has a 5% bonus point).

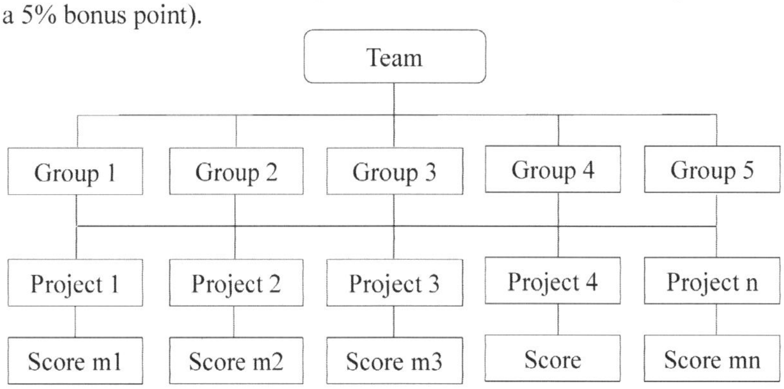

Figure 1. Framework of experimental project assessment mode based on team communication

2.1.2 Single project scoring rules.

(1) Percentage scoring. For a project with a clear number or score target, a team percentage scoring is adopted. And the basic scoring rule is the ratio of the group or team score to the target number or score multiplied by one hundred with integers are taken. Suppose the target quantity or score is *m*, the group or team score is *n*, and *int* means that the calculation result takes an integer, then the group score *s* is:

$$s = \mathrm{int}((n/m) * 100) \tag{1}$$

(2) Assessment ranking scoring. For a project that is not suitable for percentage scoring, according to the number of groups divided, the ranking and score distribution are carried out according to the completion of the group's tasks, and the score distributions of the groups with different rankings are also different. Taking the division of 7 groups as an example, from the first place to the seventh place, the score is 100 to 60 respectively. Please refer to Table 1 for details.

Table 1. Group assessment ranking and score distribution table

Ranking	1	2	3	4	5	6	7
Score	90-100	85-89	80-84	75-79	70-74	65-69	60-64

(3) Subjective scoring. For the project that has requirements on project process or results but the percentage scoring or assessment ranking scoring is not suitable for use, the subjective scoring is applied. And scoring is carried out to the completion quality or result of each group according to the project scoring standards, and the score value is 100 to 60 points. Taking the classification of 5 levels as an example, please refer to Table 2 for the group subjective scoring standard and score distribution.

Table 2. Group subjective scoring standard and score distribution table

Standards	Total conformity	Conformity	Average	More non-conformity	Nonconformity
Score	90-100	80-89	70-79	60-69	0-59

2.1.3 Single project scoring rules of groups.
If there is only one task in the experimental project, the applicable percentage scoring, assessment ranking scoring or subjective scoring rules will be adopted for group project scoring. If the experimental project has 2 or more tasks, then the average value of the scores of all tasks will be taken as the final score of the group project.

2.1.4 Final scoring rules of groups.
The average value of the scores of all the experimental projects of the group is taken as the final score of the group, namely, the individual assessment scores of all members of the group (the team leader has another 5% bonus point).

2.2. Experiment subject
The assessment mode has been applied in an experimental course of a university with 8 experimental projects, including 5 projects that participated in the assessment, and 223 students from 8 classes of 4 undergraduate majors participated in these projects

3. Experiment and Discussion

3.1. Experiment
In the course experiment, the experimental projects 1, 2 and 3 are single task projects. And project 1 is scored according to the subjective scoring rules, project 2 is scored according to the percentage scoring rules, and project 3 is scored according to the rules of the assessment ranking scoring. Project 4 and project 5 both include two tasks, which are scored respectively according to percentage scoring and assessment ranking scoring, and the average value of the scores of two tasks is taken as the final score of the project. Projects 1, 3, 4, and 5 are scored in groups, with the group score being the individual score of the student, and project 2 is scored with the team as a unit, and with the team score being the project score of the group and the individual. Finally, the performance assessment results of each project are sorted out and obtained according to the professional grades, classes, groups and individuals.

3.2. Analysis and discussion
Using the Stata15.1 statistical analysis software, the statistical analysis is carried out according to the assessment performance of professional grades, classes, groups and individuals.

3.2.1 Statistical analysis based on professional grade assessment performance.

Detailed descriptive statistics is carried out to experimental projects 1 to 5 (p1, p2, p3, p4, p5) and average scores (as), and the results are shown in Table 3.

Table 3. Detailed descriptive statistical analysis results based on each project assessment performance of professional grades

Variable	Obs	Mean	Std. Dev.	Variance	Min	Max	50%	Skewness	Kurtosis
p1	4	82.08	2.29	5.24	79	84	83	-0.78	1.97
p2	4	70.25	17.63	310.92	49	92	70	0.05	1.97
p3	4	75.75	1.37	1.88	75	78	75	1.15	2.33
p4	4	74.88	2.82	7.95	71	77	76	-1.02	2.21
p5	4	74.79	2.61	6.80	71	78	75	-0.48	2.03
as	4	76.55	4.32	18.63	71	82	76	0.08	2.00

It can be seen from the above table that the average value of p1 to p5 is between 70.25 and 82.08; the standard deviation is between 1.37 and 17.63; the variance is between 1.88 and 310.92; the minimum value is between 49 and 79; the maximum value is between 77 and 92; the median is between 70 and 83; the skewness is between -1.02 and 1.15, basically around 0; and the kurtosis is between 1.97 and 2.33, both less than 3, but close to 3. On the whole, there is no obvious abnormal data, the data are normally distributed, and the data quality is high.

3.2.2 Statistical analysis based on class assessment performance.

(1) Descriptive statistical analysis of data. After carrying out descriptive statistical analysis of data to the projects 1 to 5 (p1, p2, p3, p4, p5) and the average scores (as), the results are shown in Table 4.

Table 4. Detailed descriptive statistical analysis results based on each project assessment performance of classes

Variable	Obs	Mean	Std. Dev.	Variance	Min	Max	50%	Skewness	Kurtosis
p1	8	82.14	3.21	10.33	76	85	82	-0.92	2.92
p2	8	70.25	16.32	266.50	49	92	70	0.05	1.97
p3	8	75.48	4.17	17.42	69	80	77	-0.57	1.75
p4	8	74.62	5.87	34.50	67	84	73	0.53	2.01
p5	8	74.70	4.58	21.00	68	82	75	0.01	2.01
as	8	76.43	4.38	19.21	71	85	76	0.60	3.06

It can be seen from the above table that the average value of p1 to p5 is between 70.25 and 82.14; the standard deviation is between 3.21 and 16.32; the variance is between 10.33 and 266.50; the minimum value is between 49 and 76; the maximum value is between 80 and 92; the median is between 70 and 82; the skewness is between -0.92 and 0.60, basically around 0; and the kurtosis is between 1.75 and 3.06, basically around 3. On the whole, there is no obvious abnormal data, and the data quality is high.

(2) Normality test of data. The normality test is carried out to experimental projects 1 to 5 (p1, p2, p3, p4, p5) and average scores (as), and the results are shown in Table 5.

Table 5. Normality test results based on each project assessment performance of classes

Variable	Obs	Pr(Skewness)	Pr(Kurtosis)	adj chi2(2)	Prob>chi2
p1	8	0.12	0.35	3.69	0.16
P2	8	0.94	0.67	0.19	0.91
P3	8	0.34	0.40	1.93	0.38
P4	8	0.37	0.73	1.01	0.60
P5	8	0.98	0.73	0.12	0.94
as	8	0.31	0.28	2.68	0.26

It can be seen from the above table that the test results of the skewness and kurtosis of all the experimental projects and their average scores are greater than 0.05, accepting the null hypothesis of the normal distribution. Considering the skewness in combination with the kurtosis, on the whole, the

Prob>chi2 values of all test results are greater than 0.05, which all accept the null hypothesis of normal distribution, and the data show a significant normal distribution.

To summarize, there is no obvious abnormality in the class assessment performance data, the data show a significant normal distribution. The data quality is good, and the assessment mode is feasible.

3.2.3 Statistical analysis based on the group assessment performance.

(1) Descriptive statistical analysis of data. Detailed descriptive statistics is carried out to experimental projects 1 to 5 (p1, p2, p3, p4, p5) and average scores (as), and the results are shown in Table 6.

Table 6. Detailed descriptive statistical analysis results based on each project assessment performance of groups

Variable	Obs	Mean	Std. Dev.	Variance	Min	Max	50%	Skewness	Kurtosis
p1	28	81.96	7.12	50.63	70	95	83	-0.25	2.15
p2	28	70.25	15.55	241.82	49	92	70	0.05	1.97
p3	28	75.89	9.53	90.84	60	90	75	-0.03	1.82
p4	28	74.75	12.53	156.99	50	95	77	-0.25	2.03
p5	28	74.71	11.23	126.19	50	98	74	-0.02	2.71
as	28	76.52	6.03	36.35	63	92	76	0.23	3.05

It can be seen from the above table that the average value of p1 to p5 is between 70.25 and 81.96; the standard deviation is between 6.03 and 15.55; the variance is between 36.35 and 241.82; the minimum value is between 49 and 70; the maximum value is between 90 and 98; the median is between 70 and 83; the skewness is between -0.25 and 0.23, basically around 0; and the kurtosis is between 1.82 and 3.05, basically around 3. On the whole, there is no obvious abnormal data, and the data quality is high.

(2) Normality test of data. The normality test is carried out to experimental projects 1 to 5 (p1, p2, p3, p4, p5) and average scores (as), and the results are shown in Table 7.

Table 7. Normality test results based on each project assessment performance of groups

Variable	Obs	Pr(Skewness)	Pr(Kurtosis)	adj chi2(2)	Prob>chi2
p1	28	0.53	0.30	1.62	0.44
P2	28	0.90	0.11	2.79	0.25
P3	28	0.94	0.03	4.60	0.10
P4	28	0.53	0.16	2.57	0.28
P5	28	0.96	0.91	0.01	0.99
as	28	0.57	0.55	0.72	0.70

It can be seen from the above table that only the kurtosis of experimental project 3 is 0.03, less than 0.05, rejecting the null hypothesis of normal distribution, and other test results of experimental projects 1-5 and their average scores are greater than 0.05, accepting the null hypothesis of normal distribution. Considering the skewness in combination with the kurtosis, on the whole, the Prob>chi2 values of all test results are greater than 0.05, which all accept the null hypothesis of normal distribution, and the data show a significant normal distribution.

To summarize, there is no obvious abnormality in the group assessment performance data, the data show a significant normal distribution. The data quality is good, and the assessment mode is feasible.

3.2.4 Statistical analysis based on individual assessment performance.

(1)Descriptive statistical analysis of data. Detailed descriptive statistics is carried out to experimental projects 1 to 5 (p1, p2, p3, p4, p5) and average scores (as), and the results are shown in Table 8.

Table 8. Detailed descriptive statistical analysis results based on each project assessment performance of individuals

Variable	Obs	Mean	Std. Dev.	Variance	Min	Max	50%	Skewness	Kurtosis
p1	223	82.09	6.91	47.77	70	95	80	-0.24	2.18
p2	223	68.77	15.31	234.43	49	92	68	0.15	1.98
p3	223	75.76	9.35	87.48	60	90	75	-0.01	1.82
p4	223	74.81	12.18	148.45	50	95	77	-0.24	2.02
p5	223	74.57	10.75	115.49	50	98	73	0.04	2.68
as	223	76.20	6.06	36.73	63	96	76	0.27	3.11

It can be seen from the above table that the average value of p1 to p5 is between 68.77 and 82.09; the standard deviation is between 6.06 and 15.31; the variance is between 36.73 and 234.43; the minimum value is between 49 and 70; the maximum value is between 90 and 98; the median is between 68 and 80; the skewness is between -0.24 and 0.27 basically around 0; and the kurtosis is between 1.82 and 3.11, basically around 3. On the whole, there is no obvious abnormal data, and the data quality is high.

(2) Normality test of data. The normality test is carried out to experimental projects 1 to 5 (p1, p2, p3, p4, p5) and average scores (as), and the results are shown in Table 9.

Table 9. Normality test results based on each project assessment performance of individuals

Variable	Obs	Pr(Skewness)	Pr(Kurtosis)	adj chi2(2)	Prob>chi2
p1	223	0.13	0.00	15.95	0.00
P2	223	0.34	0.00	33.71	0.00
P3	223	0.97	0.00	70.25	0.00
P4	223	0.13	0.00	29.05	0.00
P5	223	0.79	0.34	1.01	0.60
as	223	0.10	0.58	3.02	0.22

It can be seen from the above table that the test values of the skewness of the experimental projects 1, 2, 3 and 4 are all greater than 0.05, accepting the null hypothesis of the normal distribution; the kurtosis is 0.00, less than 0.05, rejecting the null hypothesis of normal distribution. Considering the skewness in combination with the kurtosis, on the whole, the Prob>chi2 values of the test results from projects 1 to 4 are all less than 0.05, rejecting the null hypothesis of normal distribution. The skewness, kurtosis and the overall test value of the experimental project 5 and the average score are both greater than 0.05, accepting the null hypothesis of normal distribution. Project 5 and individual average score data all show a significant normal distribution.

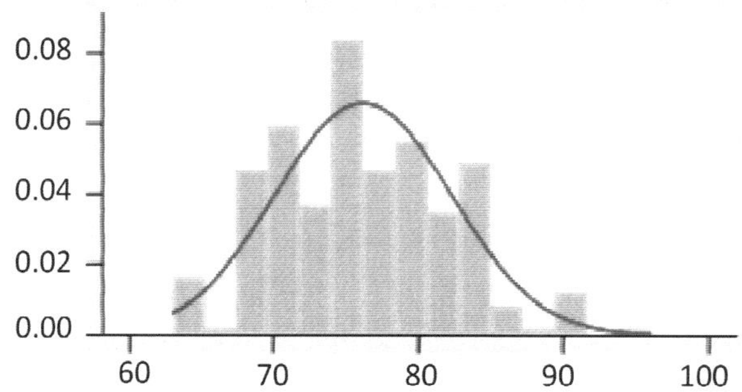

Figure 2. Histogram and normal distribution curve of the average score

On the whole, since the average score is the final assessment result of individual scores, the detailed descriptive statistics and normality test results of the average scores of individual assessment performance show that there is no obvious abnormality in the average scores of individual assessment performance, and the data show a significant normal distribution. The data quality is good, and the assessment mode is feasible.

4. Conclusions

The experimental project performance assessment mode based on team communication takes professional grades as a team, and groups in the team as the unit to complete each experimental project through communication and coordination of each group. According to the task completion of the team or groups, it adopts quantitative scoring, ranking scoring or subjective scoring and other scoring rules, and takes the project scores of the team or groups as the assessment performance of each member of the group, and takes the average score of al projects as the final assessment performance of each student (the group leader has a 5% bonus point). Judging from the actual assessment results, the main performance data all show normal distribution, and this assessment mode is practically feasible, with promotion and application value.

The shortcomings are that the experimental data in this paper come from 8 classes of 4 majors in one grade of an experimental course. The overall sample size is not enough, which may have some impact on the experimental results. In its application in the following professional grades of this experimental course and other experimental courses, the sample size of this assessment mode will be continuously increased to further improve the data quality.

Some statistical indicators in the sample data are not significant. For example, the group's performance kurtosis of project 2 and the individual performance kurtosis of projects 1-4 show a non-normal distribution, indicating that the capabilities of different majors or student groups in some projects are objectively quite different, and the distribution of people with different scores is quite different. This phenomenon may change with the increase of the sample size, but it may also be an objective phenomenon of some majors or student groups, which will not change with the change of sample sizes.

In summary, judging from the experimental results, the experimental project performance assessment mode based on team communication is feasible and worthy of application in the experimental courses suitable for grouping. In future application, the application results of this mode will be further tested through more abundant experimental data.

References

[1] Feng Xu, Yong Sun, Hongxi Zhang and Shuo Ji 2019 Exploration and practice of experimental teaching model in accord with requirements of applied talent training *Experimental Technology and Management* vol 36 (06) p 214-215+224. (In Chinese)

[2] Nan Shi, Xianhui Zhu, Pengtao Yuan, Caifeng Ma and Lin Sang 2018 Research on the Experimental Teaching Mode under the Target of Applied Talents Training *Heilongjiang Education (Theory and Practice)* (04) p 44-46. (In Chinese)

[3] Hua Qiang, Zhengwang Li and Shihui Wu 2018 Exploration of practical teaching assessment method in colleges and universities *Experimental Technology and Management* vol 35 (06) p 170-173. (In Chinese)

[4] Yancheng Yuan, Nianling Lv, Lijuan Lai and Ruixiang Yin 2019 Scientific Design and Process Management Scheme of Experimenta *Experimental Technology and Management* vol 36 (01) p 192-198+216. (In Chinese)

[5] Zhifang Yang, Zehui Chen, Xun Min, Anlin Chen, Zeling Dong, Xianlian Chen, Huan Yang and Ting Zhou 2017 Establishment and Implementation of Formative Evaluation Model of Clinical Microbiology Testing Technology Experiment Course *Continuing Medical Education* vol 31 (11) p 56-58. (In Chinese)

[6] Xihua Chen, Haining Huang, Peijie Huang 2019 An Evaluation of the Scientificity of Test Papers Based on Students' Score Analysis - Application of Data Mining and Feature Description *Journal of Beijing City University* (01) p 38-43. (In Chinese)

Journal of Physics: Conference Series **1423** (2019) 012037 doi:10.1088/1742-6596/1423/1/012037

An empirical study on the relationship between foreign trade of agricultural products and the development of cold chain logistics in Sichuan Province

Shilei Lin[1], Chong Wang[2*]

[1]School of Management, Sichuan Agriculture University, P.R.China
[2]School of Business, Sichuan Agriculture University, P.R.China
[1]739446590@qq.com, [2]wc000500@163.com

Corresponding author and e-mail: Chong Wang, wc000500@163.com

Abstract. This paper selects the data of Sichuan Province from 2001 to 2018 as the sample, which is the total trade volume of agricultural products (TA), the cold chain logistics volume of agricultural products (CD), and the cold chain logistics volume of agricultural products (CS). Through the time series of three variables, the stationary, co-integration and vector error correction models are used to study the long-term equilibrium and short-term fluctuations between the agricultural product foreign trade and agricultural product cold chain logistics development level, in order to provide suggestions for the common development of both. The results show that: First, the level of agricultural trade and the level of cold chain logistics of agricultural products continue to increase, but the overall level of the two is still insufficient. Secondly, the time series of the both are smooth sequences through the first-order difference. Through the cointegration test, there is a long-term equilibrium relationship, and the short-term relationship between the two is greater than the long-term relationship. Thirdly, the demand and supply of cold chain logistics of agricultural products is a one-way causal relationship between the trade volume of agricultural products. In the future, we must give full play to the advantages of Sichuan's characteristic agricultural products and strengthen the construction of logistics facilities and the connection between logistics nodes.

1. Introduction

Sichuan Province has rich agricultural resources and has strong comparative advantages and international market competitiveness. Strengthening agricultural foreign trade at the moment, promoting agricultural efficiency and increasing farmers' income with agricultural products exports has become an important path to transform agricultural development methods and extend industrial chains, and is also an important way to explore the development of Sichuan's export-oriented agriculture. Under the background of the "One Belt and One Road", the establishment of the Sichuan Free Trade Zone and the three major strategies for the development of the western region, Sichuan agricultural trade has ushered in tremendous development opportunities. How to seize the opportunity to more quickly make the "going out" of Sichuan's characteristic agricultural products and the introduction of high-quality agricultural products is the top priority of Sichuan's agricultural trade. In the process of agricultural product trade circulation, it will inevitably involve the issue of logistics and transportation of agricultural products. With the continuous development of agricultural products trade in Sichuan Province, the trade methods of different agricultural products are changing, the requirements for the quality of agricultural products are higher, the fresh agricultural products themselves are perishable, and the demand for cold chain logistics will be expanded, so the trade of agricultural products and There is a negligible link between the cold chain logistics of agricultural products. In recent years, the relationship and development between the two have also received increasing attention from the government and industry. "Sichuan Province's "13th Five-Year" Agricultural Products Cold Chain Logistics Development Plan" proposes to vigorously develop international cold chain logistics such as warehousing and distribution of agricultural products, improve the level of service trade facilitation, and continuously expand the international market of agricultural products cold chain logistics in Sichuan Province. The development space to realize the internationalization of Sichuan's characteristic agricultural products. No. 1 document of the central government in 2019, it was mentioned that it accelerated the promotion and supported agriculture to go global, strengthened the "One Belt, One Road" agricultural international cooperation, actively expanded the domestic shortage of agricultural products, expanded diversified import channels, and strengthened the agricultural product logistics backbone network and cold chain logistics system construction. It can be seen from the policies issued by the government and the planning of the industry that there is a very close relationship between agricultural products trade and agricultural products cold chain logistics. However, due to the late start of the construction of the cold chain logistics system in Sichuan Province, there are many defects in the construction of cold chain logistics infrastructure. The logistics information construction and technology application are not popular in Sichuan logistics enterprises. The export of agricultural products in Sichuan Province is dominated by traditional primary products. The food processing industry, the meat processing industry and the tea industry have fewer exports, the added value and technical content are lower, and the industrial chain is short. In the foreign trade of agricultural products and cold chain logistics, there is no effective integration between the two, which hinders the common development of the two. Studying the relationship between the two and exploring the path of stable and efficient development between the two will help the development of agricultural trade in Sichuan Province, so as to better build the open land of the western inland and the improvement of the level of opening up. At the same time, it also promotes the development of agricultural cold chain logistics and logistics enterprises in Sichuan Province, and achieves the goal of reducing trade time and economic cost, and enhances the competitiveness of Sichuan agricultural products in the international market. This article will use the collected yearbook data and existing research results to explore the relationship between the two, and provide countermeasures for the common development between the two.

2. Literature review and hypothetical reasoning

2.1. Literature review

By consulting relevant literature, domestic and foreign scholars have achieved more results in the study of foreign trade and logistics. Foreign scholars have studied this aspect earlier. Krugman (1991) invented the distance trade model. He quantitatively analyzed the influence of distance on international trade. His theory opened the first place for quantitative research on the relationship between logistics and international trade. The research in this area has had a profound impact. Boylan (1991) pointed out that the improvement of the current logistics system can make regional economic trade more competitive and improve its position in the international market. Egger (2005) selects more than one hundred countries to evaluate the logistics efficiency of each country based on the variables of cargo transportation time. After regression analysis, it is found that when the time of transporting goods is reduced, the logistics efficiency can be significantly improved, thus promoting each National import and export trade. In recent years, foreign scholars have focused on the impact of logistics performance on export trade. XiaoBo (2012) believes that the development of cold chain logistics in China is relatively lagging, which makes the loss of agricultural products in circulation seriously, which hinders the further development of agricultural trade between the two sides. And based on the supply chain integration of cold-chain logistics resources on both sides of the strait to build a cold chain logistics model and establish a cold chain logistics system, build a cross-strait cold chain logistics information platform, develop cold chain logistics technology, and train cold chain logistics talents. Gani and Azmat (2017) explored the role of logistics performance in international trade. The results show that overall logistics performance is significantly positively correlated with imports and exports. Martí and Luisa (2017) studied the importance of logistics performance in international trade and its trade costs, while using the gravity model to determine the importance of logistics in the export of these countries, and their findings indicate an improvement in the country. The logistics infrastructure is not only to promote their trade, but also to increase export competitiveness. Munim and Schramm (2018) used the structural equation model to empirically analyze the significant economic impacts of port infrastructure quality and logistics performance. The results show that port infrastructure can help improve logistics performance, promote maritime trade, and bring higher economic growth. .

Chinese scholars have also made great achievements in the study of foreign trade and logistics in recent years. Some scholars analyzed the relationship between the two through a time series model. Yu (2012) used the time series model to empirically analyze the relationship between modern logistics and foreign trade in Zhejiang Province. The study found that the cointegration test showed a long-term equilibrium relationship between the two, and the Granger causality test showed a one-way causal relationship between cargo throughput and total import and export. Zhao (2014) used the unit root and cointegration method to analyze the relationship between logistics industry efficiency and foreign trade dependence. The result is a long-term equilibrium relationship between logistics industry efficiency and foreign trade dependence. Development is conducive to the development of foreign trade, but in turn the development of foreign trade cannot significantly promote the development of logistics. Tang (2014) conducted an empirical analysis of the relationship between Sichuan's foreign trade and modern logistics. His research found that Sichuan's foreign trade and logistics industry have a significant relationship. Zhang and Chen (2015) combined the cointegration test, vector error correction model and Granger causality test to study the external economy of logistics facilities, and found that the completion of logistics facilities can promote the development of foreign trade.

Some scholars use the trade gravity model to analyze the relationship between the two. Wang, Dong et al. (2018) empirically analyzed the logistics performance level of China's foreign trade potential by expanding the trade gravity model based on the collected logistics performance index and its sub-indices in 51 countries and regions along the "Belt and Road". influences. Their research found that logistics performance significantly positively affected the bilateral trade volume between China and countries along the route. Wang and Wang (2018) used the gravity model to empirically analyze the impact of logistics level and export trade in Guangdong Province. The results of the study show that the improvement of logistics level can reduce the impact of distance on exports and deal with port efficiency and road transport efficiency. Increase the conclusion that the export value will increase accordingly.

Through the above research, it can be found that the development of logistics development level and the improvement of infrastructure are conducive to the development of foreign trade. Sichuan Province is China's major agricultural province, and cold chain logistics is an integral part of logistics. Among them, fresh agricultural products have a greater demand for cold chain logistics. The import and export trade of agricultural products plays an important role in the economic development of Sichuan Province. Therefore, it is of theoretical and practical significance to study the relationship between the import and export trade of agricultural products and the development level of agricultural cold chain logistics.

2.2. Hypothetical reasoning

Ma (2018)'s research shows that 70%~80% of China's urban agricultural product consumption is still circulating agricultural products through the wholesale market, and from the domestic and international environment, the application of cold chain logistics in the future agricultural product wholesale market is strengthened. And circulation is an important part of cold chain logistics. Therefore, this paper selects the circulation of agricultural products cold chain logistics as an indicator to measure the development level of cold chain logistics. Liu and Ci (2017) mentioned that China has more than 400 million tons of fresh agricultural products entering the circulation field every year, and its loss rate in logistics is as high as 25%-30%. In the current fierce competition in international trade in agricultural products, Dong and Huang (2016) believe that the quality of agricultural products is related to the competitiveness of Chinese agricultural products and the interests of 700 million farmers. Therefore, in the process of circulation of fresh agricultural products, it is necessary to ensure quality efficiently, reduce the rate of corrosion during transportation, effectively increase the circulation of cold chain logistics of agricultural products, enhance the international competitiveness of China's agricultural products and protect the interests of farmers. Based on the above analysis and the content of this study, this paper assumes that:

H1: Increasing the circulation of cold chain logistics of agricultural products has a positive impact on the import and export of agricultural products.

Fresh agricultural products are one of the main sources of nutrients needed by people in daily life. The research by Yuan and Wang (2017) shows that 70%~80% of imported fresh agricultural products are circulated in the domestic market through the agricultural product wholesale market. Fresh agricultural products basically need cold chain transportation to ensure quality and safety. Therefore, the importance of cold chain logistics is self-evident. However, the current level of cold chain logistics in Sichuan Province is still low, the cost of cold chain logistics is high, the number of refrigerated trucks and cold storage is insufficient, and the construction of cold chain infrastructure is imperfect. This also makes the development of cold chain transportation of fresh agricultural products encounter obstacles. At present, the comprehensive cold storage rate of agricultural products in Sichuan Province has only reached 50%, which has resulted in the agricultural products not being effectively protected during the import and export transportation. The quality of agricultural products has been greatly reduced, and quality and safety incidents have occurred frequently. Therefore, in the face of the short-board problem of cold chain logistics, Liu and Wang believe that (2016) should take measures to deal with cold chain logistics in a targeted manner, reduce the logistics cost of agricultural products, and promote the development of cold chain logistics, in order to improve the transportation rate of cold chain of agricultural products. It will help to better carry out the import and export trade of agricultural products. Therefore, this article assumes that:

H2: The cold chain logistics transportation volume of agricultural products has a positive impact on the import and export of agricultural products.

3. Data selection and variable setting

3.1. Data selection

Since the development of Sichuan cold chain logistics started late and China joined the World Trade Organization in 2001, the agricultural trade in Sichuan Province has been growing rapidly., this paper selects Sichuan Province 2001 by referring to the existing literature and considering the availability of Data of agricultural product import and export trade and agricultural product output in 2018 were analyzed as samples. At the same time, in the selection of agricultural product output data, the agricultural products, such as potatoes and eggs, which are not required for The total output of meat, aquatic products, fruits and vegetables, tea and milk in Sichuan Province was selected. The class does the raw data. The data is derived from the China Statistical Yearbook, the Sichuan Statistical Almanac, the Sichuan Provincial Bureau of Statistics and the data queried on the Internet.

3.2. Variable setting

TA is the total import and export trade of agricultural products in Sichuan Province, in order to represent the development level of Sichuan agricultural products import and export trade over the years. The development level of cold chain logistics of agricultural products, through consulting the literature and Drawing on the research results of predecessors, selects the evaluation indicators from the two perspectives of demand (CD) and supply (CS) to measure the development of cold chain logistics in Sichuan Province.

The circulation of agricultural products is an economic activity that realizes the transfer from agricultural production to consumption through the form of buying and selling. Of the products of Sichuan Province. According to the contents of the "13th Five-Year" Agricultural Products Cold Chain Logistics Development Plan of Sichuan Province, the cold chain circulation rate of agricultural products such as meat, fruits and vegetables, aquatic products, milk And tea is 40%, 25%, 40%, 50%, 70%,. after the five types of agricultural products are processed accordingly, the circulation of agricultural products in Sichuan Province is calculated.

The supply of cold chain logistics of agricultural products. The total amount of agricultural products transported by cold chain logistics is used to indicate the supply of cold chain logistics. At the same time, the planning also mentioned that the refrigerated transport rate of agricultural products is 55%, 40%, 65%, 55%, 60%, respectively. The output of the five types of agricultural products is also treated accordingly, and the transportation of cold chain logistics of agricultural products in Sichuan Province is obtained. Amount.

In order to eliminate the possible heteroscedasticity, the processed data are taken logarithmically, and LNTA, LNCD, and LNCS respectively represent the total agricultural trade volume of Sichuan Province, the demand for cold chain logistics in Sichuan Province, and the natural logarithm of supply.

4. Empirical part

4.1. Unit root test:

In order to prevent the occurrence of pseudo-regression, the unit root test is performed on all time series data. Because the selected sample size is small, for the reliability of the test, ADF test methods are used to test the unit root, and the AIC criterion is used. Judging the criteria to determine the stationarity of each variable sequence.

It can be seen from Table 1 that in the ADF test, the agricultural product trade volume in Sichuan Province and the original sequence of supply and demand of agricultural cold chain logistics in Sichuan Province are all unstable sequences. After the first-order difference is made to each variable sequence, ADF test is a stationary sequence. In the ADF test, the agricultural product trade volume in Sichuan Province is stable at 1%, and the supply and demand sequence of agricultural cold chain logistics in Sichuan Province is stable at 5 percent.

Table 1. Unit root test results

	Variable	(C、T、K)	T statistic	Threshold			P value
				1%	5%	10%	
	LNTA	(C、T、3)	-2.643	-3.959	-3.081	-2.681	0.107
	LNCD	(C、T、3)	-1.205	-3.887	-3.052	-2.667	0.647
ADF	LNCS	(C、T、3)	-1.187	-3.887	-3.052	-2.667	0.654
	DLNTA	(C、T、3)	-5.485	-3.959	-3.081	-2.681	0.0006
	DLNCD	(C、T、3)	-3.199	-3.920	-3.066	-2.673	0.039
	DLNCS	(C、T、3)	-3.123	-3.920	-3.065	-2.673	0.045

Note: In the inspection type, C represents the intercept term, T represents the trend term, and the number represents the lag order. Adding D to the variable indicates that the variable is first-order differential.

4.2. Cointegration test

Co-integration test can test whether there is a long-term equilibrium relationship between sequences. According to the above unit root test, the sequence of the three variables is a first-order stationary sequence, so co-integration analysis can be performed. In order to verify the trade volume of agricultural products in Sichuan Province and the co-integration relationship between the development level of cold chain logistics of agricultural products in Sichuan Province, in order to ensure the reliability of the test, two test EG two-step method and Johanse cointegration test are used for inspection, based on AIC. The criteria and SC criteria select the appropriate lag order.

(1) E-G two-step method. Firstly, the OLS regression is carried out on the three variables. The agricultural product trade volume in Sichuan Province is used as the explanatory variable, and the agricultural cold chain logistics in Sichuan Province is used as the explanatory variable to model the regression, and the following regression equation is obtained.

$$\mathrm{LNTA} = -2.474 + 6.473\mathrm{LNCD} - 4.178\mathrm{LNCS} \quad (1)$$

(7.787) (16.975) (16.879)

R^2=0.865 A-R^2=0.847

Then, the residual equation E is extracted from the regression equation (1), and the stationarity of the residual E is tested by ADF. According to Table 2, the residual sequence is stable at the level of one percent, thus indicating the trade volume of agricultural products in Sichuan Province. There is a cointegration relationship between the development level of cold chain logistics of agricultural products in Sichuan Province.

Table 2. Residual unit root test

variable	(C、T、K)	T statistic	Threshold			P value	Test results
			1%	5%	10%		
E	(C、T、3)	-2.575	-2.741	-1.968	-1.604	0.014	Stable

From the E-G two-step test, it is known that there is a cointegration relationship between the three sequences. Through the above cointegration test, the amount of agricultural products traded in Sichuan Province can be obtained, and there is a long-term equilibrium relationship between the supply and demand of agricultural cold chain logistics in Sichuan Province.

Error Correction Model Construction

According to the cointegration test, there is a long-term equilibrium relationship between the three variable sequences, but because of other factors, there will be deviations in the short term. Therefore, a modified error model is constructed to study the short-term relationship among the three. The modified error model constructed in this paper is as follows:

$$EC_{1,\ t-1} = LNTA_{t-1} + 15.897 - 3.688LNCS_{t-1} \quad (2)$$

$$EC_{2,\ t-1} = LNCD_{t-1} + 0.415 - 0.990LNCS_{t-1} \quad (3)$$

$$LNTA_t = 0.342EC_{1,\ t-1} - 34.307EC_{2,\ t-1} - 0.395LNTA_{t-1} - 0.588LNTA_{t-2} - 39.532LNCD_{t-1} - 18.398LNCD_{t-2} + 34.952LNCS_{t-1} + 22.813LNCS_{t-2} + 0.182 \quad (4)$$

R2=0.681 AIC=-1.389 SC=-0.964

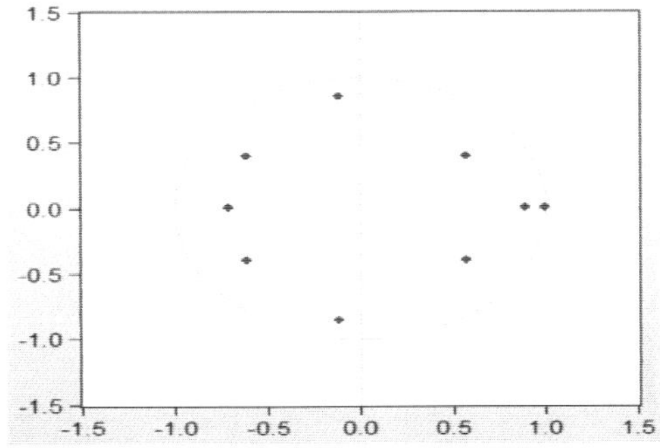

Figure 1. VEC model AR feature polynomial inverse root map

According to the regression coefficient R value is 0.681, so the overall model is more significant. The AIC is -1.389 and the SC is -0.964. The values of AIC and SC are both small, which indicates that the lag length of the model is reasonable. According to Figure 3, all estimated points are within the circle, so the estimated VEC model is stable. The coefficient of $EC_{1,\ t-1}$ is 0.342. When the model deviates from the long-term equilibrium with short-term fluctuation, the non-equilibrium state will be pulled back to the equilibrium state with the adjustment force of 0.342. It can be seen from the comparison of the coefficient coefficients. The impact of short-term relationships is greater than the impact of long-term relationships.

4.3. Impulse response analysis

Pulse analysis is used to describe the trend of influence between agricultural trade and agricultural cold chain logistics, as shown in the following figure. It can be seen that the short-term demand impact of agricultural cold chain logistics has a small initial value and a second lag. As time goes by, its contribution to agricultural trade volume begins to rise. When the demand for the second phase of cold chain logistics, that is, the circulation of cold chain logistics, increased, the trade volume of agricultural products in the third phase also increased, and the rate of increase was very fast. Then the speed began to slow down, but the trade in agricultural products was still rising slowly. the trend of. Therefore, the results of the impulse response are consistent with our hypothesis, and the assumption is true. Compared with the cold chain demand of agricultural products, agricultural trade has a greater response to the supply of agricultural products. The supply of cold chain logistics of agricultural products in the initial stage has little impact on the trade volume of agricultural products. However, as the supply of cold chain logistics of agricultural products increases in the second period, the force of agricultural products trade increases in the third period, and the effect continues. Strong sex, when the supply of agricultural products is weakened, the positive impact on the trade volume of agricultural products is still large, indicating that the increase in the supply of cold chain logistics of agricultural products can promote the increase of agricultural trade volume and its role has a sustained strong effect, the result of impulse response and Our hypothesis 2 is consistent, and hypothesis 2 is established.

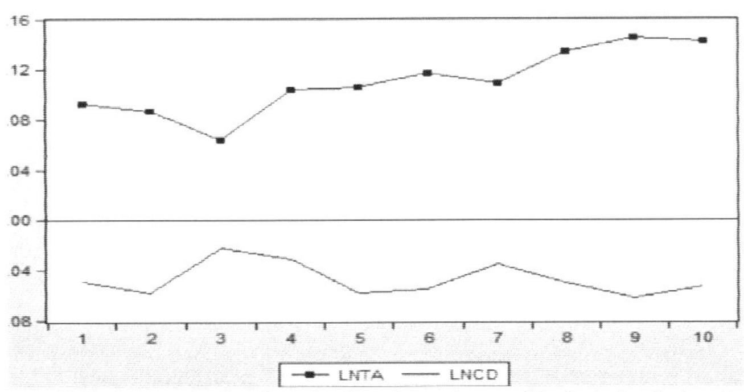

Figure 2. Pulse analysis of agricultural product cold chain logistics demand for agricultural product trade volume

Journal of Physics: Conference Series **1423** (2019) 012037 doi:10.1088/1742-6596/1423/1/012037

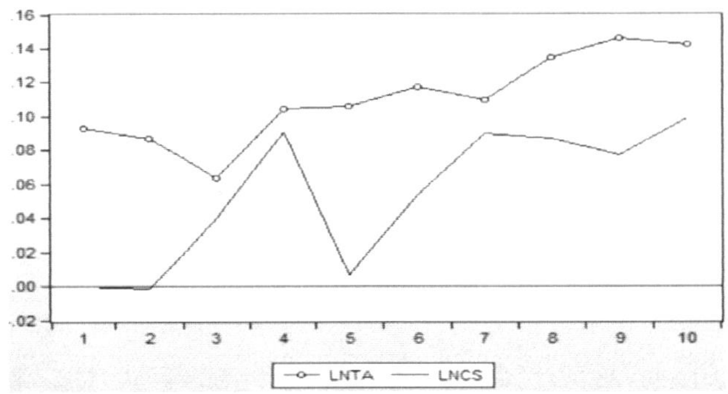

Figure 3. Pulse analysis of the supply of agricultural cold chain logistics to the trade volume of agricultural products

4.4. Granger causality test

The above test shows that there is a cointegration relationship between the variables, and both are first-order differential stationary sequences, so the Granger causality test is further carried out. As can be seen from Table 4, the demand for cold chain logistics is the Granger cause of agricultural trade, and the P value is 0.071, which is significant at the level of 10%. The supply of cold chain logistics is the Granger cause of agricultural trade, with a P value of 0.096, which is significant at the level of 10%. At the same time, the supply and demand of cold chain logistics are the Granger reasons of the other party. The P value is 0.000, which is significant at the level of one percent. Therefore, the results of the Granger causality test indicate that the increase in the demand and supply of cold chain logistics in Sichuan Province will increase the trade volume of agricultural products in Sichuan Province. The results of the causal test are consistent with our hypothesis 1 and 2. The growth of demand and supply reflects the development of cold chain logistics in Sichuan Province on the other hand. Therefore, the development of cold chain logistics in Sichuan Province can accelerate the development of agricultural trade in Sichuan Province.

Table 3. Inspection of Granger causality test results

Null hypothesis	F Statistics	P value	conclusion
Cold chain logistics demand is not Granger causality of agricultural trade	3.472	0.071	Refuse
Agricultural trade is not cold chain logistics demand Granger causality	0.076	0.928	accept
Cold chain logistics supply is not agricultural trade Granger causality	2.985	0.096	Refuse
The trade volume of agricultural products is not for cold chain logistics .Granger causality	0.065	0.938	accept

5. Conclusions and recommendations

Based on the above empirical analysis results, the relevant conclusions are as follows: (1) In the long run, there is a long-term equilibrium relationship between agricultural products trade in Sichuan Province and cold chain logistics of agricultural products. Although the total trade volume of agricultural products in Sichuan Province, the demand for cold chain logistics of agricultural products and the supply of cold chain logistics of agricultural products are all unstable sequences, and are all single-stage single sequence. (2) The improvement of Sichuan cold chain logistics level can continue to promote the development of agricultural products trade in Sichuan Province, but there is a certain lag, indicating that when the logistics level is improved, the trade volume of agricultural products will increase after a certain period of time, and agricultural products will be increased. Trade responds to cold chain logistics supply more than cold chain logistics demand. (3) From the result of Granger causality test, it is concluded that the demand and supply of cold chain logistics in Sichuan Province is a one-way causal relationship with the agricultural product trade in Sichuan Province. Even though Granger causality is not equal to the actual causal relationship, it also has reference value, which explains to some extent the continuous improvement of the cold chain logistics level of agricultural products in Sichuan Province, and will promote the development of agricultural trade in Sichuan Province in the long run. However, the development of agricultural trade in Sichuan Province is not necessarily due to the improvement of the level of cold chain logistics of agricultural products in Sichuan Province.

According to the analysis, it is known that there is a strong sustained impact between the two. Therefore, the following suggestions are proposed for the joint development of the agricultural trade in Sichuan Province and the cold chain logistics of agricultural products in Sichuan Province:

First of all, Sichuan pork, vegetables, tobacco and products have strong trade competitiveness, but there are very few deep-processed agricultural products. Therefore, we should actively promote Sichuan's high-quality agricultural products, such as tea, seeds, fruits and vegetables, and other quality products. At the same time, we should coordinate and integrate with logistics, standardize service standards, improve circulation efficiency, ensure the freshness and safety of perishable agricultural products, promote the development of agricultural trade and better coordinate domestic and international markets.

Secondly, for the cold chain logistics of agricultural products in Sichuan Province, we should vigorously develop third-party cold chain logistics, improve logistics technology, strengthen the infrastructure construction of cold chain logistics, and increase the number of refrigerated trucks and the number of cold storages. At the same time, it reduces the circulation of agricultural products and accelerates the connection between logistics nodes. Based on Chengdu and the surrounding cities of Chengdu, we will build a large-scale comprehensive multi-functional cold storage logistics center integrating meat, aquatic products, vegetables and fruits. At the same time, we must play the role of the Qing bai jiang railway transportation hub and achieve better coordination with Asia's largest railway container center, docking Rong Ou fast rail. This will establish a cross-regional agricultural product circulation base and an international commodity trading platform and it can directly connect with the Central Asian and European markets and increase the speed of the construction of port bonded logistics centers and international logistics parks. Relying on the Cheng Kun man International Channel, the market in Yunnan and Xinjiang can be connected to Europe, Central Asia and Southeast Asia, and a direct export system for transnational land transport will be built. The Tianfu big market relies on the three-dimensional transportation system to realize the circulation of agricultural products in the province for 24 hours. The construction of the transit corridor for land and sea transport promotes the smooth cross-border circulation of agricultural products.

Acknowledgment
The authors are supported by the National Natural Science Foundation of China (No. 71972136, 71602134, 71702156), Humanities and Social Sciences of Ministry of Education of China (No. 19YJC630063, 17YJC630098), and Sichuan Mineral Resources Research Center (No. SCKCZY2019-YB006).

References

[1] Krugman, Paul. Increasing Returns and Economic Geography[J]. Journal of Political Economy, 1991, 99(3):483-499.
[2] Boylan J. Logistics: Where Ends Have to Meet [J]. Journal of the Operational Research Society.
[3] Egger P. On the Impact of Transportation Costs on Trade in a Multilateral World[J]. Southern Economic Journal, 2005, 71(3):592-606.
[4] XiaoBo C. Construction of Cross-strait Cold-chain Logistics System and Development of Agricultural Products Trade [J]. Taiwan Agricultural Research, 2012.
[5] Gani, Azmat. The Logistics Performance Effect in International Trade[J]. The Asian Journal of Shipping and Logistics, 2017, 33(4):279-288.
[6] Martí, Luisa, Puertas R . The importance of export logistics and trade costs in emerging economies[J]. Maritime Economics & Logistics, 2017, 19(2):315-333.
[7] Munim Z H, Schramm H J. The impacts of port infrastructure and logistics performance on economic growth: the mediating role of seaborne trade [J]. Journal of Shipping and Trade, 2018, 3(1):1.
[8] Yazhen Yu. Empirical Research on the Relationship between Modern Logistics and Foreign Trade——Based on the Data of Zhejiang Province from 1986 to 2009[J]. International Trade Problem, 2012(01):99-107.
[9] Zhao Zhang. Research on the Dynamic Relationship between Logistics Industry Efficiency and Foreign Trade Dependence[J]. Statistics and Decision, 2014(22): 139-142.
[10] Xinzhi Tang. Research on the Relationship between Modern Logistics and Foreign Trade——An Empirical Analysis Based on Data from 2001 to 2013 in Sichuan Province[J]. Logistics Technology, 2014,33(23):245-246+259.
[11] Jin Zhang, Yiheng Chen. External Economic Research and Empirical Analysis of Logistics Facilities[J]. China Circulation Economy, 2015,29(12):46-53.
[12] Dongfang Wang, Qianli Dong, Lixin Yu. Logistics Performance of China and Foreign Countries along the Belt and Road Initiative and China's Foreign Trade Potential[J]. China Circulation Economy,

2018, 32(02): 17-27.
[13] Aihu Wang, Mengyao Wang, Yuyan He. The Influence of Guangdong Province's Logistics Level on Its Export Trade under the Background of "Belt and Road"——An Empirical Study Based on Gravity Model [J]. Industrial Engineering, 2018, 21(02): 1-8.
[14] Zengjun Ma. Review and Prospect of China's Agricultural Products Wholesale Market Development in the Past 30 Years[J]. China Circulation Economy, 2015,29(05):5-10.
[15] Genrong Liu, Yu Ci. Research on China's Agricultural Product Circulation Innovation and Its Impact on Farmers' Income[J]. China Economic Issues, 2017(03): 113-122.
[16] Yinguo Dong, Junwen Huang. Quality Measurement of China's Exported Agricultural Products—Based on Nested Logit Model[J]. China Rural Economy, 2016(11): 30-43.
[17] Lingling Yuan, Beibei Wang. Analysis on the Causes and Trends of the Growth of Import Trade of Fresh Agricultural Products in China[J]. China Circulation Economy, 2017,31(12):56-64.
[18] Jianxin Liu, Keshan Wang, Chunlin Zhang. Main Problems and Countermeasures of Ecommerce Development of Fresh Agricultural Products[J]. China Circulation Economy, 2016,30(12):57-64.

Journal of Physics: Conference Series **1423** (2019) 012039 doi:10.1088/1742-6596/1423/1/012039

Free cash flow and position-related consumption of senior management-based on empirical data of state-owned listed companies

Xing Tan[1], Shengdao Gan[2], Wei Yang[3], Furong Guo[4*]

[1] Finance department, Chengdu Wanjiang Gangli Science and Technology Co., Ltd., Chengdu, China
[2] Business School, Sichuan University, Chengdu, China
[3] Business School, Sichuan University, Chengdu, China
[4] Business School, Sichuan University, Chengdu, China

Corresponding author and e-mail: Furong Guo, guofurong@stu.scu.edu.cn

Abstract. The ultimate owner of the state-owned listed companies is the State, but the State cannot directly participate in the supervision of the managers of the company, so the owner may be "absent" in many state-owned listed companies, leading to excessive position-related consumption of the company's senior management by using the resources of the company in order to seek personal benefits. When the free cash flow is abundant, the senior management will more intend to put it into their own position-related consumption, resulting in higher agency costs. This article collects the data of A-share state-owned listed companies from 2009 to 2015 as research sample, and empirically studies the impact of free cash flow and the "Eight Regulations" on the position-related consumption of senior management. According to the research, excessive free cash flow will promote the position-related consumption of senior management of state-owned listed companies, resulting in higher agency costs; and in case the enterprise has certain free cash flow, the promulgation of the "Eight Regulations" has a restraining effect on the position-related consumption of senior management in the state-owned listed companies.

1. Introduction

Senior management's position-related consumption behavior is a reflection of agency cost caused by free cash flow, which has drawn wide concern. For a long time, the problem of position-related consumption of senior management in Chinese enterprises is relatively prominent, the ultimate owner of state-owned listed companies is the State, but the State cannot directly participate in the supervision of managers of the enterprises, and the owner's "absence" problem is serious. Due to the lack of supervision over real shareholders, serious agency problems exist in state-owned enterprises. At the same time, because of the "pay ceiling" policy, it is the good choice for the senior management of state-owned enterprises to make use of the resources of the enterprises for personal gains, and excessive position-related consumption naturally becomes a shortcut for them to gain benefits.

State-owned enterprises play an important role in the national economy and people's lives. However, along with the continuous development of economy, the drawbacks of state-owned enterprises are gradually exposed, the problem of position-related consumption of senior management in state-owned enterprises has been relatively prominent, the "sky-high wine price", "sky-high hospitality price" event and so on, which have been widely concerned by all sectors of society, reflect the problem of position-related consumption of senior management. The original intention of position-related consumption is to encourage the senior management of state-owned enterprises to work hard, create more value for the enterprises and maximize the interests of shareholders. And from the high level of position-related consumption, it can be seen that the actual results are contrary to the original intention. The position-related consumption of senior management in state-owned enterprises not only increases the cost of the enterprise, damages the interests of shareholders, but also causes a great loss to the image of the enterprise.

For this purpose, relevant government departments have issued a series of management measures to control the increasingly serious problem of position-related consumption of senior management in state-owned enterprises, among which, the most influential measure is the promulgation of the "Eight Regulations". After the promulgation of the "Eight Regulations," the Political Bureau of the CPC Central Committee started from its own, forming a good atmosphere of top-down supervision. Both the leaders and the grass-roots officers have really felt the tremendous changes during this period.

Based on the empirical analysis of China's A-share state-owned listed companies in Shanghai and Shenzhen from 2009 to 2015, this article discusses the impact of free cash flow and government regulatory policies on the position-related consumption of senior management in state-owned listed companies, with certain characteristics and innovation, which is mainly reflected in the following aspects: 1. This article discusses the impact of free cash flow on position-related consumption through empirical analysis, enriches the research sample on the correlation between free cash flow and position-related consumption, and makes more research results on the level of position-related consumption. 2. In view of the position-related consumption of senior management is hidden and difficult to quantify, quantitative analysis is difficult. This article introduces the cost behavior theory into the study of position-related consumption of senior management, divides the administrative expenses into fixed cost and variable cost, and takes the administrative expenses after deducting the amortized intangible assets which are easy to quantify in the fixed cost as the quantitative description of the position-related consumption. 3. This article studies the impact of "Eight Regulations" on the position-related consumption of senior management in state-owned listed companies under the condition that enterprise has certain fixed free cash flow, which is innovative and enriches the research sample with respect to the impact of government policies on position-related consumption, especially provides microscopic evidence on the impact of "Eight Regulations" on position-related consumption.

2. Literature Review and Hypothesis Proposal

According to agency theory, senior management and shareholders' goals are not always the same, senior management behave for their own-interest, which means, they maximize their own interests by sacrificing the interests of shareholders (Jensen, Meckling, 1976), position-related consumption is a typical case of agency cost. Position-related consumption has caused negative economic consequences to enterprises, which is a kind of agency conflict between managers and shareholders, and will bring about the reduction of enterprise value. Some cases have shown that the agency cost of position-related consumption is much greater than the benefit it brings (Hart, 2001). According to Fama (1980), certain preconditions exist when the position-related consumption becomes the agency cost between the owner and the manager. Only when the salary adjustment of the manager cannot make up for the company resources consumed by position-related consumption, position-related consumption will become a kind of agency cost, and position-related consumption will also become the opportunistic behavior of the senior management in enterprise. The existence of free cash flow increases the possibility of position-related consumption by senior management. Jianping HU, Shengdao GAN and Li WANG (2008) found that the free cash flow of enterprises has a positive correlation with discretionary income and expenditure. Because of the inconsistency of objectives and information asymmetry, there are different opinions between owners and managers in term of the use of free cash flow, while the management contract is not perfect and the manager market is not mature in China, so the senior management of enterprises tend to choose self-motivation through position-related consumption (Yinguo LIU and Chen ZHANG, 2012). By means of restricting free cash flow of enterprise, increasing debt financing (Jensen and Meckling, 1976) and cash dividend payment (Agrawal and Jayaraman, 1994), the cash flow that managers can control is reduced, the control power of managers is reduced, the agency cost between owners and managers is reduced, and the business performance is improved. Based on this, hypothesis 1 is proposed:

Hypothesis 1: Excessive free cash flow will encourage position-related consumption by senior management of state-owned listed companies, resulting in higher agency costs, free cash flow is positively correlated to position-related consumption level.

On December 4, 2012, the Political Conference of the CPC Central Committee deliberated and adopted eight regulations on improving work style and maintaining close ties with the people. These "Eight Regulations" put forward the requirements of "simplifying reception process", "not arranging banquets", "streamlining conference activities", and "strictly enforcing regulations on work and living treatment such as housing and vehicle equipping", which will inevitably have a certain impact on the behavior of the senior management of state-owned enterprises. Owners of state-owned enterprises are "absent", which makes it difficult for them to monitor and control the senior management effectively. In this case, the promulgation of the "Eight Regulations" of the central government is bound to strengthen the supervision of the senior management and weaken the self-interest behavior of the senior management. Therefore, from the perspective of free cash flow, this article studies the impact of the "Eight Regulations" policy on the position-related consumption of senior management in state-owned enterprises under the condition of certain fixed free cash flow. Based on this, hypothesis 2 is proposed:

Hypothesis 2: Under the condition of certain fixed free cash flow, the promulgation of the "Eight Regulations" has reduced the level of senior management position-related consumption in state-owned listed companies, and government supervision has played a restraining role on senior management's position-related consumption

3. Research and Design

3.1. Data Source and Sample Selection

In order to verity the hypothesis 1, this article selects the financial data of A-share state-owned listed companies in Shanghai and Shenzhen from 2011 to 2015; In order to verity hypothesis 2, this article selects the financial data of A-share state-owned listed companies in Shanghai and Shenzhen from 2009 to 2015, and selects the sample data for this article. Considering that in 2012, the Political Bureau of the CPC Central Committee deliberated and adopted the "Eight Regulations." Therefore, this article takes the data for a total of seven years as research samples, including data for three years before the promulgation of "Eight Regulations" (2009-2011), on the year of the promulgation (2012), and for three years after the promulgation (2013-2015). The relevant data source involved in this study is the CSMAR database. On the basis of the whole sample collected, the following screening was performed: 1. The samples with missing data are removed. 2. Abnormal samples whose data are obviously not comparable such as negative free cash flow, asset-liability ratio greater than 1 (insolvent), negative position-related consumption level (administrative expenses less than amortization of intangible assets) are removed.

3.2. Selection and Measurement of Variables

3.2.1 Explained variables

Position-related consumption: Based on Ang, Cole, Lin's research method and cost behavior theory, this article uses the difference of administration expenses minus the amortization of intangible assets (fixed cost) to measure the level of senior management's position-related consumption, and uses the ratio of the difference to the main business income (NPC) to quantitatively describe the level of senior management' position-related consumption, namely:

Position-related consumption level (NPC) = (administration expenses-amortization of intangible assets)/main business income

3.2.2 Explanatory variables

Free cash flow: Based on the calculation methods [31] of Standard & Poor's, Professor Copeland (1990), Lehn (1989) and other scholars, this article adopts the following calculation methods to calculate the absolute value of free cash flow:

Free cash flow = (net profit + interest expense + non-cash expenditure)-operating capital supplement-capital expenditure

At the same time, in order to eliminate the influence of enterprise size and make the enterprises of different scales comparable, the free cash flow intensity index is selected to measure the free cash flow level (FCF).

FCF=Free cash flow/total assets of the enterprise

3.2.3 Control variables
Based on the research methods of Yinguo LIU, Chen ZHANG (2012), Rong YANG (2016) and other scholars, this article introduces management pay (MP), capital structure (LEV), enterprise size (SIZE) as control variables.

The specific calculation method is as shown in Table 1.

Table 1. Variable Connotation Summary

Variable abbreviation	Variable name	Variable connotation
NPC	Position-related consumption level	(Administration expenses-amortization of intangible assets)/main business income
FCF	Free cash flow	Free cash flow/total assets at year-end
MP	Management pay	Annual total salary of senior management such as director or supervisor /main business income
LEV	Capital structure	Total liabilities at year-end/total assets at year-end
SIZE	Enterprise size	Natural logarithm of total assets at year-end

3.2.4 Model construction
According to the above variables, based on hypothesis 1, we construct the following multiple regression model to check the relationship between the free cash flow (FCF) of state-owned enterprises and the level of position-related consumption (NPC):

$$NPC=\beta_0+\beta_1 FCF+\beta_2 MP+\beta_3 LEV+\beta_4 SIZE+\varepsilon$$

To verify hypothesis 2, that is, to verify the impact of the "Eight Regulations" on the position-related consumption level of senior management in state-owned listed companies under the condition that the free cash flow remains unchanged, this article uses the ratio of the absolute number of position-related consumption level to the free cash flow of the enterprise to measure the level of position-related consumption of the senior management in the state-owned listed companies, which indicates how much will be used to increase the position-related consumption of the senior management in case of increase of one yuan of free cash flow of the enterprises. The details are as follows:

NPC/FCF=Absolute value of position-related consumption /free cash flow
= (administration expenses-amortization of intangible assets)/free cash flow

4. Empirical Analysis

4.1. An empirical analysis of free cash flow and senior management's position-related consumption

4.1.1 Descriptive statistics
The descriptive statistics of the whole samples is as shown in Table 2. The average value of the level of position-related consumption (NPC) is 0.0763, showing great difference between the two levels. The average value of free cash flow (FCF) is 0.0594, showing significant difference between the two levels. The average value of management pay (MP) is 0.0022, the lowest value is 0, the highest value is 0.0989 (at the same time, its position-related consumption level is also the highest); Average value of capital structure (LEV) is 0.5204, in which: The number of enterprises with asset-liability ratio greater than 0.5 accounted for 54.54%, and the number of enterprises with asset-liability ratio between 0 and 0.5 accounted for 45.46%. The average value of enterprise size (SIZE) is 22.6334. After logarithmic processing, the gap of enterprise size is getting narrowed.

Table 2. Descriptive statistics of main variables

Variable name	Variable abbreviation	Quantity	Minimum	Maximum	Average value	Standard deviation
Position-related consumption level	NPC	2719	0.0018	0.7934	0.0763	0.0615
Free cash flow	FCF	2719	2.73E-06	1.3504	0.0594	0.0615
Management pay	MP	2719	0.0000	0.0989	0.0022	0.0040
Capital structure	LEV	2719	0.0103	0.9961	0.5204	0.2031
Enterprise scale	SIZE	2719	18.3701	28.4820	22.6334	1.3747

4.1.2 Correlation analysis

Because of the introduction of free cash flow (FCF), management pay (MP), capital structure (LEV), and other variables, there may be multiple collinearity in explaining the dependent variable position-related consumption level (NPC). Pearson correlation analysis is used to test the correlation between the explanatory variables, and Cohen's correlation coefficient and correlation criterion in the social science field are referred to. Pearson correlation analysis was performed on the model constructed in this article, and the results are shown in Table 3:

Table 3. Pearson correlation analysis results

		NPC	FCF	MP	LEV	SIZE
NPC	Pearson correlation	1				
	Significance (two-sided)					
FCF	Pearson correlation	.039*	1			
	Significance (two-sided)	.042				
MP	Pearson correlation	.533**	.031	1		
	Significance (two-sided)	.000	.105			
LEV	Pearson correlation	-.212**	.079**	-.220**	1	
	Significance (two-sided)	.000	.000	.000		
SIZE	Pearson correlation	-.320**	.018	-.391**	.333**	1
	Significance (two-sided)	.000	.351	.000	.000	

*. Significant correlation at level of 0.05 (two-sided). **. Significant correlation at level of 0.01 (two-sided).

The Pearson coefficient and absolute value of each variable are mostly less than 0.5, which indicates that there is no highly collinearity problem between the main variables selected in this model. Therefore, the established regression model will not cause multicollinearity problem. It can be seen from the table that the free cash flow (FCF) is significantly and positively correlated with the position-related consumption (NPC) at a level of 5%, which can verity hypothesis 1 to some extent.

4.1.3 Regression analysis

This article makes a regression analysis on the relationship between the free cash flow (FCF) of the state-owned listed companies and the level of position-related consumption (NPC) of senior management. According to model 1, the following regression results can be obtained:

Table 4. Regression results of free cash flow and senior management's position-related consumption of state-owned listed companies

Variable	B	SE	P
Intercept	0.1832	0.0180	7.22E-24
Free cash flow (FCF)	0.0322	0.0161	0.0454
Management pay (MP)	7.2096	0.2672	2.94E-142
Capital structure (LEV)	-0.0221	0.0052	2.01E-05
Enterprise size (SIZE)	-0.0050	0.0008	7.34E-10
P		3.97E-212	
R^2		0.3045	
Adjustment R^2		0.3035	

In Table 6, the free cash flow (FCF) coefficient is positive (B=0.0322) and statistically significant at 0.05 level (P=0. 0454), which indicates that the state-owned enterprises will stimulate the senior management to use the enterprise funds for position-related consumption when the enterprise has a higher level of free cash flow, which fully proves the free cash flow hypothesis. Excessive free cash flow increases the agency costs of enterprises and encourages the position-related consumption of senior management. Therefore, the results support hypothesis 1 again.

4.1.4 Robustness test

The absolute value index is used to replace the relative value index in the model, the absolute values of position-related consumption (NPC), free cash flow (FCF) and management pay (MP) are used to replace the previous relative values. The regression model is still statistically significant at 0.05 level (P=0), and the free cash flow coefficient is still positive (B=0.6372), and statistically significant at 0.05 level (P=0). The regression results also support the main conclusions of this article.

At the same time, all variables are winsorized by 0.5% in this article. The regression model is still statistically significant at 0.05 level (P=0), free cash flow coefficient is still positive (B=0.0316), and statistically significant at 0.1 level (P=0.075). The regression results also support the main conclusions of this article.

4.2. An empirical analysis of senior management' position-related consumption in state-owned listed companies before and after the "Eight Regulations" were promulgated.

4.2.1 Descriptive statistics

Table 7 shows the number of samples by year. It can be seen from the table that there were 1,405 samples before the promulgation of the Eight Regulations (2009-2011), 550 samples in 2012 when the Eight Regulations were promulgated, and 1,660 samples after the promulgation of the Eight Regulations (2013-2015).

Table 5. Statistical Table of Sample Quantity by Year

Time	2009	2010	2011	2012	2013	2014	2015	Total
Number of enterprises	425	454	526	550	547	568	545	3615

4.2.2 Analysis of the position-related consumption of senior management before and after the promulgation of "Eight Regulations"

Through the analysis of the position-related consumption data of senior management in state-owned listed companies in the three periods: before the promulgation of the Eight Regulations (2009-2011), in the year of promulgation (2012) and after the promulgation of the Eight Regulations (2013-2015), as shown in Table 8:

The consumption level after the promulgation of the "Eight Regulations" is significantly lower than that before the promulgation, and the government policies play a positive role in curbing the position-related consumption of senior management in company.

Table 6. Descriptive statistics of position-related consumption

Time	Quantity	Minimum	Maximum	Average value	Standard deviation
2009	425	0.0024	409.5300	5.2794	25.0482
2010	454	0.0122	984.3778	5.6401	48.2410
2011	526	0.0060	1104.1548	5.6879	66.3271
2012	550	0.0041	157.6844	2.3330	8.2853
2013	547	0.0039	353.4061	3.7630	17.9872
2014	568	0.0140	144.6882	3.2056	10.7969
2015	546	0.0070	248.7918	3.3660	13.4666

As shown in Table 8, the average value of the NPC/FCF ratio reached a maximum of 5.6879 in 2011, and the standard deviation thereof reached a maximum of 66.3271; In 2012, the minimum value was 2.3330 and the standard deviation was 8.2853. The average value and standard deviation of NPC/FCF ratio for each year are shown in Figure 1:

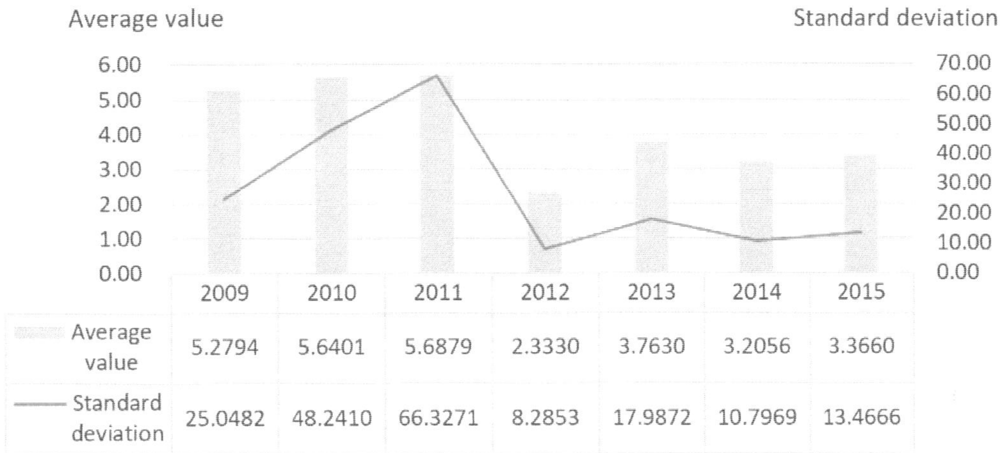

Figure 1. Average value and standard deviation of NPC/FCF ratio

The average value of NPC/FCF ratio for the three years following the promulgation of the Eight Regulations (2013-2015) was generally lower than that for the three years before the promulgation (2009-2011). In the year of promulgation (2012), due to the timeliness of the policy, it has a strong restraining effect on position-related consumption, so the decline trend is obvious. The average value of NPC/FCF ratio for three years after the promulgation of the Eight Regulations was 3.4419, compared with 5.5489 before the promulgation, marking a decrease of 37.97%. Therefore, under the condition that free cash flow remains unchanged, the level of position-related consumption of senior management in state-owned listed companies is obviously lower than that before the promulgation of the "Eight Regulations", and the government policy has a positive effect on restraining position-related consumption of senior management in state-owned listed companies.

At the same time, after the promulgation of the "Eight Regulations," the standard deviation of NPC/FCF ratio also decreased obviously, the discreteness of the sample enterprises' position-related consumption level was lower, and the range of NPC/FCF ratio was lower than that before the promulgation, indicating that the randomness of position-related consumption of senior management in state-owned listed companies was obviously weakened, and the management tended to treat position-related consumption more cautiously in order to control unnecessary position-related consumption expenditure. To sum up, the results of this study support the hypothesis 2. Under the condition that there is certain fixed free cash flow, the promulgation of "Eight Regulations" reduces the level of position-related consumption of senior management in state-owned listed companies, and government regulation plays an inhibitory role on senior management's position-related consumption.

5. Conclusions and Suggestions

Through the empirical study of state-owned listed companies from 2009 to 2015, this article discusses the impact of free cash flow and "Eight Regulations" on position-related consumption of senior management, and draws the following main conclusions:

(1) The free cash flow of state-owned listed companies is positively correlated to the level of position-related consumption of senior management. Therefore, in order to control the level of position-related consumption of senior management in state-owned enterprise, it is necessary to keep the free cash flow within a reasonable level, so as to ensure that the interests of the owners of enterprises will not be harmed.

(2) The promulgation of the "Eight Regulations" has restrained the level of position-related consumption of senior management in state-owned enterprise. Upon the promulgation of the "Eight Regulations", the level of position-related consumption of senior management in state-owned enterprise has shown a significant decrease trend, and the government's regulatory policy has played a better role in restraining such consumption.

The author believes that in order to control the consumption behavior of senior management reasonably, it is allowed to control the free cash flow (internal) and strengthen government supervision (external). For example: Control free cash flow rationally through cash dividend and stock repurchase; Control the capital structure reasonably through moderate debt, and therefore, the tax shield effect can also be obtained. Strengthen the government supervision; Improve the ability of the SASAC in performing its duties and strengthen the supervision of the owners; Strengthen the level of information disclosure.

References

[1] Michael C. Jensen, Agency Costs of Free Cash Flow, Corporate Finance, and Takeovers, American Economic Review, May 1986, Vol.76, No.2, pp.323.
[2] Tom Copeland, Tim Koller, Jack Murrin. Valuation: Measuring and Managing the Value of Companies, fifth edition [M], McKinsey and Company.
[3] Cornell, Company Value Assessment: A Tool for Effective Assessment and Decision-Making. Translated by Zhiqiang ZHANG and Chunxiang WANG. Huaxia Publishing House, 2001.
[4] James C. Van Horn et al. Modern Corporate Financial Management (11th Edition). Translated by Hao GUO. Economic Science Press, 1998.
[5] Kenneth Hankel, Yossi Livant, Cash Flow and Securities Analysis (2nd edition), Translated by Wang ZHANG, et al., Huaxia Publishing House, 2001.
[6] Hart O.Financial Contracting [J].Journal of Economic Literature, 2001, 39(4):1070-1100.
[7] Fama E. F.. Agency problems and the theory of the firm [J]. The Journal of Political Economy, 1980, 88(2):288-307.
[8] Jianping HU, Shengdao GAN. Agency Cost of Free Cash Flow: Theory and Evidence [J]. Contemporary Finance and Economics, 2009, (12): 109-114.
[9] Yinguo LIU, Chen ZHANG. Free Cash Flow and Position-related Consumption: An Empirical Study Based on Ownership and Corporate Governance [J]. Management Review, 2012, 24 (10): 18-26.
[10] Agrawal, A. and Jayaraman, N. The dividend policies of all-equity firms: a direct test of the free cash flow theory [J]. Managerial and Decision Economics, Vol. 15, 139-148.
[11] Ang J.S., Cole R.A., Lin GW.. Agency cost and ownership structure [J].Journal of Finance, 2000, 55(1):81-106.
[12] Lehn,K.,Poulsen, A., Free cash flow and stockholder gains in going private transactions[J].Journal of Finance,1989(44).
[13] Rong YANG. Eight Regulations, MP Control and Position-related Consumption [J]. Journal of Shanghai Normal University (Philosophy & Social Sciences Edition), 2016, 48 (1): 138-148.
[14] Jie MEI, Yang GE. The Effect of Policy Intervention on Position-related Consumption of State-owned Enterprises' Managers--A Pseudo-natural Experiment Based on "Eight

Regulations" in 2012 [J]. Economist, 2016 (2): 75-83.
[15] Shengdao GAN. Monographic Study on Free Cash Flow [M]. Dalian: Dongbei University of Finance & Economics Press, 2009: 8-17.

Journal of Physics: Conference Series **1423** (2019) 012040 doi:10.1088/1742-6596/1423/1/012040

The application of EDXRF on soil heavy metal analyzation in a planting area

Xiaozhe Li[1], Yi Chen[1], Liangquan Ge[1,*], Fei Li[1], Wei Tan[1], Jingru Xu[1]

[1] Chengdu University of Technology, Chengdu, China

* Corresponding author and e-mail: Liangquan Ge, glq@cdut.edu.cn

Abstract. With the increase of heavy metal concentration in soil, serious environmental problems have arisen. At present, heavy metals mainly refer to biotoxic heavy metals, including Lead, Cadmium, Mercury and Chromium, etc., which are produced in large quantities by human activities and difficult to be degraded, resulting in their loss of original activity and chronic poisoning because of the reaction with proteins and enzymes in human body. The soil concentration in the planting area exceeds the standard and is enriched in human beings, animals and plants, which directly endangers the living environment and health. In this paper, the soil in the planting area next to Xiangzhang canteens of Chengdu University of Technology was selected for field investigation, and the data analysis and risk assessment were carried out after testing the concentration of heavy metals in the surrounding soil through reasonable sampling, sample treatment and Energy-Dispersive X-ray fluorescence analysis techniques.

1. Introduction

Human activities such as mining, smelting, and agriculture cause heavy metal pollution in the soil. (Hong Tao et al.,2019) The chemical and metallurgical industries are the most important sources of metals in the environment, which are classified as "heavy metals". From the analysis of the environment of the entire urban, various human activities have brought a lot of heavy metals into the urban soil, causing the accumulation of these heavy metal elements to cause great damage to the ecological environment, especially the pollution of the soil environment(García-Sánchez et al., 2019), which threaten the safety of farmland ecological environment and the quality of agricultural products(Wu Tangfu et al., 2019). The main purpose of this study is to evaluate the soil conditions in the planting area beside the Xiangzhang canteen of Chengdu University of technology. Most of the plants in the planting area can be used. However, it is obvious that eating the vegetables planted in the contaminated soil will damage the health(Liu et al., 2019). Unlike other types of pollution, such as air pollution or ozone exposure, the individual risk associated with soil pollution largely depends on the heavy metal concentration of vegetables, and soil is the most important factor directly determining whether the amount of heavy metal in vegetables exceeds the limit. At present, the commonly used heavy metal detection methods mainly include a new electrochemical based on physical and chemical reactions, biological detection and optical detection method, but the steps are complicated and the detection cost is high. The biggest advantage of EDXRF detection technology based on atomic excitation is its high efficiency. At the same time, it can deal with the above problems simply, and it can also achieve the effect of nondestructive and rapid detection in the field.

2. Materials and method

2.1. Overview of the study area

The sampling point is located in the planting area next to the Xiangzhang canteen of Chengdu University of Technology, Chengdu, Sichuan Province. Due to the planting area next to the Xiangzhang canteen is too large, the area of 40 × 40 m^2 is selected in this experiment. The sampling interval is 10 m, and the soil below 2 cm in the surface layer is taken (no ground slide, no impactor, and no soil error from other different areas). 25 samples were taken and transferred to the plastic vacuum sample bag, numbered and sealed, about 20g each.

2.2. Experimental instrument

IED-2000T portable X-ray fluorescence analyzer was used in this experiment, which is developed by Chengdu University of Technology. Its specific parameters are as follow: HardGain=2, SoftGain=10000, ShapTime=2, TubeHv=30, Tubel=2, TestTime=100, EgA=0.0346331, EgB=0.0259799.

The dryer is DHG type constant temperature dryer. The grinder is developed by Beijing Zhonghe pioneering technology development Co., Ltd. The grinding time is 10s and the sieving scale is 200 mesh. The tablet press is a powder tablet press for Zhonghe ZHY-401B fluorescence spectrometer, with a pressure of 40t. In addition, there are several sample bags, grinding mortar and other glass instruments

2.3. Moseley's theorem

The EDXRF system is consisted of an X-ray tube, a sample chamber, a collimator, a detector, a counting circuit and a computer. When the characteristic X-ray of the substance contained in the sample is recorded by the detector, the electron hole pair will be generated immediately. The number is proportional to the energy of the incident photon, and the pulse within a certain energy length will be obtained through a series of circuits. Finally, the types and concentrations of the elements to be measured are obtained through noise reduction, background deduction and separation of overlapping peaks.

Moseley's study found that the calculating photon energy is (Mingtai Yang, et al.,2009):

$$E = h\nu = hC/\lambda$$

Where, E: energy of the X-ray photon; ν: frequency of light waves; h: Planck's constant; C: the speed of light.

The energy generated by the electron transition from the different layers is recorded as ΔE, which is released in the form of X-ray. The calculation formula is

$$\Delta E = RhC(Z - \sigma) \left(\frac{1}{n_1^2} - \frac{1}{n_2^2}\right)$$

Where, ΔE is the characteristic X ray energy; R is Rydberg constant; σ is a constant related to the excited energy level.

According to the formula, the type of elements can be obtained by the wavelength or energy of EDXRF. Since EDXRF intensity is related to the content of elements, the qualitative and quantitative analysis of heavy metal elements can be conducted accordingly.

3. Results and discussion

A total of 25 samples were measured in this experiment. After calibrating the handheld EDXRF with Zn and Cu peaks, 16 kinds of elements were obtained from the output results of the instrument: Mg, Ar, K, Sc, Ti, V, Mn, Fe, Co, Cu, Zn, As, Rb, Sr, Y and Zr. The output mode is the characteristic peak count of the element (no unit). The range of the characteristic peak area count of Mg element is 0~0, and the average value is 0.The characteristic peak area count of Ar elements ranges from 82 to 783, and the average value is 253.The characteristic peak area count of K element ranges from 425 to 802, and the average value is 570.The characteristic peak area count of Sc elements ranges from 2070 to 5834, and the average value is 3515.The characteristic peak area count of Ti element ranges from 1311 to 1855, and the average value is 1561.The characteristic peak area count of element V ranges from 186 to 541, and the average value is 395. The characteristic peak area count of Mn elements ranges from 210 to 969, and the average value is 607.The characteristic peak area count of Fe is 63070~81256, and the mean value is 70081.The characteristic peak area count of Co element ranges from 10854 to 13603, and the average value is 12065.The characteristic peak area count of Cu elements ranges from 217 to 737, and the average value is 466.The characteristic peak area count range of Zn is 659~1270, and the average value is 995.The characteristic peak area count of As element ranges from 516 to 1071, and the average value is 689.The range of the characteristic peak area count of Rb element is 2096~4388, and the average value is 2918.The area count of the characteristic peak of Sr elements ranges from 2044 to 5966, and the average value is 3108.The characteristic peak area count of Y element ranges from 567 to 1923, and the average value is 958.The characteristic peak area of Zr elements ranges from 4186 to 7091, with an average value of 5275. In this experiment, the peak area count of Fe element was the largest, far more than other elements, and the peak area count of Mg element was the smallest. The count of Co element was the most stable in 25 groups of samples, and the peak area count of each measurement had a small difference, with a maximum deviation of 5.6% from the average. In the 25 groups of samples, the count of Ar element fluctuated the most, and the peak area count of each measurement varied greatly, with the maximum deviation from the average value being 56%. Nevertheless, the overall measurement results showed high stability and reliability (Qi Zhang, 2019). The concentrations of specific elements are shown in the figure 1 below.

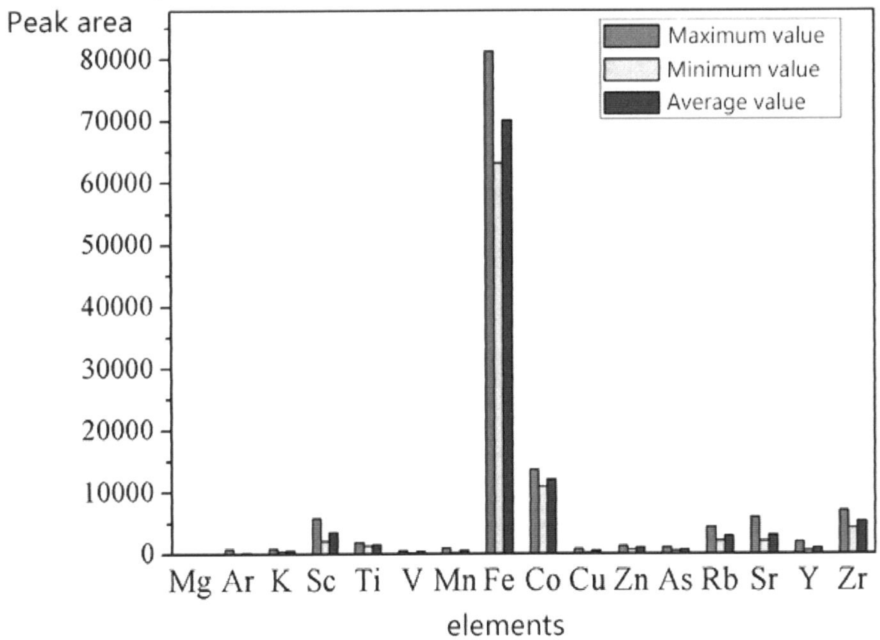

Figure 1. Statistical figure of peak area count of 16 elements

According to the purpose of this study, heavy metal elements that can be detected by EDXRF and need to be analyzed through this experiment are As, Cu and Zn.

As is calibrated according to As standard solution (elemental standard solution). In the 25 samples, the minimum value of As concentration is 8.67mg/kg, the maximum value is 17.99mg/kg, and the average value is 11.58mg/kg. At the same time, according to the soil properties of the collection site, soil PH = 6.8 is measured according to the experiment, and the As concentration requirement of the secondary standard PH between 6.5 and 7.5 according to GB15618-1995 Soil Environmental Quality Standards is less than 25 mg/kg. The concentration of the As element measured in the experiment and the national standard can be seen in the figure 2(a), and all the samples measured do not exceed the national standard.

The calibration of Cu is also carried out according to Cu standard solution (element standard solution). In 25 samples, the minimum value of Cu concentration is 3.65mg/kg, the maximum value is 12.39mg/kg, and the average value is 7.84mg/kg. At the same time, according to the soil properties of the collected land, the soil PH value is measured as 6.8. According to the GB15618-1995 Soil Environmental Quality Standards, the secondary standard PH value is between 6.5 and 7.5, and the Cu concentration required is less than 100mg/kg. The comparison between the concentration of Cu elements measured in the experiment and the national standard is shown in the figure 2 (b). All the samples measured do not exceed the national standard.

Zn is calibrated according to the Zn standard solution (elemental standard solution). In 25 samples, the minimum value of Zn concentration is 11.09mg/kg, the maximum value is 21.34mg/kg, and the average value is 16.72mg/kg. At the same time, according to the soil properties of the collected land, the soil PH value is measured as 6.8. According to the GB15618-1995 Soil Environmental Quality Standards, the secondary standard PH value is between 6.5 and 7.5, and the Zn concentration required is less than 250mg/kg. The comparison between the concentration of Zn elements measured in the experiment and the national standard is shown in the figure 2 (c). All the samples measured do not exceed the national standard.

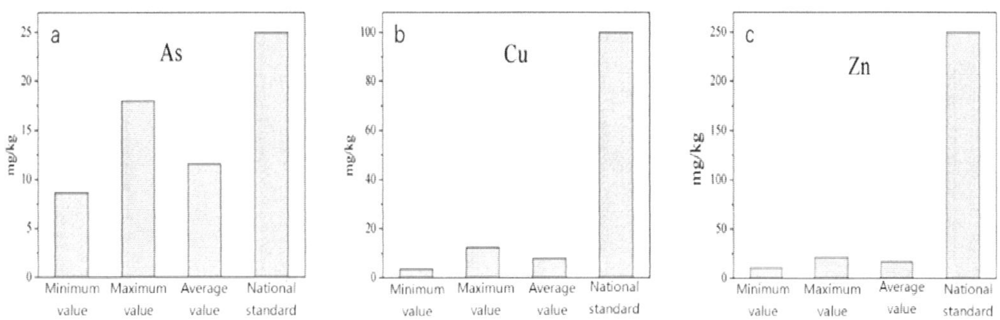

Figure 2. Comparison of the concentration of As(a), Cu(b) and Zn(c) with national standard values

4. Conclusion

According to the comparison of the experimental data with the national standard, the heavy metal concentration near the Xiangzhang canteen of Chengdu University of Technology is stable. Although it is convenient and efficient to use the EDXRF for analysis, the detection limitation is low in this experiment, and the heavy metal concentration in the soil is unavailable, such as Pb, Hg, Cd, Cr and other elements. However, during this experiment, it is found that there are more As elements in the soil. It does not exceed the National second-level soil standard, but it must also attract our attention. It can be inferred that there was groundwater under the soil near the Xiangzhang canteen years ago, and the concentration of arsenic in the natural water body ranged from 0.5 μg /L to 5,000 μg /L. Due to precipitation accumulation effect in the groundwater, in the midst of groundwater As element concentration accumulation. Therefore, it is concluded that there is likely to be groundwater infiltration under the soil years ago near the Xiangzhang canteen of Chengdu University of Technology, resulting in high as concentration in the soil. In addition to the high As concentration, the detected Fe concentration is also higher than other elements. It is speculated that there was precipitation and accumulation of iron ore in the groundwater before.

Acknowledgement
This work was funded by the National Natural Science Foundation of China (No.41774147), the National Key R&D Project (No.2017YFC0602100) and Science and technology activity project for overseas personnel in Sichuan province.

References

[1] GARCíA-SáNCHEZIE, BARRADAS V L, HILL CAPDL, et al. 2019. Effect of heavy metals and environmental variables on the assimilation of CO2 and stomatal conductance of Ligustrum lucidum, an urban tree from Mexico City. Urban Forestry & Urban Greening.

[2] Tangfu Wu, Xiangqin Wang, Changwen Li, et al. 2019. Preliminary study on farmland control technology of heavy metal Cadmium pollution in Jianghua county. Crop research. 220-222-245

[3] LIU J, ZHA F, XU L, et al. 2019. Mechanism of stabilized/solidified heavy metal contaminated soils with cement-fly ash based on electrical resistivity measurements. Measurement. 141.

[4] Mingtai Yang, Lianping Yang, et al. 2009. Similarities and differences between WDXRF spectrometer and EDXRF spectrometer. Nuclear Electronics and Detection Technology. 28(5): 1008-1011.

[5] Qi Zhang, et al. 2019. Problems and countermeasures in soil heavy metal concentration analysis. World nonferrous metals. 239-241.

[6] Tao Hong, Xiangsheng Kong, Xiangfei Yue, et al. 2019. Evaluation of soil heavy metal concentration, source and potential ecological risk in peak-cluster depression in southeast Yunnan. Environmental Science. 1-12

Data processing of pulsation signals based on local wave decomposition and teager energy

Fengli Wang[1] and Yannian Cai[2]

[1]College of Marine Engineering, Dalian Maritime University, Dalian 116026, P. R. China
[2]College of Marine Engineering, Dalian Maritime University, Dalian 116026, P. R. China

Corresponding author and e-mail: Fengli Wang, wangflsky997@sina.com

Abstract. Nonstationary and nonlinear signals are often encountered in the research and development of turbomachinery. One such example is the pulsating strain signal measured during engine ramping to find the maximum resonant strain in the application. Because the pulse signal may come from different interference sources, it is difficult to detect weak useful signal in the background of noise. In order to solve this problem, a new method based on local wave decomposition (LWD) and Teager energy is proposed. According to the local characteristics of the vibration signal, the optimal prediction operator of the transformed sample is constructed by selecting the appropriate square error minimization criterion, so that the second generation wavelet basis function can fit the local characteristics of the vibration signal. The adaptive second generation wavelet is used as the prefilter to improve the effect of LWD decomposition. Then the correlation kurtosis is used to select the sensitive internal model function (IMFs). Finally, the Teager energy operator algorithm is applied to the selected sensitive IMF to identify the characteristic frequency. The validity of the method is verified by the measured strain signal of a turbocharger turbine as an example.

1. Introduction

Nonstationary and nonlinear signals are often encountered in the research and development of turbomachinery. Sometimes the frequency of these signals changes with time. An example of this is the measurement of a pulsating strain signal in an engine ramp to find the maximum resonance of a turbomachinery blade. Because the signal frequency is proportional to the engine speed, it changes when the speed changes.

Traditionally, these signals are analyzed by FFT method [1]. But for the non-stationary signal which changes with time, Fourier transform is powerless. In view of the nonstationarity and nonlinearity of these signals, several advanced time-frequency analysis techniques are introduced and applied to the analysis of these signals. Local wave decomposition (LWD) is one of the most effective time-frequency analysis methods [2]. It is an adaptive time-frequency signal processing method. LWD decomposes the original signal into a series of band limited intrinsic mode functions (IMFs). Each IMFs component can be modulated or modulated according to the characteristics of the signal. It can well reveal the nonlinear and non-stationary information of the vibration signal. It has been successfully applied to the fault diagnosis and structural health monitoring of rotating machinery [3].

However, due to the problem of pattern mixing, LWD can not extract fault features accurately. In order to reduce mode mixing, it is important to eliminate noise before decomposition. Wavelet transform is widely used in signal denoising. The result of wavelet decomposition is related to wavelet basis function. In addition, improper wavelet transform will cover up the local characteristics of vibration signal and lose some useful original signal details. In order to overcome the above limitations, this paper introduces lifting algorithm to design adaptive wavelet function for signal denoising [4-5].

Because of the different sources of the fluctuating strain signal, it is difficult to detect the weak useful signal in the noise background if the signal processing error interferes. This paper reports the results of this investigation.

2. New method

2.1. Local wave decomposition

The local wave decomposition (LWD) method is developed under the simple assumption that any signal contains different simple intrinsic oscillation modes. Any signal can be decomposed into a finite number of IMF [2]. By definition, any signal x (t) can be decomposed as follows:

Firstly, all the local extremum of the signal is obtained, and the local mean m(t) is fitted by cubic spline, and the detail h(t) = x(t)−m(t). is extracted. H(t) is regarded as a new $x(t)$, and the above operations are repeated until h(t) satisfies the IMF condition, and then the first IMF $c_1(t)$ = h(t) is obtained. Let the residual $r_1(t)$ = x(t)−h(t) be a new signal, repeat the above process, and get other order IMFs. Therefore, we can decompose the signal into IMFs $c_1(t)$, $c_2(t)$, . . . , $c_n(t)$, and a residue, which is the average trend of.

$$x(t)=\sum_{i=1}^{n}c_i(t)+r_n(t) \tag{1}$$

For each IMF in Eq. (1), we can always have its Hilbert transform,

$$H[c_i(t)]=\frac{1}{\pi}\int_{-\infty}^{+\infty}\frac{x(\tau)}{t-\tau}d\tau \tag{2}$$

According to this definition, we can get an analytic signal,

$$x_i(t)=c_i(t)+jH_i(\mathrm{t})=\mathrm{a}_i(\mathrm{t})\mathrm{e}^{j\varphi_i(\mathrm{t})} \tag{3}$$

After Hilbert transformation of each IMF component, the original signal can be expressed as,

$$x(t)=RP\sum_{i=1}^{n}\mathrm{a}_i(\mathrm{t})\mathrm{e}^{j\varphi_i(\mathrm{t})} \tag{4}$$

2.2. Second generation wavelet

Sweldens introduces lifting scheme as a powerful tool to construct second generation wavelets [4-5]. By choosing the criterion of least square error, we get the optimal prediction operator of the transform sample, which makes the constructed wavelet function fit the transient characteristics of the original signal. The stage of the second generation wavelet transform is described as [4-6].

In the process of wavelet threshold denoising, the selection of threshold is the key, which is directly related to the denoising effect. In order to obtain the best denoising effect, this paper adopts the threshold selection scheme proposed by Pan [7].

2.3. The improved LWD

In order to overcome the shortcomings of LWD, the adaptive second generation wavelet is used as the pre-filter bank program as follows:

(1) Using lifting scheme to construct self-adaptive second generation wavelet to denoise the original signal.

(2) After wavelet de-noising, the second generation wavelet is used to reconstruct the original signal, and then the original signal is decomposed into a small number of IMFs by LWD.

Here we consider a noisy nonstationary signal with amplitude modulation,

$$x(t) = 1.2(1 + 0.2\sin(2\pi \times 7.5t))\cos(2\pi \times 30t) + 2.5\sin(2\pi \times 10t) + 0.25randn \tag{5}$$

where randn is a random number vector of normal distribution, with an average value of 0 and a standard deviation of 1.

The simulated signal in Figure 1 is analysed using LWD. The decomposition results are shown in Figure 2(a). Due to the interference of noise, the mode mixture occurs between IMFs. Thus, the decomposition results of LMD fail to represent the real characteristics of the simulated signal. To improve the results, the simulated signal is decomposed by the improved LWD, and the results are shown in Figure 2(b). It can be seen that the first IMF $c_1(t)$ has the obvious amplitude-modulated characteristic, the second IMF $c_2(t)$ is sinusoidal component.

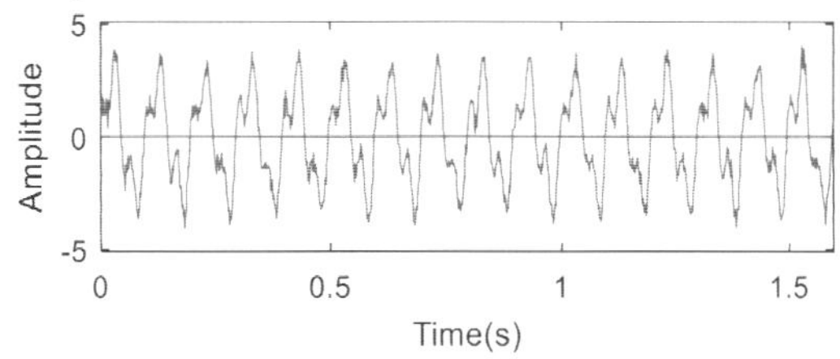

Figure 1. Waveform of simulated signal

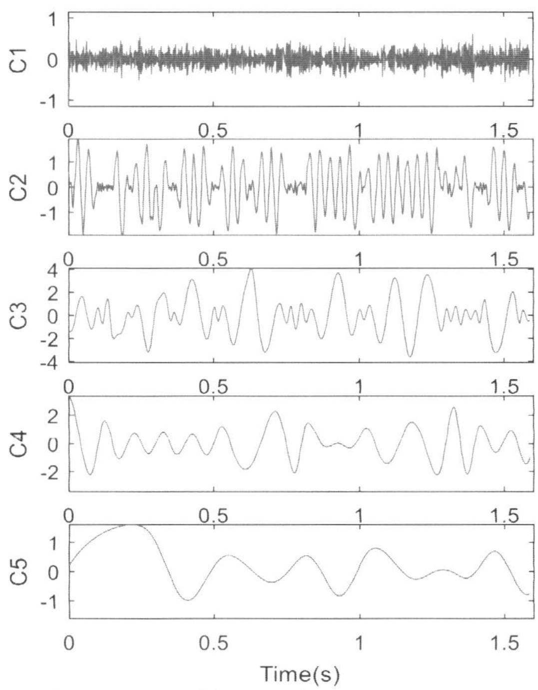

(a) Decomposition results by LWD

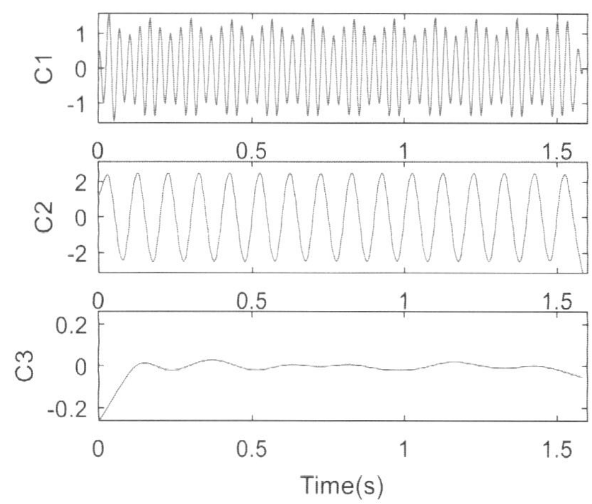

(b) Decomposition results by improved LWD

Figure 2. Decomposition results of the simulated signal

2.4. Teager energy operator

The Teager energy operator of time-varying signal is defined as [8],

$$J[x(t)] = [\dot{x}(t)]^2 + x(t)\ddot{x}(t) \tag{6}$$

where $\dot{x}(t)$ and $\ddot{x}(t)$ are the first and second derivatives of signal with respect to time t respectively.

For discrete-time signals, the Teager energy operator is transformed into discrete-time signals.

$$J[x(n)] = [x(n)]^2 - x(n-1)x(n+1) \tag{7}$$

The Teager energy operator is only suitable for narrow band signals. For complex multicomponent signals, the improved LWD algorithm can decompose the signals into IMFs.

2.5. Correlation kurtosis

Correlation kurtosis not only retains the characteristics of kurtosis, but also has the characteristics of correlation function. It is a parameter reflecting the strength of periodic pulse signal in fault signal [9]. When applied to turbine blade signal, if there is interference in the pressure distribution introduced by rotating blade, the interference is similar to the impact signal with obvious period, and the kurtosis is larger, while the correlation kurtosis of other impact signals is smaller. The correlation kurtosis is calculated as,

$$CK_M(T) = \frac{\sum_{n=1}^{N}\left(\prod_{m=0}^{M} x_{n-mT}\right)^2}{\left(\sum_{n=1}^{N} x_n^2\right)^{M+1}} \tag{8}$$

where x_n is the signal sequence, N is the number of samples in the original signal, T is the period of the pulse signal, and M is the period of shift.

3. Experimental analysis

The turbine involved is a small turbocharger installed in a gas station test unit. The output flap mode of turbine blade is a problem worthy of attention. Under the static and room temperature conditions, the natural frequency of strain measuring blade is measured as 12670Hz. The turbocharger is running on the engine, and its speed is rising and falling constantly. In this way, the recorded signal of turbine blade strain can be obtained from the excitation Figure 3 (a), which is the standardized Figure 3 (b) The peak frequency of FFT spectrum of blade strain is 13650 Hz. This is an unexpected result: the natural frequency of the first blade mode is 12670 Hz, and the expected maximum strain of the blade at room temperature and static state should appear near or on the frequency. Because of the combined use of the centrifugal stiffness and Young's modulus reduction of the turbine blade material at hot state, the order of the engine is 7, so the exciting force is relatively small, and the small strain of blade is expected to be obtained. However, the occurrence of peak strain at 13650 Hz is puzzling and requires an explanation.

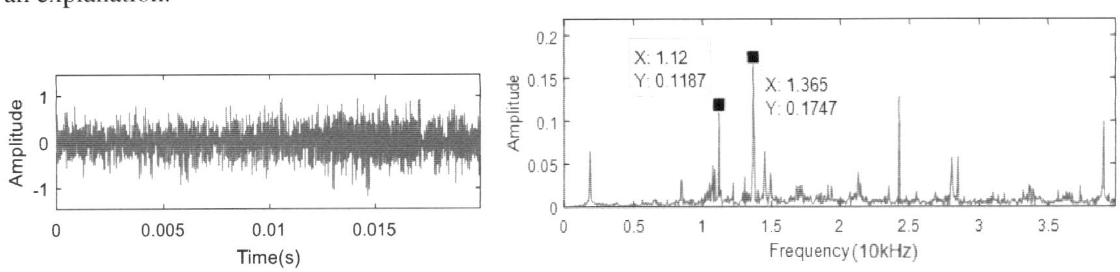

(a) Waveform of blade strain signal (b) FFT spectrum of the blade strain signal

Figure 3. Measured turbine blade strain signal

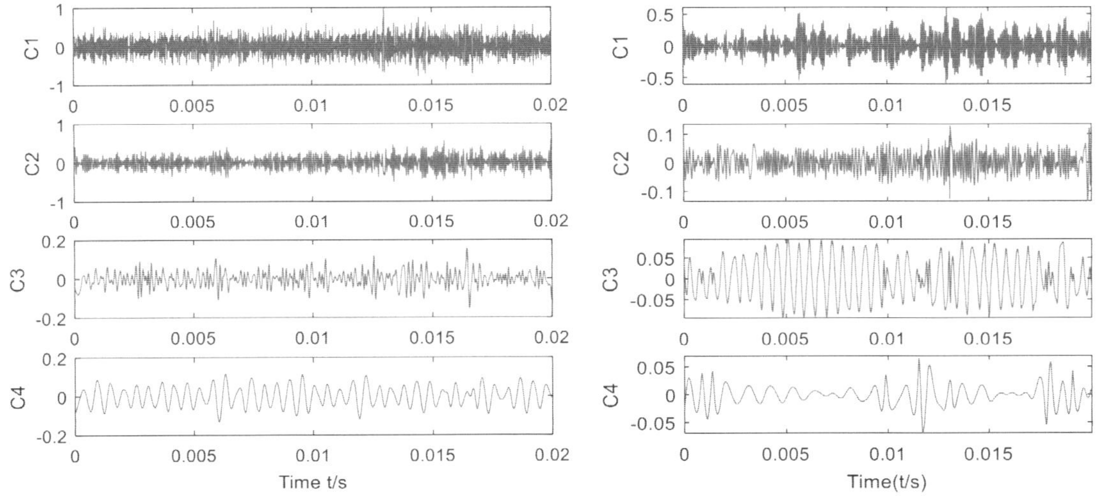

(a) The first four IMFs by LWD (b) The first four IMFs by improved LWD

Figure 4. Decomposition results of blade strain signal

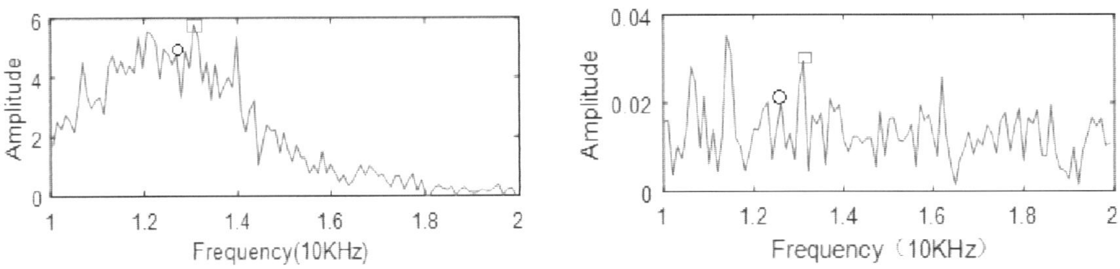

(a) Marginal spectrum of IMF2 in Figure 4(a) (b) Teager envelope spectrum of IMF1 in Figure 4(b)

Figure 5. Envelope spectrum analysis

The strain signal was decomposed by LWD, and the first four IMF decomposition results are shown in Figure 4 (a). The second IMF contains the pulse characteristic signal whose marginal spectrum is shown in Fig. 5 (a). The circular mark corresponds to 12650 Hz, and the rectangular mark corresponds to the peak value of 13070 Hz and 12650 Hz. This may be the vibration mode pursued by the blade. A higher frequency peak value of 13070 Hz is also given. The reason for this is not clear.

Therefore, the proposed method is applied to strain signal processing. Figure 4 (b) shows the results of the improved LWD decomposition. The correlation kurtosis of each IMF component is calculated according to the period of detection component. According to the maximum correlation kurtosis criterion, IMF1 was selected as the sensitive component. Then the Teager envelope analysis is applied to the IMF1, and the Teager envelope spectrum is shown in Figure 5(b). The circular mark corresponds to 12650 Hz, and the rectangular mark corresponds to 13070 Hz. It can be seen clearly that there is an obvious peak value at 12650 Hz compared with the LWD result in Figure 5(a), which further proves the value of the new method.

4. Conclusions

In order to overcome the difficulties of traditional FFT methods, a new signal processing method is proposed in this paper. In this method, a self-adaptive second generation wavelet is used to reduce the influence of noise and improve the effect of LWD decomposition. The LWD method is used to extract feature components from the original signal and separate them from the background noise and irrelevant components. The correlation kurtosis is used to screen out sensitive feature components, effectively eliminate the interference components and capture the feature information. Teager energy enhances the detection of the characteristic frequency of weak impact. Experiments are carried out on a turbine wheel to verify the effectiveness of the FFT method. The results show that this method can identify the first vibratory mode of the blade from the weak strain gauge signal, which is difficult for the traditional FFT method.

Acknowledgement
This research was supported by the Fundamental Research Funds for the Central Universities (3132019330).

References

[1] H. Hackenberg, A. Hartung 2015 *Proceedings of ASME Turbo Expo 2015*. p 1.
[2] N.E. Huang, Z. Shen, S.R. Long, et al. 2003 *Proceedings of the Royal Society of London*. Series A, Vol.459, p 2317 .
[3] Q. Gao, C. Duan, H. Fan, et al. 2008 *Mech. Syst. Signal Process.* Vol 22 p 1072.
[4] W. Sweldens 1996 *Appl. Comput. Harmon. Anal*. Vol 3 p 186.
[5] W. Sweldens 1997 *SIAM Journal on Mathematical Analysis*.Vol 29 p 511.
[6] R. L. Claypoole, M. D. Geoffrey, W Sweldens 2003 *IEEE Transactions on Image Processing*. Vol 12 p 1449.
[7] Q. Pan, L. Zhang, G. Dai, H. Zhang 1999 *IEEE Transactions on Signal Processing*. Vol 47 p 3401.
[8] R. Maragos, E. Kaiser, T. Quatieri 1993 *IEEE Trans Signal Processing*. Vol 41 p 3024.
[9] G. McDonald, Q. Zhao, M. Zuo 2012 *Mechanical systems and Signal Processing*. Vol 33 p 237.

Journal of Physics: Conference Series **1423** (2019) 012042 doi:10.1088/1742-6596/1423/1/012042

Performance of Bundle Protocol (BP) for deep space communications with long link disruptions

Ding Wang[1], Hongzhi Yu[2]

[1] Northwest Minzu University, Lanzhou, China
[2] Northwest Minzu University, Lanzhou, China

Corresponding author and e-mail: Ding Wang, wangding212@qq.com

Abstract. Near Earth missions and deep space missions will continue to be a major future space mission. A long link disruption is inevitable in space communications because of spacecraft rotation, planetary bodies and limited relay capability. A few works have been done with delay/disruption tolerant networking (DTN) technology for deep space communications and provided feasibility for its adoption in LEO- space and deep space missions. However, not much work has been done to fully evaluate the performance of DTN in such an environment, especially in the presence of custody and long link transfer. In this article, we present an experimental performance evaluation of DTN architecture and protocol stack with DTN custody transfer and long link break underneath bundle protocol (BP), in typical LEO-satellite and deep space communication infrastructures accompanied. The experiment was conducted by performing realistic file transfers over a PC-based test-bed.

1. Introduction

Bundle protocol is a protocol layer of DTN architecture [1]. It designed to work at the application layer of the DTN architecture. The protocol data unit communicated by this DTN bundle protocol is called Bundles and each bundle consist of two or more blocks of protocol data.[2] The node component that executes the programs of the BP and offers the BP services is known as the bundle protocol agent (BPA). The convergence layer adapter (CLA) utilizes the service of the underlying network protocol layers to send and receive the bundle on behalf of the BPA. [3]

Each DTN node keeps custody of bundles it forwards to the adjacent node until it receives a positive ACK from the adjacent node is so called store and forward mechanism. If the data is lost or if the node receives negative ACK, it re-forwards the bundle again. The custodian bundle is released when the node receives a notification that some other node or the particular destination node accepted the custody of the same bundle; or the bundle is deleted explicitly due to lifetime expiration. [3] So by this mechanism BP provided reliable transfer of data even in a highly stressed environment. Some of the other services provided by bundle protocol are error-free data delivery, completeness in delivery, fragmentation and extensibility[4].

Custody transfer is a good way to improve the reliability of network. Especially, it can help deliver a message to a moving responsibility destination. Protocol will request an acknowledgment to enhance the reliability. A message with custody transfer will not be deleted until it can be received by another node with custody transfer. Node will hold a message and is so called a custodian. Some custodians hold one message, some custodians own a message or message fragment. Protocol requests custody transfer to perform and deliver acknowledgment when host system can move the message to other nodes.

2. Experimental setup and configuration

This work focuses on performance evaluation of DTN in a LEO-satellite environment. LEO-satellite communication architectures have been well described in literature. For detail, refer to[5]. A PC-based Space Communication and Networking Testbed (SCNT) was built at Lamar University to emulate a relay-type of LEO-satellite and deep space communications infrastructure for the proposed experimental evaluation of DTN. Previous research [6] shows that the evaluation results obtained from the SCNT have generality and the test-bed can effectively evaluate the realistic performance of a protocol. A network emulation package of Linux OS, netem[7], was adopted to simulate the satellite channel conditions such as link outage, propagation delay, packet corruption, and packet loss. (http://www.linuxfoundation.org/collaborate/workgroups/networking/netem)

Table 1 illustrates a block diagram of SCNT. The test-bed consists of a Linux-based file source personal computer (PC) (*tx.lamar.edu*), file relay PC (*relay.lamar.edu*), file destination PC (rx.lamar.edu), and the SLS, which connects three file PCs.

Table 1. Experimental Factors and Configuration for Deep Space

Protocol Layering

ION BP/LTPCL/UDP

(Note: with a fixed block varying buddle size)

Custody transfer

Have custody transfer	Have not custody transfer

Delay

300second

Link Breaks

10 min	15 min	30 min	1 hour	2 hour	4 hour	8 hour

Ratio of Channel Rate

1:1

BER

0	10^{-6}	10^{-5}

File Size

10k	100k	1M	10M

3. Experimental results and discussions

In this section, we present and discuss the performance evaluation results of the LTP-based DTN protocol stack for deep space communication. A particular attention is on impact of BP custody transfer option.

3.1. BP/LTPCL with Custody Transfer vs BP/LTPCL without Custody Transfer with 1MB File Size

In Figure 1, a comparison of goodput performance for the BP/LTPCL with custody transfer vs BP/LTPCL without custody transfer to transmit a 1 Mbyte file over deep space channels is presented at three BERs with respect to no link outage and long link outages ranging from 10 minutes to 8 hours.

A direct observation is that for all the experimented transfer configurations (i.e., with custody transfer and without custody transfer) and bit error rate (0, 10-5, and10^{-6}), the transmissions without a link outage involved show significantly higher goodput than those with a link outage involved. While the goodput without a link break involved ranges from 3,000 bytes/s to 1,000 bytes/s, the goodput with a link break of 1hour involved merges and approaches the same value. This is reasonable because a long link break dominates the short file transmission time, resulting in very poor goodput performance, provided that a high speed data channel rate was configured for transmission of a 1 M byte file during the experiment.

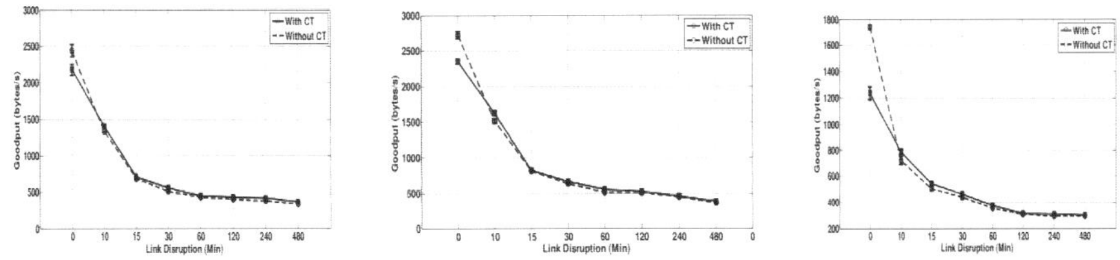

(a) (b) (c)

Figure 1. Goodput comparisons for the LTP-based DTN protocol to transmit a 1Mbyte files over deep space channels with different links disruption and BER. (a) BER=0 (b) BER=10^{-6} (c) BER=10^{-5}

It is also observed that the performance is significantly different among two different transmission modes with custody transfer and without custody transfer for all the configurations when no link break is involved. In comparison, their performance merges and shows no significant differences for all link break durations varying from 2 hour to 8 hours.

As another observation, for all two transmission modes, the goodput performance without a link break decreases when either a packet bit error rate of 10^{-5} and/or $10^{-6}$are introduced. This can also be easily explained. Data loss causes retransmission of data packets (although in a deferred mode), creating additional file transmission time thus degradation of goodput. When bit error rate is involved, the forward channel rate may be too low to handle the transmission of returning ACKs effectively, resulting in late arrival or frequent losses of ACKs and consequentially performance degradation of the protocol.

The key reason for performance degradation when custody transfer is added to bundle communication over LTP is that it doubles the number of bundles that must be sent: for each original data bundle that is sent – via LTP reliability – in the forward direction, a small custody acceptance signal bundle is sent – again via LTP reliability – in the reverse direction. The total number of bytes is not doubled, because custody signals are much smaller than most original data bundles, but transmission of a custody signal causes an additional report segment and report ACK segment to be transmitted, so the LTP overhead is doubled.

Custody transfer does hop-by-hop reliable transmission while TCP does end-to-end reliable transmission. A single custody transfer hop may span multiple bundle forwarding hops, because not all nodes are required to take custody; the next custodian might not be an immediate neighbor in the network. A node that has taken custody of a bundle will typically re-forward the bundle if no custody acceptance signal has been received (for that bundle) from any other node prior to expiration of a "countdown timer" for that bundle.

The custody transfer system does not do any integrity checking (such as a checksum) on received bundles, but convergence-layer protocols will typically discard any received protocol data unit that is determined to be corrupt. A node would typically never receive the bundle transmitted by another node, because the convergence layer would have discarded it. If the convergence layer adapter did pass the corrupt bundle up to BP at the node, the bundle protocol agent could check its integrity and detect the corruption. In that event it might theoretically send back a custody refusal signal, but there really isn't any custody transfer signal reason code that signifies to retransmit; reasons for custody refusal tend to be shortages of storage or forwarding resources.

From the above discussions, custody transfer really isn't designed to be a NACK-based retransmission system. In fact, the only event that causes retransmission by a bundle custodian is expiration of a countdown timer. In general, custody transfer isn't designed to be a full-featured reliability system at all. The specification imposes very few solid requirements on the implementer.

3.2. BP/LTPCL with Custody Transfer vs BP/LTPCL without Custody Transfer with 10MB File Size

In Figure 2, a comparison of goodput performance for the BP/LTPCL with custody transfer vs BP/LTPCL without custody transfer to transmit a 10 Mbyte file over deep space channels is presented at three BERs with respect to no link outage and long link outages ranging from 10 minutes to 8 hours. It illustrates a comparison of goodput performance for the transmissions with custody and without custody at three different bit error rates of 0, 10^{-5}and10^{-6}.

We observe that for all three BERs, the transmissions without custody transfer show significantly higher goodput than those with custody transfer and the performance advantage is more significant when no link break is involved. The biggest performance advantage is observed at a packet corruption rate of 0, and it gradually decreases with the increase of link break. The advantage drops significantly when a 10^{-6} error rate is introduced and is further decreased when the bit error rate is further increased to 10^{-5}.

As we see, the goodput difference between custody transfer and no custody transfer at an error rate of 0 decreases from 3000 bytes/s to 1000 bytes/s and then to 500 bytes/s when the link disruption is increased from 0 to 10 minutes and then to 480 minutes. In comparison, the goodput difference without custody transfer decreases more significantly, from 2500 bytes/s to 800 bytes/s and then to only 440 bytes/s.

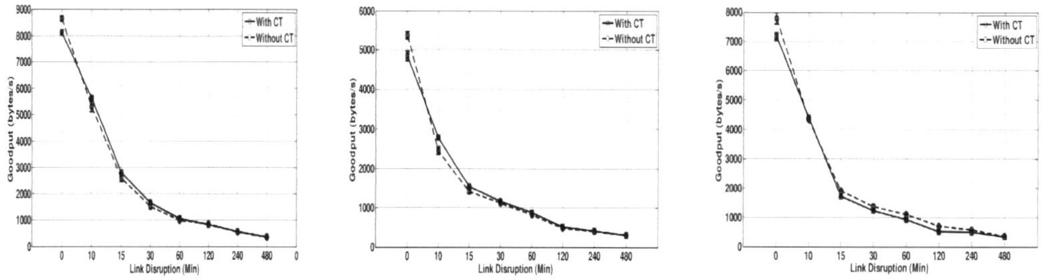

(a) (b) (c)

Figure 2. Goodput comparisons for the LTP-based DTN protocol to transmit a 10Mbyte files over deep space channels with different links disruption and BER. (a) BER=0(b) BER=10^{-6} (c) BER=10^{-5}

The comparison results indicate that the impact of deep space channel with custody and without custody transfer on the goodput performance of LTP-based DTN decreases along with the increase in link disruption. The impact decreases more significantly when link disruption (around 10 minutes) is occurred over the channel in comparison to no disruption. In other words, the impact of link disruption (with a rate around 10 minutes ~120 minutes) on the goodput performance of LTP increases for the transfer mode without custody. The impact increases more significantly over a deep space channel.

Figure 2 illustrates a comparison of goodput performance for the transmissions with bit error rates of 0, 10^{-}5and10^{-6} at different file sizes for both transmissions with custody transfer and without custody transfer. For both transfer configurations, the transmissions with a bit error rate of 0 show significantly higher goodput than those with a bit error rate of 10^{-5}, and the performance advantage is more significant for without custody transfer. This tells us that the impact of the experimented bit error rate (around 10^{-5}) on the goodput performance of LTP is more significant without custody transfer than that with custody transfer.

3.3. BP/LTPCL with Custody Transfer vs BP/LTPCL without Custody Transfer without Break with Respect to File Size

In Figure 3, we compare the averaged goodput performance for the LTP-based DTN protocol to transmit the 10Kbyte 100Kbyte 1Mbyte and 10Mbyte file over deep space channels experiencing without long link disruption.

In Figure 3(a), the goodput performance is compared among two different transfer modes with bit error rate of 0. An obvious observation is that for an exponential increase in the different file size, there is an almost exponential increase in goodput. In fact, each time when the file size becomes ten times than before, the goodput increases to roughly 2.5 times. The goodput performance varies in the range of 9000 bytes/s to 400 bytes/s.

We also observe that the goodput is actually about the same between custody transfer or without custody transfer when the file size is the same except at 10Mbyte, regardless of the changes of bit loss rate, and transfer configuration (i.e. custody transfer or without custody transfer). This happens because a big file size of 10-1000kbyte is so small that they do not lead the transmission into steady step. Therefore, the goodput of the LTP-based DTN transmissions is not most adversely affected by custody transfer and without custody transfer. In other words, for a transmission over deep space channels with no link disruptions, there is not much difference between custody transfer and without custody transfer in goodput performance of transmission except for a large file size (≥10Mbyte).

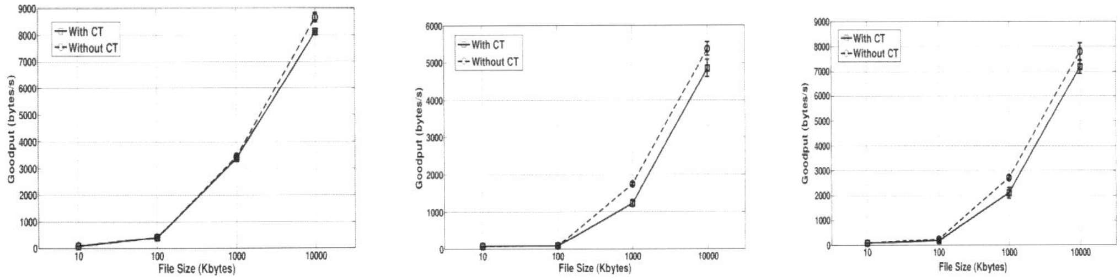

(a) (b) (c)

Figure 3. Goodput comparison for the LTP-based DTN protocol to transmit a 10Kbyte 100Kbyte 1Mbyte and 10Mbyte file over deep space channels with respect to no link outage and long link outages involved. (a) BER=0 (b) BER=10^{-6} (c) BER=10^{-5}

The three different bit error rates have little noticeable goodput differences for big file size, especially at 1Mbyte and 10Mbyte, but they have a small disparity when the file size is around 1Mbyte~10Mbyte. This indicates that the experimented bit error rate from 0 to 10^{-6} has little impact on the performance of the protocol when the link outage is zero but it has no impact at all when the file size is around 100Kbyte or smaller because a small file size has less effect on the goodput than the experimented bit error rates.

For the transmissions with BER involved (i.e. with BERs of 10^{-6} and 10^{-5}), the comparisons show very similar results, as in Figure 3 (a), we see a roughly exponential decrease in goodput along with the exponential increase in link outage duration. We observe that the performance of two different transfer mode configurations is quite similar except at big file size. This indicates that the custody transfer configuration has no impact at all on the performance of the protocol for a small file.

4. Conclusion

For deep space communications, BP/LTPCL protocol is very effective in data transmission in presence of an extremely long link delay of 5 minutes of one ways delay. The comparison results are different depending on the transmission condition. BP/LTP without custody transfer has performance advantage over with custody transfer when no break or over short break is involved for transmission of a big file. The comparisons with custody transfer and without custody transfer shows very similar performance for a long link break channel and/or small file sizes. BP/LTPCL without custody transfer to transmit big size file with long break is very effective. According to this study, custody transfer really isn't designed to be a NACK-based retransmission system, and it does not show performance advantage in deep space communication.

Acknowledgment

The author would like to acknowledge Scott C. Burleigh at Jet Propulsion Laboratory (JPL) of California Institute of Technology for his support in configuring ION and conducting the experiment and many significant discussions. This paper is supported by "the Fundamental Research Funds for the central universities of Northwest Minzu University" (Grant No.31920190093)

References

[1] Ding Wang. "Performance of Licklider Transmission Protocol (LTP) in LEO-Satellite Communications with Link Disruptions" IEEE ICCICC16 Proceedings
[2] Burleigh, S. and K. Scott. 2007. Bundle protocol specification. Delay tolerant networking group.
[3] Warthman, Forrest. 2003. Delay-Tolerant Networks (DTNs): A tutorial. http://www.dtnrg.org/docs/tutorials/warthman-1.1.pdf (accessed June 2009).
[4] www.ietf.org/rfc/rfc5050.txt (accessed September 2010).
[5] R. Wang, B. Gutha, and P. K. Rapet, "Window-based and rate-based transmission control mechanisms over space-Internet links," IEEE Transactions on Aerospace and Electronic Systems, vol. 44, no. 1, January 2008, pp. 157-170.
[6] R. Wang, S. Burleigh, P. Parik, C-J Lin, and B. Sun, "Licklider Transmission Protocol (LTP)-based DTN for cislunar communications," IEEE/ACM Transactions on Networking, vol. 19, no.

Journal of Physics: Conference Series **1423** (2019) 012042 doi:10.1088/1742-6596/1423/1/012042

2, April 2011, pp. 359-368. [Online]. Available: http://ieeexplore.ieee.org/stamp/stamp.jsp?tp=&arnumber=5671516.

[7] Netem: network emulation package of Linux OS, [Online]. Available: http://www.linuxfoundation.org/collaborate/workgroups/networking/netem

Study on data-driven PEMFC humidity mechanism soft-sensing model

Baohua Tan[1,2] **, Hongbin Chen**[3]**, Rui Quan**[4]**, Shuhai Quan**[5]**, Yuanye Zhou**[4]

[1] School of Science, Hubei University of Technology, Wuhan, China
[2] Hubei Engineering Technology Research Center of Energy Photoelectric Device and System, Hubei University of Technology, Wuhan, China
[3] Department of Respiratory Medicine, Renmin Hospital of Wuhan University, Wuhan, China
[4] School of Electrical and Electronic Engineering, Hubei University of Technology, Wuhan, China
[5] The New Energy Detection and Control Technology Research Centre, Wuhan University of Technology, Wuhan, China

Corresponding author and e-mail: Yuanye Zhou, 85802102@qq.com

Abstract. Proton exchange membrane fuel cell (PEMFC) has a special closed structure, whose humidity cannot be measured directly. This paper selects the diffusion model as the modelling and analysis basis of the humidity mechanism soft-sensing based on the experimental research object. After the study of the basic principles of the diffusion model, a one-dimensional data-driven proton exchange membrane fuel cell humidity mechanism soft-sensing model is analysed and completed. At last, the actual experimental data drives the humidity mechanism soft-sensing model and the predicted output error based on the humidity mechanism soft-sensing model can be controlled within 3%. The output results verify the effectiveness of the humidity mechanism soft-sensing model.

1. Introduction

Proton exchange membrane fuel cell (PEMFC) is essentially an "inverse" device for water electrolysis [1-2]. As the "fuel", hydrogen enters and reaches the anode of the battery through a gas channel on the bipolar plate. As the "oxidant", oxygen or air enters and reaches the cathode of the battery through the air channel on the bipolar plate. Through the action of the catalyst, the hydrogen and oxygen molecules form water by electrochemical reaction respectively, and at the same time, the reaction release electrical energy and heat. The product of fuel cell is water, heat and electric energy, so that it will not pollute the environment, and it is a very clean energy.

When the fuel cell is operating, it needs to work under a certain humidity condition. The water content of the PEM polymer film will directly affect the voltage output. The proton conductivity of the polymer PEM depends on its internal water content, and the unhealthy humidity status of the film, which includes “dry” and “flooded”, will have negative effects on the performance of the fuel cell [3].

This paper selected the diffusion model as the modeling and analysis basis of the humidity mechanism soft-sensing based on the experimental research object [4]. After studying the basic principles of the diffusion model, then a one-dimensional data-driven proton exchange membrane fuel

Journal of Physics: Conference Series **1423** (2019) 012043 doi:10.1088/1742-6596/1423/1/012043

cell humidity mechanism soft-sensing model was analyzed and completed. At last, the actual experimental data drove the humidity mechanism soft-sensing model. And the predicted output error based on the humidity mechanism soft-sensing model can be basically controlled within 3%. And which verify the effectiveness of the humidity mechanism soft-sensing model.

2. Humidity mechanism model experiment and soft measurement analysis

The operating principle of proton exchange membrane fuel cell is shown in figure 1. It has a special closed structure, whose humidity cannot be measured directly. As the only product of electrochemical reaction, the fuel cell generates water at the cathode by combining hydrogen and oxygen under the action of catalyst [5]. The water in PEMPC is usually in the form of gas and liquid. In addition to the water generated at the cathode through the electrochemical reaction, it should also contain the water brought in by the humidified reaction gas input at the cathode and anode. Proton exchange membrane is the core element of PEMFC. It provides a pathway for protons to migrate and transport from the anode to the cathode. However, proton transport process is carried out in the form of hydrated ions, so the membrane should have the right amount of moisture humidity.

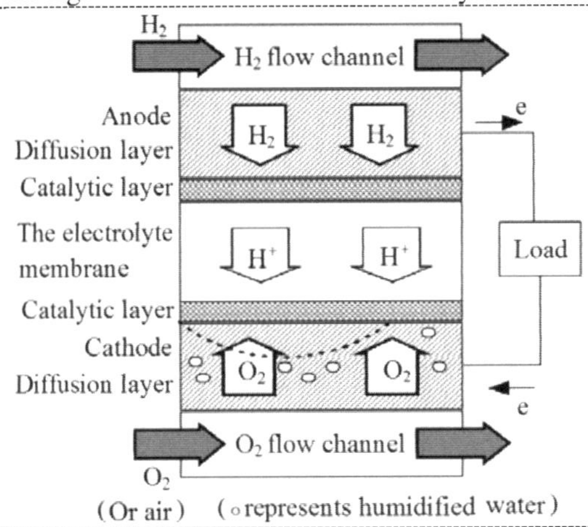

Figure 1. The operating principle of proton exchange membrane fuel cell

Currently, there are many researches on water management models of proton exchange membrane, but most of them are realized around convection and transfer model, dust flow model, statistical mechanics model and diffusion model [6]. And all these models assume that the fuel cell system is stable, adiabatic, and one-dimensional.

In the preliminary work of this paper [7-8], four kinds of humidity mechanism models of various proton exchange membranes were analyzed and compared. And typical assumptions were compared, and conclusions and deficiencies of each model were analyzed. The results show that the diffusion model assumes that water is in equilibrium in the gas phase and the membrane phase, the membrane is uniformly wet and completely saturated with water, and the water dissolved in the membrane is mainly generated in the cathode by the electroinduced drag migration and electrochemical reaction. It is consistent with the actual situation of proton exchange membrane fuel cell operating state. Therefore, the research object in this paper is more suitable for applying the diffusion model.

2.1. *Hypothesis of humidity mechanism model*

In this paper, according to the basic idea of diffusion model, the basic mechanism model of water management of proton exchange membrane fuel cell is established by taking the diffusion model as the prototype mechanism. In all concrete analysis, some established hypothesis condition should be followed, which is shown in table 1.

2.2. Soft-sensing model analysis of humidity mechanism

In this paper, the rectangular coordinate system is used to build the soft-sensing model. In the proton exchange membrane, the direction from anode to cathode is defined as the X direction, and all parameters are set to change in the X direction, thereby a one-dimensional moisture diffusion mechanism model is established.

Table 1. Hypothesis conditions of humidity mechanism model

No.	Postulated conditions
1	The proton exchange membrane is completely hydrated.
2	The capillary phenomenon is ignored.
3	Regardless of the configuration of water, both liquid and gaseous water are considered as water molecules in gaseous form.
4	There is electroneutral in the membrane.
5	Fuel agent and oxidant gas are ideal pure gas.
6	The solubility of H2 and O2 in the flow channel and gas diffusion layer in the liquid water is neglected.
7	The bidirectional nature of water flow is not considered, and the water flows in one direction.
8	The battery temperature inside the battery is constant and evenly distributed.
9	The thickness of the diffusion layer is not considered.
10	The pressure of anode and cathode are uniform and equal distribution.

From the water balance process of proton exchange membrane fuel cell, and according to the electroinduced drag migration, concentration difference reverse diffusion and pressure migration, there are:

$$W_a^{in} - W_a^{out} = -W_{drag} + W_{diff} + W_{pres} \tag{1}$$

Where, W_a^{in} is anode humidifying water; W_a^{out} is the Water which expelled from the anode in gaseous form; W_{drag} is the water of the electroinduced drag migration; W_{diff} is concentration difference reverse diffusion water; W_{pres} is the water coursed by pressure migration.

In the electroinduced drag migration process, there is:

$$W_{drag} = n_d \frac{iA}{F} \tag{2}$$

If the activity of water is denoted as α_w, the electroosmosis coefficient can be calculated as:

$$n_d = \begin{cases} 0.0049 + 2.02\alpha_w - 4.53\alpha_w^2 + 4.09\alpha_w^3 & \alpha_w \le 1 \\ 1.59 + 0.159(\alpha_w - 1) & 3 \ge \alpha_w > 1 \\ 16.8 \times \dfrac{2.5}{22} & \alpha_w > 3 \end{cases} \tag{3}$$

In the concentration difference reverse diffusion process, there are:

$$\begin{cases} W_{diff} = D_w \dfrac{\partial C_m^w}{\partial x} A \\ W_{press} = n_d \dfrac{i}{F} - D_w \nabla C_m^w - C_m^w \dfrac{K_m}{\mu} \nabla P_m \end{cases} \tag{4}$$

The diffusion coefficient D_w is:

$$D_w = \begin{cases} [(0.0049 + 2.02\alpha_w - 4.53\alpha_w + 4.09\alpha_w^3)\times 10^{-6}]\exp^{\left[2416\left(\frac{1}{303}-\frac{1}{T}\right)\right]}, \alpha_w \le 1 \\ \left\{\left[1.59 + 0.159(\alpha_w - 1)\right]\times 10^{-6}\right\}\exp^{\left[2416\left(\frac{1}{303}-\frac{1}{T}\right)\right]}, 3 \ge \alpha_w > 1 \\ [16.8\times 10^{-6}]\exp^{\left[2416\left(\frac{1}{303}-\frac{1}{T}\right)\right]}, \alpha_w > 3 \end{cases} \tag{5}$$

Take ρ_m as the density of the membrane, and the unit is $kg \cdot cm^{-1}$; EW is the molar mass of the membrane, and the unit is $g \cdot mol^{-1}$, then the water concentration in the membrane is expressed as:

$$C_m^w = \frac{\rho_m}{EW}\lambda \tag{6}$$

Set λ as the water content of proton exchange membrane, then the water content of proton exchange membrane can be expressed as:

$$\lambda = 0.043 + 17.81\alpha_w - 39.85\alpha_w^2 + 36.0\alpha_w^3 \tag{7}$$

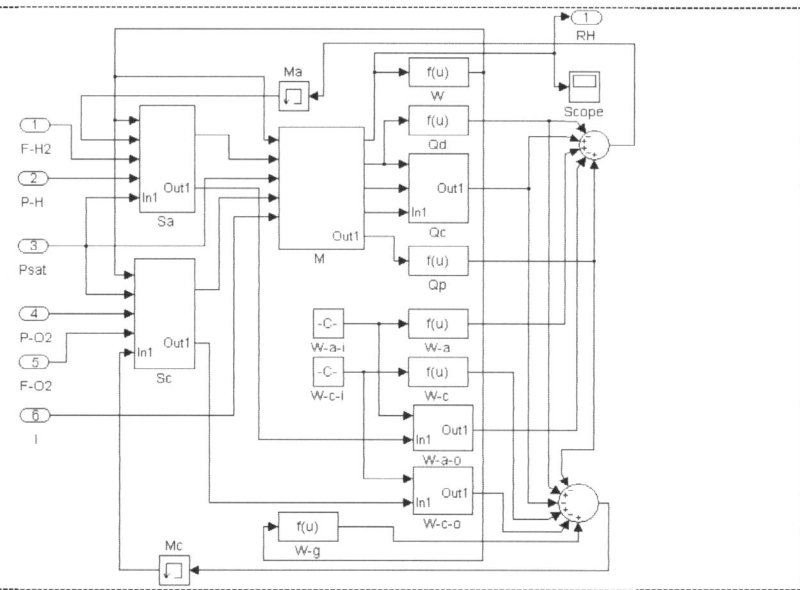

Figure 2. Basic humidity mechanism model

3. Humidity mechanism model and results analysis

According to the previous analysis and research of this paper, the basic humidity mechanism model of fuel cell is finally determined, as shown in figure 2.

Where, "F-H2" is the hydrogen flow rate, "P-H2" is the hydrogen pressure, "P_{sat}" is the saturated vapor pressure of water, determined by the reaction process,"P-O2" is the air pressure,"F-O2"is the air flow rate, and "I" is the working current. By adjusting the values of each input, the average water content output value "RH" described by the humidity mechanism model can be obtained. In this paper, when conducting soft-sensing research based on the mechanism model, the values of relevant parameters are shown in table 2.

Journal of Physics: Conference Series **1423** (2019) 012043 doi:10.1088/1742-6596/1423/1/012043

Table 2. The values of relevant parameters

No.	The parameter name	The parameter value
1	The hydrogen flow rate	0.15 l/s
2	The hydrogen pressure	700bar
3	The air pressure	4 standard atmospheric pressure
4	The air flow rate	0.75 l/s
5	The running current value	30A

The FuelCon Evaluator-C test platform is used as the hardware platform, the TrueData-EIS analyzer is used as the test tool, and the FuelWork software is used as the test software. According to the time sequence, the test data is obtained, and then the simulation data of the mechanism model is compared and analyzed. The experiment data comparison table is shown in table 3.

Table 3. The comparison table of the measured output data and the test data

Time(s)	Test data	Output data	Time(s)	Test data	Output data
0.00	0	0	3.60	0.530	0.536
0.20	0.156	0.134	3.80	0.538	0.536
0.40	0.195	0.167	4.00	0.531	0.536
0.60	0.225	0.200	4.20	0.526	0.536
0.80	0.242	0.219	4.40	0.535	0.536
1.00	0.264	0.245	4.60	0.542	0.536
1.20	0.291	0.283	4.80	0.546	0.536
1.40	0.324	0.333	5.00	0.532	0.536
1.60	0.363	0.371	5.20	0.538	0.536
1.80	0.395	0.400	5.40	0.529	0.536
2.00	0.431	0.421	5.60	0.540	0.536
2.20	0.458	0.443	5.80	0.536	0.536
2.40	0.491	0.476	6.00	0.524	0.536
2.60	0.509	0.486	6.20	0.536	0.536
2.80	0.526	0.513	6.40	0.525	0.536
3.00	0.541	0.526	6.60	0.534	0.536
3.20	0.538	0.532	6.80	0.538	0.536
3.40	0.536	0.536	7.00	0.530	0.536

According to table 3, the soft-sensing simulation results of the humidity mechanism model are obtained, as shown in figure 3. Where, the dotted line is the output value of the mechanism model, and the solid line is the actual test data.

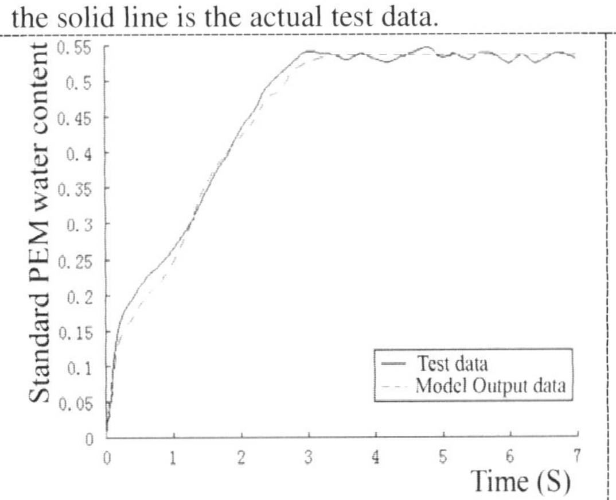

Figure 3. Soft-sensing simulation results of humidity mechanism model

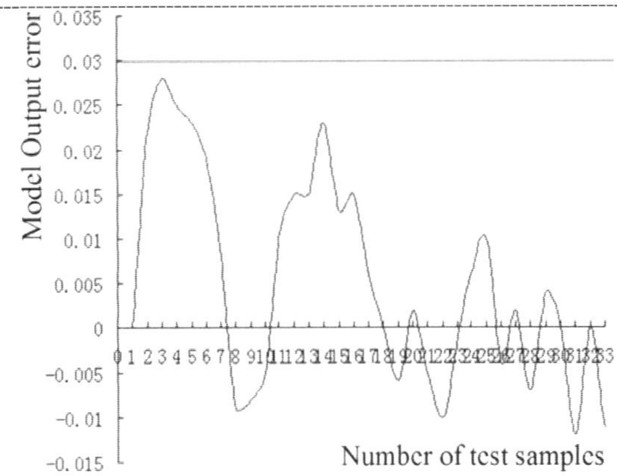

Figure 4. The Soft-sensing simulation test results error curve

Shown as table 3 and figure 5, after the fuel cell operating, the output results of mechanism model based on soft-sensing consistent with the test data curve. The error is within 2%, and most of them in less than 1%. However, the local data has appeared in the large deviation, especially in the early stages of beginning, whose maximum error closed nearly to 3%. The soft-sensing simulation test results error is shown in figure 4. This may be due to the uneven distribution of the gas at the beginning. And the final output of humidity is only 0.532, which may be caused by the small test current used in the mechanism model soft-sensing analysis.

4. Conclusions

In this paper, the humidity mechanism soft-sensing establishment and simulation results of a data-driven proton exchange membrane fuel cell are analyzed and studied. The specific works are as follows:

(1) According to the experimental research object, the diffusion model is selected as the modeling and analysis basis of the humidity mechanism soft-sensing.

(2) According to the basic principles of the diffusion model, a one-dimensional humidity mechanism soft-sensing model was designed and completed.

(3) According to the analysis of actual experimental data, the humidity mechanism soft-sensing model was driven and the predicted output error based on the humidity mechanism soft-sensing model can be basically controlled within 3%. The analysis results of measured data verify the effectiveness of the humidity mechanism soft-sensing model.

However, the design of this paper is under a lot of assumptions and conditions, the actual proton exchange membrane fuel cell humidity mechanism will be more complex and difficult, and worthy of being studied and explored continually.

Acknowledgement

The relevant research work of this paper was supported by the national "863" program for energy saving and new energy vehicles (2008AA05Z105), National natural science foundation of China(51106116), National natural science youth fund(61004018/F030102),Hubei province natural science fund(2010CDB05803)and Hubei university of Technology doctor start fund(BSQD2014063).

References

[1] ZHANG Liyan, QUAN Shuhai. Fuel cell system modeling and optimization control [M]. Beijing: the electronics industry press, 2011.

[2] QUAN Rui, TAN Baohua, QUAN Shuhai, Study on Fault Tree Analysis of Fuel Cell Stack Malfunction[C], Proceedings of 2010 International Conference on Measuring Technology and Mechatronics Automation (ICMTMA2010), March 13-14, 2010, Changsha, China, pp: 579-582.

[3] TAN Baohua,QUAN Shuhai,ZHANG Liyan,Study on Membrane Water Content Soft-sensing technology Method based on internal resistance in PEMFC[J]. International Journal of Digital Content Technology and its Applications,2012,Vol.6,No.11,pp 205-211.

[4] PAN Lideng, LI Dazi, MA Junying. Principles and applications of soft-sensing technology[M]. Beijing: China electric power press, 2009.

[5] LUO Liangqing. Study on online test software system of fuel cell internal resistance and health status monitoring [D]. Wuhan: wuhan university of technology, 2011.

[6] CAO Xiaolei, Study on the Soft Measurement for Proton Exchange Membrane Fuel Cell Humidity [D]. Wuhan: Wuhan University of Technology, 2011.

[7] TAN Baohua,QUAN shuhai,Cao Xiaolei,Study on PEMFC Humidity Measurement Based on Fuzzy Sets[J],Applied Mechanics and Materials,2014, Vol. 556-562,pp: 2986-2989.

[8] TAN Baohua. Study on Humidity Soft-sensing of Vehicular PEMFC[D]. Wuhan: Wuhan University of Technology, 2014.

Journal of Physics: Conference Series **1423** (2019) 012044 doi:10.1088/1742-6596/1423/1/012044

Research on the optimization of express delivery route of logistics base under the constraint of carbon quota

Zhigao Liao[1], Zongqi Deng [2], Kerong Jian[3], Yi Guo[4]

[1] Graduate Department, Guangxi University of science and technology, 545006
[2] School of Economics and Management, Guangxi University of science and technology, 545006
[3] Office of Financial Affairs, Guangxi University of science and technology, 545006
[4] School of Economics and Management, Guangxi University of science and technology, 545006

Corresponding author and e-mail: Zhigao Liao, liaozhigao@126.com

Abstract. This paper considers the route selection of express delivery under the constraint of carbon emission. Two kinds of path optimization models, not con carbon emissions and under the constraint of carbon quota, are constructed respectively to optimize the distribution route of express delivery stations, and this paper further analyses the impact mechanism of carbon quota on the optimization decision-making of distribution route. Finally, carbon emission policy recommendations are obtained.

1. Introduction

There are three main sources of carbon emissions. The transportation industry's carbon emissions are second only to the manufacturing industry, while the automobile's carbon emissions have accounted for 25% of the total carbon emissions. Foreign research shows that in recent 10 years, in addition to the carbon emissions of transportation industry is still increasing year by year, the carbon emissions of other industries and fields have been effectively curbed, and transportation industry has become the most important factor causing the world climate change. Since the reform and opening up, China's economy has been developing rapidly, and the corresponding carbon emissions in China are also increasing. In 2005, the carbon emissions generated by road transportation in China have reached 21%. In 2015, China issued the first 10-year action program of manufacturing power strategy "made in China 2025", which requires that the added value of unit carbon emissions in 2025 should be reduced by 40% on the basis of 2015.

At present, most of the optimization of logistics distribution path only considers the factor of the lowest total cost, but to develop the logistics industry healthily and efficiently, it is not enough to optimize the distribution path network only from the economic point of view. It is also necessary to add the factor of carbon emission into the optimization of logistics distribution path network, so that the total cost and carbon emission in the distribution process can be minimized or balanced This is also the requirement of energy conservation and emission reduction and the development of green logistics in China. As a big energy consumer, the development of low-carbon logistics is not only the requirement of energy conservation and emission reduction, but also the requirement of sustainable development.

Therefore, it is of great significance to study the optimization of logistics distribution path under carbon emission constraints. In recent years, some foreign scholars have added the influence of carbon emissions to the research of vehicle distribution path optimization. Cairns [1] studies the impact on the environment when the grocery store delivers goods to the door, transforms the objective function from the distribution distance to the carbon emission and optimizes it, but does not consider the impact of speed changes and other factors. Andrew Palmer [2] calculated the impact of speed on vehicle fuel consumption with the help of driving conditions, indirectly confirmed the impact of vehicle flow and speed on carbon emissions, and constructed e-vrp model to optimize vehicle distribution path. Figliozzi [3,4] added the constraints of time window, took the sum of vehicle, distance, time and carbon emission cost as the objective function, banjaafar[5] incorporated the carbon emission constraints into the logistics distribution system, established the strict carbon quota constraint model, carbon trading constraint model, and elaborated the carbon quota and carbon trading constraint.

China's research on the optimization of distribution path under low-carbon constraints started late. Gao Minji [6] established the optimization model of low-carbon logistics distribution network based on cloud computing. The model uses multi-modal transport, cloud logistics platform, and re plans the path, making the distribution path change to the direction of low-carbon emissions, completely changing the traditional distribution network, and realizing green logistics. Lin Diansheng et al [7] established a low-carbon logistics distribution centre location model based on the uncertainty of demand. The model fully considered the carbon emissions of each link related to the distribution centre location decision in the distribution network to avoid the result error. Li Shuangyan et al. [8] established a path optimization model for split transportation of low carbon vehicles to minimize the total carbon emissions in the transportation process. In this paper, through the establishment of the path optimization model of carbon limit constraints, the comparison and analysis of the path selection and cost changes, starting from the reality, it is of theoretical and practical significance to really guide enterprises to choose the low-carbon distribution path and achieve the best overall and environmental benefits of the distribution path.

2. Optimization model of distribution path without considering carbon emission

2.1. Model hypothesis

When choosing the distribution route, we usually consider the necessary principles and some influencing factors. From the economic point of view, this paper constructs a distribution route optimization model with the objective function of minimizing the total transportation cost. Finally, in order to facilitate the establishment of model analysis, and make the following assumptions:

(1) Model 1 is mainly constructed to compare with the later carbon constraints, so storage cost and other costs are not considered in the total cost of this paper;

(2) The loading capacity of each vehicle is equal;

(3) All distribution vehicles move at a constant speed;

(4) Each express station has and only has one vehicle for delivery.

2.2. Model parameter description

V refers to the collection of all express delivery sites (nodes), including logistics base. i and j are any two nodes, and S represents the starting point (Logistics base), D represents the destination (Express Station), and the logistics base number is 0.N is the number of express delivery stations, and M, number of distribution vehicles. c_i, transportation cost of unit goods in node i, and f is the fixed cost per kilometre of each vehicle. Q is freight capacity of each vehicle and q_i, the quantity of goods at express delivery point, where i= 1, 2,…,n. A: the number of stations distributed by each vehicle. d_{ij}: represents the distance between node i and node j.

Decision variables is

$$x_{ij}^{k}=\begin{cases}1 & \text{Goods are transported from node i to node j by vehicle k,}\\ 0 & \text{Otherwise}\end{cases}$$

$$y_{i}^{k}=\begin{cases}1 & \text{Vehicle k completes the cargo transportation of node i}\\ 0 & \text{Otherwise}\end{cases}$$

2.3. Mathematical model construction

The objective function is

$$MIN\,Z=\sum_{k=1}^{M}\sum_{i=0}^{N}\sum_{j=0}^{N}\mathrm{f}x_{ij}^{k}d_{ij}+\sum_{i=1}^{N}c_{i}q_{i} \tag{1}$$

$\sum_{k=1}^{M}\sum_{i=0}^{N}\sum_{j=0}^{N}\mathrm{f}x_{ij}^{k}d_{ij}$ represents the variable cost in transportation cost and $\sum_{i=1}^{N}c_{i}q_{i}$ is the fixed cost in transportation cost, subject to

$$\sum_{i=1}^{N}y_{i}^{k}q_{i}<Q \quad k=1,2...M \tag{2}$$

$$\sum_{k=1}^{M}y_{i}^{k}=1 \quad i=1,2...N \tag{3}$$

$$\sum_{i=1}^{N}x_{ij}^{k}=y_{j}^{k} \quad j=1,2..N,k=1,2..M \tag{4}$$

$$\sum_{j=1}^{N}x_{ij}^{k}=y_{i}^{k} \quad i=1,2...N,k=1,2..M \tag{5}$$

$$\sum_{i=1}^{N}y_{i}^{k}\le A \quad k=1,2..M \tag{6}$$

$$x_{ij}^{k},y_{i}^{k}\in\{1,0\} \quad \forall i,\forall j\in V,\forall k\in M \tag{7}$$

Model 1 does not have the limitation of carbon constraint, but chooses the corresponding path with the minimum cost as the objective function. Equation (1) is the objective function, constraint (2) is the load limit of each vehicle, constraint (3) indicates that each node has a vehicle distribution, constraint (4) and (5) ensure that each node can only enter and exit once, and constraint (6) indicates the limit of the number of stations for each vehicle distribution , constraints (7) are all 0-1 variables. In fact, this model is equivalent to the model of finding the shortest distribution path.

3. Optimization model of distribution path under carbon limit

3.1. Model hypothesis

In model 2, based on Model 1, carbon limit constraints are added. Carbon limitation is a kind of emission reduction tool regulated by the government. It mainly relies on administrative means to limit the carbon emission ceiling of enterprises to force enterprises to reduce carbon emissions. This kind of carbon emission reduction restriction is relatively easy for the government to implement, the management cost is very low and the operation is very simple. The government only needs to regulate the relevant laws and regulations, and formulate a reasonable carbon emission ceiling, so it is easy to promote. However, for enterprises, this means mainly relies on administrative means, which cannot help enterprises to save energy and reduce emissions in the long run. Enterprises often reduce emissions at the expense of their own interests. Administrative means can only achieve the effect of energy conservation and emission reduction in the short term, and cannot improve the enthusiasm of enterprises in energy conservation and emission reduction, so the acceptability of enterprises will be low.

3.2. Model concept

First of all, the government stipulates the upper limit of carbon emission for an enterprise. Under the administrative means, in order to achieve the emission reduction goal, the enterprise will be forced to choose the distribution route with low carbon emission to achieve the government's emission reduction goal. According to the effect of emission reduction in a certain period of time, the government can choose to readjust the carbon emission ceiling, or give some enterprises some preferential policies to help them better achieve the emission reduction goals.

3.3. Model parameter description

Model 1 has given some specific model parameters, which are not repeated here, only the parameters of the newly added model are given.

e represents the carbon emission factor, indicating carbon dioxide emission per unit distance of goods, and L, upper limit of carbon dioxide emission, actual carbon dioxide emissions of enterprises and T, the total weight of the distribution vehicle when carrying the goods.

3.4. Mathematical model construction

The carbon emission of enterprises shall be lower than the carbon emission limit stipulated by the government:

$$\sum_{i=0}^{N}\sum_{j=0}^{N}\sum_{k=1}^{M} ex_{ij}^{k}\mathrm{t}d_{ij} \le L \tag{8}$$

The actual carbon emission e of the enterprise is

$$E=\sum_{i=0}^{N}\sum_{j=0}^{N}\sum_{k=1}^{M} ex_{ij}^{k}td_{ij} \tag{9}$$

Model 2 adds a carbon emission constraint (8) on the basis of model 1, so the distribution path optimization model under the carbon quota constraint is as follows: the objective function is (1), and the constraint conditions are (2) to (8). The objective function is still the lowest total cost, and the constraints (2) to (7) are the same as model 1.

4. Example

Take the logistics distribution data of Nanning Suning logistics base and 22 express delivery stations as an example. Due to the limited space, the distance data is not listed, only according to the model 1, according to the size of the total cost, eight distribution routes are listed, which are low cost distribution schemes in Nanning Suning logistics base. Based on this, the carbon quota model is analysed.

The upper limit of carbon emissions is not constant. Policy makers can adjust the upper limit of carbon emissions appropriately according to the size, complexity, overall emission objectives and operation of enterprises. When the upper limit of carbon emission l becomes smaller, the enterprise's optional distribution path will become less. When l increases, the enterprise's optional distribution path will become more. If we want to implement a carbon limit policy, decision-makers can only adjust the size of L to determine the degree of easing of the policy.

This paper discusses the impact of different circumstances on the selection of distribution routes and the change of cost from two aspects, i.e. the initial carbon quota L is free ($\alpha = 0$) and the need for auction ($0 < \alpha < 0.3$).

(1) In the case of free issuance of initial carbon quota, the impact of initial carbon quota L and transaction price P2 on line selection and cost.

In this paper, we use the data of 8 kinds of distribution schemes calculated by the model (1) to analyse these route distribution schemes under the constraint of carbon trading again, and compare the cost differences before and after, so as to provide reference for enterprises. Take P2 = 40, L= 50 as an example to analyse the cost changes of these eight distribution schemes. The results are shown in Figure

1. The total cost 1 is the simple transportation cost calculated by the model 1, and the total cost 2 is the total cost after adding the carbon transaction cost under the carbon transaction constraint. It can be seen from this that in scheme 6 and later distribution schemes, only the distribution cost of carbon trading constraints exceeds the total cost before. Therefore, the distribution routes of enterprises will also produce corresponding changes, and the distribution schemes have changed from 8 to 5.

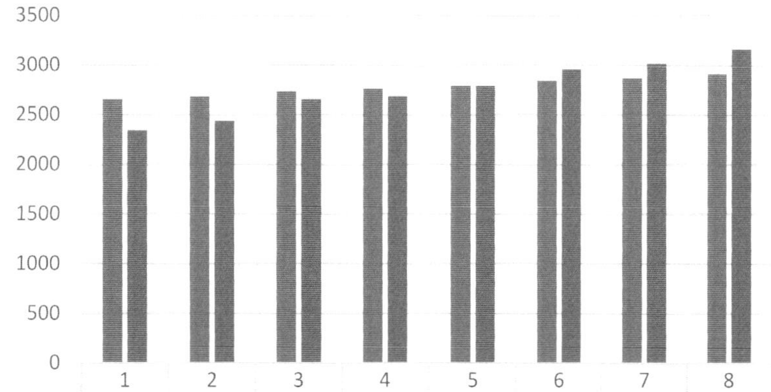

	1	2	3	4	5	6	7	8
Total cost 1	2648.5	2680.5	2735.7	2756.9	2790.5	2842.7	2866.3	2910.9
Total cost 2	2340.46	2434.08	2655.05	2684.54	2786.83	2952.95	3017.09	3160.62

Figure 1. Total cost change of distribution scheme under carbon constraint

Under the constraint of carbon trading, both the initial carbon quota L and the trading price P2 can be regulated, while different L and P2 have different effects on the selection of distribution scheme. Here, the discussion range of the initial carbon quota L is 25-50, and the trading price P2 is 10, 20, 30, 40, 50 and 60 respectively to compare the distribution scheme 1, and draw a figure 2. Because the carbon emission of this distribution scheme is 42.29903, the six curves in Figure 2 intersect at the place where the initial carbon quota is 42.29903, with 42.29903 as the critical value, and the two sides of the intersection are:

(1) E-L > 0 is the left part of the critical value. When the actual carbon emission in the distribution process of Nanning Suning logistics base is greater than the initial carbon emission L, then the carbon cost P2 (E-L) > 0 becomes the penalty cost, and with the reduction of the initial carbon quota and the increase of carbon trading price, the total cost will be higher.

(2)E-L < 0 is the right part of the critical value. When the actual carbon emission in the distribution process of Nanning Suning logistics base is less than the initial carbon emission L, the remaining carbon quota can be sold to other enterprises. At this time, the carbon cost P2 (E-L) < 0 can offset the transportation cost. And the larger the initial carbon quota and transaction price, the greater the offset transportation cost.

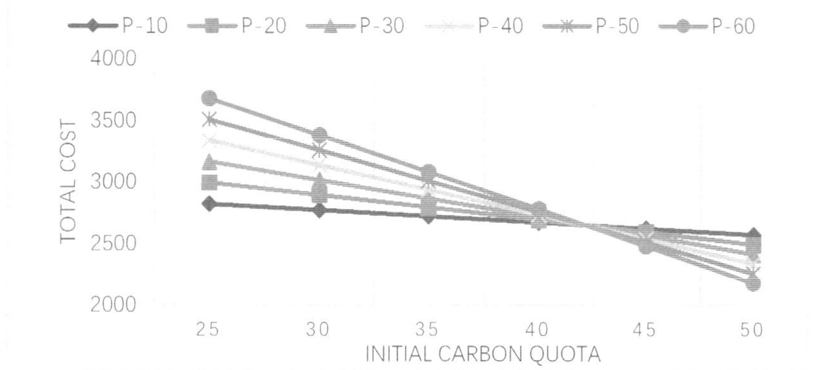

Figure 2. Total cost changes under different initial carbon quotas and transaction prices

5. Conclusion

5.1. Path optimization without considering carbon constraints
Because in the modelling, the total cost is selected as the solution goal, the model is the optimal distribution path without considering the carbon constraints, but the optimal distribution path strength is optimized in the perfect situation. In reality, enterprises can not only choose one distribution path scheme. This model is mainly compared with other models.

5.2. Path optimization under carbon limitation
From the perspective of policy makers, this paper gives a reference method to determine the carbon emission ceiling. By adjusting the carbon emission ceiling, the policy makers can implement more strict or loose emission reduction policies. From the perspective of the enterprise, the distribution scheme that the enterprise can choose changes with the change of the carbon emission upper limit. When the carbon emission upper limit is reduced, the distribution scheme that the enterprise can choose will be reduced. On the contrary, when the carbon emission upper limit is increased, the distribution schemes that the enterprise can choose will also be increased correspondingly.

Acknowledgement
This work is supported by Western project of NSFC (19XGL025), Guangxi Philosophy and Social Sciences Project (17FJY007) and Innovation Project of Guangxi Graduate Education (YCSZ2014203).

References
[1] Cairn S. Home delivery of shopping: the environmental consequences[D]. ESRC Transport Studies Unit/University College London.1999.
[2] Palmer A. The development of an integrated routing ang carbon dioxide emissions model for goods vehicles[D]. Cranfield: Univ. of United Kingdon,2007.
[3] Miguel F M. A. Vehicle routing problem for emissions minimization[J]. Transp. Res. Rec: J. Transp. Res. Board, 2010, 2179:1-7.
[4] Miguel F M. A. The impacts of congestion on time-definitive urban freight distribution networks CO2 emission levels: Results from a case study in Portland, Oregon[J]. Transp. Res. Part C, 2011, 19(05):766-778.
[5] Banjaafar S, Li Yanzhi, Daskin M. Carbon footprint and the management of supply chains: insights from simple models[J]. Automation Science and Engineering, IEEE Transactions on, 2013, 10 (01):99- 116.
[6] Gao min. low carbon logistics distribution network path optimization design based on cloud computing [J]. Journal of Harbin Business University (NATURAL SCIENCE EDITION), 2018, 34 (05): 560-563. (In Chinese)
[7] Lin Diansheng, Zhang Zhiyong, Wang Jiaxin, Liang Xi, Shi Yongqiang. Study on the location of low carbon logistics distribution center under uncertain demand [J]. Control and decision: 1-9 [2019-04-29].Https: //doi.org/ 10.13195 /j.kzyjc. 2018.0646. (In Chinese)
[8] Li Shuangyan, Wang Zhongwei, Zhang Dezhi. Low carbon vehicle path optimization based on split transportation [J]. Journal of Railway Science and engineering, 2018, 15 (07): 1871-1878. (In Chinese)

Journal of Physics: Conference Series **1423** (2019) 012045 doi:10.1088/1742-6596/1423/1/012045

Application research of data mining algorithm in big data environment

Zenghui Zhao[1]

School of Intelligent Science and Information, Xi'an Peihua University, Shaanxi 710000, China

Corresponding author and e-mail: Zenghui Zhao, 296883619@qq.com

Abstract. The Hadoop platform forms a complete large-scale ecological distribution system, including HDFS, MapReduce, HBase and other subsystems. This paper analyzes the parallel processing of Hadoop platform and applies it in the field of data mining algorithms. In order to obtain better algorithm efficiency, a K-Modes clustering algorithm based on big data platform is proposed. It uses cluster mode to replace the central node. The mining process uses naive Bayes to improve mining efficiency. The experimental results show that it has better adaptability, saves time and improves the efficiency of the algorithm.

1. Hadoop platform

Hadoop platform is a distributed processing cluster platform composed of common commercial servers, which can process massive data resources stored in parallel. Since the open source of Apache Hadoop system, at the beginning, there were only HDFS, MapReduce, HBase and other subsystems. At present, Hadoop platform has formed a complete large-scale ecological distribution system.

MapReduce consists of two phases: Map and Reduce, through which parallel computation and stored procedures are realized. In addition, MapReduce framework provides a universal interface for parallel programming, which allows users to invoke the interface according to their needs and quickly complete the data program processing of big data parallelization. MapReduce realizes the data processing through the input method of key value data, and realizes the data processing, and automatically completes the partition data and scheduling data management. In the process of parallel program computing storage, MapReduce completes processing scheduling and allocation of computing resources by analyzing and processing the parallel framework. In addition, it classifies the input/output data and performs relevant degree scheduling to monitor the degree state. In this way, it can realize the collection, sorting and synchronization of computing nodes.

2. Research on data mining algorithm

2.1. Study on clustering data mining algorithm

2.1.1. Clustering algorithm

K-modes clustering algorithm is a kind of unsupervised learning method, very adaptive. The principle of the algorithm is as follows: assume that the sample is, and the number of attributes in the data set is, and then the difference value of the sample can be obtained:

$$d(X,Y)=\sum_{j=1}^{m}\delta(x_j,y_j)$$

$$\delta(x_j,y_j)=\begin{cases}0,(x_j=y_j)\\1,(x_j\neq y_j)\end{cases}$$

Q is represented as a cluster center matrix, $Q_l=[q_{l,1},q_{l,2},\ldots,q_{l,m}]$, Then W sample matrix can be divided:

$$w_{i,l}=\begin{cases}0,i\in l\\1,i\notin l\end{cases}$$

The objective function of k-modes clustering algorithm is:

$$P(W,Q)=\sum_{l=1}^{k}\sum_{i=1}^{n}\sum_{j=1}^{m}w_{i,l}\delta(x_{i,j},q_{l,j})$$

Where, k represents the number of clusters, n represents the number of samples, and m represents the number of attributes.

The basic principle of k-modes clustering algorithm is to first divide the samples into highly similar class clusters, but the similarity between different classes of clusters is very low. Assumes that the number of samples is N, the algorithm from the first N samples freely choose K samples, as a cluster center, then, in turn, to calculate the distance between each sample and cluster center, maximum sampling the similarity of samples, to be classified as group clustering, clustering cluster formation, after iteration, until after all kinds of the center of the cluster stability can be concluded that K clusters. The basic flow is shown in table 1.

Table 1. Basic flow chart of k-modes algorithm

Input:	data source to be clustered, cluster central set defined for the first time
Output:	clustering results
Map()function	
1.	First, define the Map node. Divide the first round of clustering center into the initial Map nodes with DistributedCache algorithm;
2.	for each i(sample)in m(The data set) Solve the difference value between i and the cluster center, and mark the smallest difference value as i
3.	In formula (2), the target value is accumulated and calculated
4.	for each h(attribute) in i(sample) Count the mode of each attribute and write it to the mode matrix
5.	Output mode matrix and final objective function.
Reduce()function	
1.	Store all the results from the map () function
2.	The value of each element in the same position in the cumulative mode matrix is formed into a new matrix, and the value of the objective function is calculated twice.
3.	Select the value according to the principle of minimum difference value and update the cluster center;
4.	Returns the target function value and enters it into the Main() function
Main()function	
1.	If the result of two loops is the same, the algorithm terminates.
2.	If the target values of the two loops are different, continue with the Map() function

Journal of Physics: Conference Series **1423** (2019) 012045 doi:10.1088/1742-6596/1423/1/012045

2.1.2. Algorithm instance

Clustering algorithm k-modes application practice, to analyze the sample analysis of ants looking for food, the ant is through its tentacles and touch feet to detect the distance between food and its own, so through the ant touch feet and tentacles for the attribute analysis of ant colony food search and food distance relationship, the data is shown in table 2.

Table 2. Partial data table

The serial number	Antenna length(mm)	distance(mm)	Touch the foot length(mm)	distance(mm)
1	10	30	3	50
2	13	50	6	70
3	12	60	8	90
4	15	80	3	100
5	21	100	5	120
6	22	120	10	150

Then divide it into 2 groups, that is, k=2.

Firstly, the objects in the table are grouped in the form of 3 groups and allocated to Mapper nodes. The group number corresponds to the node number of Mapper, and then the Cost variable Cost=-1 is calculated through Reduce function.

Then select any data from the group to obtain the clustering center, respectively:

CostA= (length of tentacles =10, distance =30, length of contact feet =3, distance =50)

CostB= (antenna length =21, distance =100, contact length =3, distance =100)

According to the algorithm flow in table 1, the above data is calculated and processed, and the final result is output, so that the clustering center can be obtained. After the above iteration, it can be concluded that the final node center is:

CostANew= (length of antenna =12, distance =60, length of touch foot =8, distance =90)

CostBNew= (length of tentacles =15, distance =80, length of contact feet =4, distance =100)

2.2. Classification algorithm based on Hadoop platform

The main process of naive bayesian classification algorithm of Hadoop platform is as follows: the user will send a request job to the Hadoop platform, and the Hadoop platform will randomly assign a job number after receiving the job. The background job server will then process the job. After analyzing the job structure information, it will be assigned to the task server of relevant job nodes for processing. All training sets and intermediate processing results in the process of job processing are stored in the HDFS file set. After processing the training set data, the classifier will return relevant result values to the corresponding subsequent task nodes. When all JobTracker has finished, the algorithm is finished.

2.2.1. Naive bayesian classification algorithm

The basic theory of naive bayes classification algorithm is bayes' theorem, and the mathematical model of bayes' theorem is as follows:

$$P(A \mid B) = \frac{P(B \mid A)P(A)}{P(B)}$$

In the above equation, the probability of P (A) only when A occurs is also called prior probability. The posterior probability is P (A|B), representing the probability of A occurring under the condition of B; Similarly, P (B|A) is the probability of B occurring under A condition.

When two events A and B are independent of each other, P(AB)=P(B|A)P(A).

Training and testing are important steps in the process of classification algorithm. First of all, after the prior probability is calculated, the posterior probability is obtained by using the bayesian formula. In the classification of object attributes, the sample classifier should be used for probability calculation, so as to obtain the category of the maximum probability. The specific process is as follows:

If we define $C_1, C_2, ..., C_n$, with n classes, thus the data set sample X to be classified can be obtained:

$$P(C_i | X) > P(C_j | X) \quad 1 \le j \le n, j \ne i$$

The naive bayesian classification algorithm directly assigns the unclassified sample X mentioned above to the class with the highest posterior probability.

When $P(X | Y_i)P(Y_i)$ reaches the maximum value, the conditional probability reaches the maximum value, and the calculation formula is as follows:

$$P(X | Y_i) = \prod_{k=1}^{n} p(x_k | Y_i)$$

In the formula, P(X) can be omitted as a fixed value. P(Y) is obtained by calculating the ratio of the training record. The attribute set $X = \{x_1, x_2, ..., x_n\}$ contains n attributes.

The posterior probability calculation formula is as follows.

$$P(Y_i | X) = \frac{P(Y_i) \coprod_{k=1} P(x_k | Y_i)}{P(X)}$$

2.2.2. Test results

Experimental data were derived from the data set in UCI: Pokerhand and Skin Segmengtation were shown in table 3.

Table 3. Original data table

The serial number	The data set	Sample size	dimension	Number of categories
1	Pokerhand	1010884	10	11
2	Skin Segmengtation	1246183	5	3

Part of the data in the data set is randomly taken as the test set, and the average value is taken as the result output after running the result for several times. The number of test sets was gradually increased, and the time required for the execution of naive bayesian algorithm was analyzed, as shown in table 5, as the speedup ratio (parallelization processing time on Hadoop platform/stand-alone processing time - 1). The analysis of table 4 shows that large-scale data on Hadoop platform has classification efficiency.

Table 4. Experimental results schedule

The serial number	Number of test sets (10,000)	Speed up than
1	1	0.46
2	5	0.72
3	10	0.81
4	25	1.21
5	50	1.42
6	75	1.61
7	90	1.82
8	100	1.89

3. Conclusions

Hadoop platform forms a complete large-scale ecological distribution system, including HDFS, MapReduce, HBase and other subsystems. The algorithm proposed in this paper can effectively improve the efficiency of the algorithm by replacing the clustering center with cluster mode, and it has better self-adaptability. The effective implementation of classification mining based on naive bayes algorithm can effectively save the algorithm time.

Acknowledgement
Fund Program: University-level Scientific Research Program of Xi'an Peihua University in 2019, "Research on the construction of industry-university-research platform based on artificial intelligence" (No. PHKT19003).

References

[1] Li shuang, Chen ruirui, Lin nan. Hadoop framework k-means clustering algorithm for big data mining [J]. Computer engineering and design, 2018, 39(12):3734-3738.
[2] Chen e x. application research of data mining algorithm based on Hadoop [J]. Journal of bohai university (natural science edition), 2018, 39(03):274-280.
[3] Chen Juan. Algorithm analysis and application based on Hadoop Internet of things data mining [J]. Computer times, 2018(06):29-31+34.
[4] Wang 100. Study on clustering k-means algorithm based on Hadoop platform [J]. Computer and telecommunications, 2018(04):18-20.
[5] Hao yanni, tian weili. Application of data mining algorithm based on Hadoop in wine information data analysis system [J]. Computer application, 2017, 37(S1):72-74+79.
[6] Hong bo, lv yanxia, huang lei. Research and implementation of data mining algorithm based on Hadoop framework in big data environment [J]. Electronic design engineering, 2017, 25(07):41-44.
[7] Liu baolong, su jin. Double MapReduce improved Canopy-Kmeans algorithm [J]. Journal of xi 'an university of technology, 2016, 36(09):730-737.
[8] Wang hao, shi wei, li huan. Research on meteorological data mining based on bayesian network under Hadoop [J]. Electronic devices, 2016, 39(04):841-846.
[9] Bu renzi. Study on parallelization of dbik-means clustering algorithm based on Hadoop platform [J]. Journal of xinyu university, 2016, 21(04):115-118.
[10] Li yanghui. Study on k-means clustering algorithm based on MapReduce [D]. Xi 'an university of technology, 2016.
[11] Wang lei, zhang yongjian, jia jipeng, niu xiaoguang, nie changlong. Data mining method of public building energy consumption based on Hadoop [J]. Computer system application, 2016, 25(03):34-42.
[12] Ma bin, Yin lifeng. A parallel naive bayesian network public opinion classification algorithm based on Hadoop platform [J]. Modern book information technology, 2015(02):78-84.

Journal of Physics: Conference Series **1423** (2019) 012046 doi:10.1088/1742-6596/1423/1/012046

The design of micropower field data collector based on MSP430 microcontroller

Chun Jiao[1], Ming Tang[2]

Xi'an Peihua University, Xi'an, China

Corresponding author and e-mail: Chun Jiao, 2024988445@qq.com

Abstract. Field data collectors are widely used in industrial control and information monitoring. According to the design requirements of micropower consumption and reliability, combined with MSP430 microcontroller, this paper expounds the design scheme of a new type of micropower field data collector, and discusses the overall design of the system and the implementation method of main functional units.

1. Introduction

In many application fields such as industrial measurement and control, automation control, and information monitoring, it is often necessary to collect and process various types of status information on the monitoring site to achieve effective on-site monitoring and control functions. In a typical application environment, it is often necessary to use network transmission technology to set up a measurement and control network in the field environment, and transmit relevant data to the remote central station of the measurement and control system for subsequent analysis and processing. Therefore, how to achieve data acquisition, storage and transmission in the field measurement and control environment with lower hardware cost, lower power consumption, smaller size and higher reliability is a problem that application engineers often need to consider and solve in actual design.

2. The System Design

The field data collector is a typical microcontroller application system. The system can be used as a local measurement node of the distributed measurement and control network to collect various kinds of on-site information (including various status information and image information) in real time, and can store and process useful monitoring data. As shown in Figure 1, through the various sensors, measurement unit, functional units, built-in control program and various functional programs in the system, the functions of information monitoring, acquisition and storage at the measurement and control site can be completed.

The data collector is used in the field of measurement and control, so the system is a battery-powered application system. Micropower design is an important part of the overall design of the system. At the same time, the overall technical architecture, hardware and software design of the system also need to meet the requirements of cost, volume and reliability.

The system's micropower design includes intrinsic low power design, operational power analysis, and power management [1]. The low-power circuit structure and devices with essentially low power consumption should be preferred. At the same time, it is necessary to comprehensively analyze the power consumption of each device in the operation of the system, and seek a task arrangement that can minimize the average power consumption of the system. Workflow optimization and intelligent power management for micropower design can be achieved through a combination of software and hardware. At the specific software design level, the control program and various functional programs of the system are designed based on the interrupt program structure to achieve refined power management.

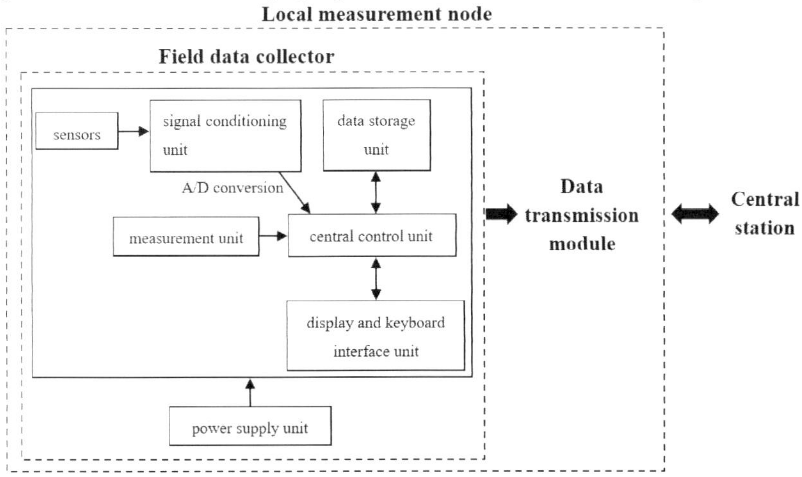

Figure 1. The hardware block diagram of the data collector

As shown in Figure 1, all kinds of measured information from the measurement and control site can be converted into electrical signals by the corresponding sensors, amplified, shaped and filtered by the signal conditioning unit, and input to the central control unit after A/D conversion.

The 16-bit microcontroller MSP430F5419 with outstanding micropower characteristics is selected as the central control unit of the system. The chip has many peripheral function modules [2], which is powerful, which helps to improve the system's advancement and integration, and reduces the design difficulty.

The data storage unit is used to store various types of on-site monitoring data from the forward channel. According to different data types and storage time intervals, data storage chips of different forms and storage capacity can be selected.

The display and keyboard interface unit provides a Chinese human-computer interaction interface for the operation of the system, and realizes functions such as operation setting, function menu display, prompt and status information display. The system adopts OLED12864 liquid crystal display module with resolution of 128×64, supports multiple interfaces, uses low voltage power supply, and has the advantages of small size, good shock resistance and wide viewing angle.

The power supply unit supplies power to both the analog and digital circuit parts of the system at the same time, provides a stable power supply voltage, and cooperates with the system control program to implement power partition management. At the same time, it has a high power output quality to meet the requirements of output efficiency, reliability and electromagnetic compatibility.

3. The Forward Channel

As shown in Figure 1, the sensors, the signal conditioning unit and the measurement unit are all forward channels in the system. On the one hand, various types of physical quantities (temperature, humidity, pressure, flow, etc.) in the field environment can be converted into electrical signals through various types of sensors. The output signals of the sensors need to be amplified, shaped and filtered by the signal conditioning unit. After A/D conversion, they become digital signals and are input to the central control unit. On the other hand, in the actual design, according to the requirements of system integration, cost, volume, etc., it is also possible to directly select a specific functional module that can be used for secondary development as a measurement unit in the system, such as a blood oxygen saturation module capable of measuring blood oxygen saturation and heart rate or a dedicated image monitoring module. The measurement unit composed of a specific functional module is capable of outputting a digital signal or a digital image of a certain format, which is directly received by the central control unit. Some sensors integrated with signal conditioning and A/D conversion function can also directly output digital signals, which can be directly selected in combination with cost and volume requirements. The design method of directly selecting the digital sensor or the integrated measuring module eliminates the signal conditioning unit in the system, and the output digital signal does not require analog-to-digital conversion, which improves the system integration and reduces the design difficulty.

Due to cost, integration and volume constraints, in many cases it is still necessary to design a signal conditioning unit separately for the output signal of the sensor. The signal conditioning unit performs signal amplification, noise filtering and shaping on the sensor output signal, and the electrical signal (usually a voltage signal) with a small amplitude of the sensor output is amplified or adjusted without distortion, and is finally converted into a voltage signal that can be directly sampled by the A/D conversion module. The signal conditioning unit mainly includes signal amplifying circuit, filtering circuit and voltage reference circuit to realize signal amplification, filtering and shaping functions. The structure of the signal conditioning unit should be adapted to the signal form of the output from the front stage sensor.

The signal conditioning unit of this system should meet the following design requirements:

(1) The amplification of the sensor output signal can be completed, and the useless signal outside the effective signal band can be filtered out;

(2) It can provide the required voltage reference and adjust the sensor output signal to the full-scale range of the A/D converter module;

(3) Its input and output have less influence on the front stage sensor and the A/D conversion module of the latter stage;

(4) Its accuracy, sensitivity, linearity and frequency response range are able to meet system requirements;

(5) It uses single-supply and low-voltage power supply with low static and dynamic power consumption.

Taking a common piezoresistive sensor as an example, the output signal generally includes two parts: a common mode signal and a differential mode signal. Moreover, the equivalent resistance of the sensor changes with the measured physical quantity. Therefore, the signal amplifying circuit should have a large input resistance, and should have strong common mode rejection and differential amplification capability, that is, have a high common mode rejection ratio. The signal amplifying circuit should also have a lower input offset current and a smaller temperature drift to further reduce the effect of sensor bridge resistance variation on accuracy.

In summary, the signal amplifying circuit of the system should use an amplifier with a non-inverting input differential structure. Specifically, a non-inverting input differential structure composed of two high-quality operational amplifiers can be employed. This structure is advantageous for increasing the input impedance of the amplifying circuit itself, and the differential structure has a strong suppression capability for the common mode signal output by the sensor. The use of two parallel operational amplifiers also facilitates mutual compensation for drift. In the specific implementation method, an instrument-level amplifier with an adjustable gain and a non-inverting input differential structure can be selected, which is advantageous for reducing design difficulty and improving system integration. When there is a higher flexibility requirement for the signal amplification range, a digital potentiometer can be used instead of the adjustable resistor to achieve programmable gain amplification.

The filter circuit is used to filter out unwanted signals outside the frequency band of the wanted signal, and the high frequency harmonics introduced by the power supply circuit or electromagnetic radiation can be pre-filtered before sampling to eliminate noise. In order to simplify the design, an active filter circuit composed of an operational amplifier of the same phase structure and an RC network is used in the system. This form of filter circuit has a high input impedance and a certain gain and buffering effect.

The gain range G of the signal amplifying circuit in the signal conditioning unit can be calculated by the following equation.

$$G = \frac{\text{full scale } voltage \text{ range of the amplified signal}}{\text{voltage range of measured signal} \times \text{sensor's sensitivity} \times \text{passband gain of the filter}}$$

4. The Central Control Unit

This system uses TI's new 16-bit microcontroller MSP430F5419 as the central control unit. The MSP430F5419 integrates four independent serial interfaces, 14 channels of high-precision A/D conversion module, 128KB of program memory and 16KB of RAM [2], enabling efficient data acquisition without the need for a dedicated A/D converter chip. It has an RTC real-time clock module and an on-chip CRC check module based on the CRC-CCITT standard, which provides technical guarantee for further improving the reliability of data storage and transmission.

The MSP430F5419 has five low power modes. Depending on the low power mode, you can choose to keep the data in RAM or turn it off directly. Each of its internal functional modules can be turned off by software operation when idle. With the micropower characteristics of the MSP430F5419, the main control program and each function program of the system can be designed based on the interrupt structure, which can make MSP430F5419 in a low-power sleep mode during idle time and be wakened up by various interrupt request signals, and perform functions such as data acquisition and data storage in interrupt service routine.

5. Data Collection

The MSP430F5419 microcontroller integrates an A/D converter module ADC12, which provides 14 external channels for 12-bit A/D conversion with a sampling rate of 200ksps. The ADC12 module includes an ADC core with sample/hold function, sample and conversion timing circuitry, conversion memory logic, an internal reference level generator and a variety of clock sources. And there are four A/D conversion modes of single channel single, single channel repetition, sequence channel single and sequence channel repetition. These internal functions of the ADC12 module can be individually set as needed, providing a high degree of flexibility for data acquisition of the system.

Since the system needs to continuously collect multiple measured signals at the measurement and control site, the useful frequency range of each measured signal should be combined to determine the highest frequency component to be retained. The sampling frequency suitable for the system is then determined according to the requirements of Nyquist's sampling law. In order to retain more frequency components and detailed information of the measured signal as much as possible, the sampling frequency of the system can be appropriately increased to ensure that the measured signal can be restored without distortion according to the collected data.

The data acquisition program of this system adopts the conversion mode of sequence channel repetition. In this mode, the sampling order of each channel can be flexibly set as needed. The sampling control signal, clock source, and reference level also need to be set in the data acquisition program.

Writing the acquired data to the data storage chip requires a certain write time. In order to ensure that the data acquisition and data storage processes do not affect each other, two dedicated buffers of the same size should be opened in the RAM area of the MSP430F5419. During the data acquisition process, the MSP430F5419 temporarily stores the collected data in these two buffers. When the buffer is full, the data of this buffer is written to the memory chip, and the other buffer continues to temporarily collect the data. By switching between the two buffers and combining the page data writing function of the data storage chip, the collected data can be stored in a timely and reliable manner, which effectively improves the system reliability.

The specific process of the data acquisition program is shown in Figure 2. In order to effectively reduce the average power consumption of the system, the data acquisition program uses an interrupt program structure. That is, during the A/D conversion process, MSP430F5419 is kept in the low power sleep mode. At the end of each A/D conversion, the A/D conversion interrupt flag ADC12IFG is automatically set to wake up MSP430F5419 and process the acquired data in the interrupt service subroutine. The trigger signal for each start of sampling is generated by the on-chip hardware timer of MSP430F5419, and its timing time corresponds to the sampling frequency. The data acquisition program based on the interrupt structure in the system reduces the running time and the unnecessary waiting state of MSP430F5419, and achieves the purpose of accurately controlling the running power consumption.

The flow chart of data acquisition program is shown in Figure 2.

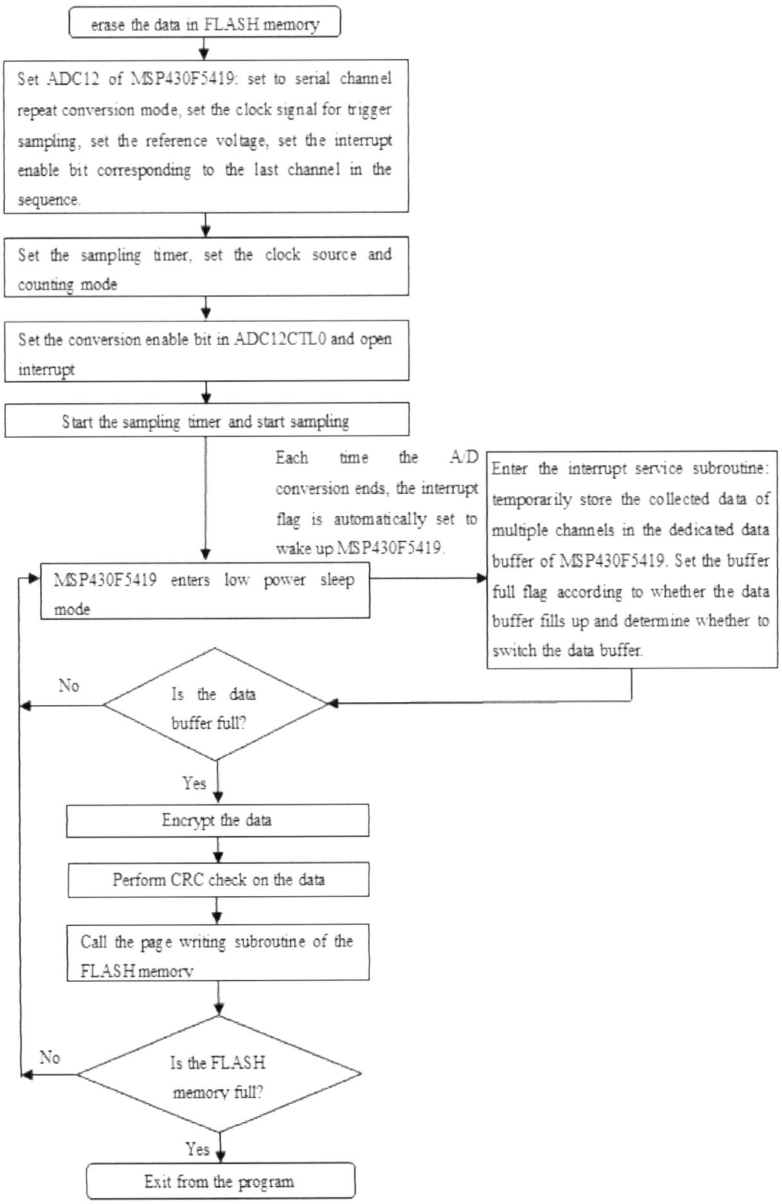

Figure 2. The flow chart of data acquisition program

6. Conclusions

This paper combines the design principles of micropower consumption and reliability to design a micropower data collector suitable for various industrial control and information monitoring sites. The data collector is powered by battery, and can realize real-time monitoring, collection and storage of various types of information in the measurement and control field.

Acknowledgement

This work was supported by Special Research Project of Shaanxi Provincial Department of Education (18JK1085).

References

[1] He Limin. MCU minimum power system design outline V0.6[J]. Electronic Technology Application, 1999, (10): 4-5, 22. (In Chinese)
[2] TEXAS INSTRUMENTS. MSP430F543x, MSP430F541x Mixed-Signal Microcontrollers. TEXAS INSTRUMENTS data sheet, 2018.
[3] Ma Youkun. Design Considerations for Operational Amplifiers in Active Filter Design[J]. Telecommunications Technology, 2001(4): 103-105. (In Chinese)
[4] Atmel Corporation. 64-Mbit, 2.7V Minimum Serial Peripheral Interface Serial Flash Memory AT25DF641. Atmel datasheet, 2010.
[5] Maxim Integrated Products. 16μA IQ,1.2A PWM Step-Down DC-DC Converters MAX1556/MAX1556A/MAX1557. Maxim data sheet, 2011.

Journal of Physics: Conference Series **1423** (2019) 012047 doi:10.1088/1742-6596/1423/1/012047

Study on application of entropy averaging method in airborne gamma-ray spectrum for local disturbance elimination

Chao Xiong [1], Jiankun Zhao [1], Hexi Wu [1], Liangquan Ge [2,*]

[1] School of Nuclear Science and Engineering, East China University of Technology, Nanchang 330013

[2] Applied Nuclear Techniques in Geosciences Key Laboratory of Sichuan Province, Chengdu 610059[2]

*Corresponding author and e-mail: Liangquan Ge, glq@cdut.edu.cn

Abstract. This paper briefly describes the basic principle and the entropy averaging method for local disturbance elimination in airborne gamma-ray spectrum. This method was employed to process t boundary of different lithologies and the identification of metal ore points. The results showed that applying the entropy averaging method to eliminate the interference of the airborne gamma-ray spectrum data, and selecting a larger window width, achieve more significant results compared with the ground gamma ray spectrum.

1. Introduction

Airborne gamma-ray spectrum measurement that installed on an aircraft by an airborne gamma-ray detector is used to measure gamma ray from radioactive material in the surface medium and atmospheric radioactive material. We can explore shallow surface radioactive minerals and metal minerals through this method. Due to the randomness of the decay of radioactive materials, the radioactive element content has statistical fluctuation characteristics. In addition, formation of geological bodies are also random in the long process of geological evolution. At the same time, airborne gamma-ray spectrum measurements are susceptible to interference from factors such as air humidity, radon concentration in the air, topography, and flying height, which makes the measured data a random superposition of the real information of the radioactive level of the surface medium and the interference noise information. How to reduce the measurement statistical noise and improve the measurement accuracy is a new development direction in data processing in recent years.

Many scholars have tried various ways to eliminate the errors caused by statistical fluctuations. Eliminating local interference of gamma spectral data has been proven to be a necessary and effective method, and it is proved by experiments that the entropy averaging method is better than the multi-point slewing method. (Chen X, 2011), Liu et al.(Liu J, 2010) used the entropy averaging method to eliminate the local interference of the ground gamma-ray spectrum profile data of geological mapping in shallow coverage area. The results show that using the method can eliminate the random interference of surface vegetation and local soil enrichment. Xu (Xu S, 2011) compared a variety of methods to eliminate random interference of ground gamma-ray spectrum data. However, there are some significant differences between airborne gamma-ray spectrum and ground gamma spectrum survey. For instance, the number of measuring points on the airborne gamma-ray spectrum line is large; the single line covers a wide area and the point distance is fixed; due to the influence of terrain and other factors, the ground gamma-ray spectrum cannot be evenly distributed on the line, or the point distances are completely equal. In summary, the method of using the entropy averaging method to eliminate random interference of airborne gamma-ray spectrum data has not been reported.

2. The principle of entropy

The concept of entropy averaging is derived from the development of physics, which was originally used to measure the disorder degree of a thermodynamic system. Claude Elwood Shannon first applied the concept of entropy to information theory in 1984. So entropy is also called Shannon entropy. In information theory, entropy is the concept of measuring uncertainty. That is, the higher the entropy, the larger the amount of information, and the smaller the amount of information. In the airborne gamma-ray spectrum measurement, the random variable X is set as the information source, where $X \in \{a_1, a_2, \cdots, a_k\}$. The frequency Pk at which the variable X appears is as follows:

$$p_k = P(a_k)\ (i = 1, 2, \cdots, K)$$

Hence each line can be recorded as a finite number of independent repeat tests.

$$X = \begin{Bmatrix} a_1 & a_2 & \cdots & a_k \\ P_1 & P_2 & \cdots & P_k \end{Bmatrix}$$

This process can be expressed as a equation between X and Pk:

$$H(X) = -\sum_{k=1}^{k} P_k \log P_k$$

Where H (X) is called information amount or information entropy about the source. In general, information entropy can be calculated if the probability of a source is known. But because people often have difficulty knowing the probability of random variables

appearing pk=p(ak)(i=1,2,…,K),, therefore entropy is actually a mathematical expectation of the sum of the bit quantities of random variables and the probability of sequential occurrence multiplied. If the mathematical expectation of the function f(x) of x is:

$$\sum_{k=1}^{K} p(a_k) f_m(a_k) = c_m, m = 1, 2, \cdots, M$$

thus we can estimate the distribution pattern $(p_1, p_2, \cdots, p_k)$of X, and then find the solution of the X distribution to calculate the information entropy.

In 1957, E.T.Jaynes proposed the principle of maximum entropy. The core idea of it is to make full use of only a small amount of information to judge the distribution of unknown information without adding speculation (Jaynes E.T, 1957). It can be known from the principle of maximum entropy that the estimation problem of probability distribution has in fact been transformed into the optimal solution problem of the planning scheme: $H(X) = -\sum_{k=1}^{k} P(a_k) \log P(a_k)$

that is, the probability distribution of $P = (p_1, p_2, L, p_k)$ when the entropy is maximized. At the same time, the following constraints are met:

$$\begin{cases} \sum_{k=1}^{K} p(a_k) f_m(a_k) = c_m, (m = 1, 2, \cdots, M) \\ \sum_{k=1}^{K} p(a_k) = 1, p(a_k) \geq 0 \end{cases}$$

The optimal solution obtained under this constraint is the probability distribution that maximizes the system information entropy.

3. Entropy averaging function of airborne gamma-ray spectrum data

Since the airborne gamma-ray spectrum measurement has statistical fluctuations, and the measured data is easily superimposed with the interference information, thus uncertainty appears. Such uncertain attributes can be described by information entropy as follows:

$$h_i = -kq_i \log q_i$$

where qi is the probability of describing the discreteness of a certain point data. The measurement process of a single airborne gamma-ray spectrum point can be considered as an independent repeat test. Independent repeat trials are uniformly distributed and can be considered as equal probability events. Since the uncertainty of the equal probability event is the largest, we can assume that the entropy of the measured values of the airborne gamma-ray spectrum is the largest. At this time, the larger the value of the sharp change point in the window, the smaller the influence on the entropy averaging. Here the expression of qi can be written in the following form:

$$q_i = \frac{|A_i - A_0|}{\sum_{i=1}^{n} |A_i - A_0|}$$

where A0 represents the actual measured value of the airborne gamma-ray spectrum, and A1 represents the measured value when the entropy is maximum.

The steps of extracting anomaly information of airborne gamma-ray spectrum data based on entropy averaging method are as follows:

(1) Select n consecutive airborne gamma-ray spectrum measurement points from a certain measurement line to form a measurement window. The actual values of the measurement points are set to A1, A2, A3....An;

(2) Select the point in the data set of the window that can make H (A0) obtain the maximum value. Since the equal probability event has the largest uncertainty, it has the largest entropy. To make H (A0) take the maximum value, A0 should be the most equal distance from other observations falling within the window;

(3) Solve the following equations

$$\begin{cases} A_H = A_0 + \sum_{i=1}^{n} (A_i - A_0) h_i \\ k = \dfrac{1}{\log(n-1)} \\ q_i = \dfrac{|A_i - A_0|}{\sum_{i=1}^{n} |A_i - A_0|} \end{cases}$$

where AH is the entropy averaging value of n observation points in the current window, and AH is recorded in the center of the window;

(4) Move the window with one measuring point as the reference in the direction of the measuring line, and repeat steps (1)-(3) to complete the calculation of the single measuring line;

(5) Repeat steps (1)-(4) with the line as the calculation unit to complete the calculation of all the measurement areas.

In order to visually reflect the effect of entropy averaging on raw data processing, the coefficient of variation is introduced into the statistics of airborne gamma-ray spectrum line data, that is, the coefficient of variation (CV) is used to characterize the degree of line change after different parameters are processed.

Because the ore-bearing areas are in the region of high coefficient of variation (Sun, 2009), the coefficient of variation of gamma energy spectrum can reflect the changing trend of uranium, thorium and potassium in the survey area, so it can be used to study the spatial distribution law of radioactive elements, general exploration and prospecting. The calculation formula is as follows:

$$CV = \frac{\sqrt{\frac{1}{n-1} \sum_{i=1}^{n} (X_i - \overline{X})^2}}{\overline{X}}$$

where $\overline{X}$ is the average value of a certain element in the survey area; n is the total number of points in the survey area; Xi is the element content value of a certain measuring point in the survey area.

The practical significance of the coefficient of variation in the measurement data of airborne gamma-ray spectrum is: Taking the average of the arithmetic points of the measured points in a single line as the reference, obtain the coefficient of variation of the measuring points, and further compare the degree of dispersion of the line data. The larger the coefficient of variation, the more intense the data jitter and the greater the instability.

Application to airborne gamma-ray spectrum data

The flight data of 1:5000 scale in a metal mine survey area in Inner Mongolia was selected as the data source. The data was measured by the AGS863 airborne gamma-ray spectrometer developed by Chengdu University of Technology. The geological map of the survey area is shown in Figure 1. The area is mainly composed of granite, sandstone and sedimentary rocks, and there are metal ore points (lead-zinc ore, iron ore) in the southeast of the survey area. The measured full spectrum of the gamma-ray spectrometer is shown in Figure 2, and the area of the survey area is about 300 square kilometers, with 21 lines and 6705 points. The measured nuclide statistics are detailed in Table 1.

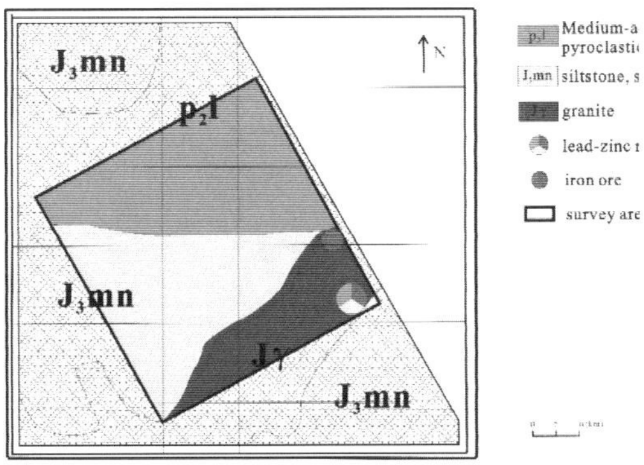

Figure 1. Geological map of the survey area

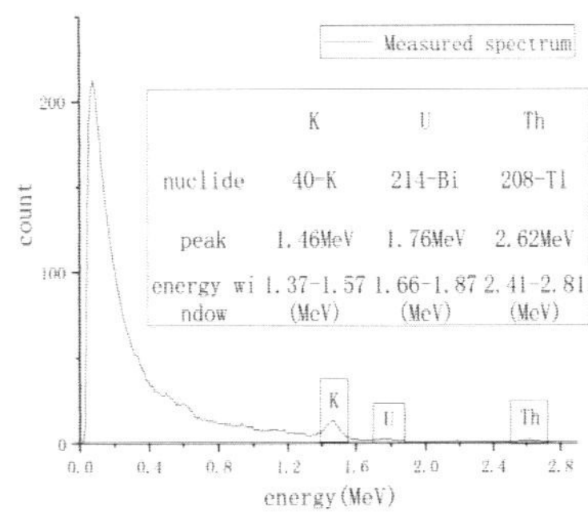

Figure 2. Measured airborne gamma-ray spectrum by AGS863

Table 1. Measurement data statistics table

Statistical parameter \ Indicator name	K (%)	U($\mu g/g$)	Th($\mu g/g$)	TC(cps)
average value	3.59	3.97	14.33	20.26
Median	3.23	3.32	11.76	17.52
Standard deviation	1.47	3.19	7.79	9.46
variance	2.15	10.15	60.74	89.56
Minimum value	0.04	0.049	0.26	0.41
Maximum value	19.87	37.20	57.19	107.32

Select a line in the survey area, and use the 5 -point, 9 -point, 13 -point, 17 -point and 21 -point entropy averaging methods to process the line and calculate the coefficient of variation after processing. The comparison of the total count rate and coefficient of variation of a section of a survey line is shown in Figure 3.

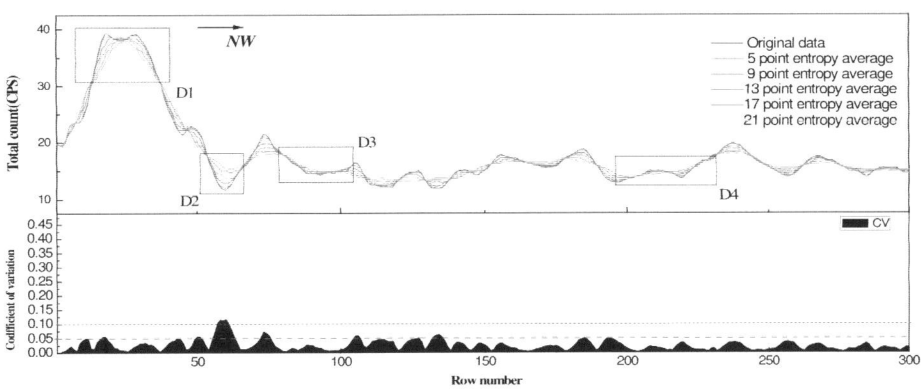

Figure 3. Comparison of different entropy mean and coefficient of variation for the total count rate of a survey line

Journal of Physics: Conference Series **1423** (2019) 012047 doi:10.1088/1742-6596/1423/1/012047

It can be seen from the upper part of Figure 3 that the raw data can maintain the general trend of the content of the line elements after being processed by the entropy averaging method of different width windows, and the coefficient of variation is mostly below 0.05, and has the following characteristics:

In the region where the coefficient of variation is gentle (D3, D4), the results obtained by the entropy averaging method are basically consistent with the original data of the line, and the corresponding coefficient of variation is also less than 0.25, which is a weak variation. In the low-valued area D2, the data after the five-point entropy averaging is basically consistent with the original data form, and as the window increases, the low-value trend of D2 gradually decreases, and the lowest point increases significantly. The coefficient of variation here is the maximum of the whole line, reaching 0.12, which is a strong variation, indicating that the window size of the entropy averaging method has a greater impact here. The overall value of the D1 region is roughly the same as the original data, but with the increase of the entropy averaging window, it is controlled by the sliding average algorithm, and the peaks are suppressed by overlapping; while the coefficient of variation is weakly less than 0.6.

As can be seen from the above, the size of the window width directly determines the effect of the entropy averaging algorithm. According to the experience of ground gamma spectrum measurement, we select 7-point, 15-point, 30-point entropy averaging method to process the raw data, and extract the uranium, strontium and potassium anomaly information of the airborne gamma-ray spectrum. The results are shown in Figure 4 to Figure 6:

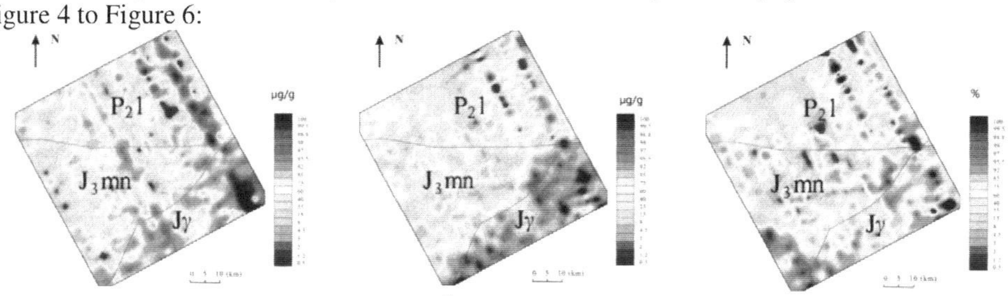

a b c

Figure 4. 7-point entropy averaging method (a, uranium; b, thorium; c, potassium)

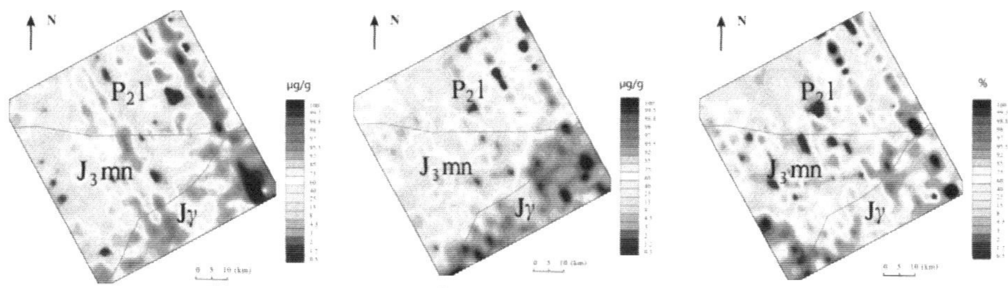

a b c

Figure 5. 15-point entropy averaging method (a, uranium; b, thorium; c, potassium)

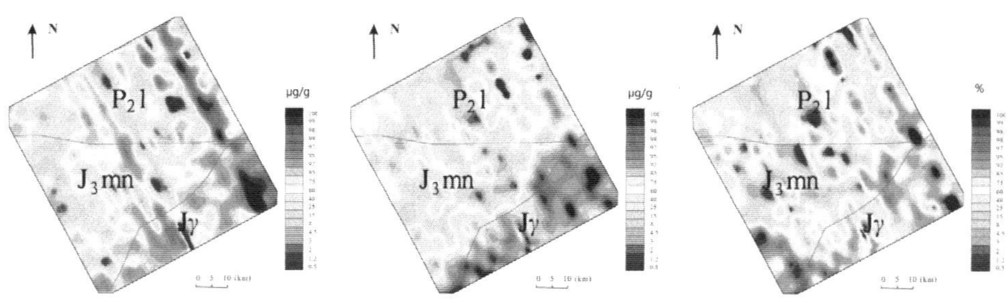

a b c

Figure 6. 30-point entropy averaging method (a,uranium; b, thorium; c, potassium)

It can be seen from Figure 4 to Figure 6 that when n=7 and n=15, the entropy averaging does not substantially eliminate the interference effect on the high, medium and low value regions of the airborne gamma-ray spectrum profile. Liu Jinghua (Liu, 2010). studied the ground gamma spectrum measurement and found that using the entropy averaging method can eliminate local interference well and has less influence on anomaly information when n=5. The reason for this difference is the difference in the measurement carrier. Since the airborne gamma-ray spectrum system records a full spectrum of data per second, and the aircraft is flying at a high speed (speed 70-120 m/s), the area covered per unit time is much larger than the area of ground gamma spectrum measurement, which results in the distribution density of the sampling points of the airborne gamma-ray spectrum measurement is much higher than density of the ground gamma ray spectrum. Therefore, airborne data reflects far more radioactive information than ground data. However, the average gamma-ray spectrum entropy window width is much negligible compared to the airborne gamma-ray spectrum data, which requires increasing the entropy averaging window width to increase the data capacity within the window. When the window width n is chosen to be 30, it is found that the anomaly range and lithology distinction are the most consistent with the geological background shown in Figure 1. Therefore, it is determined that n=30 is used as the entropy averaging method to extract the window width value of the airborne gamma-ray spectrum anomaly information.

4. Conclusion

The entropy averaging of airborne gamma-ray spectrum is an effective interference cancellation method, for which data can maintain the general trend of the content of the line elements after being processed by the entropy averaging method of different width windows. After repeated tests, it is determined that n=30 is used as the entropy averaging method to extract the window width value of the airborne gamma-ray spectrum anomaly information.

Acknowledgement

This work is supported by the National Natural Science Foundation of China (No. 41804114), National Key R&D Project (No.2017YFC0602100) and Scientific research project of jiangxi education department (GJJ170451)

References

[1] Chen Xin.2011.Application research of entropy average method in radioactive data processing[J], (02):13-15.

[2] Liu Jing-Hua et al. 2010. Study on the application of entropy average method in ground gamma spectrometry for local disturbance elimination, Journal of Geophysics –Chinese edition[J], 53(8). 1961-1967

[3] Xu Shi. 2011. Research on Ground Gamma Spectroscopy Measurement Random Interference Cancellation Method[D],21-30

[4] Jaynes ET. 1957. Information theory and statistical mechanics. Physical Review[J], 106:620–630.

[5] Sun Zhongren.2009. Study on the coefficient of variation for the statistical work of specimen magnetic parameters[J],45(1).65-69

The design of sports energy expenditure measuring device based on three-dimensional motion sensor

Chun Jiao[1], Ming Tang[2]

Xi'an Peihua University, Xi'an, China

Corresponding author and e-mail: Chun Jiao, 2024988445@qq.com

Abstract. The effectiveness of the measurement method based on accelerometer for human sports energy expenditure has been confirmed by research in recent years. In this paper, combined with the overall design requirements of micropower consumption and miniaturization, this paper describes the design scheme of a portable measuring device that uses three-dimensional accelerometer to measure sports energy expenditure. The general design idea of the measuring device and the implementation method of main functional units are discussed in detail.

1. Introduction

Daily physical exercise or training can improve human health. Moderate physical exercise can reduce the early mortality rate of adults by 20% to 30%, which can reduce the incidence of chronic diseases such as coronary heart disease, type 2 diabetes and stroke by 50% [1]. Exercise can also slow muscle decline, maintain bone health and promote child growth. In the process of exercise health management, whether it is to study the relationship between exercise and health or disease, or to conduct sports intervention or conduct personalized sports guidance research, objectively and accurately assess the energy consumption during human exercise is a key issue.

In recent years, with the development of microelectronics and computer technology, more and more researchers have begun to use accelerometers to study the assessment of sports energy expenditure. Researchers have demonstrated through a large number of scientific experiments that the integral value of body motion acceleration versus time is linear with energy or oxygen consumption. That is, the accelerometer output is highly correlated with the energy consumption [2][3][4][5]. This provides a specific theoretical basis for the use of accelerometers to assess human exercise energy consumption.

The three-dimensional accelerometer can simultaneously measure the vertical, front-back and lateral human motion accelerations, and has obvious advantages in sensing motion type and exercise intensity, and can more accurately reflect the actual movement of the human body. The reliability and validity of using three-dimensional accelerometers to assess energy consumption has also been confirmed by more and more studies [6-9].

2. The System Design

As shown in Figure 1 and Figure 2, the measuring device of sports energy expenditure is a battery-powered portable single chip application system. The measuring device can collect, process and store human motion information in the motion field, monitor the motion state, and calculate the sports energy consumption generated during the human body motion. At the same time, the sports intervention function can be completed by the control program and function programs built in the measuring device.

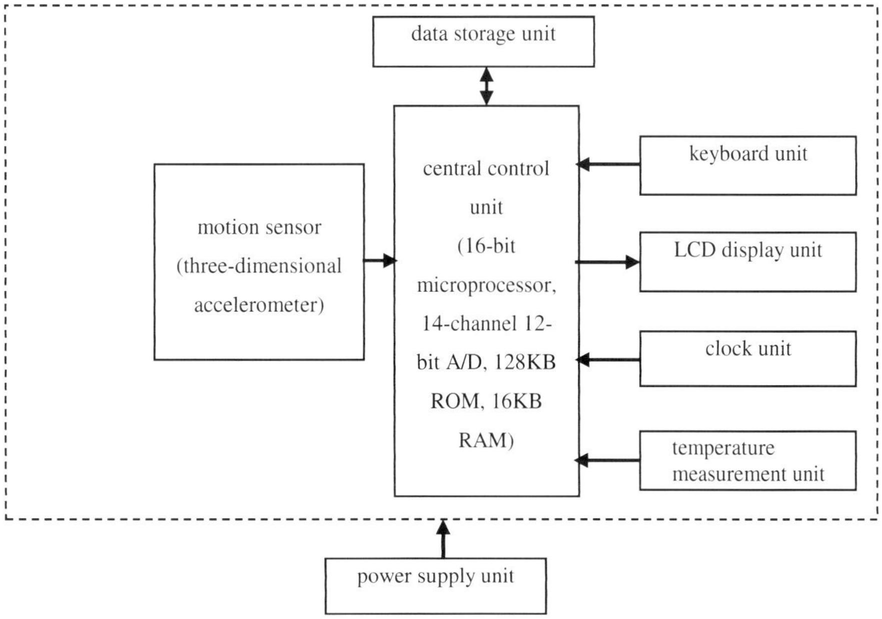

Figure 1. The hardware structure diagram of the measuring device

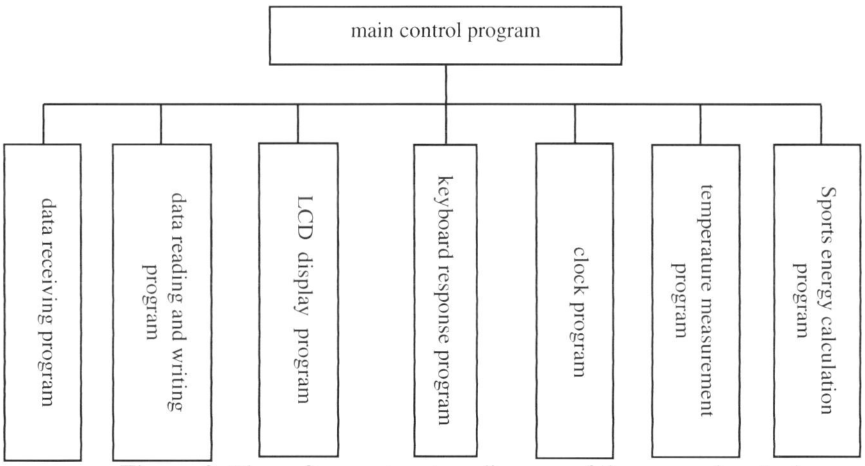

Figure 2. The software structure diagram of the measuring device

Since the measuring device needs to continuously monitor and record the motion information at the sports site, its field usability has high requirements on the overall micropower consumption design of the system. That is, the entire system should have low static power consumption and low, controllable dynamic power consumption to extend the use time of the measuring device. At the same time, the measuring device should be small in size and light in weight, which is convenient for the athlete to wear and does not affect the normal exercise. Therefore, micropower consumption and miniaturization design are important contents and difficulties of this system. Component selection, circuit design, and software design in the system should be closely related to micropower consumption and miniaturized design.

3. The Central Control Unit

The central control unit of the system uses TI's new 16-bit microcontroller MSP430F5419. Compared with the earlier models of the MSP430 series of microcontrollers, MSP430F5419 integrates more functional modules. The integration is further improved, the micropower characteristics are more prominent, the functions are more powerful, and the cost is significantly reduced.

MSP430F5419 integrates four independent serial interfaces, a 12-bit precision multi-channel A/D conversion module, 128KB of program memory and 16KB of RAM, enabling efficient data acquisition without the need for a dedicated A/D converter chip [10]. It has an RTC real-time clock module that provides available time coordinates for the system and provides reliable time stamp for the orderly storage of relevant data. The system does not require an additional clock chip, which improves system integration. MSP430F5419 also provides an on-chip CRC check module based on the CRC-CCITT standard, which provides technical guarantee for further improving the reliability of data storage and transmission.

MSP430F5419 features outstanding micropower consumption with five low-power modes that can be woken up from low-power modes in as little as 5μs. Its hardware and software architecture are designed for low-power applications. Its internal RAM controller can choose to retain RAM data or shut down RAM directly, depending on the needs of the different low-power modes. Each of its internal functional modules can be turned off by software operation when idle. At the same time, MSP430F5419 interrupts and subroutine calls have no hierarchical restrictions. It is beneficial to design the control program and function programs based on the interrupt structure, and further reduce the system power consumption from the software design level.

4. The Motion Sensor

ADI's ADXL345 chip is a motion sensor that monitors the three-dimensional acceleration of human motion. ADXL345 features threshold monitoring and motion acceleration measurement with a maximum range of ±16g and a bandwidth of 1.6KHz. It can output a 16-bit two-complement digital signal that can resolve tilt changes of less than 1.0° and can sense the presence or absence of motion [11].

Because ADXL345 has built-in A/D converter, it can directly output the digital signal in the complement form, which eliminates the signal conditioning unit that realizes the amplification, shaping and filtering functions in the system, which simplifies the system design and saves the board space. The amplitude range of the human body motion acceleration signal is -12g to +12g, and the frequency range is 0Hz to 20Hz. ADXL345 is selected as the three-dimensional acceleration sensor to meet the measurement range required by the system.

At the same time, the power supply range of ADXL345 is 2.0V to 3.6V, support SPI serial interface, so its power supply range and interface form can be matched with MSP430F5419, the two can be directly connected. As shown in Figure 3, when connected in SPI mode, both chips work in master-slave mode. MSP430F5419 is used as the master to provide the clock signal and ADXL345 is the slave.

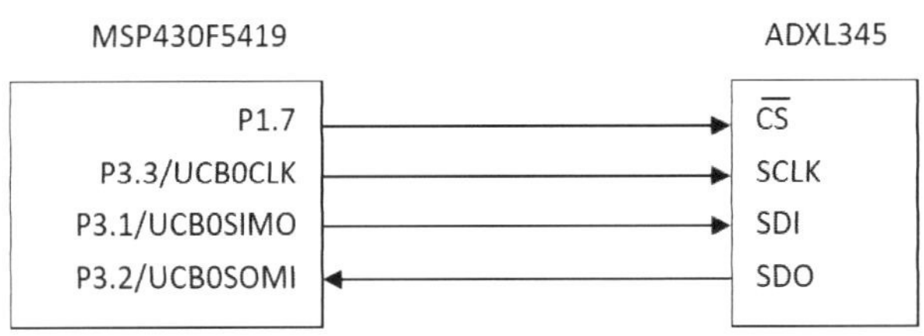

Figure 3. The connection diagram between MSP430F5419 and ADXL345

ADXL345 features outstanding micropower consumption. It consumes only 23μA in its measurement mode and quiescent current in standby mode is only 0.1μA. ADXL345 features a low power mode and a sleep mode. When static is detected, ADXL345 automatically switches to a low power, low sample rate mode. At the same time, ADXL345 integrates 32 FIFO buffers on-chip, with four modes of operation: bypass mode, FIFO mode, flow mode and trigger mode, which can effectively reduce the load of MSP430F5419. ADXL345 has a complete interrupt mechanism and provides two interrupt output pins, INT1 and INT2, which can be used to set the interrupt output in combination with different operating modes. In this system, ADXL345's low-power mode, interrupt mechanism, and interrupt-based control program and data receiving program are combined to achieve motion state monitoring and motion acceleration measurement with lower power consumption.

In order to further reduce the error and improve the system accuracy, in the sports energy calculation program, the three axial acceleration outputs of ADXL345 should be temperature compensated in combination with the actual ambient temperature value measured by the temperature measurement unit in the system.

ADXL345 is available in an LGA package that is optimized for miniaturization and its volume is only 3mm x 5mm x 1mm.

5. The Data Storage Unit

Considering storage capacity, power consumption, volume and cost factors, the system selects ATMEL's FLASH data storage chip AT25DF641. Its operating voltage is 2.7V to 3.6V, and the standby current is only 25μA. Its storage capacity reaches 8M bytes [12]. AT25DF641 supports SPI serial interface, which saves space and reduces switching noise during high-speed data transmission.

Since the supply voltage and interface form of AT25DF641 can be matched with MSP430F5419, the two chips can be directly connected. When connected using the SPI serial interface, both chips work in master-slave mode. MSP430F5419 is used as the master to provide the clock signal, and AT25DF641 is the slave. In order to further increase the storage capacity, the system simultaneously connects two AT25DF641 chips through one SPI serial interface. The enable pins of each AT25DF641 chip are controlled by different pins of MSP430F5419 to realize time division multiplexing of two AT25DF641 chips.

The flow chart of the data storage program is shown in Figure 4.

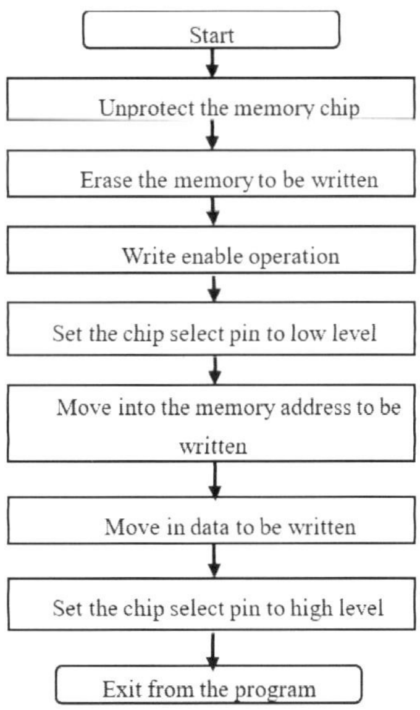

Figure 4. The flow chart of the data storage program

This system uses the MISTY1 encryption algorithm to design a special data encryption program. The information segment of each frame of data can be data-encrypted, and the information segment and the data segment of each frame of the stored data are subjected to CRC check, which further increases the reliability and security of data storage.

The flow chart of the data encryption program is shown in Figure 5.

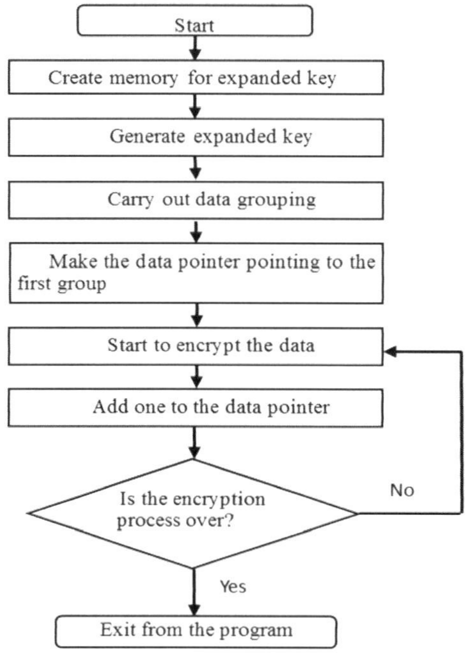

Figure 5. The flow chart of the data encryption program

6. The Power Supply Unit

The power supply unit is an important part of the hardware design of the system. The output quality of the power supply unit will directly affect the accuracy and reliability of the system.

In this system, the design of the power supply unit must meet the following requirements:

(1) It can output high quality power supply voltage. The output ripple is small, the electromagnetic radiation is low, and the voltage regulation effect is good;

(2) The input voltage range is wide. It can meet a wide range of changes in battery voltage from the input;

(3) The power supply efficiency is high. It can effectively reduce unnecessary energy loss;

(4) Miniaturization. There are few external components and no large capacitance or large inductance.

The power supply unit of this system uses Maxim's DC-DC converter MAX1556. Its quiescent current is only 16μA, its output current can reach 1.2A, and its conversion efficiency can reach 97% [13]. The 1.8V, 2.5V, 3.3V or adjustable output voltage can be selected via the pin. Its input voltage range is 2.6V to 5.5V, which is very suitable for powering the system with lithium ion battery or 2 to 3 alkaline batteries. MAX1556 is also available in a tiny 10-pin TDFN (3mm x 3mm) package.

7. Conclusions

In this paper, with the design principle of micropower consumption and miniaturization, the design scheme of the measuring device for assessing sports energy expenditure based on three-dimensional motion sensor is expounded. The measuring device can collect, process and store human motion information in real time on the sports site, and can quantitatively evaluate the human sports energy consumption. It is an intelligent instrument suitable for family or individual sports health management.

Acknowledgement

This work was supported by Shaanxi Natural Science Fund Project (2018JM7037025).

References

[1] Department of Health , Physical Activity , Health Improvement and Prevention. At least five a week: Evidence on the impact of physical activity and its relationship to health. 2004.
[2] Meijer, G.A., Westerterp, K.R., Koper, H. Assessment of energy expenditure by recording heart rate and body acceleration[J]. Medicine and Science in Sports and Exercise,1989,21:343-347.
[3] C.V.C. Bouten, K.T.M. Koekkoek, M. Verduin. A triaxial accelerometer and portable data processing unit for the assessment of daily physical activity[J]. IEEE Trans. Biomed. Eng.,1997,44:136-147.
[4] Scott E., Crouter•James R, Churilla. Estimating energy expenditure using accelerometers[J]. Eur J Appl Physiol,2006,98:601-612.
[5] Troiano, Richard P., Berrigan, DAVID. Physical Activity in the United States Measured by Accelerometer[J]. Medicine & Science in Sports & Exercise, 2008,40:181-188.
[6] Juliette Hussey, Kathleen Bennett, et al. Validation of the RT3 in the measurement of physical activity in children[J]. Journal of Science and Medicine in Sport,2009,12:130-133.
[7] Meredith A. Perry, Paul A. Hendrick, et al. Utility of the RT3 triaxial accelerometer in free living: An investigation of adherence and data loss[J]. Applied Ergonomics,2010,41:469-476.
[8] Tanaka, Hidenori; Kawamata, Tomoyuki; et al. Evaluation of the Physical Activity of Anesthesiologists in the Operating Room During Daily Work Using a Triaxial Accelerometer[J]. Archives of Environmental & Occupational Health, 2015.70:77-80.
[9] ROWLANDS, ALEX V.; YATES, THOMAS; et al. Sedentary Sphere: Wrist-Worn Accelerometer-Brand Independent Posture Classification[J]. Medicine & Science in Sports & Exercise,2016.48:748-754.
[10] TEXAS INSTRUMENTS. MSP430F543x, MSP430F541x Mixed-Signal Microcontrollers. TEXAS INSTRUMENTS data sheet, 2018.
[11] Analog Devices,Inc. 3-Axis,±2g/±4g/±8g/±16g Digital Accelerometer ADXL345. Analog Devices data sheet, 2009.
[12] Atmel Corporation. 64-Mbit, 2.7V Minimum Serial Peripheral Interface Serial Flash Memory AT25DF641. Atmel datasheet, 2010.
[13] Maxim Integrated Products. 16μA IQ,1.2A PWM Step-Down DC-DC Converters MAX1556/MAX1556A/MAX1557. Maxim data sheet, 2011.

Full multi-dimensional dynamic comprehensive optimization stochastic simulation is the sustainable trend of future petroleum reservoir description

Xiaonan Xiao

Xiamen University Tan Kah Kee College, Xiamen, 363105, China

Corresponding author and e-mail: Xiaonan Xiao, xiaoxn@xujc.com

Abstract. Being an important and sustainable approach in future petroleum reservoir description as well as a new frontiers subject in information optimization decision making of today's oil/gas field development, multi-information stochastic dynamic comprehensive optimization simulation bears profound research significance and broad application prospect. The present paper is an introduction to theoretic performance and characteristics of multi-dimensional stochastic dynamic comprehensive optimization simulation, to classification, steps of model establishment, frequently-adopted approaches and problems to be considered in stochastic simulation. The paper, based on the author's years of profound study and practice, also indicates the trend and latest application progress of stochastic simulation.

1. Theoretical performance and characteristics of full-dimensional dynamic comprehensive optimization stochastic simulation

Scientific and rational petroleum reservoir description is an important part of scientific development and reconstruction of oil and gas reservoirs. Its purpose is to deepen the understanding of oil and gas reservoirs through dynamic comprehensive analysis and prediction of oil and gas reservoirs based on scientific theories, methods and optimization methods. Furthermore, the adjustment and perfection of oilfield development plan should be done well to develop oilfields scientifically and rationally from the beginning.

In the future, most of China's onshore oil fields will enter a high water-cut period, and some will enter a very high water-cut period. Moreover, with the deepening of the degree of exploitation, the relationship between underground oil and water will become more and more complex, and it will be more and more difficult to stabilize and adjust the production potential of oilfields. However, at present, the proportion of reserves and production of old oilfields still accounts for more than 70% of the total land reserves and production. Therefore, how to make a deep petroleum reservoir description of the old oil fields in the future and find out the relative enriched areas of the remaining oil to the maximum extent has become the primary task of the future petroleum reservoir development research [1]. However, in petroleum reservoir description, especially in the study of reservoir heterogeneity and spatial distribution of petroleum reservoir attribute parameters, it is found that under the condition of given limited data, the models established by traditional interpolation methods often have been limited without considering the spatial correlation of the estimated values. Therefore, in the future reservoir description, special attention should be paid to the overall dynamic correlation and dynamic comprehensive optimization of the results, and the reservoir property space should be described in a comprehensive and dynamic manner. In order to overcome the shortcomings of traditional methods, we should pay special attention to the stochastic simulation method and means of multi-dimensional dynamic synthesis optimization under the condition of insufficient information of future petroleum reservoir description and distribution of underground reservoirs.

We should conduct in-depth search, excavation, dynamic filtering and optimization of many complex information flows in the objective reality of underground oil and gas reservoirs, and then obtain a series of practical random models for accurate description of all-round multi-information reservoirs and oil and gas. Therefore, it provides a new theory, method and approach of intelligent full-dimensional simulation for stochastic information optimization modeling and computer simulation application software development of various complex petroleum reservoirs. So people will have a deeper understanding of petroleum reservoirs. Thus because we can determine the distribution of oil and gas reservoirs and their remaining oil more effectively in this way, the well pattern development mode can be determined more reasonably and the maximum development benefit can be obtained with less investment. The above-mentioned in-depth research on the series optimization theory, method, model and intelligent application software development of petroleum reservoir description in the future will play a very positive role in promoting the development of the future dynamic development analysis of oilfields and the industrialization of digitalization, informatization, computerization and decision-making analysis of oil and gas field benefits by vigorously tapping the petroleum reservoir potential and improving the overall recovery of various types of reservoirs [2].

2. Classification, steps and methods of the full multi-dimensional dynamic comprehensive optimization stochastic simulation

The full multi-dimensional dynamic synthesis optimization stochastic simulation is based on in-depth search and full exploitation of various complex information flows, and uses modern stochastic analysis, modern information quantification analysis and stochastic simulation in an all-round and multi-perspective manner to establish an optional A stochastic dynamic comprehensive optimization model with equal probability and high precision reflecting the spatial distribution of variables.

2.1. Classification of the full multi-dimensional dynamic comprehensive optimization stochastic simulation

According to different classification criteria, stochastic simulation of full multidimensional dynamic synthesis optimization has different classification. According to whether the simulation results are faithful to the constraints, stochastic simulation of full multi-dimensional dynamic synthesis optimization can be divided into conditional simulation and unconditional simulation. The model generated by unconditional simulation only needs to reproduce the correlative structure of spatial dynamic distribution of reservoir parameters, while the model generated by conditional simulation requires not only to reproduce the correlative structure of spatial dynamic distribution of reservoir parameters, but also to be consistent with the known well observation position.

The stochastic simulation of the full multi-dimensional dynamic comprehensive optimization stochastic simulation can also be divided into discrete and continuous types. The discrete stochastic simulation of full multi-dimensional dynamic synthesis optimization mainly describes the geological characteristics of discrete nature in an all-round way. The continuous stochastic simulation of full multi-dimensional dynamic synthesis optimization mainly describes the spatial distribution of continuously changing reservoir parameters in an all-round way.

2.2. Steps of the full multi-dimensional dynamic comprehensive optimization stochastic simulation

The discrete stochastic simulation of the full multi-dimensional dynamic comprehensive optimization stochastic simulation is closer to geological dynamic interpretation and more suitable for reservoir dynamic simulation of discontinuous and large-scale heterogeneity, while the continuous stochastic simulation of full multi-dimensional dynamic synthesis optimization is more suitable for spatial dynamic distribution simulation of continuous rock physical parameters. Therefore, the steps of stochastic modeling should include two stages of simulation of two models:

In the first stage, the full multi-dimensional dynamic comprehensive optimization stochastic simulation method is used to optimize stochastic simulation of large-scale heterogeneity related to geology and sedimentary based on omni-directional seismic data and logging data [3].

In the second stage, on the basis of the simulation results in the first stage, the full multi-dimensional dynamic comprehensive optimization continuous simulation method is used to simulate all kinds of physical parameters of different sedimentary belts [4-5]. At this stage, the parameters of different sedimentary zones are different, so we should determine the variation of variables and the relationship between variables and spatial structure based on core data. Thus we obtain the relevant dynamic simulation parameters. In summary, the first stage of simulation provides important geological morphology and spatial structure, while the second stage of simulation provides the dynamic change law of rock physical parameters in a small range.

2.3. Methods of the full multi-dimensional dynamic comprehensive optimization stochastic simulation
Stochastic simulation of the full multi-dimensional dynamic comprehensive optimization provides many important means and methods with great development prospects for reservoir description in the future. These methods include Boolean dynamic random simulation, sequential Gauss random simulation, sequential indicator random simulation, truncated Gauss random simulation, probability field dynamic random simulation, dynamic random annealing simulation, fractal dynamic random domain simulation, Markov dynamic random field simulation, and mosaic process dynamic random simulation, etc. All the above methods have their own applicable conditions and suitable types. In practical application, we should choose according to the specific conditions.

2.4. Problems to be considered in the full multi-dimensional dynamic comprehensive optimization stochastic simulation

2.4.1. The problem of the main heterogeneity of reservoirs in the full multi-dimensional dynamic synthesis optimization stochastic simulation.
In the full multi-dimensional dynamic comprehensive optimization stochastic simulation, the most influential heterogeneous feature must be considered first. If the phenomenon studied is a mixture of different objects and statistical population, the geometry should be simulated first. Then the distribution of reservoir attributes parameters is simulated in geometric form.

2.4.2. The problem of Joint Simulation of Multiple Variables
In many cases, it is important to reproduce the spatial interdependence of several variables. In stochastic reservoir simulation, the relationship between porosity, permeability and oil saturation is often considered. Most simulation methods based on stochastic functions can be used to simulate several variables jointly. However, due to the complexity of practical problems, it is difficult to deduce and simulate cross-covariance on computer. Usually, in practical applications, it is necessary to simulate the important or the most relevant variables (principal variables) first, and then sample them with conditional distribution.

2.4.3. Ergodicity problem in the fully multi-dimensional dynamic synthesis optimization stochastic modeling
The purpose of stochastic modeling for the full multi-dimensional dynamic comprehensive optimization is to generate statistics that reflect and fit the original data. The key to this problem is to simulate the degree to which the statistical loyalty entity realizes the original data statistics. The given model statistics are obtained from the sample statistics. The limited sample size often makes these statistics rough. Therefore, it is not necessary and impossible to reproduce each statistic accurately through each implementation. The difference between the simulated statistics and the statistic of a given model is called "ergodic bias". A random function is said to be ergodic if its corresponding statistics tend to a certain value when the study area becomes larger. Therefore, as long as the random function is stationary and ergodic, and the simulation field is large enough, then any realized statistics of the random function can accurately reflect the parameters of the model.

2.4.4. Selection of the full multi-dimensional dynamic synthesis optimization stochastic model
When we build the selection criteria of the full multi-dimensional dynamic synthesis optimization stochastic model, we should pay attention to the following factors. First, the realized statistics should be as close as the model statistics. Second, how is the subjective aesthetic evaluation of the entire model? Third, can we optimize an implementation that reflects the data and features of the model that was not imported before the simulation? Fourth, Can we dig all kinds of complex information that can truly reflect the objective reality of underground oil and gas reservoirs? Whether it is from a subjective aesthetic or based on actual data, the choices made must be based on actual data. In this way, the research phenomenon can be better reflected.

3. New progress in application of the full Multi-dimensional dynamic comprehensive optimization stochastic simulation
The full-dimensional dynamic comprehensive optimization stochastic simulation has become a new and important means of petroleum reservoir description. Compared with the traditional interpolation method, it has great advantages and has been widely used in petroleum reservoir description [6].

In recent years, the introduction of some new theory and methods (such as fractal theory, genetic coefficient method, neural network technology, etc.) not only enriches the stochastic simulation theory, but also broadens the application range of stochastic simulation and improves the application effect of stochastic simulation. With the development of reservoir description technology, the continuous improvement of geostatistics and the development of computer technology, the full-dimensional dynamic comprehensive optimization stochastic simulation method has played an increasingly important role in petroleum reservoir description.

In fact, the full Multi-dimensional dynamic comprehensive optimization stochastic simulation in petroleum reservoir description has become a new frontier cross-cutting subject in information optimization decision-making in oil and gas field development. So the author has made a thorough study of this topic. In his study, for the first time, a series of modern information quantification methods and stochastic information optimization methods have been fully used [7]. He creatively proposed a complete information optimization decision-making idea. According to the characteristics of many types of oil reservoirs and complex structures in China's onshore oilfields, he studied how to extend the information flow in the general information space of oil and gas reservoirs in onshore high water cut oilfields to the non-homomorphism information flow in the generalized measure and functional information space [8-9]. The author made full use of modern information processing and quantification methods such as omnidirectional multi-information stochastic dynamic optimization analysis theory and method, stochastic control theory and method, stochastic signal optimal filtering theory and method, optimal information entropy method and multi-factor Fuzzy information search method [10-12]. He conducted in-depth search, mining, dynamic filtering and optimization research on many complex information flows in the objective reality of deep stratigraphic oil and gas reservoirs. He provides a batch of omni-directional and multi-information optimization decision models and intelligent application optimization software for further improving the multi-factor and multi-information optimization method in petroleum reservoir description and continuously enhancing the comprehensive information decision-making level and benefit in oil and gas field development.

This research provides stochastic simulation and application optimization software for distribution description of underground petroleum reservoirs, theoretical description and method of remaining oil distribution, and scientific evaluation method of oil and gas reservoirs, etc. After being adopted by Changqing Petroleum Exploration Bureau, Liaohe Petroleum Exploration Bureau, Central Plains Petroleum Exploration Bureau, Yanchang Oil Mine Administration Bureau, Jilin Oilfield Branch, Xinjiang Oilfield Branch, Yumen Oilfield Branch and Northwest Sinopec Company, remarkable economic benefits have been achieved. According to incomplete statistics, the annual cumulative production of crude oil will be increased 191170 tons and natural gas will be increased 7009×10^4 for our country. It can be predicted that the in-depth study of the full Multi-dimensional dynamic comprehensive optimization stochastic simulation will provide a series of new theories, new means, new methods and new models with far-reaching significance and broad application prospects for the sustainable development of reservoir description in the future.

References

[1] Mosisa Aga. Edgeworth Expansion of the Parametric Bootstrap tstatisic for Linear Regression Processes with Strongly Dependent Errors [J].Journal of Statistical Theory and Applications,2015,14(1):52-59.

[2] Obiora-Ilouno H O,Mbegbu J I.A Bootstrap Approach to Error-Reduction of Nonlinear Regression Parameters Estimation [J].IOSR Journal of Mathematics (IOSR-JM),2015,11(1):45-51.

[3] Bello Oyedele Adeshina,Bamiduro,Timothy Adebayo.Bootstrap Nonlinear Regression Application in a Design of an Experi-ment Data for Fewer Sample Size [J]. International Jounal Of Research (Ijr),2015,2(2);1425-1441.

[4] Anwer Fitrianto,Ng Mei Cing. Empirical Distributions of Parameter Estimates in Binary Logistic Regression Using Bootstrap [J].Int.Journal of Math. Analysis,2014,8(15);721-726.

[5] Lorelied A Santos,Erniel B Barrios.Small Sample Estimation in Dynamic Panel Data Models;A Simulation Study[J].American Open Journal of Statistics,2011(1):58-73.

[6] Andreja Radman Matija Gredicak, Ivica Kopriva. Predicting Antitumor Activity of Peptides by Consensus of Regression Models Trained on a Small Data Sample [J]. International Journal of Molecular Sciences, 2011(12): 8415 - 8430.

[7] Rui wang, Stephen w Lagakos . Robert j Gray. Testing and interval estimation for two-sample survival comparisons with small sample sizesand unequal censoring[J].

Journal of Physics: Conference Series **1423** (2019) 012049 doi:10.1088/1742-6596/1423/1/012049

Biostatistics ,2010, 11(4): 676-692.
[8] Park J,Shintani M.Testing for a unit root against transitionalautoregressiv emodels[J]. Vanderbilt University Department ofEconomics Working Papers, 2010.53-62.
[9] Choi C, MohY. How useful are tests for unit-rootin distinguishingunit-root process from stationary but non -linear process[J].Econometrics Journal ,2007(10) :82-112.
[10] Rodrigues P M. Properties of recursive trend-adjusted unit root tests [J]. Economics Letters ,2006 (91):413 - 419.
[11] Caner M, Hansen B E. Threshold Autoregression with a Unit Root [J]. Econometrica 2001(69):1555-1596.
[12] Efron Bradley. bootstrap methods: another look at the Jack Knifi[J]. the annals of statistics , 1979,7(1):1-26.

Journal of Physics: Conference Series **1423** (2019) 012050 doi:10.1088/1742-6596/1423/1/012050

Design and implementation of a high-precision turbidity measuring instrument of a mine

Xueming Wang [1, 2], Baoyuan Chen[1], Libin Zhang[1, 2]

[1] CCTEG Changzhou Research Institute, Tiandi (Changzhou) Automation Co., Ltd. Changzhou, China, 213015
[2] Tiandi (Changzhou) Automation Co., Ltd., Changzhou, China, 213015

Corresponding author and e-mail: Xueming Wang, wangxueming@cari.com.cn

Abstract. In this paper, the principle of turbidity detection based on light scattering principle is expounded, and a design method of high-precision turbidity measuring instrument in coal mine is proposed. The software and hardware design of such turbidity measuring instrument are introduced. The measuring instrument adopts integrated circuit design such as imported photodiode and high-precision operational amplifier. The MATLAB simulation software tool is used to formulate the curve of the turbidity measurement data, and the turbidity measurement value is compensated and corrected to improve the turbidity measurement accuracy. Furthermore, the measuring instrument adopts low-power design, the OLED display technology in particular, to replace the traditional digital tube or LCD liquid crystal; the anti-interference design is applied in the digital interface communication design, and the electromagnetic compatibility test is very suitable for the electromagnetic environment in the coal mine

1. Introduction

Turbidity refers to the degree of hindrance of the solution to the passage of light, and is an important parameter that can indirectly reflect the use of equipment in the production process of water companies [1]. Especially in China's coal mining enterprises, the turbidity of mine water is generally large, which makes the management of mine water treatment difficult, and can not guarantee the stability of the effluent, which makes it more difficult to reuse the mine water. With the continuous improvement of coal mine automation level in recent years, underground hydrological monitoring, on-line monitoring of underground water quality, on-line monitoring of mine water quality, on-line monitoring of downhole equipment/cooling water, etc. [2] have been introduced, so a mining turbidity is needed. The detection technology detects the turbidity of underground fire water, downhole equipment cooling water, and downhole spray water, thereby indirectly understanding the use of equipment and adopting different water treatment measures to achieve remote monitoring or further improve the level of automation management.

There are many methods for detecting turbidity, such as transmitted light, ratio method and scattered light type. The transmitted light type has poor detection accuracy for low turbidity concentration; the ratio method is to combine the projection and scattering methods, but the structural design. It is more difficult to implement; the scattering method is relatively easy to implement, and the measurement linearity is good. It is the preferred measurement method for high-precision performance products abroad. This paper mainly uses the light scattering principle detection method. In order to ensure the design accuracy, the imported high stability is selected. As a light source, the high-intensity op amp and analog-to-digital converter are used to amplify the micro-signal to improve the measurement accuracy. In software data processing, the MATLAB software tool is used to fit the measurement data curve to further improve the turbidity. The measurement accuracy is compared with the American HACH and the domestic mainstream brand turbidity meter. The designed turbidity measuring instrument has high measurement accuracy, good stability and electromagnetic compatibility, and rich output interface. It is suitable for use in the complex electromagnetic environment of coal mines. .

1 optical turbidity detection principle

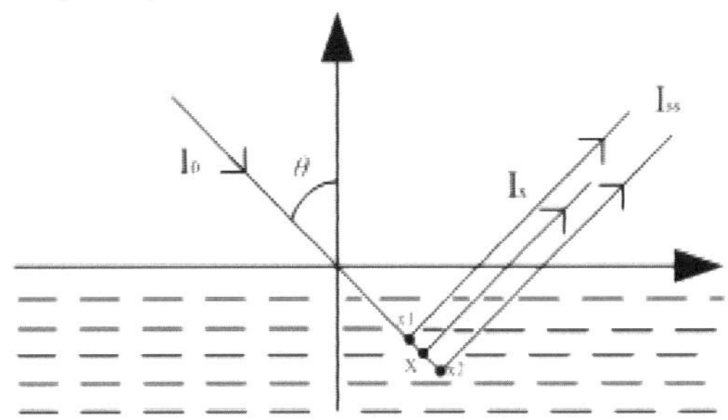

Figure 1. Block diagram of light scattering turbidity measurement

As shown in Fig. 1, a light source is disposed in the turbidity optical detecting probe, and a silicon battery photodetector is mainly used for receiving 90° scattered light. The turbidity probe is infiltrated in the solution to be tested, the surface of the lens is in contact with the solution to be tested, and the light source emits a fine parallel light having a light intensity of I0 into the solution [3], and has an angle θ with the Y axis, which is reached according to the Beer-Lambert law. The light intensity Ix at B is:

$$I_X = I_0 e^{-K_1 Tx\sin\theta} \tag{1}$$

(1) where T is the turbidity value, I0 is the light intensity of the light source emitted by the light source, and K1 is the absorption coefficient. The light hits the particle and emits a plurality of scattered light, wherein 90° of the scattered light is received by the photodetector through the light decay, and the total scattered light Iss at x1~x2 is:

$$I_{SS} = K_{SS} I_0 T \Delta x \tag{2}$$

Where Kss is the scattering coefficient, $\Delta x = x_2 - x_1$

It can be seen that the scattered light intensity Iss is proportional to the turbidity T and Δx, and the measured turbidity value in the water can be measured indirectly by measuring the scattered light intensity.

2. Design of the mine water quality turbidity measuring instrument

The light scattering turbidity detecting and measuring instrument mainly receives the light intensity signal through the front end silicon battery, and the silicon battery mainly functions to convert the received light intensity value into a collectable electric signal, and the electric signal realizes signal amplification processing through the precision signal conditioning circuit. The amplified electrical signal enters the analog-to-digital conversion chip (A/D) and is converted into a digital signal. The digital bus is used to read the data between the microprocessor and the A/D chip, and the microprocessor calculates and compensates the collected data. Calculate and display the final calculated result on the liquid crystal.

2.1. The hardware design of the Water quality turbidity measuring instrument

The hardware design of the turbidity meter mainly includes the following parts: photoelectric signal conversion and acquisition conditioning circuit, display and communication circuit, power supply circuit, microprocessor and peripheral circuit.

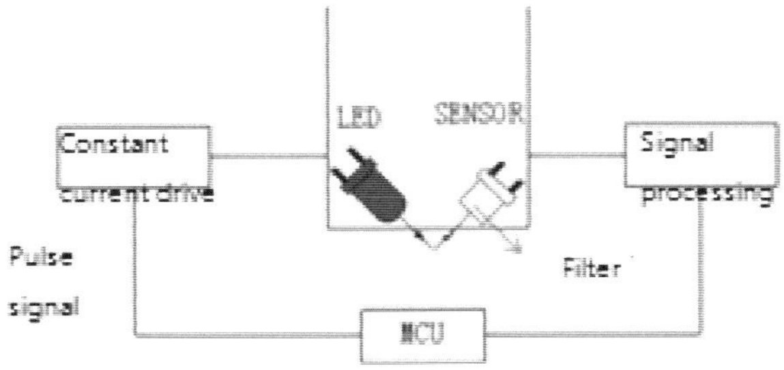

Figure 2. Turbidity meter hardware design block diagram

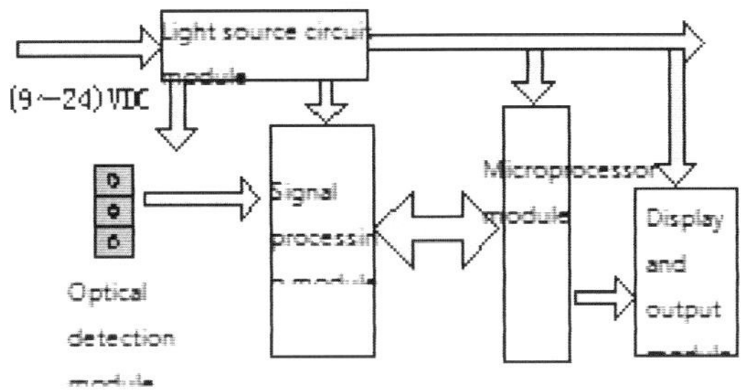

Figure 3. Main components of the turbidity meter hardware module

Design of light source and signal conditioning circuit

(1)Light source circuit design

The infrared light emitting part adopts the imported infrared LED reflective diode with a wavelength of 860 nm. The light intensity of the light emitting diode is linear with the driving current [4]. The magnitude and stability of the light intensity directly affect the strength of the received signal, and the measurement accuracy of the measuring instrument. The influence is large, so the high-precision constant-current driving circuit is used in the design. In order to ensure the stability of the reference voltage, the micro-power high-precision voltage reference chip LM285-2.5V is selected, and the temperature drift coefficient of the chip is 20ppm [5], which greatly improves the design. The stability of the voltage reference with environmental fluctuations improves the accuracy of the constant current source and reduces the influence of temperature on the turbidity measurement error.

(2) Signal conditioning and processing circuit design

As shown in FIG. 4, the silicon photodiode D13 converts the received optical signal into a weak electrical signal [6]. The diode is designed to operate in a light guide mode and a zero bias state, and convert the received light intensity signal into a photocurrent signal. It can effectively prevent the reverse bias current, meet the weak light intensity signal conversion, improve the turbidity measurement accuracy, and there is a certain drift and error in the data amplification process. In order to improve the measurement accuracy, a compensation capacitor is designed in the design process. C1 reduces drift and suppress current noise.

The latter stage modulates the weak signal, mainly using a current (I) to voltage (V) conversion circuit. Since the generated photocurrent signal is very weak, normally between pA and μA, the latter integrated operational amplifier chip is required to have Very low bias current, ultra-low noise, high stability [7], this paper uses a high-precision operational amplifier TLV2372 chip as a signal conditioning amplifier chip [8], the photo-diode generated weak photocurrent signal through the TLV2372 The chip realizes high-precision signal amplification of the current signal to the voltage signal, and the amplified signal voltage amplitude is between 0 and 150 mV, and the voltage signal is converted and processed by the ADS1100 analog-to-digital converter of the PGA programmable amplifier integrated in the latter stage. The microprocessor interacts with the A/D chip through the I2C serial bus, as shown in Figure 4.

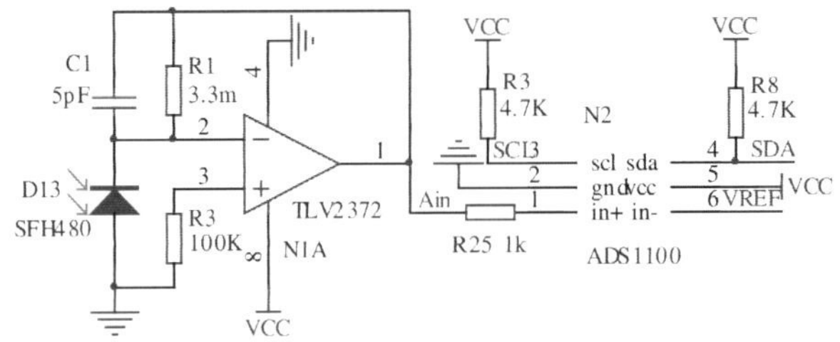

Figure 4. Silicon cell optical signal receiving conversion circuit

The A/D analog-to-digital converter in the circuit uses a 16-bit self-calibrating high-precision analog-to-digital converter ADS1100, which supports software programmable maximum 8 times gain signal amplification [9]. The internal amplifier amplifies the pre-stage differential signal and then the voltage signal. Adjusted to the A/D acquisition allowable range, the effective measurement accuracy of the voltage measurement by the experimental test converter can reach more than 12 bits. The ADS1100 chip based on serial I2C digital bus communication has strong communication anti-interference ability [6]. The microprocessor uses the average and median double filtering algorithm to process the read A/D chip data, and then performs software compensation on the measurement result. .

(3)Display and communication circuit design

In order to meet the special environmental requirements of poor underground coal mine light, the measuring instrument adopts self-illuminating OLED dot matrix liquid crystal screen, and adopts Qingda photoelectric 128×64 dot matrix liquid crystal screen HGS128647, which has built-in SSD1305 controller [10], through serial The bus realizes data reading and writing for the microprocessor, and the implementation is simple and reliable. The operating temperature range of the liquid crystal display is wider than that of the conventional ordinary liquid crystal, and the working environment temperature range is -40+85 ° C. Since the self-luminous display is adopted, the display needs to be lit. The dot matrix is energized, the other parts are not powered, and the power consumption of the liquid crystal is about 30 mA when the LCD is normally displayed. The low power consumption design is very suitable for the intrinsically safe power supply and transmission requirements of the coal mine.

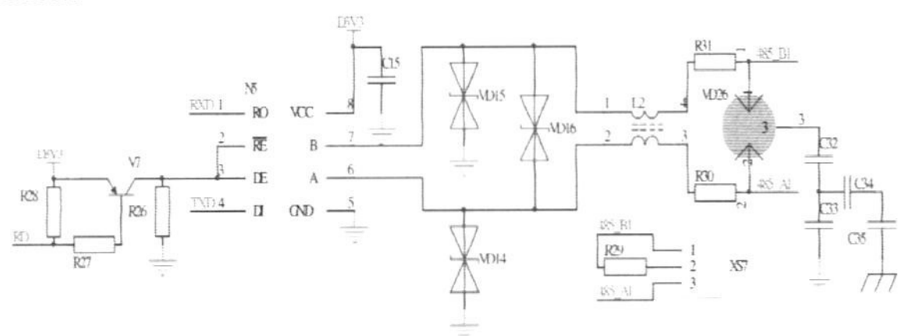

Figure 5. RS485 bus interface circuit

The RS485 communication bus interface circuit of the measuring instrument adopts the MAX3072 chip design [11]. As shown in Figure 4, the V7 triode circuit in the circuit is mainly to let the measuring instrument be in the state of receiving data when it is powered on, in order to improve the anti-electromagnetic interference capability of the bus communication and reach the coal mine. The level 2 electromagnetic radiation required by the measuring instrument design, the 2-stage pulse group resistance and the DC power supply and the signal port level 2 surge (impact) immunity test, to achieve reliable protection of the interface, select three TVS on the bus A, B line The SMBJ17CA implements transient voltage suppression clamp protection, and uses a common mode choke coil to enhance the suppression of common mode interference. The discharge tube is mainly used for overvoltage, lightning and electrostatic protection [12].

The current output interface of the measuring instrument adopts the special current output chip AD5410 integrated circuit design, and the sound and light alarm circuit adopts the dual triode drive control design, as shown in Fig. 6.

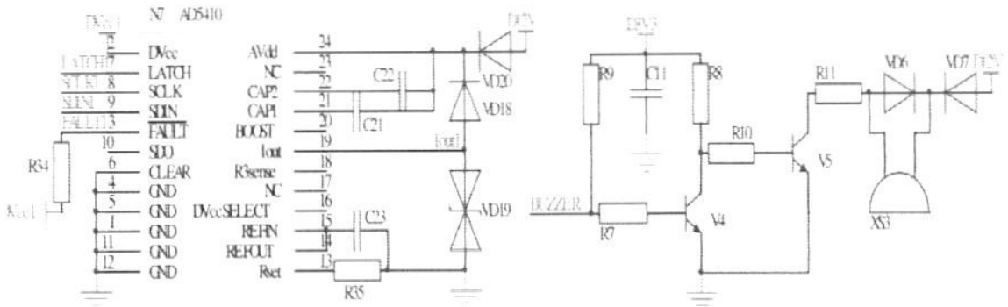

Figure 6. Current output, alarm control circuit

(4)Microprocessor and peripheral circuit design

The measuring instrument microprocessor adopts C8051F340, which is the mixed signal system core SoC [13] of Cygna Company of USA. It is packaged in a small OEM 48-pin LQFP with a supply voltage range of 2.7-3.6 V. This design is powered by a 3.3V power supply. . At the same time, the processor chip has high integration, and the I/O port supports programmable adjustment. It can be adjusted according to actual conditions during PCB layout. The processor supports online simulation and debugging, supports C2 simulation interface and offline program download, and is convenient to use. . The internal FLASH of the processor integrates the power-down storage rewritable data storage unit [14]. In order to ensure reliable storage data, the peripheral circuit adopts EEPROM memory chip CAT24C02 (N6) to realize double backup of stored data to ensure reliable and secure stored data. .

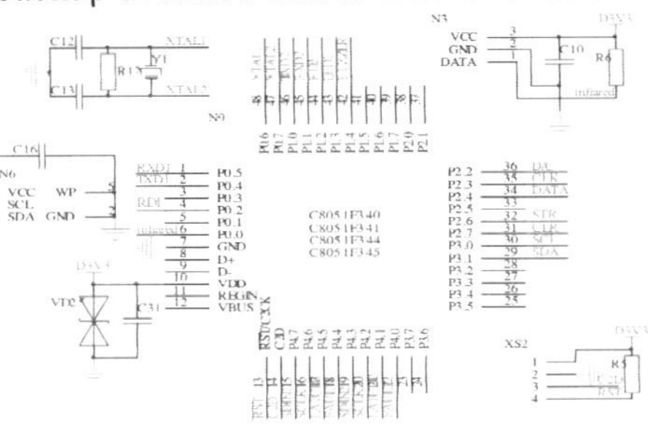

Figure 7. Microprocessor and peripheral interface circuit

2.2. The software design of the Water quality turbidity meter

The software program design of turbidity meter mainly includes the following aspects: 1) basic parameter configuration of single chip system; 2) A/D analog-to-digital converter configuration, signal acquisition and filtering; 3) data compensation and processing calculation (MATLAB simulation) The calculation formula performs curve fitting on the filtered data; 4) OLED liquid crystal display communicates with the bus; 5) signal and alarm output. The main software flow chart is shown in Figure 8.

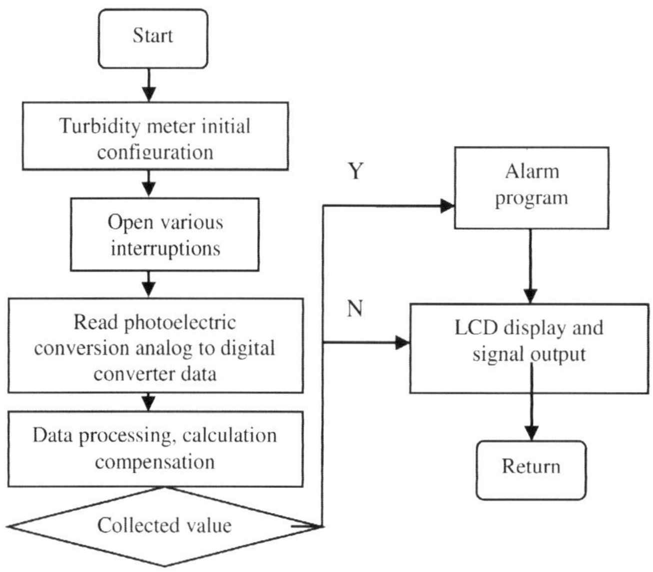

Figure 8. Overall software design block diagram of the turbidity meter

3. Measurement data analysis and MATLAB curve fitting

The turbidity measuring instrument designed in this paper is configured to calibrate and verify the measurement accuracy and error by configuring the standard turbidity solution of Formazin. Six kinds of concentration solutions are arranged in the range of 0-1000 NTU, and 0 and 1000 NTU are used before the measurement. The solution is initialized to two points of the measuring instrument, and then the measured measuring instrument is placed in other concentration solutions to measure the data. Under ideal conditions, the light intensity signal should be linear with the turbidity concentration. According to the two-point calibration method, the linear relationship is calculated as: y= x/15.004 (where x is the intensity of the light intensity signal and y is the turbidity concentration). The measurement error is found in the experiment, the maximum measurement error is at 400NTU concentration point, and the maximum error value is 44.68NTU. In order to improve measurement accuracy and reduce measurement error, the received light intensity signal value and standard are obtained by MATLAB software tool [15]. The relationship curve of the solution concentration value is fitted, and the light intensity signal value data corresponding to the concentration of the standard solution is corrected and compensated, and the fitting formula is y= 0.00000064134x2+0.057443x-1.98 (where x is the intensity value of the light intensity electric signal, y is Turbidity concentration), the internal processor of the turbidity meter is processed according to the compensation relationship, and then the measuring instrument designed in this paper is compared with the American HACH turbidity meter TSS Portable and a domestic brand turbidity meter. The test data is as shown in the table. According to the experimental results, the measurement error of the measuring instrument designed in this paper is about 3%, while the error of HACH turbidity meter is about 2%, and the measurement error of domestic turbidity meter is about 3.8%. After the measurement instrument, the measurement error is significantly reduced, which can greatly improve the turbidity measurement accuracy. The measurement data of the original 400NTU standard solution maximum error point measuring instrument is 408.84 NTU, which is wrong compared with the standard solution. The difference is reduced to 8.84 NTU. The test results show that the measurement accuracy and error of the measuring instrument designed in this paper are obviously better than the performance of the domestic mainstream products tested, but there is still a certain gap compared with imported HACH and other measuring instruments.

Table 1. Comparative analysis of turbidity measurement data of measuring instrument

Serial number	standard solution (NTU)	standard solution	Compensation before the two points method measurement and error (NTU)	
1	0	0x0000	0.00	0.00
2	200	0x0d44	216.34	16.34
3	400	0x1a01	424.68	24.68
4	600	0x24ad	625.77	25.77
5	800	0x2fab	813.32	13.32
6	1000	0x3a9c	1000.00	0.00

Serial number	Measured value and error after compensation(NTU)		HACH turbidity meter measurement and error(NTU)		Measurement value and error of a mainstream domestic turbidity meter(NTU)	
1	-1.98	-1.98	0.00	0.0	2.02	2.02
2	200.49	0.49	200.2	0.2	202.54	2.54
3	408.84	8.84	403.5	3.5	409.83	9.83
4	593.89	-6.11	602.4	2.4	609.53	9.53
5	791.50	-8.50	806.7	6.7	789.82	-10.18
6	1005.27	5.27	1005.8	5.8	1012.3	12.30

4. Conclusion

In this paper, a mine water turbidity measuring instrument is designed through the analysis of the working principle of optical turbidity detection. The measuring instrument is mainly based on the principle of light scattering. The measurement error is minor (the accuracy is within 3%), and the front end adopts silicon photoelectric. As a photoelectric conversion component, the diode adopts high-precision operational amplifier circuit and A/D analog-to-digital converter design. It adopts low-power OLED liquid crystal display for human-computer interaction and anti-interference design of digital interface RS485 bus to meet complex electromagnetic interference in coal mine. In terms of the environment and data processing, the MATLAB software simulation tool is used to fit and compensate the measurement data curve, which effectively improves the accuracy of turbidity measurement. Compared with measuring instruments of other brands and the performance of the domestic measuring instruments, the measurement error of the measuring instrument has been greatly improved, but there is still a certain gap between the measurement error of foreign HACH products. Thus, further improvement and optimization in front-end silicon photodiode selection, light source and signal conditioning circuit are needed to improve the performance of the measuring instrument. The turbidity measuring instrument mentioned in this paper has been tested on site in Shenhua Shendong, China Coal Group, Zaozhuang Group and other mines. The work is stable and reliable. The mine water turbidity measuring instrument provides a reliable online monitoring for the water quality of underground equipment. Means, high real-time and effective guidance for daily maintenance of water treatment equipment have greater application and promotion value.

Acknowledgment

Fund Project: China Coal Science and Technology Group Tiandi Science and Technology Innovation and Entrepreneurship Fund Special Project (2018-TD-QN019)

Journal of Physics: Conference Series **1423** (2019) 012050 doi:10.1088/1742-6596/1423/1/012050

References

[1] Mao Weidong, Zhou Rulu, Guo Zhongquan. Technology and Application of Zero Emission Treatment in Coal Mine Water[J].Coal Science and Technology, 2017, 45(11): 205-210.
[2] He Feilong. Characteristics of mine water quality in Lulusu coal mine and suitable water treatment technology [J]. Coal Science and Technology, 2018, 46 (S2): 255-258.
[3] Wang Xueming. Study on turbidity measurement and nonlinear compensation method for double-scattered light water quality[J].Electronic World,2016,(07):123-124.
[4] Wang Li. Research on scattering underwater turbidity measurement method [J]. Foreign Electronic Measurement Technology, 2016, 31(9): 27-30.
[5] National Development and Reform Commission, National Energy Administration. The 13th Five-Year Plan for the Development of the Coal Industry [EB/OL]. [2018-02-04].
[6] He Xuwen, Qian Dayi, Tan Yuanbin, et al. Water quality characteristics of high turbidity mine water[J]. Journal of University of Science and Technology Beijing, 2008(08): 844-847.
[7] Wang Zhidan, Chang Jianhua, Zhu Chenggang, et al. Design of a new wide-range turbidity measuring instrument[J]. Measuring Instruments and Microsystems, 2016, 35(05): 77-79.
[8] Song Qimin, Lu Minggang. Range and nonlinear study of scattered light turbidity measurement[J].Journal of Shanghai University:Natural Science Edition,1997,3(5):564-569
[9] Based on ADS1100 high-precision intelligent dynamometer [J]. Yan Ziyong, Liao Wenliang, Wang Huitang, Chen Wenzhao. Foreign electronic measurement technology. 2006(06):34-35
[10] Sheng Qiang. Scattering light turbidimeter and signal processing technology research. Shanxi: Taiyuan University of Technology, 2007.
[11] Lan Yu, Zhang Shunxing. Design of a direct-fired photoelectric turbidity meter[J]. Foreign Electronic Measurement Technology, 2013, 32(01): 53-55.
[12] Electromagnetic Compatibility Analysis and Design of Coal Mine Safety Monitoring Equipment [J]. Wang Xueming, Zhang Libin, Jiang Ze, Zhu Qianwei. Industrial and Mining Automation. 2009(05):30-31
[13] Li Wei, Wang Bo, Wei Qiming, et al. Design of Reference Current Source Using Piecewise Linear Compensation[J].Microelectronics,2009,39(05):602-605.
[14] Liu Chang, Wang Bingfeng. Intelligent Turbidimeter Based on AT89S52 Single Chip Microcomputer[J]. Instrument Technology & Measuring Instruments, 2007(12): 59-61.
[15] The Application of Matlab in Higher Mathematics Teaching[J]. Computer and Digital Engineering. 2013(08):25-26.

Journal of Physics: Conference Series **1423** (2019) 012051 doi:10.1088/1742-6596/1423/1/012051

Application of geometric flow and bandelet coefficient combining method to image fusion

Mantian Zhong

Department of Education, Luoding Polytechnic, Luoding, Guangdong, 527200

Corresponding author and e-mail: Mantian Zhong, 409490472@qq.com

Abstract. Based on the theory of Bandelet transformation, we use Beamlab200 toolbox and Matlab software, adopt new construction of geometric flow and Bandelet coefficient combination method, apply new introduction of smaller absolute value rules, and take advantage of usage of computer simulation to experiment image fusion to explore the coefficient combination method and new introduction of smaller absolute value rules for geometric flow and Bandelet and test effect of their application to image fusion. The result after the experiment proves that this method is superior than the Beamlab 200 toolbox method in terms of running time, convergence speed, standard deviation, signal-to-noise ratio, gradient and image effect. The method can be applied to repair damaged images and others according to the results.

1. Introduction

The domestic and traditional sampled tensor product wavelet does not have translation invariance, which is only representing singular features of isotropic. Taking a digital image as an example, when the outline and edge of an image are represented, the points on the boundary are isolated, in which continuity or smoothness along the edge direction is not utilized. In order to overcome the shortcomings of wavelet analysis, foreign scholars Pennec and Mallat proposed the Bandelet transformation in 2000. Bandelet transformation is an edge-based image fusion method, and is also an image fusion method based on the maximum absolute value fusion rule, which can adaptively track the geometric regular direction of the image. The two scholars believe that during the image processing task, if the geometric regularity of the image can be known in advance and fully utilized, the approximation performance of the image transformation method will be improved. In this study, we improved the image fusion method based on the maximum absolute value fusion rule, by introducing smaller absolute value rules, constructing a geometrical flow and by using Bandelet coefficient combining method for image fusion experiments, then obtained a good result [1].

2. Bandelet transformation theory

2.1. Geometric regular images and geometric flow

Let the support area of image $f(x, y)$ be $[0,1]^2$, and the only continuous uninterrupted edge in $[0,1]^2$ is denoted by Γ. The continuous edge Γ in the square region has a finite geometric length. For convenience of representation, let the continuous edge be $[a,b]$ and $[a,b]$ be the projection interval of Γ on the x axis, and Γ can have the form like (1). [3]

$$\Gamma = \{(x, \gamma(x)) \mid a \leq x \leq b\} \tag{1}$$

Let $\Omega = [0,1]^2 - \Gamma$, $C^{\alpha}(\Omega)$ represents the α-order Hölder space. The image $f(x, y)$ is called α-order geometric regular image if $f(x, y) \in C^{\alpha}(\Omega)$, $\gamma(x) \in C^{\alpha}[a,b]$. $f(x, y)$ is smooth in other areas except for a segmented smooth edge line. Considering that the edges of any natural image have a certain width, so we can correct $f(x, y)$

$$f = \tilde{f} * h \tag{2}$$

Where $\tilde{f} \in C^{\alpha}[a,b]$, $\Omega = [0,1]^2 - \Gamma$ .is a fuzzy kernel in a region, and it is a tightly supported one that has not yet been determined [3].

2.2. *Optimal geometric flow*

For the needs of the experiment, the coding of several geometric flow images should be considered in two directions: bit rate and distortion. Inside the square area S, it should be the best geometric flow direction, that is, we should take the smallest Lagrangian function as the formula (3).

$$(f_{\theta}, R) = \left\| f_{\theta} - \tilde{f}_{\theta} \right\| + \lambda T^2 (R_g + R_b) \tag{3}$$

Where $\tilde{f}_{\theta}$ is a one-dimensional reconstructed signal, and it is also a Bandelet coefficient that must be uniformly quantized; T is a uniform quantization threshold, R_g is the number of bits necessary to encode the geometry stream image, R_b is the number of bits required for the Bandelet coefficient, and it is the geometrically streamed image that is encoded and consistently quantized. λ is the Lagrange multiplier factor. It is concluded that $\lambda = 5/28$ is taken in the computer simulation experiment according to Pennec's conclusion,.

We can divide the rounded corners $[0, \pi)$ into $L^2 - 1$ at equal angles for the small squares of $L \times L$, that is, the possible value of θ is as illustrated in the formula (4).

$$\theta = \frac{k\pi}{L^2 - 1}, k = 0,1,2,\cdots, L^2 - 2 \tag{4}$$

However, if the experiment encounters a situation where there is no geometric flow, use $\theta = Inf$ as a feature tag to indicate that Bandeletization is not implemented here.

Consistent uniform quantization is required for Bandelet coefficients. Here we can use X to represent Bandelet coefficients, T is a uniform uniform quantization threshold, so the quantized value $Q(X)$ is calculated by the formula (5).

$$Q(X) = \begin{cases} 0, |X| \leq T \\ \operatorname{sgn}(X)\left(q + \frac{1}{2}\right)T, qT \leq |X| \leq (q+1)T \end{cases} \tag{5}$$

3. Geometric flow and Bandelet coefficient combining and smaller absolute value rules

Image fusion is as following after geometric flow and Bandelet coefficient combining rules are given.

(1) The two original images are geometrically registered.

(2) In region Ω_i, re-sampling the sample of the calculated image.

(3) Calculate the geometric flow $G_j(i)$, $(j = 1,2,\cdots,N)$ of each region and its corresponding Bandelet coefficient. N represents the total number of original images. $G_j(x,y,i)$ is the Bandelet coefficient of pixel (x,y), and it is for the of the j th image of original image.

(4) Fusion rules are as follows.

①Combine geometric flow images with smaller absolute value fusion rules

$$G_F(i)=\begin{cases}G_1(i), G_1(i)\le G_2(i)\\ G_2(i), G_1(i)> G_2(i)\end{cases} \tag{6}$$

②For Bandelet coefficients, use smaller absolute value fusion rules

$$G_F(x,y,i)=\begin{cases}G_1(x,y,i), |G_1(x,y,i)|\le |G_2(x,y,i)|\\ G_2(x,y,i), |G_1(x,y,i)|> |G_2(x,y,i)|\end{cases} \tag{7}$$

In formulas (6) and (7), let $G_F(i)$ be the geometric flow of the region Ω after fusing the two images, and use $G_F(x,y,i)$ to represent the Bandelet coefficient, which is the coefficient at the pixel point (x,y) of the fused image after fusing the two images.

(5) The fused image is obtained by the geometric flow $G_F(i)$ and the Bandelet coefficient $G_F(x,y,i)$, through reconstruction by the inverse of the Bandelet transformation.

4. The simulation experiment

4.1. Image effect comparison

In order to compare the effects of image fusion using Beamlab200 toolbox method, geometric flow and Bandelet coefficient combining method, the simulation experiment was carried out with aid of Matlab7.01 software. The images of experimental results are shown below in Figure 1, Figure 2, Figure 3. Figure 4.

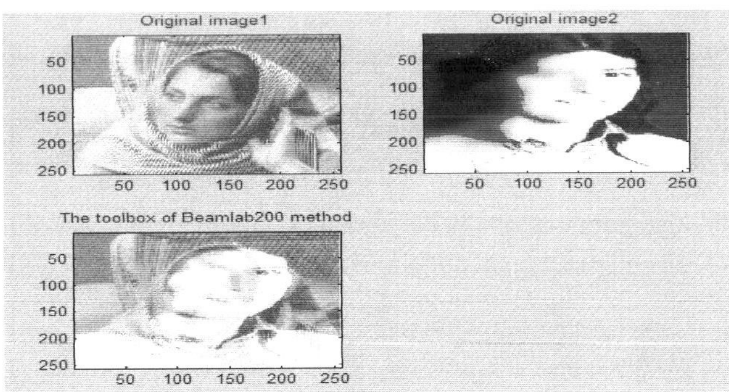

Figure 1. Beamlab200 toolbox fusion
Different image fusion based on Beamlab200 toolbox

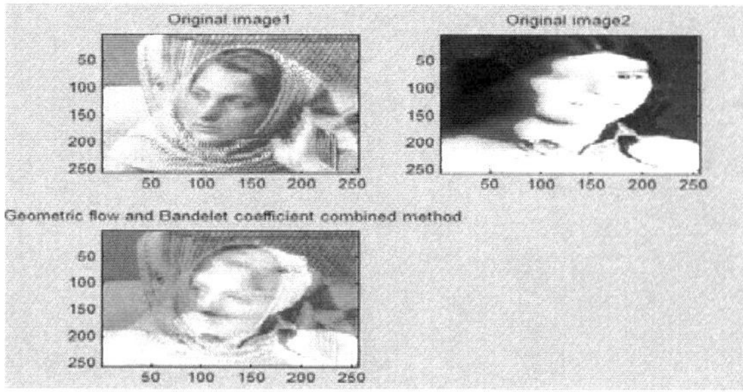

Figure 2. Different image fusion based on Geometric flow and Bandelet coefficient combining method

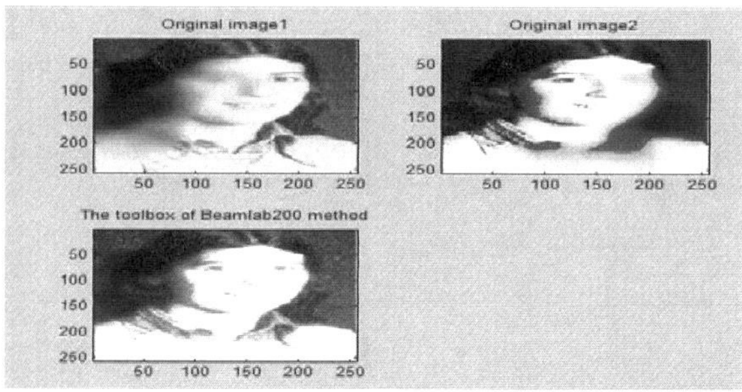

Figure 3. Half-images fusion based on Beamlab200 toolbox

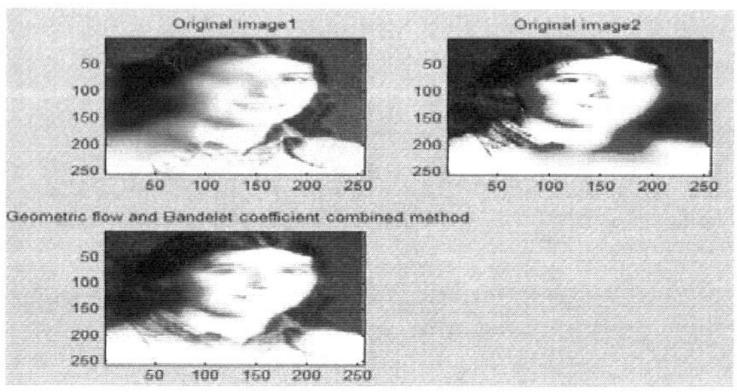

Figure 4. Half-images fusion based on Geometric flow and Bandelet coefficient combining method

4.2. Convergence speed comparison

In order to compare the convergence speeds of Beamlab200 toolbox method, geometry flow and Bandelet coefficient combination method respectively, the simulation experiment was carried out with aid of Matlab7.01 software. The images of experimental results are shown below in Figure 5.

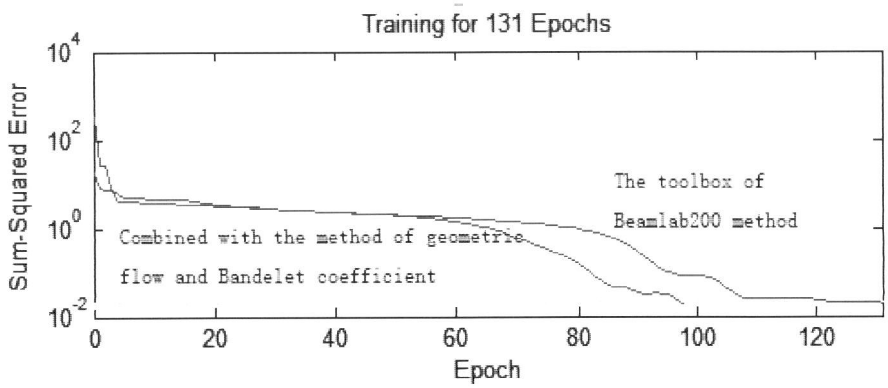

Figure 5. Convergence speed of the two methods

4.3. Parameter comparison

In order to compare the performance parameters of Beamlab200 toolbox method, geometric flow and Bandelet coefficient combining method respectively, the simulation experiment was carried out with aid of Matlab7.01 software. The images of experimental results are shown below in Table 1.

Table 1. Comparison of performance parameters of the two methods

Image fusion method name	Standard deviation	Signal to noise ratio (db)	The number of iterations	The average gradient	Entropy	Running time(s)
Beamlab200 toolbox method	71.4153	38.6529	131	19.6584	7.7583	593
Geometric flow and Bandelet coefficient combing method	69.2548	40.0058	96	18.5389	7.7695	386

5. Conclusion

Through the computer numerical simulation experiment, the results are as follows:

First, the simulation results prove the effectiveness of the geometric flow and Bandelet coefficients combing method, and the new introduction of new smaller absolute value rules. Through comparison between Figure 1 and Figure 2 as well as between Figure 3 and Figure 4, it proves that we can process the edge texture more clearly, display the stereo features of the face more realistically, and make it is more close to the original image color by using this method. Second, the simulation results prove the convergence of the geometric flow and Bandelet coefficients combining method and the new introduction of new smaller absolute value rules. According to the comparison presented in Figure 5, we converged more than 30 steps in advance by using this method, validating convergence effect of adopting this method. Third, the simulation results prove the advantage of the geometric flow and Bandelet coefficients combing method, and the new introduction of smaller absolute value rules, According to the comparison presented in Table 1, the standard deviation after using the method is slightly smaller, the signal-to-noise ratio is slightly larger, the average gradient becomes smaller, the entropy value is slightly smaller, and 200 seconds of running time is saved, which shows advantage.

We met the following conditions when perform the computer simulation experiments. First, the value of threshold T with uniform, quantization, and even nature is the key. T=50 in the experiment, and the general value is T=40-60. If it exceeds this range, the fused image will not work properly, and the convergence speed will slow down more than 100 steps.

Second, the value of N is also very important. The value of N in the experiment is N=215-335. If the value exceeds this range, the convergence speed will be slower by 50 steps or more and the effect of fused image is not good.

Third, value of λ and other coefficients can make emergence of better image fusion effect and ideal convergence speed after many experiments.

Fourth, the image pixel values are linear before and after the fusion, and the pixel values are in corresponding to position. If we use non-linear values like quadratic function, the fusion effect is not good, and the convergence speed is slower more than 100 steps.

References

[1] E Le Pennec, S mallat, Sparse Image Representation with Bandelets [DB/OL], Sub Submitted to IEEE Trens on Image Processing [Z], http://www.camp.polytechnique.fr/~mallat/papers, 2003.

[2] Erwan Le Pannec, Stephane Mallat. Bandelet Image Approximation and Compression. SIAM Journal of Multiscale Modeling and Simulation [Z], sol.4.no.3, pages991039, 2005.

[3] Yan Jingwen, Qu Xiaobo. The Super Wavelet Analysis and Application [M]. Beijing: National Defence Industry Press, 2008

[4] Bowles et al, translation by Rui Guosheng, Cheng Zheng Xing, Wang Wen. Wavelet and Wavelet Transformation introduction [M]. Beijing: Publishing House of Electronics Industry

Journal of Physics: Conference Series **1423** (2019) 012052 doi:10.1088/1742-6596/1423/1/012052

An improved SNIP denoising algorithm applied in EDXRF

Liangquan Ge[a, b], Hui Li[b], Chuanfeng Tang [b], Fei Li[a, b,*], Kai Zhu[b], Jin Yang[b]

[a] Applied Nuclear Technology in Geosciences Key Laboratory of Sichuan Province, China
[b] College of Nuclear Technology and Automation Engineering, Chengdu University of Technology, China

Corresponding author and e-mail: Fei Li, lifeimvp@sina.com

Abstract. Based on the phenomenon of peak height reduction in existing denoising algorithms, this paper modifies the traditional SNIP algorithm, proposes an improved SNIP algorithm and applies it to EDXRF denoising. Two evaluation parameters RMSE and PCC are used to compare the improved SNIP algorithm with the traditional SNIP algorithm. The experimental results show that the improved SNIP algorithm has great advantages both in similarity and relevance, and the results of the improved SNIP algorithm are more consistent with the characteristics of the original data peak than the original SNIP algorithm, which is beneficial to the qualitative and quantitative analysis of the energy dispersive X-ray fluorescence spectrum.

1. Introduction

The energy dispersive X-ray fluorescence spectrum is the characteristic X-ray peak and continuous background formed by a series of effects produced by the interaction between photons and substance. Before EDXRF can be used for qualitative and quantitative analysis of the samples to be measured, it is necessary to denoising, background subtraction and overlapping peak separation. Among them, denoising is one of the key links of EDXRF, which has a direct impact on the analysis results. The algorithms commonly used for denoising are the least square method [1], the wavelet transform method [2], the Fourier transform method [3], and so on. The experimental results show that these algorithms can effectively eliminate the statistical fluctuations, and can largely retain the energy spectrum information while eliminating the noise. SNIP algorithm commonly used in background subtraction can effectively suppress the influence of the substrate, and has a strong weakening effect on the influence of background factors such as background peak and scattering ray peak [4]. At the same time, these algorithms also have the disadvantage of reducing the spectral peak height, which is not conducive to the subsequent quantitative analysis. And most of these algorithms are gamma spectrum denoising, but few of them are used in X-ray spectrum denoising, because the peaks of X-ray spectrum are relatively dense and the matrix effect can not be avoided. In this paper, we improve the SNIP algorithm and apply it to X-ray spectrum denoising. The experimental results show that it can effectively improve the phenomenon of energy spectrum peak height reduction.

2. Material and Method

2.1. Principles and disadvantages of the traditional SNIP algorithm

The idea of snip algorithm applied to background subtraction of gamma energy spectrum was first proposed by Ryan et al. [5-7].

The actual steps to deduct the background of gamma energy spectrum are as follows:

1) carry out logarithmic transformation (LLS transformation) for each count value, and the transformation formula is as follows:

$$y(i) = ln[ln(x(i) + 1 + 1) + 1] \tag{1}$$

Where i is the corresponding count, $x(i)$ is the channel address, and $y(i)$ is the operation result. The natural logarithm operator in the formula is mainly used for processing high count peaks.

2) iterate over the energy spectrum data according to the SNIP algorithm, and the formula is as follows:

$$y_r(i) = min\left[y(i), \frac{y(i-m)+y(i+m)}{2}\right] \tag{2}$$

Where m is the artificially set parameter, which is related to the filter window width W (W=2m+1). In the traditional SNIP algorithm, it is 4 times of the peak area half height width (FWHM). The smaller value of $y(i)$ with the current channel address i and the mean value of channel address $y(i \pm m)$ at both ends of the filter window is taken as the result $y_r(i)$.

3) when all $y_r(i)$ are obtained, the background spectrum of the full spectrum can be obtained by inverse LLS operation.

The traditional SNIP algorithm usually uses a single window width when filtering and stripping the energy spectrum. The filter width is determined by the width of the spectrum peak, and the width of the spectrum peak generally increases with the increase of energy. Choosing different window width will lead to different results. For example, the window width suitable for the low energy band of gamma peak may cause the net count in the high energy band to be deducted as the background, while the window width suitable for the high energy band of gamma peak may cause the background in the low energy band to be treated as the net count without being completely deducted [8]. Therefore, using a single filter width will reduce the effect of the algorithm.

2.2. The existing improved SNIP algorithm and its shortcomings

At present, there are some shortcomings in the existing improved SNIP algorithms. Their differences lie in the calculation method of filter window width:

1) the width of the initial filter window is 3 times of the maximum value of the half height width of the energy spectrum, and then decreases to 1 [9];

2) the filter window width of each peak area is the channel value corresponding to the peak width, and the peak area of the non-omnipotent peak is 0 [10];

3) according to the characteristics of Gauss distribution, the left and right boundaries of all-around peaks are redetermined, the approximate peak boundary is determined by the first derivative method, and then the energy spectrum is stripped [11];

4) the peak width is obtained by using the functional relationship between peak width and energy, and the adaptive filter width is obtained [12].

The above methods are mainly focused on the improvement of filter width, but the improvement of filter width has no effect on the situation that the peak value of SNIP algorithm decreases after iterations.

2.3. Improved SNIP algorithm

The purpose of this paper is to find the functional relationship between the peak decline ratio of energy spectrum, the channel address (energy) and the number of iterations, and then replace it to achieve peak reduction without affecting the algorithm effect, so as to effectively avoid the situation that the peak decline ratio will gradually increase with the increase of energy after multiple iterations of SNIP.

After several iterations of SNIP algorithm, compared with the decline ratio of the original data to the processed data, the fitting is performed (see Figure 1). The energy spectrum address (energy) is used as the abscissa and the descent ratio as the ordinate. The fitting curve conforms to the exponential function distribution, so the exponential fitting is adopted.

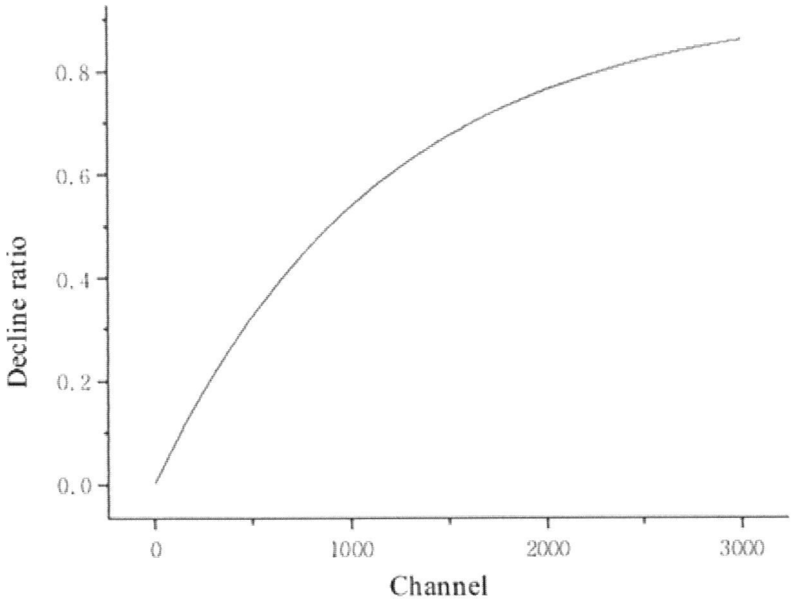

Figure 1. The change of decline ratio with channel

It can be seen from Figure 1 that the decline proportion is more and more close to 1 with the increase of channel address (energy). The general EDXRF spectrum channel address is generally 1024 or 2048, the decline ratio is 40% to 70%, and the peak shape will have obvious distortion. After multiple iterations, the decline ratio is fitted (see Figure 2 and figure 3). Among them, the number of iterations is abscissa and the descending scale is ordinate.

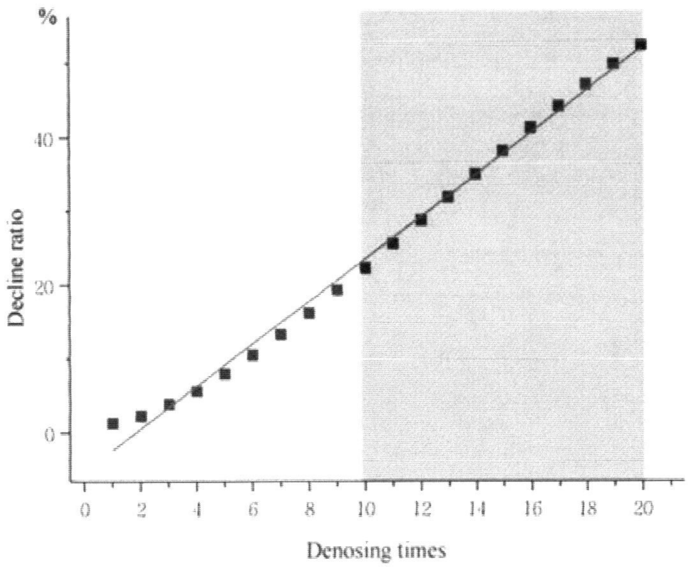

Figure 2. The change of decline retio with denoising times

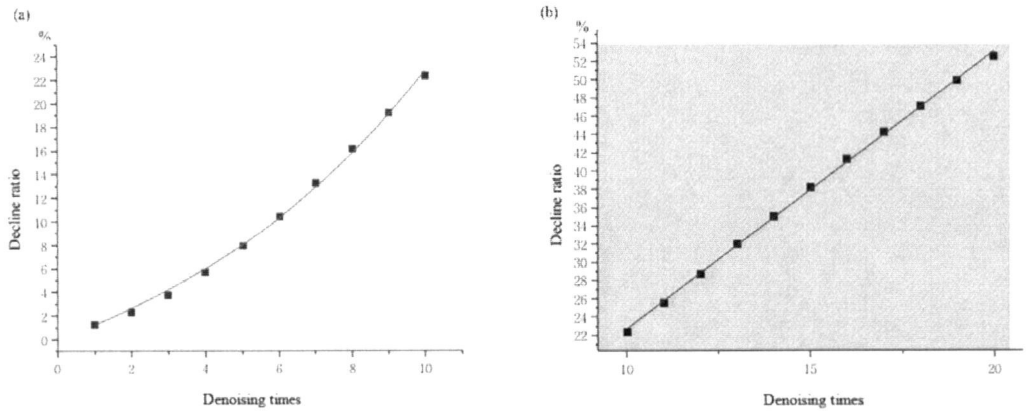

Figure 3. (a) The amplication ares with 0-10 of denoising times, (b) The amplication area with 10-20 of denoising times

The comparison shows that if the number of iterations is less than 10, the influence factor is more in line with the exponential relationship, and if the number of iterations is more than 10, it is more in line with the linear relationship.

After comprehensive consideration, the following formula is obtained:

$$f(i)' = A(M)\big(1 + B(i)\big)f(i)$$

The iterative influence formula: $A(M) = \begin{cases} -9.605241(1 - 1.12927^{M}) & M \leq 10 \\ -7.65184 + 3.03598M & M > 10 \end{cases}$

The channel address influence formula: $B\big(f(i)\big) = 0.9303(1 - \exp(-8.6101\mathrm{e} - 4 * \mathrm{f}(i)))$

Where i is the energy spectrum channel address, $f(i)$ is the original signal, $f(i)'$is the signal after denoising, and M is the number of iterations.

3. Results and discussion

The evaluation parameters selected for the improved SNIP algorithm are root mean square error (RMSE) and Pearson correlation coefficient (PCC). RMSE is related to the similarity of the two signals, while PCC is related to the correlation. The new improved SNIP algorithm proposed in this paper is used to reduce the noise of energy spectrum, and its denoising effect is compared with that of traditional SNIP algorithm.

A set of X-ray fluorescence data was iterated for 15 times, and then the average RMSE and PCC were compared and tested successively (see Table 1).

Table 1. The comparison between the traditional SNIP algorithm and the improved SNIP algorithm

传统 SNIP 算法			改进 SNIP 算法	
迭代次数	RMSE	PCC	RMSE	PCC
1	140.1989	0.9962	35.2699	0.9938
2	215.0684	0.9928	84.7045	0.9942
3	256.1275	0.9862	96.48	0.9902
4	280.0102	0.9766	105.2399	0.9819
5	294.8753	0.9647	118.7876	0.9704
6	304.6651	0.9511	136.3097	0.9567
7	311.4143	0.9362	156.3736	0.9416
8	316.2471	0.9205	178.5169	0.9256
9	319.8148	0.9042	203.0683	0.9089
10	322.5162	0.8874	230.4694	0.8918
11	324.6074	0.8704	244.6847	0.8746
12	326.2595	0.8535	259.3096	0.8573
13	327.5879	0.8366	273.4805	0.8401
14	328.6718	0.8198	287.3485	0.8231
15	329.5669	0.8032	301.111	0.8062

RMSE: root mean square error

PCC: Pearson correlation coefficient

It can be concluded from the Table 1 that after different iterations, the RMSE obtained by the improved SNIP algorithm is far smaller than that obtained by the traditional SNIP algorithm, and the PCC obtained by the improved SNIP algorithm is also significantly larger than that obtained by the traditional SNIP algorithm. It can be concluded that the improved SNIP algorithm is superior to the traditional SNIP algorithm in both similarity and correlation. With the increase of iterations, RMSE and PCC tend to be stable.

The traditional SNIP algorithm and the improved SNIP algorithm are used to reduce the original data (Figure 4. (a)) 5 times respectively (see Figure 4). It can be seen intuitively from the Figure 4 that with the increase of iteration times, the peak height obtained by the traditional SNIP algorithm drops more and more obviously. Even after the fifth iteration, the weak peak is almost lost, and the feature peak is also greatly affected, which is not conducive to the extraction of the feature peak information. Although the improved SNIP algorithm reduces the peak height to some extent, it has little influence and retains the weak peak relatively completely, which is beneficial to the subsequent qualitative and quantitative analysis.

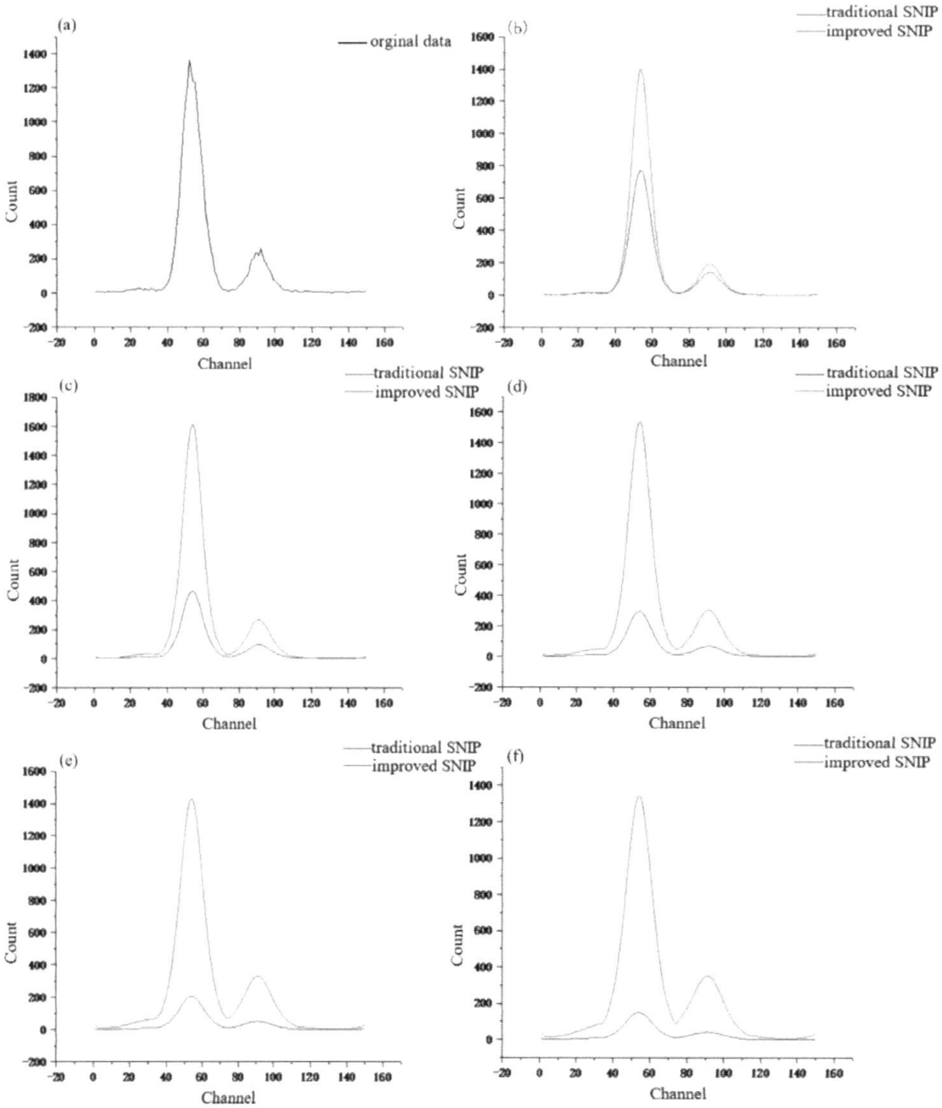

Figure 4. (a) The orginal data, (b) The result of denoising once, (c) The result of denoising twice, (d) The result of denoising three times, (e) The result of denoising four times, (f) The result of denoising five times.

4. Conclusion

In this paper, a new and improved SNIP algorithm is proposed and applied to EDXRF denoising simultaneously with the traditional SNIP algorithm. RMSE and PCC of energy spectrum processed by the two methods are calculated respectively for comparison. The results show that after several iterations, the improved SNIP algorithm is obviously superior to the traditional SNIP algorithm in terms of peak retention, making the results more consistent with the characteristics of the original data peak. Therefore, the denoising of energy spectrum can make the characteristic peak of the element more obvious, and the recognition effect can be improved obviously, which is more conducive to accurately finding weak peak and heavy peak.

Acknowledgement

This work was funded by the National Natural Science Foundation of China (No.41774147), the National Key R&D Project (No.2017YFC0602100) and Science and technology activity project for overseas personnel in Sichuan province.

References

[1] ZHU Meng-Hua, LIU Liang-Gang, QI Dong-Xu, YOU Zhong, XU Ao-Ao. Least square fitting of low resolution gamma ray spectra with cubic B-spline basis functions. Chinese Physics C. 2009, 33(1):24-30.

[2] Davide Bianchia, Claudia Lenauera, Gerhard Betzb, András Vernes. A wavelet filtering method for cumulative gamma spectroscopy used in wear measurements. Applied Radiation and Isotopes. 2017,120:51-59.

[3] M.J. Safari, F. Abbasi Davani, H. Afarideh, S. Jamili, E. Bayat. Discrete Fourier Transform Method for Discrimination of Digital Scintillation Pulses in Mixed Neutron-Gamma Fields. IEEE Transactions on Nuclear Science. 2016, 63(1):325-332.

[4] WU Hexi, LIU Qingcheng, YANG Bo, LIU Yujuan. SNIP algorithms applied to analyze y-ray spectra of natural radioactivity nuclides. He Jishu/Nuclear Techniques. 2010, 33(7):513-516.

[5] Omer M. Analysis of SNIP algorithm for background estimation in spectra measured with LaBr3: Ce detectors [C]. Zero-Carbon Energy Kyoto 2012, Tokyo. Springer: 245-252.

[6] Miroslav Morhác. An algorithm for determination of peak regions and baseline elimination in spectroscopic data. Nuclear Instruments and Methods in Physics Research Section A Accelerators Spectrometers Detectors and Associated Equipment. 2009, 600(2): 478 - 487.

[7] WU Hexi, LIU Qingcheng, YANG Bo, et al. SNIP algoritms applied to analyze γ - ray spectra of natural radioactivity nuclides[J]. Nuclear Techniques, 2010, 33(7): 513-516.

[8] LIU Yonggang. Research on decomposition method of γ spectrometer[D]. Beijing: China University of Geosciences, 2011.

[9] LONG Bin, FENG Tiancheng, SU Chuanying, et al. a self-adaptive method for the clipping of scatter background of γ spectrum[J]. Nuclear Electronics & Detection Technology, 2013, 33(10): 1293-1296.

[10] ZHANG Yupeng, XIAO Xiaoling, ZHANG Xiang, et al. Research on elimination background by extending clipping window of snip algorithm[J]. China Energy and Environment Protection, 2017, 39(3): 143-146.

[11] WANG Yiming, WEI Yixiang. Baseline elimination method for γ - ray spectra based on improved snip algorithm[J]. Nuclear Electronics & Detection Technology, 2012, 32(12): 1356-1359.

[12] GAO Yan, ZENG Guoqiang, GU Min, YAN Lei, LUO Mingtao, HOUYang, ZHU Zhu ,TANG Wei. Development of adaptive width SNIP algorithm in nuclides identification instrument[J]. Nuclear Techniques,2019,42(6):060401.

Journal of Physics: Conference Series **1423** (2019) 012053 doi:10.1088/1742-6596/1423/1/012053

Design of the gordon euryale seed automatic shelling machine

Junyan Zhang

Jiangsu Union Technical Institute Suzhou Industrial Park Branch, Suzhou, China

Corresponding author and e-mail: Junyan Zhang, 108645502@qq.com

Abstract: In this study, Aiming at the inefficiency of the manual shelling of Gordon Euryale seed, a new automatic seeding Sheller with simple structure and easy operation was developed to improve the efficiency. According to the shape and size of the Gordon Euryale Seed and the material characteristics of its hard shell, the mechanical structure of the machine were designed. It included the feeding mechanism, shelling device and separating device. Experiments showed that it had met the requirements of rapid shell peeling. The shell peeling rate was over 98% and the whole kernel rate was over 97.5%. It had reliable operation, fast speed and low failure.

1. Introduction

As one of the Eight Immortals in water, Gordon Euryale is commonly known as chicken head rice, and also known as ginseng in water, which is not only delicious but also has medical value. The shell of Gordon Euryale is hard and astringent and not suitable for eating, and should be peeled off when eating. At present, some existing Gordon Euryale shelling machine cannot completely shell and the pulp of Gordon Euryale is crushed. The quality of Gordon Euryale after processing is very poor, which affects the shelling amount and the taste of the Gordon Euryale. In addition, the shell is directly spilled on the ground, causing the environment of the site to be dirty, messy, poor, and very troublesome to clean up. Therefore, most growers still peel by hand. It is difficult to peel the hard shell by hand, and it takes a long time to peel the pulp, resulting in low output, high labor cost and high price [1].

In order to improve the market sales volume and reduce the price of Gordon Euryale, many experts and enterprises have conducted research on the shelling machine [2], but there are few researches on the technology of Gordon Euryale automatic shelling machine. Only Yan GU and Guanghe XIAO [3] applied for the patent of "A Gordon Euryale Shelling Machine" to peel off the surrounding and rear shells of Gordon Euryale, but the structure is complicated. Xiaoyan ZHANG [4] disclosed A Safer Peanut Shelling Machine, Xieqing XU, Honghui RAO, Tao LI [5] et al published the paper Design and Test of a Fully Automatic Lotus Seed Shelling and Peeling Machine. Mingquan LAI [6] invented A Fully Automatic Lotus Seed Cutting and Shelling Machine. Using these machines to shell and peel step by step, the operation is complicated, the sanitary conditions are poor, and the shelling quality is not high. In view of the above situation, this design proposes a shelling machine that can quickly separate the shell and the pulp without damaging the pulp of Gordon Euryale, which can achieve automatic shelling of Gordon Euryale instead of manual work.

Graph classification No. S226.4 Document identification code: A.

2. Overall structure design and working principle of Gordon Euryale automatic shelling machine

2.1. Overall structural design

The overall structure of the Gordon Euryale automatic shelling machine consists of three parts: feeding device, shelling device and separating device. The feeding device is composed of a bin, a ring groove, and supporting members. The shelling device is composed of a servo lifting motor, a ring cutter holder, a ring cutter and an extrusion module. The separating device is composed of a chute, a vibrating screen, a shell box and a kernel storage box. The simple map of the overall structure is shown in Figure 1. The overall structure can be divided into upper, middle and lower layers, and the layers are connected and fixed by support frames and bolts. The uppermost layer is the bin mechanism, servo motor, and ring cutter holder lifting mechanism, and the middle layer is the ring groove, ring cutter, electric cylinder, and extrusion module mechanism; The lowermost layer is the separation mechanism, vibrating and screening net, shell box, and kernel storage box, etc..

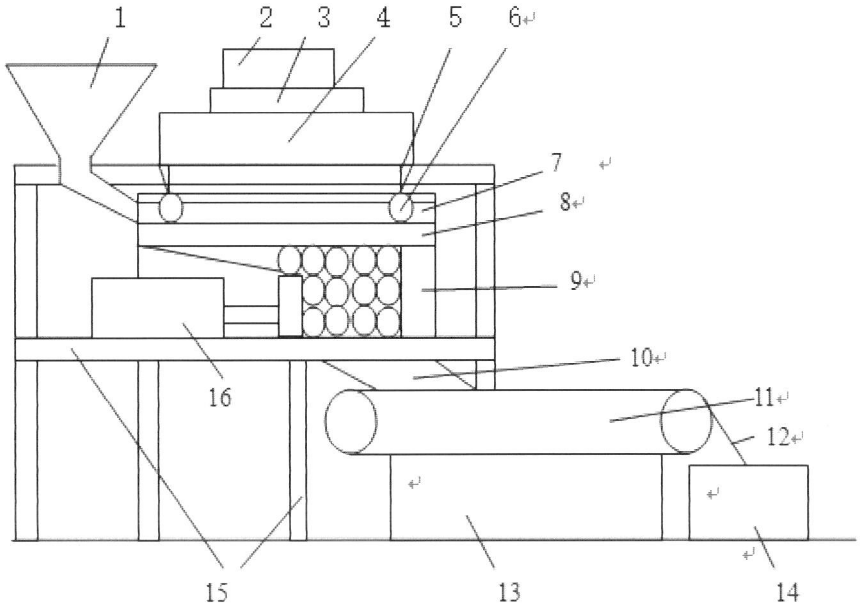

1. Bin 2. Servo lifting motor 3. Connector 4. Ring cutter holder 5. Ring cutter
6. Gordon Euryale to be processed 7. Ring groove 8. Movable baffle 9. Fixed extrusion module
10. Chute 1 11. Vibrating screen 12. Chute 2 13. Kernel storage box 14. Shell box
15. Support frame 16. Movable extrusion module

Figure 1. A simple map of Gordon Euryale automatic shelling machine

2.2. Operating principle

The shelling of Gordon Euryale is completed in two processes. First, cut with a ring cutter, and then extrude and separate it. Wash the approximately round granular Gordon Euryale with shell (raw materials to be processed) and pour it into the bin in an even and orderly manner, start the automatic shelling machine, guide the raw material uniformly and orderly into the groove, and fix the raw material when the concave groove is filled. The servo motor drives the ring cutter holder to move down. Precisely control the cutting depth of the ring cutter, and delay for a period of time after cutting the opening. The movable baffle can be opened automatically. Drop the Gordon Euryale with shell that has been opened into the extrusion device module, which continuously separates the shell and the pulp of Gordon Euryale by extruding and kneading. The extrusion force can be detected by the pressure sensor and fed back to the system to adjust the extrusion pressure and speed. Finally, the separated shell and pulp are fed into the vibrating screen through the chute, and the pulp is sent to the kernel storage box by vibration, screening and filtration, and the shell of Gordon Euryale is separated into the shell box. Finally, the kernel and the shell of Gordon Euryale are separated.

2.3. Analysis of physical properties of Gordon Euryale

For parameters closely related to this design, such as the geometric dimensioning, weight and shell thickness of Gordon Euryale with shell, shell breaking force, geometric dimensioning of kernel, etc., are calculated, providing an important basis for the structural design of the automatic shelling machine. Take 5 specimens of Gordon Euryale with shell randomly for testing, measure the diameter of Gordon Euryale with a vernier caliper, separate the shell by a sheller with digital display. After shelling, record the shelling force once for each specimen, and record the size of the kernel. See Table 1 for the statistical table.

Table 1. Structural dimensions and mechanical properties of Gordon Euryale

Number / Parameter	1	2	3	4	5	Average
Shell size / mm	20.11	18.52	19.42	19.54	20.13	19.544
Shelling force /N	28.52	24.30	25.43	24.82	29.21	26.456
Kernel size /mm	10.01	9.52	9.96	9.58	10.12	9.838

3. Design of key parts mechanism

3.1. Design of feeding device

The feeding mechanism mainly comprises a bin, a feeding trough, a vibration device and a groove. Place Gordon Euryale in the bin, and feed Gordon Euryale to the groove of the feeding trough through the inclined leak hole of the bin. The eccentric mechanism vibrates to make Gordon Euryale uniformly and orderly enter the shelling device. The gap between the leak hole and the groove is 2mm, to prevent Gordon Euryale from being caught in the gap during the vibration when Gordon Euryale falls into the groove.

3.2. Design of shelling device

The shelling device mainly comprises a guide trough, a ring cutter mechanism, an extrusion device, a discharge port, etc.. Gordon Euryale reaches the bottom through the bin hopper. The sensor detects the Gordon Euryale and turns on the vibrating motor. The Gordon Euryale is uniformly and orderly guided into the ring groove under the action of slight vibration until the entire ring groove is filled and stops vibration. The groove is tightened, and the Gordon Euryale is fixed by the tightening clamping force. Under the action of the pulse signal, the servo motor controls the precise positioning of the ring cutter to be directly above the Gordon Euryale. The ring cutter opens and rotates, and slowly approaches the Gordon Euryale to be processed for cutting. The first process ends. The outer dimensions of the ring groove and the center distance of the two-ring cutter are determined according to the shell size of Gordon Euryale, which meets the requirements of shelling of Gordon Euryale. The structure diagram is shown in Figure 2.

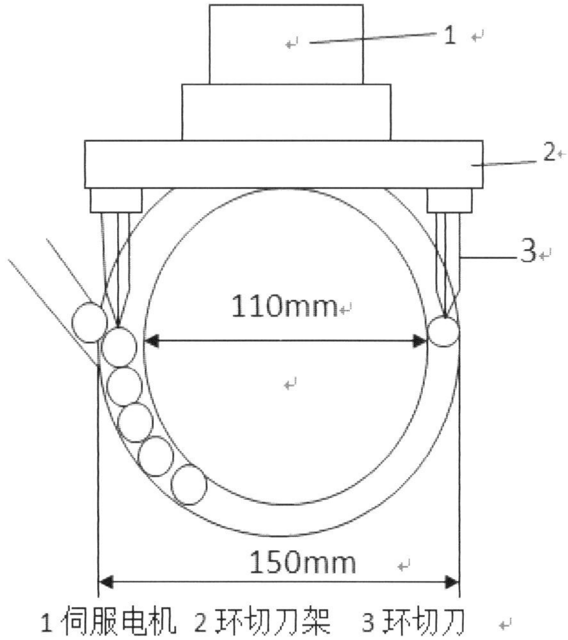

Figure 2. Structure diagram of the shelling device

According to the external characteristics of Gordon Euryale with shell, the sharp knife-shaped ring cutter is selected to cut the shell of Gordon Euryale. When the cutter touches the surface of the Gordon Euryale shell, the force analysis is performed, as shown in Figure 3.

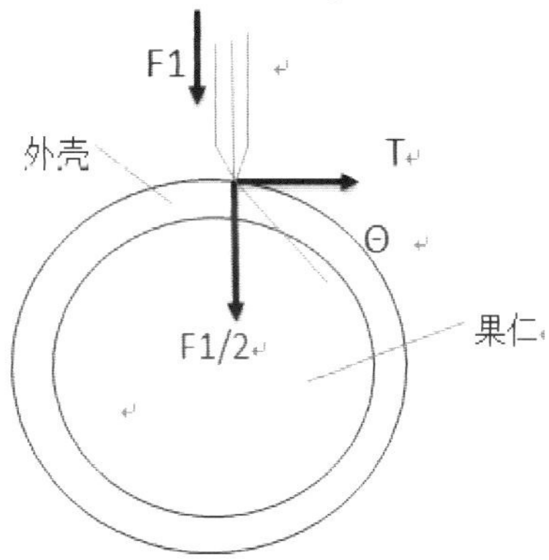

Figure 3. Force analysis diagram of Gordon Euryale with shell

The positive pressure of Gordon Euryale with shell during cutting is F1=PS, where, F1 is the shelling pressure, S is the contact area between the cutter and the Gordon Euryale shell, and P is the compressive strength of Gordon Euryale shell. The axial force on the Gordon Euryale shell is $T=F1/2 \cot\theta$

Where, T is the axial force exerted by the cutter on the shell, and θ is the angle of the blade. When the positive pressure F1 on the Gordon Euryale is greater than the critical value of F2 (the binding force on the Gordon Euryale), the shell cracks. When the cutter moves, the axial force T shears the shell of Gordon Euryale. In order to achieve a better incision effect, the cutting force of the cutter to the Gordon Euryale shell is set to 29N and the cutting depth is set to 5mm according to the above analysis results of physical properties of Gordon Euryale.

After cutting treatment, the Gordon Euryale with shell enters the extrusion device for shell breaking treatment. The fixed extrusion module baffle is fixed, and the movable extrusion module is opened to the maximum, waiting for the Gordon Euryale after opening to fall into the extrusion area. The Gordon Euryale with shell entering into the extrusion area is taken as the object for stress analysis below. In the figure, force analysis is conducted on the eight Gordon Euryale with shell as a whole. The force diagram of the separation of the shell and the kernel is shown in Figure 3.

$$G=F4\ (G=mg) \tag{1}$$

$$F5=F-f2-F3-F2 \tag{2}$$

Where, G is the total gravity of Gordon Euryale, m is the total mass of 8 Gordon Euryale, F4 is the support force of the support plate to the 8 Gordon Euryale, and the two forces are a pair of equilibrium forces in the vertical direction. F5 is the shell breaking force, F is the extrusion force of the movable extrusion module on Gordon Euryale, f2 is the total friction force of the support plate to all Gordon Euryale, F3 is the pressure of the fixed baffle to the Gordon Euryale with shell, F2 is the binding force of the Gordon Euryale with shell. When F5 is greater than 0, the shell can be separated, but to completely separate the shell and the kernel, according to the previous analysis of the shell breaking force, it has been tested that the breaking effect is optimal when F5 is set to 28N. See Figure 4.

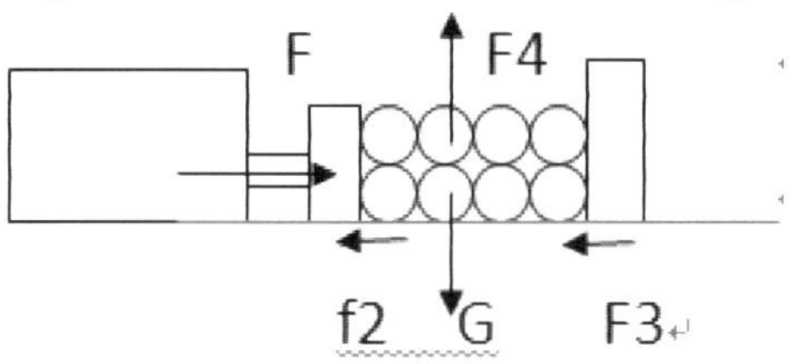

Figure 4. Force diagram of the separation of the shell and the kernel

3.3. Design of separating device

The separating device is mainly composed of a blowing port, an air flow channel, a fan, a vibrating screen, a kernel storage box, a shell box, etc.. In the separation process, the shell and the pulp are separated by controlling the wind speed of the fan to generate an air flow channel according to the difference of the specific gravity and the wind receiving area between the shell and the Gordon Euryale. The specific gravity of the pulp of the Gordon Euryale is relatively large, and the wind receiving area is small, so it is not easy to be blown away by the air flow, but slides down into the vibrating screen because of its own weight. The larger shell that cannot be blown away by the fan also slides down into the vibrating screen together with the kernels. Through slight vibration, screening and filtration, the kernels of Gordon Euryale are sieved down to the kernel storage box below, and the larger shell of Gordon Euryale is sent to the shell box through chute 2.

4. Design of control system

The control system requires PLC (programmable controller) as the controller and the touch screen HMI as the upper computer to realize the control of the vibration motor, the shelling motor, the ring cutter and the fan, receive the signal of the force sensor, and control the cutting force and extrusion force after analog signal processing; The touch screen monitors the running status of each motor in real time, gives real-time feedback, displays the cutting force and extrusion force in real time, and displays fault and alarm information at the same time. The touch screen uses TPC7062KS to transmit data to PLC (programmable controller) and PC (computer) through USB data cable. Mitsubishi PLC's GX software programming is used to achieve the control of feeding and vibration motor, start and stop of the fan, system extrusion pressure and precise positioning control of the ring cutter. PLC collects the input information of the system through various sensors, and compare the deviation between the output value and the given value after logical operation and PID algorithm, thus achieving the final control function.

5. Analysis of test results

The test result indicators are selected as: Stripping rate, kernel breaking rate, processing efficiency. The test sample is Gordon Euryale, which has a long growth period. During the test, 30 jin of raw materials are taken, and comparative analysis is conducted with the manual shelling of 5 skilled workers. This test mainly studies the two main factors: ring shear force and extrusion force. With the stripping rate and the damage rate as the test index, the stripping rate is $\delta = (M1 - M2) / M1 \times 100\%$, where M1 is the total mass of Gordon Euryale; M2 is the mass of unshelled Gordon Euryale. The damage rate is $\eta = (M3/M4) \times 100\%$, where, M3 is the mass of the kernels of Gordon Euryale that has been shelled; M4 is the mass of damaged kernels. At present, the standard stripping rate of the shelling machine industry is ≥ 98.00%, and the damage rate is ≤4.50% [7]. Add 15kg of Gordon Euryale each time, and weigh the mass of the shelled kernels, the mass of the unshelled and the mass of the damaged kernels. According to the analysis of test results [8], the processing speed of the automatic shelling machine is 60kg per hour, the striping rate is above 98%, the rate of perfect kernel is above 97.5%, the unshelling rate is below 2%, the kernel breaking rate is below 2.5%, the processing quality and the manual shelling quality is the same, the cleanliness of the processing environment is even higher than manual shelling, the processing efficiency is nearly 20 times faster than manual processing. The machine is relatively small, easy to maintain, long life, easy to carry and change the site, which is suitable for family farmers to purchase and use.

6. Conclusions

A fully automatic Gordon Euryale shelling machine [9] has been developed. The shelling machine has a reasonable structure, simple shape, easy to carry, and suitable for family use. Key technology: the shelling rate of Gordon Euryale is over 98%, and the kernel breaking rate is below 2.5%. The quality of processed kernels is the same as that of manual shelling. There is no damage and raw material waste. This device can completely replace manual shelling, with reliable work and low cost, and is favored by growers [10].

Acknowledgement

Project source: Key Subject of High-skilled Talents in Suzhou City, Jiangsu Province GJNZ201817
Junyan ZHANG: 1978/12, Xuzhou, Jiangsu Jiangsu Union Technical Institute Suzhou Industrial Park Branch Master of Engineering Associate Professor, Senior Engineer
Research direction: Mechanical design, PLC, intelligent control, etc.

References

[1] Hengyin ZHU, Jincheng HE, Wenxi FANG, Dapeng YE, Shihua LIANG, et al. Design and Experiment of Small Fresh Lotus Seed Shelling Machine [J]. Transactions of the Chinese Society for Agricultural Engineering, 2017, 33(7): 28-35.
[2] Chuanxiang ZHENG. Design and Test of Lotus Seed Shelling Machine [J]. Transactions of the Chinese Society for Agricultural Machinery, 2003, 34(5): 106-108.
[3] Yan GU, Guanghe Xiao. A Gordon Euryale shelling machine [P]. China: CN201810960001.9, 2018-10-23
[4] Xiaoyan ZHANG. A Safer Peanut Shelling Machine [P]. China: CN201820000239.2, 2018-10-12.
[5] Xieqing XU, Honghui RAO, Tao LI, et al. Design and Test of A Fully Automatic Lotus Seed Shelling and Peeling Machine [J]. Transaction of the Chinese Society of Agricultural Engineering, 2014, 30(13): 28-34.
[6] Mingquan LAI. Lotus Seed Stripping Device and Stripping Machine [P]. 922493, 2007-07-18.
[7] Wangping WANG, Yiteng TAN. Development of Fresh Lotus Seed Shelling and Peeling Machine [J].

Food & Machinery, 2014, 30(1): 118-122.
[8] Xingyuan HUANG, Weiming XIONG, Yuhai JING, et al. Rolling-cut Lotus Seed Shelling Machine and Its Experimental Research [J]. Food & Machinery, 2012, 28(4): 139-142.
[9] Ying WANG, Fang LU, Jiahui FENG, Zhanwei WANG, et al. Design and Test of a Peanut Seed Shelling Machine [J]. Journal of Agricultural Mechanization Research, 2018, (4): 68-72.
[10] Hua ZHAO, Tao WANG, Zhuoliang HE. Design of Rubber Fruit Shelling Device [J]. Food & Machinery, 2016, 32(3): 119-121.

Application of small sample BP neural network in quantitative analysis of EDXRF

Fei Li[a, b], Zhuoyao Tang[b], Liangquan Ge[a, b,*], Nanxing Wu[b], Feng Cheng[b], Kun Sun[b]

[a] Applied Nuclear Technology in Geosciences Key Laboratory of Sichuan Province, China
[b] College of Nuclear Technology and Automation Engineering, Chengdu University of Technology, China

* Corresponding author and e-mail: Liangquan Ge, glq@cdut.edu.cn

Abstract. Quantitative analysis of EDXRF is affected by matrix effect, randomness of nuclear radiation, interaction of elements, statistical fluctuation of radiation detection process, etc. Its algorithm needs to consider many factors, and the established functional relationship is often a complex non-linear function. Therefore, effective quantitative analysis method has always been the key research direction of spectral resolution technology. In this paper, we use MATLAB software to study the effect of BP neural network in quantitative analysis of EDXRF by establishing the non-linear relationship between the counting rate of each element and the content of a single element in a small sample. The experimental results show that the small sample neural network can establish a stable structure and be applied to quantitative analysis of EDXRF, but the number of samples restricts the accuracy of prediction results, most of which can only guarantee 20% to 30% relative error.

1. Introduction

In EDXRF quantitative analysis, the sample is stimulated by X-ray to produce characteristic X-ray and scattering ray. These characteristic X-rays and scattered rays enter the detector directly. Through the pre-circuit, linear pulse amplifier, single-channel or multi-channel pulse amplitude analyzer, the energy of various rays and their corresponding nuclides or elements are determined. The signals generated by certain energy rays are analyzed to determine the activity or content of nuclides. Therefore, the objective of quantitative analysis is achieved. In this process, spectral analysis techniques are needed, including spectral smoothing, peak searching, peak location determination, element identification, peak boundary path determination, background subtraction, peak area calculation, overlapping peak decomposition and content calculation. In 2019, Li Fei et al. [1] reviewed the related algorithms of spectral smoothness, background subtraction and overlapping peak decomposition in XRF spectral analysis technology. In EDXRF quantitative analysis, there are many elements in the sample to be measured, among which the elements except the elements to be measured are collectively called matrix. Matrix effect can lead to considerable errors in EDXRF analysis. The development of matrix correction can be divided into three stages: the research period of empirical correction equation, the establishment and maturity period of basic parameter method and theoretical α coefficient algorithm, and the development period characterized by automation and intellectualization

[2]. At present, the existing matrix correction methods mainly include: theoretical influence coefficient method, basic parameter method, empirical coefficient method, experimental non-standard sample calculation, experimental correction method and so on. But these algorithms are complex and depend on a specific system. For different analysis objects, different algorithms have different analysis accuracy, thus introducing different errors.

Psychologist W. S. McCulloch and mathematical logician W. Pitts first proposed the mathematical model of simulating human neurons (MP model) in 1943, which opened the age of studying artificial neural networks. Artificial neural network does not need to consider the complex relationship between the concentration and strength of the elements to be measured. In recent 20 years, it has been widely used in EDXRF spectral data processing algorithms. Guo Panlin et al. [3] quantitatively analyzed the X-ray energy spectrum of Pt-Pd alloy by using artificial neural network and least square regression method respectively. The relationship between sample concentration and strength was obtained. The fitting degree of neural network was 0.1. Compared with the traditional least square method, the accuracy of neural network in non-linear fitting was relatively higher. In 2012, Xu Lipeng et al. [4] proposed an improved BP neural network model based on principal component analysis (PCA). The model was used to quantitatively analyze Fe and Ti elements in unknown geological samples and compared with chemical analysis. The experimental results show that the relative error between the predicted value and the chemical analysis value of the improved BP neural network is less than 3%, and the BP neural network has a good prediction effect. In 2015, Wang Jun et al. [5] proposed a hybrid algorithm of genetic algorithm (GA) to optimize BP neural network (GA-BP). The content of titanium and iron in five types of ore samples in Panzhihua mining area was quantitatively analyzed through overall prediction and classification prediction experiments. The experimental results show that the relative error of 76.7% samples is less than 2% by comparing the predicted value of GA-BP neural network with the chemical analysis value, which shows the effectiveness of this method in the correction of matrix effect. Principal Component Analysis (PCA), as a chemometric method, has a good effect in the application of EDXRF, but PCA's ability to deal with non-linear relations and generalization is inferior to that of neural networks. In 2017, J. J. Okonda et al. [6] compared PCA with BP network when using EDXRF to analyze trace bio-metals in soft tissues. In the experiment, the R2 value of the neural network ranges from 0.890 to 0.941, and the R2 value of PCA ranges from 0.680 to 0.871, which shows that the generalization ability of the neural network is better than that of PCA.

The accuracy of the results of the neural network strongly depends on the training samples. The larger the amount of data, the more accurate the prediction results are. However, in the actual production, there are occasions when the number of samples is insufficient. At this time, the evaluation of the reliability of the neural network is a key problem. This paper mainly studies the prediction ability and accuracy evaluation of small sample BP neural network in EDXRF.

2. Material and Method

2.1. The Principle of Neural Network

BP neural network is a kind of feed-forward neural network with error back-transferring (see Figure 1). BP network establishes the relationship between input nodes and output nodes through initial weights and thresholds, and obtains the initial output of the network through the forward propagation of data along the network. The specific steps are as follows:

①Input: x1,x2,x3……xn are n inputs for neurons, while the total input was expressed by X=[x1,x2,x3……xn]T.

②Network weights and thresholds: The network weights and thresholds can be adjusted in the neural network, which is also an important part of the characteristic regulation of the neural network, and plays a key role in the training and learning process of the neural network. w11, w12, w13……w1n represents the weight of the network representing the proportion of the input; b, a scalar of 1*1, represents the threshold representing a network weight whose input is constant to 1.

③Summation unit: Weighted summation of input signals: $\text{net} = \sum_{k=1}^{n} x_k w_{i,k} + b$.

④Transfer function: As the main way of neuron processing, transfer function calculates the function of sum and unit to get the output of neuron.

⑤Output: The final output is processed by weighted summation and transfer function.

Then, on the premise of defining the expected output, the network corrects the weights and thresholds layer by layer according to the error function between the calculated value and the expected value, so that the error function decreases along the negative gradient. Learning stops until the network error is less than the expected error or the maximum number of training times is reached. The network will record the current weight matrix and threshold matrix for effective calculation and prediction of data to be processed. [7]

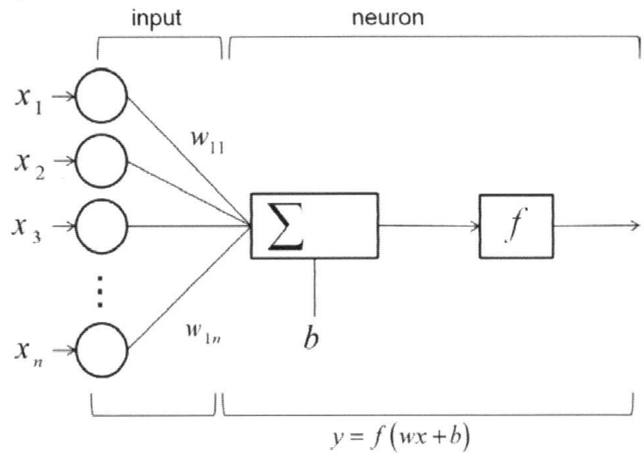

Figure 1. BP neural network

2.2. Experimental Design

The purpose of this experiment is to study the effect of small sample BP neural network in dealing with non-linear problems in EDXRF quantitative analysis. Using BP neural network, the non-linear relationship between the counting rate of each element and the content of a single element is established. The function relationship that needs to be determined can be expressed as follows:

$$C_i = f(w_1A_1 + b_1) + f(w_2A_2 + b_2) + ...f(w_iA_i + b_i) + ...f(w_nA_n + b_n)$$

In the above formula, Ci represents the content of an element. Ai represents the counting rate of an element. w and b represents the weight and threshold respectively. f (·) represents the activation function. The tangent S-type function is used in this experiment; n represents the number of elements to be processed, and the number of experimental data is 6.

In this experiment, three neural network models of target elements (Fe, Cu, Pb) were established to determine the relationship between element content and the counting rate of each element. Input experimental samples for training, the network adjusts and corrects the weights and thresholds of the above function relations continuously until the error reaches below the preset error, then stop training. At this time, BP neural network can be identified as stable, and error analysis and prediction can be carried out. The specific steps of the experiment are as follows:

①The EDXRF data are sorted and grouped for training set samples of neural network training, including the counting rate and standard content data of 15 groups of samples of 6 elements. The prediction set after network training is detected, including three sets of counting rate data of six elements and corresponding actual content data, for comparison with the predicted output value.

②The BP neural network is constructed by MATLAB. The counting rate data of each reference element is taken as input vector, the content data is taken as target output, the initial parameters of the network are set, and the training of the neural network is carried out.

③After the network training is completed, the element counting rate imported into the prediction set is compared with the actual content data, and the error is analyzed and evaluated.

Fifteen groups of standard counting rates and standard content data of six elements Zn, As, Fe, Ti, Cu and Pb were selected as training set samples of the neural network before using MATLAB to construct the neural network (see Table 1 and Table 2).

Table 1. Counting rate of each element in training sample

	Standard Counting Rate of Elements					
Number	Zn	As	Fe	Ti	Cu	Pb
1	3.87634	0.376302	199.647	0.333595	3.31702	9.57401
2	5.36762	0.077832	126.214	0.292757	1.74736	10.1929
3	3.3076	0.159167	76.1615	0.443485	1.44727	22.2033
4	1.36034	0.17394	404.633	0.22188	2.65149	4.89463
5	0.7455	1.64159	483.343	0.00222	6.61986	2.35262
6	1.03209	4.69056	320.782	0.318703	5.19784	2.64562
7	16.5059	0.313031	144.679	0.132521	1.83673	11.1681
8	10.3162	0.292433	167.151	0.165875	1.86819	10.0682
9	6.22	0.483652	149.821	0.324392	1.59367	11.9402
10	3.17609	0.399014	153.92	0.39988	1.62931	6.7775
11	13.2366	0.414737	168.244	0.050899	2.70346	12.0419
12	11.3321	0.591143	142.52	0.259835	0.896827	10.3352
13	5.46654	0.231044	203.736	0.457959	1.88377	8.35871
14	3.64284	0.41217	246.55	0.053573	3.30441	6.18569
15	1.7316	0.518939	213.743	0.43542	3.17259	5.31308

Table 2. Element Content of Training Sample

	Standard content of each element (except Fe content in units of % other units are micrograms per gram)					
Number	Zn	As	Fe	Ti	Cu	Pb
1	1.72	21	3.63	20.4	98	155
2	2.36	16.3	2.46	19.4	20	187
3	1.27	11.4	1.40	12	26	380
4	0.26	40	7.21	64	58	77
5	0.1	144	8.83	40	552	42
6	0.16	390	5.66	53	314	39
7	8.27	24.3	3.14	31.5	21	236
8	5	25	3.36	33	25	172
9	2.62	19	2.92	26	22	226
10	1.33	21.4	2.95	25.4	24.7	182
11	5.83	29	3.30	32	19	240
12	5	21.6	2.88	28.5	21.6	195
13	2.45	27.4	3.72	33	31	152
14	1.53	37	4.51	41	38	115
15	0.4	32	3.81	27.4	61	68

The counting rate of another 10 groups of samples is selected as the prediction set of the neural network (see Table 3). The output of the network operation is compared with the real value, and the error magnitude and reason are analyzed.

Table 3. Element Counting Rate of Prediction Set

	Counting rate of each element (counting rate)					
Number	Zn	As	Fe	Ti	Cu	Pb
1	3.56897	0. 054897*	236.4	0.23597	2.46358	4.89463
2	2.65847	0.077832	135.68	0. 5948*	1.5798	3.6897
3	1.5987	0.19894	89.564	0.0857	1.44727	21.568
4	1.2867	0.15896	386.59	0.18697	1.4589	3.5897
5	0.7455	1.64159	498. 6*	0.00222	6.61986	2.35262
6	0. 6359*	5. 1648*	331.548	0.218703	6. 92649*	2. 03256*
7	14.5889	0.28978	120.265	0.132521	1.76458	10.5898
8	8.6976	0.189747	145.654	0.12354	1.86819	10.0682
9	5.49	0.35978	126.56	0.35947	1.2564	10.2659
10	2.8975	0.2945	112.56	0.37584	1.2359	5.4777

The data with a * sign in the table indicates that the value is outside the training sample range.

3. Results and Discussions

The experiment uses MATLAB software to realize the neural network operation, imports the data of training set samples, takes the standard counting rate of each element as Input Data, and takes the standard content as Target Data. Then the neural network model is constructed for each element quantitative analysis in turn. In the network window, the type of neural network is selected as Feed-forward BackProp (BP Neural Network). Only one hidden layer, one input layer and one output layer are set. The neurons in the hidden layer are set to eight. An input and target are selected. The activation function is selected as TANSIG (tangent S function), and the neural network model is generated (see figure 2).

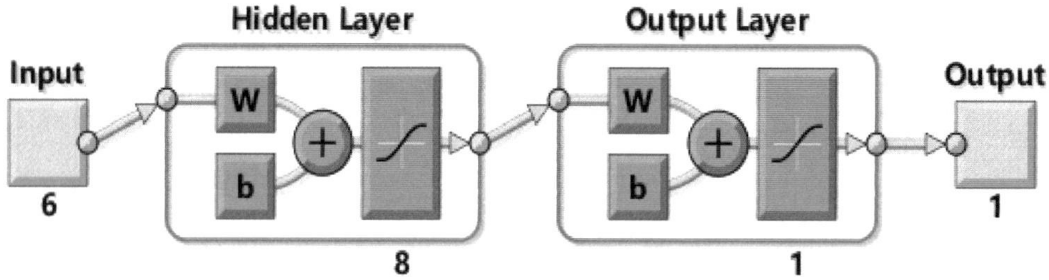

Figure 2. Neural Network Model

Some network parameters can be initialized by train window. Through repeated training experiments, the training step length is 1000, the learning time is 120 seconds, the maximum error number is 6, and the weight change rate is 0.001. At this time, the training effect is the best.

When the network is trained, the input data (i.e. counting rate data) are classified. Most of the data are used as training group for network training and learning, and the functional relationship between counting rate and content is grasped. A small number of data are extracted as test and validation groups, which are used to test and verify the functional relationship after the network completes training. Finally, the total (all) training data are fitted, and the operating chart and fitting chart of Fe, Cu and Pb are obtained respectively.

Journal of Physics: Conference Series **1423** (2019) 012054 doi:10.1088/1742-6596/1423/1/012054

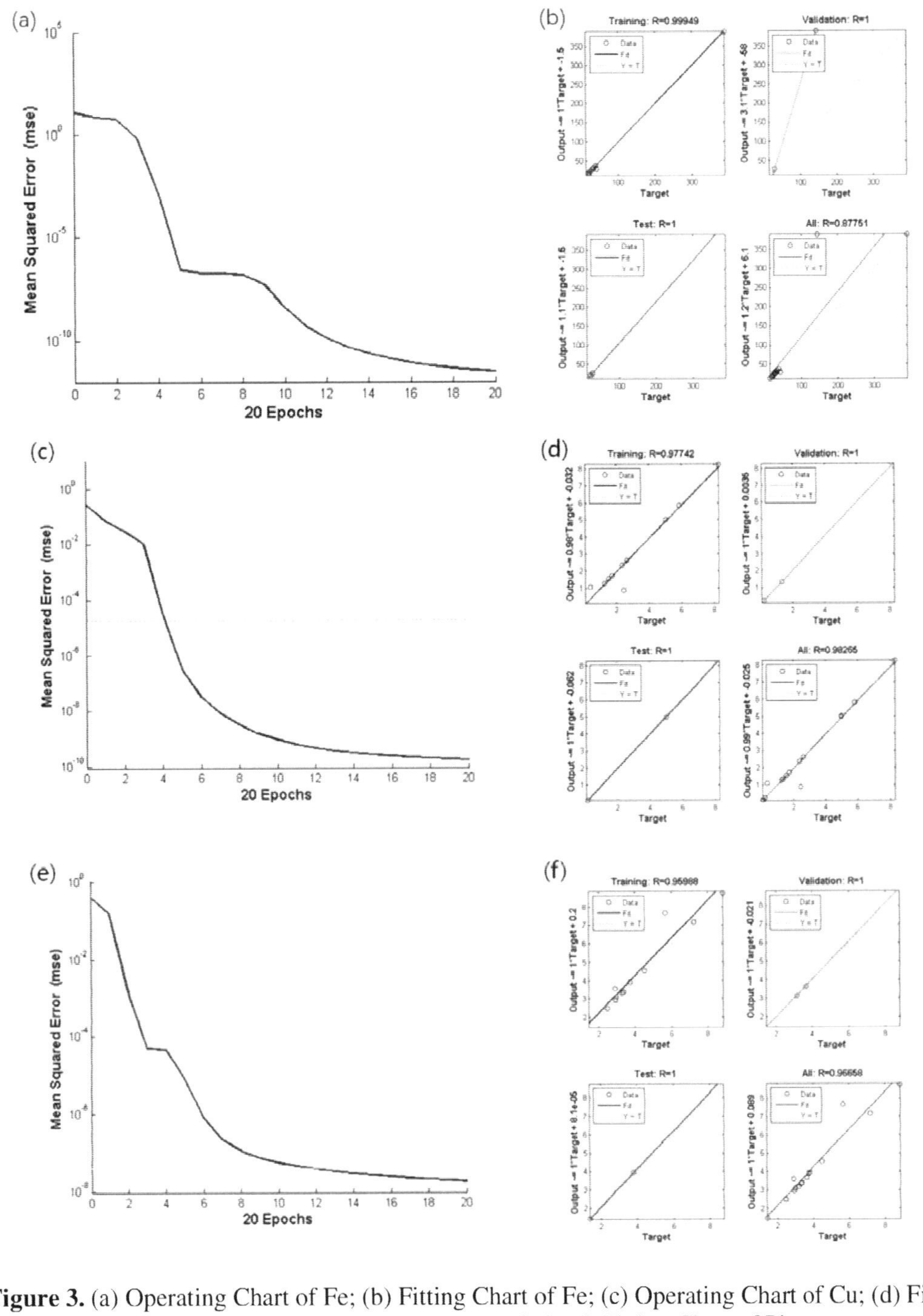

Figure 3. (a) Operating Chart of Fe; (b) Fitting Chart of Fe; (c) Operating Chart of Cu; (d) Fitting Chart of Cu; (e) Operating Chart of Pb; (f) Fitting Chart of Pb;

The prediction set data of Fe, Cu and Pb are simulated in turn and compared with the actual data to make a comparison error table. The results are as follows:

Table 4. Data Error Table of Fe, Cu, Pb Prediction Set

	Fe Prediction Error Analysis Table		Cu Prediction Error Analysis Table		Pb Prediction Error Analysis Table	
Number	absolute error	absolute error	absolute error	absolute error	absolute error	relative error
1	-0.90435	15.3019	37.9205	37.9205	-0.90435	-0.1957
2	-0.75999	15.6572	3.3605	3.3605	-0.75999	-0.2969
3	-0.45971	-9.5444	6.6806	6.6806	-0.45971	-0.2873
4	0.045206	10.1086	12.5479	12.5479	0.045206	0.0067
5	0.79023	-1.9446	0.0080129	0.0080129	0.79023	0.0827
6	-4.3299	-11.0009	-265.4928	-265.4928	-4.3299	-1.0260
7	-0.28057	1.2599	-3.0464	-3.0464	-0.28057	-0.1087
8	-0.32935	31.4553	-2.1854	-2.1854	-0.32935	-0.1272
9	-0.91627	-108.1182	-2.0254	-2.0254	-0.91627	-0.5763
10	0.53373	105.1649	-1.1202	-1.1202	0.53373	0.2765

From the error analysis table, it can be seen that the prediction errors fluctuate between 0.6% and 60%. There are five groups of data errors below 20%, three groups of data errors above 30%, and two groups of data errors between 20% and 30%. The overall accuracy of Pb element content prediction is general. From the error analysis table, it can be seen that the prediction error fluctuates between 0% and 90%. Six sets of data errors are less than 20%. Three sets of data errors are more than 30%. Only one set of data errors is between 20% and 30%. The overall accuracy of Fe element content prediction is relatively poor. From the error analysis table, it can be seen that the prediction error fluctuates between 0% and 90%. Six sets of data errors are less than 20%. Three sets of data errors are more than 30%. Only one set of data errors is between 20% and 30%.

In the operation of neural network, the overgeneralization of input data may lead to the error of final prediction results. This overgeneralization refers to the upper and lower bounds of the training set data used in the training process of the neural network. When the network training is finished and the new data is predicted and simulated, if the input data is not within the upper and lower bounds of the training set, the generalization problem will arise. If this generalization problem occurs on multiple input data, the accuracy of the final prediction results of the network will be greatly reduced. The effect of neural network prediction and simulation depends on the range of original data. The neural network used in this experiment has some limitations in the application of actual prediction and measurement data.

4. Conclusion

In this paper, the quantitative analysis of EDXRF based on small sample BP neural network algorithm is studied. Through experiments and data analysis, it is considered that the application of neural network in quantitative analysis of EDXRF can be realized and has certain advantages. However, when the number of samples is relatively small, there are some drawbacks. Because the amount of data involved in fitting is insufficient, it is easy to lose some of the original features of the data after training. In addition, small samples can easily lead to overgeneralization of the neural network, resulting in great error in experimental prediction. In practical applications, neural networks often rely too much on the data to be processed. Although small samples can reduce the time of data pre-processing, they can not guarantee the accuracy of analysis in EDXRF.

Acknowledgement

This work was funded by the National Natural Science Foundation of China (No.41774147), the National Key R&D Project (No.2017YFC0602100) and Science and technology activity project for overseas personnel in Sichuan province.

References

[1] Li Fei, Ge Liangquan, etc. 2019. Recent developments on XRF spectra evaluation.Applied Spectroscopy Reviews.

[2] Gi ang, Zhuo Shangjun, Le Guohui. 2011. Energy dispersive X-ray fluorescence spectroscopy. Science Press.

[3] Guo Panlin, Wang Jiqing, etc. 1999. Application comparison of artificial neural network and least squares regression in quantitative analysis of XRF. Nuclear technology. P725-730.

[4] Xu Lipeng, Ge Liangquan, etc. 2013. Application of EDXRF analysis based on PCA-BP neural network in determination of iron and titanium in geological samples. Spectroscopy and spectral analysis. P1392-1396.

[5] Wang Jun, Liu Mingzhe, etc. 2015. BP neural network optimized by genetic algorithm is used to predict the content of ferrotitanium in EDXRF. Atomic energy science and technology. P1143-1148.

[6] Okonda, J.J., Angeyo, K. H.. A nested multivariate chemometrics based calibration strategy for A nested multivariate chemometrics based calibration strategy for direct trace biometal analysis in soft tissue utilizing Energy Dispersive X-Ray Fluorescence (EDXRF) and scattering spectrometry. Applied Radiation and Isotopes. P49–56.

[7] Yang Huijun, Feng Keming, etc. 2019. Improved GPT2w model based on BP neural network and its global precision analysis. Systems Engineering and Electronic Technology. P500-508.

Journal of Physics: Conference Series **1423** (2019) 012055 doi:10.1088/1742-6596/1423/1/012055

Structural design of anthropomorphic robot vision system

Junyan Zhang

Jiangsu Union Technical Institute Suzhou Industrial Park Branch, Suzhou, China

Corresponding author and e-mail: Junyan Zhang, 108645502@qq.com

Abstract. Anthropomorphic robot is one of the latest research directions in the field of intelligent robot research. It is an intelligent robot with similar shape and humanity. It has mobile function, operation function, sensing function, memory and autonomy, and can realize human-computer interaction. Robot technology has always been a hotspot of technology that people pay attention to. Because of the diverse development of robots, machine vision systems are gradually being established. The machine vision system realizes the functions of collecting images, analyzing images, acquiring information, processing images and others by simulating the human eye. It is a comprehensive technology integrating automation and computer. The article deeply studies the hardware composition and image processing technology of machine vision system and gives an example study on the application of machine vision in various fields. It also analyzes the current development status of machine vision system in China and looks forward to the bright future of machine vision.

1. Introduction

With the rapid development of science and technology, the development of robots has been nearly a hundred years old. In the medical field, medical robots assist or even replace humans to diagnose and operate patients. In the industrial sector, robots replace humans for inspection and packaging of products. In addition, robots have also been used in the military field, such as helping soldiers observe the battlefield and assist shooting. Among them, the machine vision system uses the digital camera to collect various kinds of information, analyzes and processes the image data through the central processor, and obtains useful information in the picture to help people complete the tedious and monotonous work. In the development of machine vision technology, it also keeps pace with the trend of social development. From the popular PS to the current face recognition technology, it gradually moves toward artificial intelligence, adding more convenient channels for people's lives. This article will be on the machine vision system. The hardware and software are described separately.

2. Design principles

Autonomous Mobile Robot (AMR) refers to an autonomous robotic system with athletic ability, which combines the characteristics of anthropomorphic robots and autonomous robots. Due to its wide and attractive application prospects, the research of autonomous humanoid robotics involves many subject areas such as robotic mechanical structure design, artificial intelligence, computer vision, motion control, planning decision-making, image processing, pattern recognition, information fusion, etc. There are a lot of challenges in the research, so the autonomous humanoid robot technology has become one of the focuses of the current research in the field of robotics. One of the key technologies of autonomous anthropomorphic robots is how to make them robust to autonomous navigation. Autonomous anthropomorphic robots can only have a purposeful movement when they know where they are and where the various targets are. At the same time, they can detect changes in the environment and their own state in real time, and adjust their behavior accordingly to effectively accomplish specific tasks. Therefore, in order to achieve autonomy, the first task of anthropomorphic robots is to obtain an understanding of the working environment and its own state, that is, to realize the perception of the robot, which is one of the three basic elements recognized by the robot (ie, perception, planning, execution).

Among them, the visual sensor can provide the most abundant sensory information compared with other robot navigation and positioning sensors. It is also a passive sensor, which is closest to the human environment perception mode, and its price is relatively moderate. Scientific research and statistics show that about 75% of the information that humans receive from the outside world comes from the visual system. Vision sensors also have the advantage of being able to work in high-rise urban environments or indoor environments, which is not possible with GPS receivers only; it can work with a large amount of highly dynamic occlusion, which is It is difficult to achieve based on laser, ultrasonic and other distance sensors; there is no accumulation of inertia sensor error with working time. At the same time, theoretical and technical advances in related fields such as computer vision and pattern recognition have also promoted the application of visual sensors in autonomous humanoid robots, such as target recognition, scene recognition, and self-positioning. The application of vision to autonomous humanoid robots has attracted the participation of a large number of researchers around the world, and has also achieved a lot of research results, such as IEEE Transaction on Robotics, IJCV, RAS, CVIU and other famous journals and ICRA, IROS, ICCV, etc. A large number of papers have been published at mainstream conferences such as CVPR and ECCV. Vision sensors have become an indispensable sensor for autonomous humanoid robots, and computer vision research and anthropomorphic robot research have also achieved good integration. Among various vision sensors, the omnidirectional vision device can obtain the panoramic information of the surrounding area of the robot in one image because of the horizontal angle of view of 360°. Through image processing, analysis and understanding, the target recognition and construction of the robot can be realized. Graphs, self-positioning, etc., have been used more and more in various anthropomorphic robots. This chapter will detail the history, development and application of omnidirectional vision in the following sections, especially in the application and challenges of autonomous anthropomorphic robots.

3. Hardware structure

The machine vision system gives the robot a view and can quickly analyze, resolve, and process information on complex images. The application based on this robot has been broadened, reducing the time for humans to handle cumbersome work and making our lives more convenient and efficient. In the pharmaceutical industry, the production process of drugs requires strict management control and quality control, and the machine vision system can realize this series of cumbersome operations and maintain the safety of patients. In laser processing, laser processing technology combined with machine vision technology makes processing more precise and reduces processing costs. The machine vision system consists of five parts: light source, lens, camera, image acquisition card and image processing system.

The light source is one of the important structures of the machine vision system. There are different requirements for the type of light source in different situations. Generally, it is a heat radiation power source, a gas discharge light source, a solid discharge light source, and a laser. LEDs are the most commonly used among light sources, and LEDs are divided into many clocks because of their different shapes. For example, when we perform microscope illumination, we use a ring light source to highlight the three-dimensional information of the object. Use a backlight to detect scratches on transparent objects, highlighting the contours of the object. Not limited to the LEDs mentioned above, the appropriate light source can be selected in different scenarios to capture valid picture information.

The lens is an important component of the image acquisition system of the machine vision system. It is like the human eyeball to observe the distance and range of the object under test. Its types are standard, telecentric, wide-angle, telephoto and close-up. People choose according to the object distance, shooting range, aperture, focal length and other conditions.

The camera is the core component of image acquisition. The light signal is converted into an electrical signal by a photosensitive element, and the conversion of light and digital signals is realized, and the original image is collected. Cameras are also divided into a variety of ways depending on the classification method. According to the chip classification, the most common types are CCD camera and CMOS camera. They are digital cameras for photoelectric conversion. However, due to different chip types, the processing power is also different. CMOS technology is more energy efficient than CCD. More efficient, but more expensive.

The image acquisition card is an indispensable part of the image processing part. In the processing of video information, the image acquisition technology is essential to digitize the image and store it in the form of a data file in the hard disk of the computer. Image acquisition generally relies on a large number of photosensitive elements. For example, when the resolution is 1 080×720, there are 1 080 small rectangular rectangles on the photosensitive element and 720 small rectangles on the longitudinal side, waiting for the light intensity to be collected. Since the sensitivity of the light intensity received by the photosensitive elements is different, so that their resistance changes, the effects they produce will also differ to obtain the initial image. The larger the pixels that make up the picture, the clearer the image is, forming what people call high-definition pictures. In the machine vision system, the image processing system performs a series of analysis and processing through the image data obtained by the image acquisition card to obtain the desired information. First, the image information is collected by an image acquisition card, and the optical signal is converted into an electrical signal. The digital image is transmitted to the processor for grayscale processing, turning the part we are interested in into white, and the uninteresting part turning black, thus distinguishing between what we need and what we don't need. However, after obtaining the binarized image, there will be some noise around the target object, which affects the value of the edge of the subsequent judgment object. In order to eliminate these noises, we use image etching and expansion processing/filtering processing to corrode the target object, and then use the convolution operation to amplify the target, remove noise, and achieve boundary detection and image enhancement. After comparing with the database to obtain accurate information of the target in the field of view, based on this, the judgment and operation of subsequent execution.

4. Omnidirectional vision system design

Various designs of omnidirectional mirrors satisfying single-view imaging. The common feature of these mirrors is that the mirror shape is simple, the profile curve can be accurately described by mathematical analytic, and the processing difficulty is relatively small, and the omnidirectional configuration using these mirrors is relatively small. The main drawback of the vision system is that the obtained panoramic image has a large barrel distortion, and the resolution of the target imaging decreases rapidly as the distance from the visual system increases, and the imaging of the distant scene is small. A typical image captured by a hyperbolic vision system using a hyperbolic mirror in a 12m x 7m RoboCup medium group competition venue is shown in Figure 1. This omnidirectional vision system does not use football robots for a wide range of target recognition and accurate measurements for applications such as the RoboCup medium group robot soccer game. In response to the above-mentioned shortcomings of the single-view omnidirectional vision system, the researchers proposed a mirror design method that satisfies the specific imaging resolution requirements of various parts of the environment around the omnidirectional vision system, and also designed three imaging resolutions. The mirror with the same rate is the horizontal mirror, the vertical mirror and the angle mirror, which can realize the horizontal direction (perpendicular to the mirror rotation axis), the vertical direction (parallel to the mirror rotation axis) and the specular incident ray angle. The imaging resolution on the top is constant. None of these three mirrors can be used to construct an omnidirectional vision system that satisfies single-view imaging. This section first introduces the design of horizontal mirrors and vertical mirrors, and on this basis, designs a new combination mirror for the NuBot omnidirectional vision system.

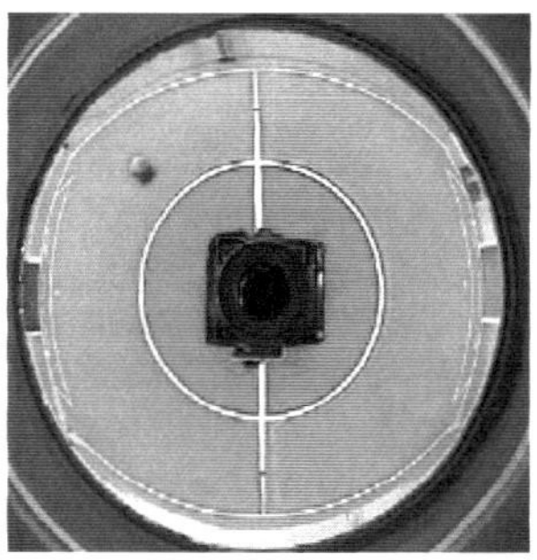

Figure 1. Typical panoramic image captured by an omnidirectional vision system composed of a hyperbolic mirror

The horizontal equal-mirror mirror refers to a mirror surface that satisfies the distance relationship between the horizontal scene and the corresponding points in the formed image when the omnidirectional vision system is installed vertically. The mirror image is shown in Figure 2.4. F (t) is a function of the specular profile, and the point P (d, 0) on the horizontal plane to the optical axis with a distance d is imaged on the camera CCD far from the optical axis x. According to the principle of the angle of incidence, the angle of incidence is equal to the angle of exit, $\varphi = \psi + \theta$, so $\tan(\varphi + \theta) = \tan(\psi + 2\theta)$, and because 'tan θ = F (t), tan ψ = x / f, where f is the focal length of the camera, so there is

$$\frac{d-t}{F}=(\frac{x}{f}+\frac{2F^{'}}{1-F^{'2}})/(1-\frac{x}{f}\times\frac{2F^{'}}{1-F^{'2}})$$

zF + 2 F − z= 0, because the mirror is a convex mirror, take its positive root

$$F^{'}=\frac{-1+\sqrt{1+z^{2}}}{z}$$

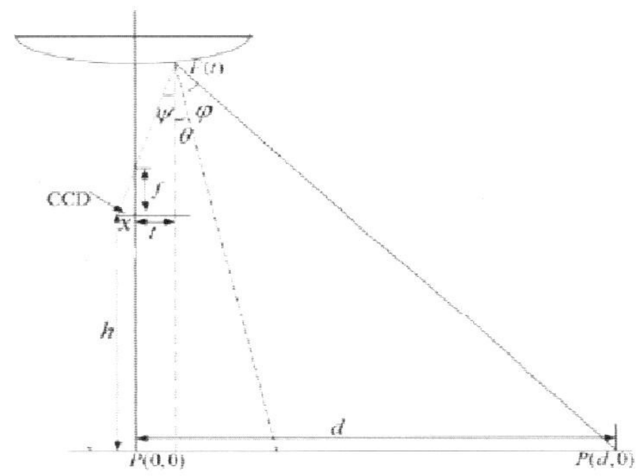

Figure 2. Schematic diagram of horizontal equal mirror design

5. Calibration of the omnidirectional vision system

The calibration of the vision system refers to determining the parameters of the mapping relationship between the two-dimensional image points on the imaging plane of the camera from the three-dimensional space point to the vision system. The vision system can only be applied to computer vision tasks such as visual perception measurement, environmental 3D reconstruction, visual system's own motion estimation, etc., and the accuracy of calibration directly determines the application effect. The omnidirectional vision system is no exception, and since the panoramic image obtained by the omnidirectional vision system has lower resolution and severe distortion compared to the ordinary fluoroscopic image, the calibration problem becomes more important. Since the omnidirectional vision system that satisfies single-view imaging has good characteristics, that each point in the panoramic image uniquely corresponds to an incident ray from the environment on the omnidirectional mirror surface, as shown in Figure 2.1, so based on imaging the omnidirectional visual calibration method of the model has been extensively studied and many achievements have been made. The existing calibration methods can be roughly divided into two categories: methods for prior knowledge of known environments, and methods that do not require the use of environmental prior knowledge. The first type of method requires three-dimensional coordinates of a series of spatial points in the environment in which the omnidirectional vision system is located, and the image points corresponding to these spatial points should be automatically extracted (or manually extracted), which we call These spatial points are control points. These control points are often obtained by setting a calibration block in the environment. The relationship between the three-dimensional space coordinates of the control points and the two-dimensional image coordinates of the corresponding image points can be used to recover the internal and external parameters in the imaging model of the omnidirectional vision system. Since the three-dimensional space coordinates and two-dimensional image coordinates of the control points need to be determined, the calibration process is cumbersome. When there is no control point in the environment, this method requires several sets of lines parallel to each other in the known space, using the intersection of the conic curves imaged by these lines in omnidirectional vision (parallel lines in the panoramic image) the nature of the vanishing point is the internal parameter information of the omnidirectional vision system. The calibration idea is mainly based on the unified imaging model of various single-view omnidirectional vision systems proposed by Geyer and Danilidis. A detailed description of the model can be found in the literature. During the calibration process, the relative position between these lines and the relative position between the line and the omnidirectional vision system are not known. Ying and Hu derived a linear projection quadratic curve to provide three independent constraint equations for the internal parameters of the omnidirectional vision system, and a spherical quadratic projection curve to provide two independent constraint equations for the internal parameters of the system. Furthermore, a new calibration method for omnidirectional vision system based on geometric invariants is proposed, and it is concluded that the method based on spherical projection invariants can obtain more robust calibration results. Wu et al. also proposed a linear single-view omnidirectional vision system calibration method, which can simultaneously calibrate all omnidirectional vision systems except the main point by imaging information in at least three lines in a panoramic image. Internal parameters. The second type of method, also known as the self-calibration method, calibrates the omnidirectional vision system by using the polar line geometric constraint relationship between the pairs of points corresponding to multiple images. The polar geometry of various single-view omnidirectional vision systems is based on their respective imaging models. For a detailed derivation, see. The advantage of this method is that there is no need to use calibration blocks, and there is no need to know any prior information of the environment. The disadvantage is that it is necessary to be able to obtain corresponding pairs of points between images robustly and accurately.

6. Camera parameter automatic adjustment algorithm for image entropy

In digital cameras and home DVs, researchers have proposed some parameter adjustment mechanisms to improve imaging effects, such as changing the aperture or shutter time for automatic exposure, auto white balance, and auto focus. In some multi-slope response cameras, the researchers adjusted the response curve through automatic exposure control to adapt the camera's dynamic response range to different lighting conditions. But these methods are based on the camera hardware level, because we can't operate the internal hardware of most cameras (except some special hardware-supported cameras) of the robot vision system, so we can't use these methods. This paper attempts to adjust some parameters supported by the camera through external software to improve the imaging effect of the vision system under different lighting conditions. Some related research is currently in the RoboCup medium group. The RoboCup Medium Group Robot Soccer is a standard test platform for robot vision related issues. Although the latest rules of the game have undergone some major changes, such as turning the yellow/blue goal into a net like a human race, the colored columns have been cancelled, but the game environment is still largely color-coded. The ultimate goal of RoboCup is that the robot soccer team can defeat the human world champion. The robot will need to be able to compete in a highly dynamic environment or even an outdoor environment, so how to make the designed vision system robustly recognize the color coding. The goal remains a research focus for RoboCup researchers. In addition to adaptive color segmentation methods, color online learning algorithms, and color recognition-free target recognition algorithms, some researchers have also attempted to implement visual sensors by adjusting camera parameters. Robustness. Reference defines the camera parameter adjustment problem as an optimization problem and uses genetic algorithms to minimize the distance between the actual color value of the pixel and the theoretical color value in the artificially selected image region to obtain the optimal solution to the problem. Since the theoretical color value is used as the reference value, the influence from the ray condition can be eliminated, but the method requires manual selection of some special image regions as reference information. Reference designed a set of PID controllers that adjust camera parameters such as gain, aperture, and two white balance channels based on the pixel color values of the white areas that are always visible in the imaging of the omnidirectional vision system. Reference designed a PI controller to adjust the exposure time to adjust the color value of the reference green area to the desired color value. Reference proposes to automatically set camera parameters such as exposure time, gain and white balance of the omnidirectional vision system based on the luminance information of the entire image and the information of the known black and white regions in the image, but the method requires Each black and white swatch is placed in advance on the field, so it can only be used for offline adjustment calibration before the game. These methods require some reference colors in the camera parameter adjustment process, thus limiting their application in many other applications.

Using an omnidirectional vision system to obtain a series of panoramic images at different exposure times and gains in indoor and outdoor environments and calculating the image entropy of all images according to equation, it is possible to analyze how the image entropy changes with camera parameters. The indoor environment is a standard 18m x 12m RoboCup medium-sized group competition venue. The lighting conditions are not only dependent on the indoor lighting, but also may be greatly affected by natural light through a large number of glass windows. The outdoor environment contains two blue swatches and some elements of the indoor environment, such as a green carpet, two orange football balls and black obstacles. The experiments for all omnidirectional vision systems in this chapter will be performed in both environments. Since the lighting conditions of the two environments are completely different, and the dynamic response range of the cameras used is limited, two sets of omnidirectional vision systems with different aperture settings (the lens aperture can only be manually adjusted) are used in indoor and outdoor environments. (i.e. two soccer robots) to conduct experiments. In indoor environment experiments, the exposure time ranges from 5ms to 40ms, and the gain ranges from 5 to 22. The time of the experiment is night, so the lighting conditions are only illumination, not affected by natural light. In outdoor environment experiments, the exposure time ranges from 1ms to 22ms, and the gain ranges from 1 to 22. The weather is cloudy, and the experiment time is noon. The minimum adjustment steps for the two parameters are 1ms and 1 respectively. During the experiment, a panoramic image is captured after each set of camera parameters is set.

Journal of Physics: Conference Series **1423** (2019) 012055 doi:10.1088/1742-6596/1423/1/012055

Considering that all images corresponding to the image entropy on the ridge curve are suitable for robot vision, by defining a certain search path, the two-dimensional optimization problem can be converted into a one-dimensional optimization problem. Since the RoboCup mid-group competition is a highly dynamic and color-coded environment, high exposure times and low gains can cause real-time performance degradation of the system, while low exposure times and high gains can cause more noise in the image, which in turn causes color Segmentation and other image processing results deteriorate. Therefore, it is necessary to make a compromise between the exposure time and the gain, and neither of the two values should be too high. Depending on the parameter range of the camera used, the search path is defined as "exposure time = gain" (since the exposure time is in milliseconds, the gain is unitless, so it is only equal in value), and the maximum image entropy is searched along this path. The camera parameters corresponding to the image entropy are the optimal parameters under the current environment and current lighting conditions. According to the imaging characteristics of omnidirectional vision, the robot itself is imaged in the central portion of the panoramic image. Therefore, in practical applications, the robot can determine whether the robot has entered a new working environment or whether the lighting conditions in the current environment have changed by calculating the brightness average of the central region of the image. If the increase in the mean exceeds a certain threshold, the robot assumes that the ray condition becomes more intense, so the optimization of the camera parameters will follow the search path and toward the exposure time and gain reduction. Similarly, if the decrease in the mean exceeds the threshold, the optimization process proceeds along the search path and toward the exposure time and gain increase. In the experiment, the threshold was set to 20. During the optimization process, the mean value of the brightness in the center of the image does not need to be recalculated, and the change does not need to be considered. After the optimal camera parameters are obtained, the brightness average is recalculated and saved, and compared with the new value to determine whether adjustment and optimization of the camera parameters are needed. Since the lighting conditions do not change all the time, this judgment only needs to be performed once every certain time.

When the robot is located at different positions in the field, the content contained in the image is different, and the maximum image entropy on the search path changes as the position of the robot changes. Therefore, once the robot finds that the camera parameters need to be adjusted, it will stop on the field until the camera parameter adjustment process is completed. Moreover, according to the discussion in the following section 3.4, the camera parameter adjustment can be completed in a few hundred milliseconds. In the optimization process, the surrounding environment of the robot can be approximated as static, and the image entropy only changes with the camera parameters. The change, so the maximum image entropy and the optimal camera parameters can be easily searched, and the optimization of the camera parameters is not affected by the surrounding environment of the robot. This paper will also verify the camera parameters optimized in the static environment through experiments. It is also an ideal parameter in a highly dynamic environment. During the optimization process, a new set of camera parameters is set to the camera, and the robot acquires a new panoramic image and calculates its image entropy according to equation. The new image entropy is used to compare with the image entropy obtained in the previous optimization step to determine if the maximum value has been reached. This iterative process continues until the maximum image entropy is searched. On how to choose new parameters, the algorithm uses variable step size techniques to speed up the search process. When the current image entropy distance Max E is not large, the optimization step size is set to 1, that is, the change of exposure time is 1 ms, the change of gain is 1; when the current image entropy distance ( )cMax E is large, the optimization step size can be Set to 2 or 3. Depending on the camera characteristics and the different requirements of the vision system for different applications, the camera parameters need to be adjusted and optimized along different search paths. For example, different cameras have widely different gain ranges. Some cameras have gains from 0 to 50, and some have gain ranges from 0 to 4000. Therefore, the search path can be defined as "exposure time = α * gain" (which is also only numerically equal), and the parameter α can be determined after analyzing the influence of the camera gain on the image. If the camera's parameter range is similar to that of the camera used in this paper, in some cases, the signal-to-noise ratio of the image is required to be high, but the real-time performance of the vision system is not too high. The search path can be defined as "exposure time = α * gain". Where $\alpha > 1$; if other occasions require the camera to output the image as fast as possible, and the image noise requirements are not strict, then the search path can be defined as "exposure time = α * gain", where $\alpha < 1$.

7. Machine vision applications

In the military field, machine vision systems play a very important role in aviation, aerospace, and tracking. The Air Force mainly uses drones to complete air missions, such as targeting and accurate annihilation. In the case of high altitude and movement, by matching the infrared sensor, the picture information is returned in real time, and the target object of the database is matched. The similarity is extremely high, and a strong infrared signal is detected, and the enemy information such as illegal nuclear weapons is judged in the area, to accurately strike the target object. Reduce personnel losses and improve operational accuracy. When performing the high-altitude

Journal of Physics: Conference Series **1423** (2019) 012055 doi:10.1088/1742-6596/1423/1/012055

reconnaissance mission of the drone, the high-resolution image acquisition and analysis processing helps us to monitor the surrounding environment at any time and prevent the enemy from invading while observing the daily operation of the enemy.

In the industrial field, product quality inspection of repetitive machinery consumes a lot of labor costs, and the processing speed is slow, which is affected by disturbance factors such as weather, physical state and working time. The robot vision system is not affected by various interference factors and has an excellent processing system to quickly identify the quality defects of large-volume products in a few seconds, which is fast and accurate. It is calculated that the industrial robot vision sorting system can complete the workload of 500 workers a day, greatly improving the efficiency of production and liberating the labor force. Industrial vision robots can continuously learn, and each robot can quickly refresh knowledge through the network, continuously learn and consolidate, reduce error links and improve accuracy, which is equivalent to having numerous reproducible senior inspectors. Not only in the inspection work, the factory intelligent robot can also use the machine vision positioning to quickly locate the machine on the assembly line, quickly analyze the material and perform processing operations.

In the civilian field, the scope of application of machine vision recognition technology is very extensive, and there are deep contacts in face verification, security, and transportation. We can see it on the computer in the security room. The computer obtains the license plate number of the vehicle entering and leaving the vehicle through the captured picture and compares whether the vehicle is the license plate number of the community, and calculates how long the vehicle enters and exits the area. It simplifies the vehicle entry and exit process to a certain extent, providing a safe and fast access for vehicles to enter and exit. When entering the airport check-in, the automatic ticket checking machine compares the face with the ID card photo, quickly and accurately judges, saves labor costs, and speeds up airport security. In addition, the camera is used to detect suspects in real time, and the photos taken by various cameras are compared with the photos of the prisoners in the database to analyze the similarities, so that criminals can be identified and arrested in the vast sea of people.

8. Conclusion

The omnidirectional vision system has become one of the more important sensors for autonomous anthropomorphic robots and has extensive attention and research on it. This paper focuses on how to improve the robustness of omnidirectional vision system and its application in target recognition and self-positioning. According to the software system of the anthropomorphic robot vision system designed in this paper, the image captured by the dual camera is better realized, and the modular structure with easy expansion is provided for the realization of algorithms such as image processing, ranging and recognition in the future, better software and hardware platform.

References

[1] Liu Huacheng. Design of Robot Vision System [J]. Science and Fortune, 2013(12):11-12.
[2] Du Junxi, Huang Xinhan. Design of Chess Robot Vision System [J]. Electronic Technology Application, 2007, 33(9): 133-136.
[3] Guan Bing, Gu Jinan, Chen Sijie, et al. Vision System Design of Cylinder Workpiece Loading and Unloading Robot [J]. Mechanical Design and Research, 2012, 28(1): 18-21.
[4] Zhao Zhengang, Lü Jianping. Design of Monocular Infrared Target Tracking Robot Vision System [J]. Journal of Optoelectronic Technology, 2008, 23(6): 19-23.
[5] Shen Yige, Qi Guo. Design of Picking Robot Vision System Based on DSP and μC/OS-II [J]. Journal of Agricultural Mechanization Research, 2017(06): 229-233.
[6] Li Lei. Research on anthropomorphic robot system design and visual navigation control [J]. Graduate, 2003.
[7] He Xuejian, Dong Guanqiang. Design of Monocular Vision System for Wheeled Anthropomorphic Robot [J]. Mining Machinery, 2005(02): 77-78.
[8] Qi Di, Li Xin. Design of Soccer Robot Vision System Based on Real-time Multi-target Tracking [J]. Mechanical & Electrical Engineering, 2001(06): 40-42.
[9] Zhu Xi, Li Yongxin, Chen Sheng, et al. Design of Soccer Robot Vision System Based on Multi-target Real-time Search--Robocup Series Research [J]. Automation and Instrumentation, 2004(2).

Journal of Physics: Conference Series **1423** (2019) 012056 doi:10.1088/1742-6596/1423/1/012056

Automatic classification of tongue texture color and tongue coating color based on BP neural network

Kang Zhang[1], Heling Zhang[2], Mingfeng Zhu[3]

[1]Jiangxi University of Traditional Chinese Medicine, Nanchang, China

Corresponding author and e-mail: HelingZhang, 21846498@qq.com

Abstract. Tongue diagnosis is an important part of the traditional Chinese medicine diagnosis, and it is one of the main criteria for clinical diagnosis of TCM. Tongue texture color and tongue coating color are the main research contents of tongue diagnosis. This paper first uses computer technology to achieve separation of tongue texture and tongue coating on the tongue image. According to the position information, the tongue texture and tongue coating template are discriminated, and the color information of tongue texture area and tongue coating area are obtained respectively. Finally, the automatic classification of tongue texture color and tongue coating color is realized through neural network.

1. Introduction

Tongue diagnosis is an important part of the four diagnosis of traditional Chinese medicine and one of the main bases of clinical diagnosis of traditional Chinese medicine. Tongue texturecolor and tonguecoating color are the main research contents of tongue diagnosis.The tongue is divided into two areas: tongue texture and tongue coating. Tongue color refers to the color of tongue quality, and coating color refers to the color of tongue coating. The tongue is the muscle vein of the tongue, located in the smooth area of the tongue. Tongue coating is a thin coating on the back of the tongue, which is located in the middle of the tongue.

At present, FCM fuzzy clustering algorithm is commonly used to separate tongue texture and tongue coating. In this paper, the algorithm is improved, which can automatically determine the number and location of clustering centers, improve the accuracy of segmentation and operation speed.

At present, the research on tongue texture color and tongue coating color mainly focuses on the quantitative analysis of color, and the automatic classification of tongue texture color and tongue coating color mainly uses statistical methods.Neural network imitates human brain and adopts adaptive algorithm. It has strong fault tolerance, self-learning, self-organizing function and induction ability. Therefore, this paper uses BP neural network to classify the color of tonguetexture and tongue coating.

2. Segmentation of tongue texture, tongue coating and background

FCM fuzzy clustering algorithm is used to realize automatic segmentation of tongue quality, tongue coating and background. The number and position of the tongue image clustering center can be determined by the peak position and the number of peaks of the chromaticity histogram of the tongue image. The specific algorithm is as follows:

(1)Convert Original tongue image from RGB space to HIS space to obtain hue histogram, as shown in Figure1.

(2)The high frequency component of the signal corresponds to the mutation region of the signal. Therefore, Figure 1 is filtered by Butterworth low-pass in frequency domain.It can remove the small and medium burr of the histogram and make the histogram smooth. The filtering results are shown in Figure 2.

(3)Obtain the number and position of the peak points in Figure 2, where the number of peaks is the number of clustering centers, and the position of peaks is the initial position of clustering centers. The FCM clustering segmentation effect is shown in Figure 3.

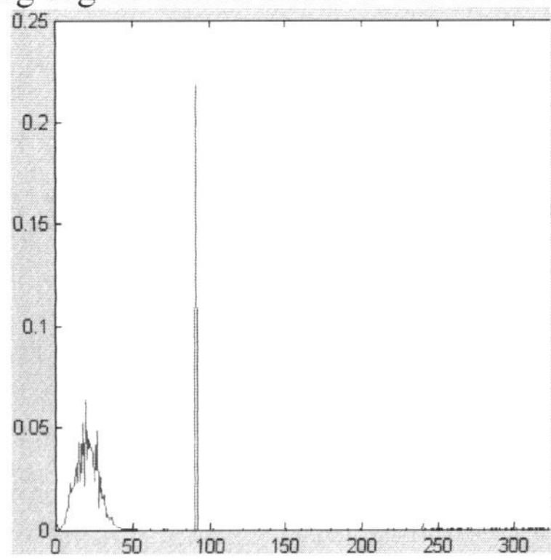

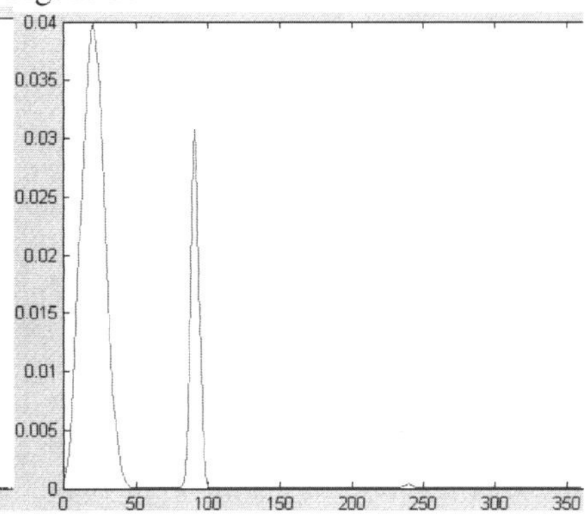

Figure 1. Hue histogram of tongue image **Figure 2.** Low pass filtering results of Fig1

Figure 3. FCM clustering segmentation effect

3. Automatic recognition of tongue texture template, tongue coating template and background template

After FCM clustering segmentation, there are three gray levels, 255,150 and 0, which correspond to three different regions. So we need to determine which of the three gray levels corresponds to the tongue quality, tongue coating and background. According to the feature that tongue coating is surrounded by tongue quality, this paper realizes the automatic recognition of tongue quality and tongue coating.

The identification steps are as follows:

(1)Get the template of grayscale 0,150 and 255. The original tongue image is shown in Figure4 (a), and the FCM clustering segmentation image is shown in Figure4 (b). In this paper, E0 is used to represent the gray level 0 template, as shown in Figure 4 (c), E150 is used to represent the gray level 150 template, as shown in Figure 4 (d), and e255 is used to represent the gray level 255 template, as shown in Figure 4 (e).

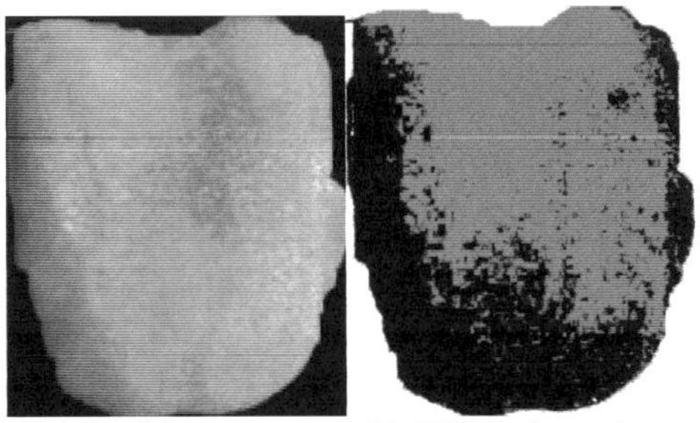

(a) Original image (b) FCM clustering segmentation effect

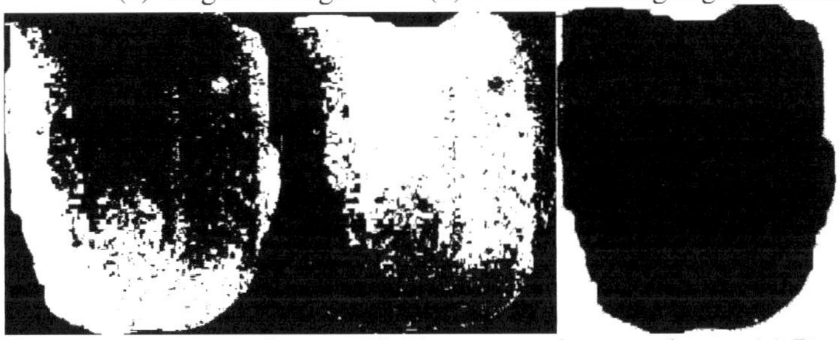

(c) Tongue texture template (d) Tongue coating template (e) Background template

Figure 4. The process of separating three gray level templates

(2)Judge background template.

According to the value of the background template in the upper left corner of the image must be 1, and the value of the tongue quality and tongue coating template in the upper left corner must be 0, the background template can be found in three templates.

If e255 (1, 1) = = 1 background = e255; else if e0 (1, 1) = = 1 background = e0; else if e150 (1, 1) = = 1 background = e150

(3)Judgment of tongue texture and tongue coating template

The remaining two gray level templates are used to determine which gray level corresponds to the tongue quality, and the other gray level corresponds to the tongue coating. Take Figure 4 as an example. In the second step, we can find that the background template is e255.

The binary morphological operations are carried out on e0 and e150 respectively to remove the small connected components, only the largest connected components are retained, and the set of coordinates of all pixels in the largest connected components is obtained. Because the tongue coating is sometimes at the root of the tongue, we can judge the position information of two gray levelstemplate from three directions. When one template is surrounded by another template in any two directions of the left, right and lower three directions, it can be judged that the template is tongue coating template, and the other template is tongue quality template.

4. Automatic classification of tongue texture and tongue coating color on tongue images using BP neural network

Due to the limited number of tongue image samples, this paper does not classify all tonguetexture colors and tongue coating colors. In this paper, the color of tonguetexture is divided into red, purple, light, deep purple and light red. The color of tongue coating is divided into white and yellow.

The BP neural network of tonguetexture color classification has three inputs. The input data are the values of H, S and I in the tongue region, which represent the information of hue, saturation and brightness in the tonguetexture region. The network has three output terminals: y3y2y1, y3y2y1 = 000(red), y3y2y1 = 001(deep purple), y3y2y1 = 010(light red), y3y2y1 = 011 and y3y2y1 = 100(light red).

The number of samples of each type of five tongue colors is 8, including 5 samples of each type of training set and 3 samples of each type of test set.

The mean square error of the network is set to net.trainparam.goal=1e-3, and the number of iterations is net.trainparam.epochs=10000.The effect of mean square error iteration is shown in Figure 5. It can be seen from the figure that when the iterative operation is more than 2000 times, the mean square error is very small.

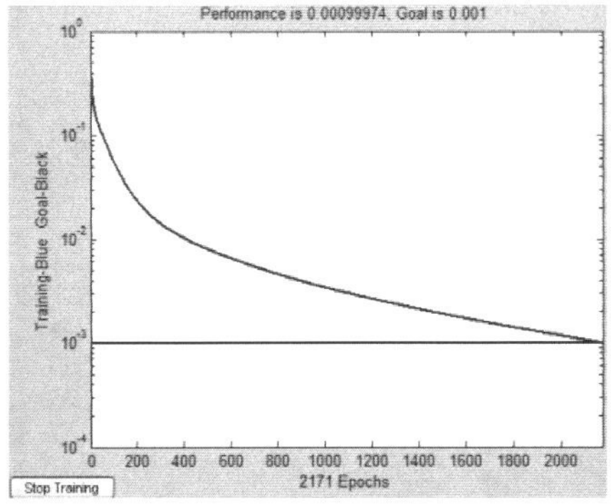

Figure 5. Mean square error iteration effect

The BP neural network of tongue coating classification is established. There are three input nodes in the input layer of the network. The input data are the values of H, S and I in the tongue coating area, which represent the information of hue, saturation and brightness in the tongue coating area. The output layer has an output terminal Y1, when Y1 = 0, it represents moss white, Y1 = 1, it represents moss yellow.

The number of samples of each type of two kinds of tongue coating color is 30, including 20 samples of each type of training set and 10 samples of each type of test set. The mean square error and the number of iterations of the network are the same as those of the tonguetexture color classification network.

After obtaining the network through the training set, the network is verified through the test set. Set T as the ideal output result (the first three columns 0 represent white moss, the first four to six columns 1 represent yellow moss), is shown in table1,A is the experimental output,is shown in table2.Mean square error equal tosum ($(t-a)^2$) =0.094.

Table 1. Ideal output T.

	white moss Example 1	white moss Example 2	white moss Example 3	yellow moss Example 1	yellow moss Example 2	yellow moss Example 3
T	0	0	0	1	1	1

Table 2. Experiment output result A.

	white moss Example 1	white moss Example 2	white moss Example 3	yellow moss Example 1	yellow moss Example 2	yellow moss Example 3
A	0.00342	0.00428	-0.00413	0.931	1.043	1.052

5. Conclusions

In this paper, the color information of tongue texture and tongue coating area are automatically obtained by computer, and the color information is used as the input of BP neural network, and the tongue texture color and coating color are used as the output to realize automatic classification of tongue texture color and coating color, and obtain high accuracy. However, as the number of tongue image samples is not enough, it has not been realized for all kinds of tongue texture color and coating color. In the follow-up study, the number of samples will be increased to further improve the recognition accuracy and the number of classification types.

Acknowledgement
Jiangxi Provincial Health and Family Planning Commission for Traditional Chinese Medicine Research (2017A233)

References

[1] ZhaoZhong-Xu,WangAi-ming,Shenlan-sun,Journalof Beijing University of Technology, 1999,12(4);480.
[2] Tan wu-xing,Li bing,Yue Xiao-Qiang.Journal of University of Science and Technology of China, 2010.8 vol.40.
[3] Prasanna Sahoo,Gurdial Arora.Pattern Recognition Letters,2006,27(6):520-528.
[4] Wang LU,Cai Zi-Xing.Mini-Mico Syetems,2005,25(10):1744-1777.
[5] Zhou yue,Yang jie,Shen li,JOURNAL OF BIOMEDICAL ENGINEERING,2004;21(6):917-920.
[6] YANG Yong,XIE Gangsheng.Journal of East China Geological Institute,2003,26(4):24-38.

Journal of Physics: Conference Series **1423** (2019) 012057 doi:10.1088/1742-6596/1423/1/012057

Dynamic modelling and simulation of 9.0 m double-tamper mechanism for asphalt paver

J Sun, C Liu, Y H Zhou

Mechanical and Electronic Department, Xuzhou Institute of Technology, Xuzhou 221018, China

Corresponding author and e-mail: J Sun, Sunjian751@163.com

Abstract. Compound motion of double-tamper mechanism has impact force on the ironing plane frame, which affects the quality of the load when the Asphalt Paver running. The dynamic model of the double-tamper mechanism is established by adopting multi-body dynamics, the virtual prototype model of 9.0m Asphalt Paver is established and simulated through the analysis of the principle of the Asphalt Paver. The impact regularity of the vibrating beam inertia force and moment is obtained, and that could provide a reference for the design of double-tamper mechanism for asphalt paver and the adjustment of construction parameters.

1. Introduction

The screed is the main device of the asphalt paver which plays a very important role in the construction. It achieves modeling, leveling and compaction of paving roads [1]. The screed with double-tamper mechanism has been widely applied in the market due to the characteristics of high level, high density and high operating efficiency. The double-tamper Mechanism in the front of the screed is driven by the rotation of the vibrating beams, and its rotational frequency is commonly 0 ~ 25 Hz. The inertia force impacting on the screed is very large. In some early researches, the movement of the tamper mechanism is simplified as simple harmonic vibration which didn't take the vibration of the vibrating beams into account[2]. Few studies [3] have investigated the theoretical calculation based on the method of complex vector. The process of this method is trivial and only the movement parameters of the tamper mechanism can be calculated. This paper established the dynamics model and the virtual prototype model of the double-tamper according to the analysis of the Structure and principle of the double-tamper Mechanism. The impact regularity of the vibrating beam inertia force and moment is obtained which could provide a reference for the design of double-tamper mechanism for asphalt paver and the adjustment of construction parameters.

2. Work principle of double-tamper mechanism

The first paragraph after a heading is not indented. Double-tamper mechanism for asphalt paver mainly composed of the eccentric shaft, eccentric sleeve, main vibrating beam and auxiliary vibrating beam. Eccentric shaft is supported on the frame by the bearings and driven by a hydraulic motor. The initial compaction of the paving mixture is realized through the vibration beam move up and down which is caused by the rotation of the eccentric shaft. The main and auxiliary vibrating beams are considered as the slab and girder structure. The top of the beams is grid, and the beams are hanged on

Journal of Physics: Conference Series **1423** (2019) 012057 doi:10.1088/1742-6596/1423/1/012057

the eccentric shaft, the structure is shown as shown in figure 1. There are only some extended parts which are parts of the double-tamper mechanism in the figure 1.

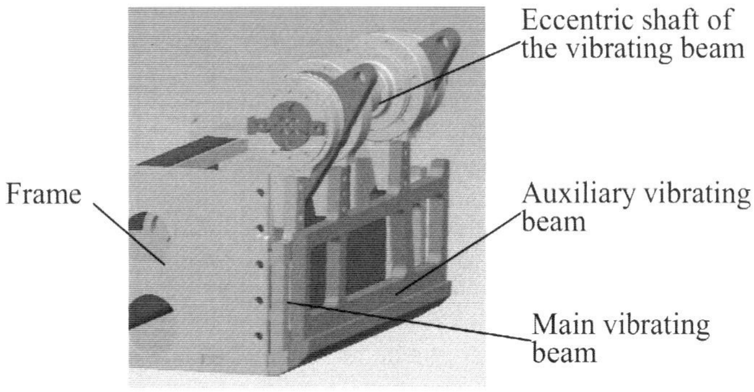

Figure 1. Structure type of double-tamper mechanism

The eccentricity between the eccentric part of the eccentric shaft and the rotating center of the mechanism is e_1, and the eccentric sleeve is installed on the eccentric shaft. The eccentricity between the outer circle of the eccentric sleeve and the inner hole center of the eccentric sleeve is e_2. The total eccentricity e is the vector sum of the shaft and sleeve eccentricity, which is calculated by the following formula [4].

$$e = \overline{OP} = \sqrt{e_1^2 + e_2^2 - 2e_1e_2\cos\alpha} \tag{1}$$

Where e_1 is eccentricity of eccentric shaft, e_2 is eccentricity of eccentric sleeve, α is rotation angle of eccentric sleeve relative to the eccentric shaft.

A variety of different combinations of amplitudes can be obtained by adjusting the eccentricity of the eccentric shaft and eccentric sleeve, also different phase difference between the vibrating of the main and auxiliary vibrating beams. Thus a better inertia force can act on the flatiron-box which can better improve the road paving quality [5].

3. Simulation model establishment of double-tamper mechanism

In this article, Adams is used to build a virtual prototype model of 9.0m double-tamper mechanism for simulation analysis. At present, ADAMS is widely used in simulation analysis of mechanical system, including powerful constraint database, force database and machine part database, and powerful solver with numerical analysis function. Based on Lagrange equation in Dynamics of Multi-body System, ADAMS can automatically establish dynamic equation of mechanical system, quickly and accurately solve various kinematics and dynamics problems of mechanical system. The results of simulation analysis can be displayed in the form of curve and graph [6].

3.1. Fundamental theory

Adams can use the method of Lagrange equation according to mechanism model, use Cartesian coordinates of center of mass of rigid body i and the Euler angles of the position of rigid body as generalized coordinates, which expressed as an equation is $q_i = [x, y, z, \psi, \theta, \varphi]_i^T$, $q = [q_1^T, q_2^T, \ldots, q_n^T]^T$ .To each rigid body is described by six generalized coordinates. The algebraic equation of constraint, external force equation and self-defined Differential-Algebraic Equations of the system are as follows.

$$\left.\begin{array}{l}\phi(\dot{q},q,t)=0\\ F(\dot{u},u,q,f,t)=0\\ diff(\dot{u},u,q,f,t)=0\end{array}\right\} \quad (2)$$

Where q is generalized coordinates, u is differentiation of each generalized coordinates, f is external force of system and constraint.

Let $y=[q,u]^T$ be state vector, then the equation of the system can be expressed as

$$G(y,\dot{y},t)=0 \quad (3)$$

3.2. Model establishment

In this paper, Pro / E is used to establish the three-dimensional digital model and entity assembly relationship of the double-tamper mechanism. Each quality parameters of each component are as table 1.

Table 1. Parameters of System

parameters of the base part			
r_1 /mm	2.5	r_2 /mm	3
b_1 /mm	6.27	b_2 /mm	24
m_1 /kg	5.1	m_2 /kg	90.07
m_3 /kg	71.53	m_4 /kg	7.84
J_2 /kg.mm^2	3.55x10^6	J_3 /kg.mm^2	2.68x10^6
parameters of the added part			
m_1 /kg	5.1	m_2 /kg	75.04
m_3 /kg	49.16	m_4 /kg	7.84
J_2 /kg.mm^2	3.01x10^6	J_3 /kg.mm^2	1.94x10^6

By using the interface module MECHANANISM/Pro, the data transmission between Pro/E and ADAMS can be realized. Add the gravity restrain on the whole mechanism and the rotational restrain between the eccentric shaft and the main and auxiliary vibrating beams [7]. Then add the bushing retrain to imitate the elastic coupling. In practice, the main and auxiliary vibrating beams can be simplified as them to two planes fixed on the earth in the dynamic model by applying planar constraint between the vibrating beams and the panels. The dynamic model is shown in Fig.2.

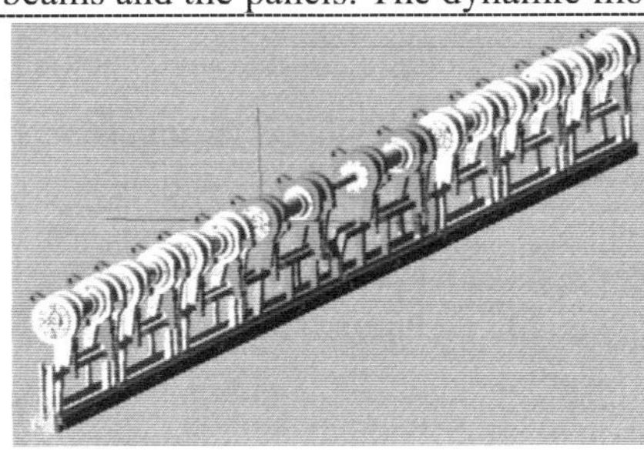

Figure 2. Simulation model

4. Dynamic simulation of double-tamper mechanism

4.1. Analysis of inertia force under different vibration frequency

The inertial force and moment of inertial can be obtained by measuring the force of rotational constrain between the eccentric shaft and the bearing. Take the amplitude of the main vibrating beam as 5mm, the amplitude of the auxiliary vibrating beam as 6mm, the simulation is carried out at different vibration frequencies 10,15,20,25 Hz, and carry out the simulation.

4.2. The inertial force curve of the right base of the screed in x and y direction are shown as follows.

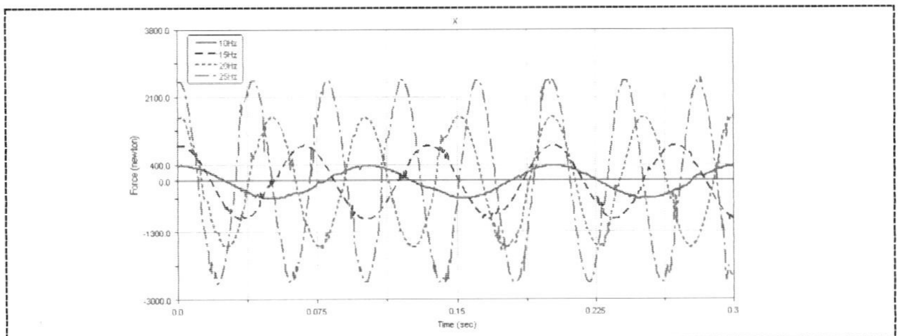

Figure 3. Iinertia force in X direction

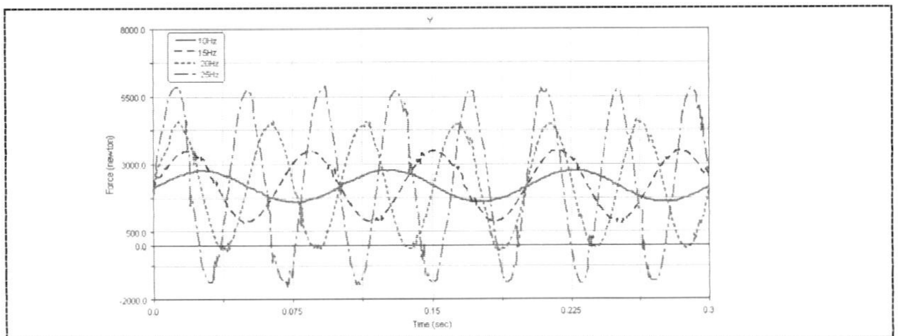

Figure 4. Iinertia force in Y direction

As shown in the figure, the frequency of tamper mechanism has a greater impact on inertia force. Take the inertial force in X direction for example. While the frequency is respectively 10Hz, 15HZ, 20HZ, 25HZ, the maximum of the inertial force in X direction is 275.7N, 622.2N, 1145.3N, 1827.6N. It is shown that the inertia force of the vibrating beams increases greatly with increasing of the vibration frequency, and the inertia force approximately increases k^2 times when the vibration frequency increases k times. The transverse inertia force is large, which will lead to transverse vibration of the flatiron-box. In the previous analysis, we only consider the vibration in the vertical direction, and neglect transverse vibration of the flatiron-box, that is not correct, it is not able to fully reflect the working status of the screed.

4.3. Plotting the effective value of the inertia force of each parts of the double-tamper mechanism

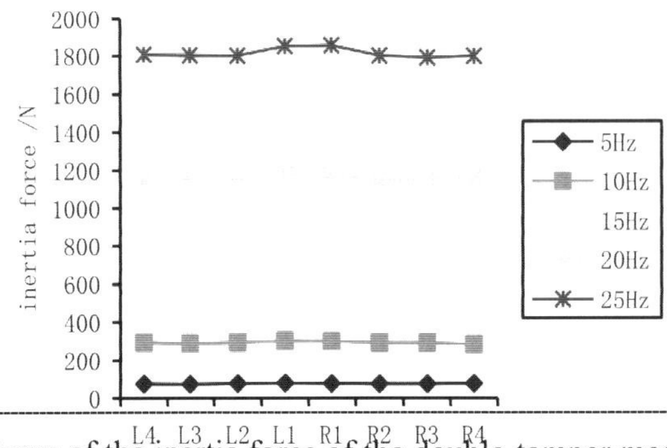

Figure 5. Curve of the inertia force of the double-tamper mechanism at different frequency

It is shown that the lager the speed the larger increasing extent of the inertia force and has a bad evenness. We should take low-frequency vibration in the precondition of meeting the pavement compaction conditions. The vibrating mechanism has a small effect on the screed at a low frequency. The vibrating mechanism has a significant effect on screed at a high frequency.

4.4. Analysis of inertia force under different vibration amplitudes

Take the amplitude of the main vibrating beam as 5mm, the amplitude values of the auxiliary vibrating beam are 0,3,6,9,12 mm, at the condition of different frequencies 5,10,15,20,25 Hz, take the maximum of the inertial force and moment to draw graphics, the results is shown as follow.

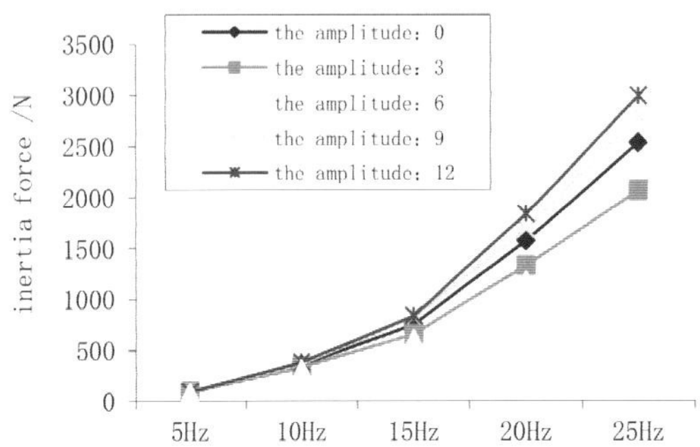

Figure 6. Inertia force of right base part of the tamper mechanism

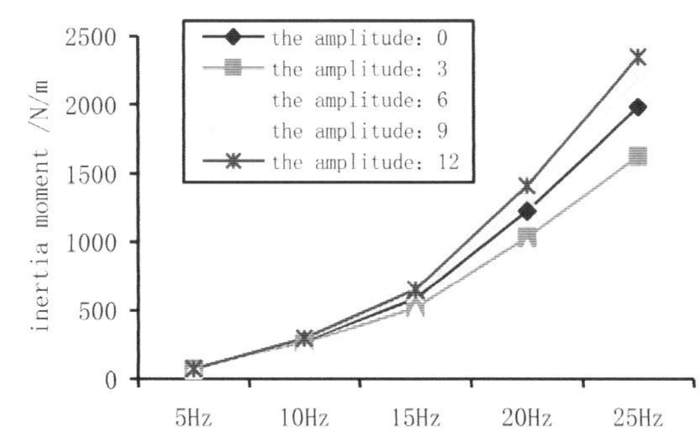

Figure 7. Moment of Inertia of right base part of the tamper mechanism

As shown in the above figure, the change of the inertia force and the moment is small while the amplitude of the main vibrating beam is 5mm, the amplitude of the auxiliary vibrating beam changes from 3mm to 6mm. while the amplitude of the auxiliary vibrating beam is 9mm or 12mm, the inertia force and inertia moment is to the peak. The result of the simulation and theoretical analysis are in good agreement.

4.5. Analysis of inertia force under different phase angles between adjacent vibrating beams

This paper carries the simulation by Taking the phase difference α between the left and right base part as 60 ° and the phase difference β between the base part and the added part as 0° ,60° ,90° ,150° ,180° . The result of inertial force of each part are shown as follows.

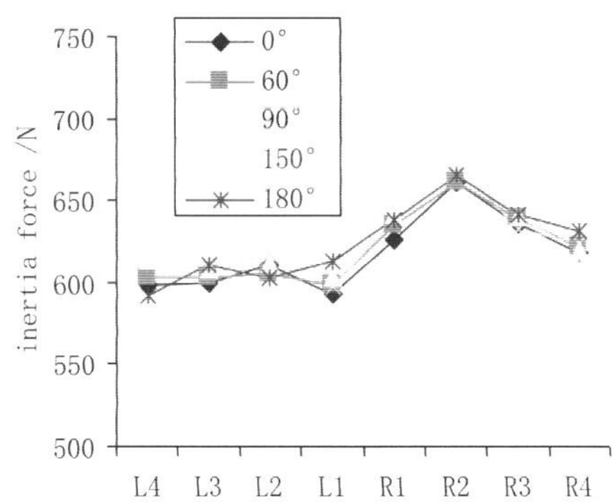

Figure 8. The inertia force of each double-tamper mechanism with different phase

The mean value and variance of the value of each parts of the double-tamper mechanism are shown in table 2.

Table 2. Mean value and variance

Phase	0°	60°	90°	150°	180°
Mean value / N	618.11	620.52	627.19	625.66	624.82
Variance	23.03	22.42	24.49	23.07	23.92

The driven moment to maintain the rotation of the 9.0m double-tamper mechanism can be obtained by the measure of the added motor.

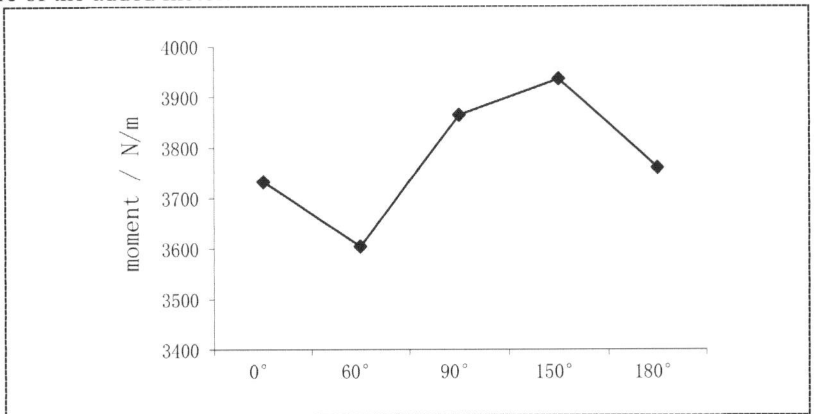

Figure 9. driven moment with different phase

It is shown that the inertia force is small and the while the phase β is 0° or 60°, the inertial force is lager and has small changes at 90°、150° and 180°. The inertial force has a good evenness at 60°, however the inertial force has a bad evenness at 90° and 180°. The driven moment is small at 60°, but lager at 90° and 150°. In order to have a little effect on the screed, so we should set the phase between the base part and the added part as 60° in practice.

5. Conclusions

In this paper, the multi-body dynamic model of tampering mechanism of asphalt paver is established based on Pro/E and ADAMS. The dynamics characteristic of the mechanism is simulated. The virtual prototype model is established according to the real machinery, compared with previous simplified calculation model. This paper considers the space structure and mass distribution of each component, avoiding product the real machinery, and improving the design efficiency. This method is very suitable for the research on complex mechanism.

Acknowledgement

The research work is supported by Major Program of Natural Science foundation for Colleges and Universities of Jiangsu Province (18KJA460008), the "Summit of the Six Top Talents" Program of Jiangsu Province (GDZB-128), and Xuzhou applied basic research project (KH17012).

References

[1] Gang L, Pu-yue T, and Zhong-yu X 2005 *Journal of agricultural machinery* vol 36 p 34-37.
[2] Qing X 2006 *Road Machinery & Construction Mechanization* vol 3 p 15-19.
[3] Guo-pu Z 2009 *Road Machinery & Construction Mechanization* vol 6 p 35-38.
[4] Chao Y, Jianrun Z and Tongmeng S 2018 *JOU RNAL OF VIB RATION AND SHOCK* vol 8 p167-171.
[5] Guo-pu Z 2009 *Road Machinery & Construction Mechanization* vol 6 p35-38.
[6] Tianhong L, Xinfu G 2010 *2010 IEEE ICICTA* p800-803.
[7] Yuxiang C, Mengyan Z and Yong C 2012 *China Mechanical Engineering* vol 8 p996-1000.

Journal of Physics: Conference Series **1423** (2019) 012058 doi:10.1088/1742-6596/1423/1/012058

BP neural network study on the quality evaluation system of innovation and entrepreneurship education in local universities

Wenfeng Huang[1], Bowen Liu[1], Xingjie He[1,*], Yan Zhou[1], Binbin Ning[1], Liang Wang[1]

[1] Chengdu University of Technology, China

*Corresponding author and e-mail: Xingjie He, 345325319@qq.com

Abstract. When the country vigorously advocates the goal of cultivating innovative and entrepreneurial engineering talents, the quality evaluation of innovation and entrepreneurship education has naturally become a part of the higher education system. However, there are few studies on the quality evaluation of innovation and entrepreneurship education in engineering majors in China. Firstly, through literature research, combined with the current development trend of quality evaluation of innovation and entrepreneurship education at home and abroad, BP neural network-based evaluation method was selected. Secondly, according to the three-element theory of education, namely, the three factors of educators, educators and education, and comprehensive research, the content of quality evaluation of innovation and entrepreneurship education in engineering majors in colleges and universities is determined, that is, from professional links, teaching links and students. Levels to build evaluation indicators. After establishing the evaluation index system, this paper uses Chengdu University of Technology as a research sample to issue questionnaires for engineering students in Chengdu University of Technology to obtain data for use, and to analyze and analyze the data by AHP, so as to obtain the weight of each indicator. Finally, based on the data collected by the questionnaire, the BP neural network data fitting and model training and optimization are carried out to determine the feasibility of the quality evaluation index of innovation and entrepreneurship education in engineering majors in colleges and universities, and to establish BP neural network evaluation model for engineering majors. The innovation and entrepreneurship education quality evaluation system provides new research methods and research ideas, enriches the connotation and breadth of the quality evaluation system of innovation and entrepreneurship education in engineering majors, and makes up for the shortcomings and vacancies in related fields.

1. Introduction

With the implementation of the national strategy of innovation and entrepreneurship driven development, the cultivation mode of university talents has also undergone significant changes. In April 2002, the pilot work symposium on entrepreneurship education in national ordinary universities held by the ministry of education indicated for the first time that innovation and entrepreneurship education should focus on the process and the cultivation of comprehensive quality of innovation and entrepreneurship, and put forward that innovation and entrepreneurship education is a concept [1]. The

13th five-year plan and the report of the 19th national congress of the communist party of China also put forward the call to "stimulate the vitality of innovation and entrepreneurship, promote mass entrepreneurship and innovation" and "stimulate and protect entrepreneurship, and encourage more social subjects to engage in innovation and entrepreneurship" [2]. In addition, a series of policy documents issued by the General Office of the CPC Central Committee and The State Council, as well as conferences and speeches on related topics, have pushed the wave of innovation and entrepreneurship to the forefront of The Times, and innovation and entrepreneurship education in colleges and universities has attracted much attention.

In 1945, scholars in Harvard University proposed that the joint cultivation of students' knowledge and innovation ability should be emphasized. The educational philosophy of the universities of Cambridge and Oxford in Britain is to "explore, explore and develop students' potential ability and stimulate individual creativity", which shows the importance of innovation and entrepreneurship education [4]. In the national defense education act of 1958, the U.S. government proposed to enhance international competitiveness by enhancing innovative education [5]. In 1981, "besson entrepreneurship annual conference" was successively held by American Besson College, university of Washington and other universities [6]. In 1989, UNESCO held "international seminar on education for the 21st century", which pointed out that entrepreneurship education should focus on cultivating students' pioneering ability and innovative spirit [7]. In 2010, the European Union stated in its education report that to strengthen the reform of entrepreneurship education, it is necessary to develop entrepreneurship education with an open and cooperative attitude, encourage member states to continue to formulate special entrepreneurship education development strategies, and encourage cooperation between schools and enterprises in the field of entrepreneurship education [8]. By 2017, nearly 2,000 colleges and universities in the United States have set up courses related to innovation and entrepreneurship education, including primary, high school, university and postgraduate education, and formed a relatively complete education and teaching system. More than 75% of colleges and universities in the United Kingdom have set up courses related to innovation and entrepreneurship education [9]. In addition, in Australia, innovation, entrepreneurship education teaching faculty got the high attention, innovative entrepreneurship education instructor is mostly entrepreneurs with higher education, they have professional theoretical foundation and rich experience in actual combat, in other developed countries such as Japan, Singapore, innovation, entrepreneurship education has also been incorporated into the education system in [10].

Relevant scholars, experts, researchers mostly committed to innovative entrepreneurship, method, model of research and analysis, evaluate the quality of college engineering specialty innovative entrepreneurship education is little, so to establish a set of perfect innovation entrepreneurship education quality evaluation system to help the engineering innovation business of development become the urgent needs of quality education, this paper adopts literature analysis, questionnaire investigation, literature analysis, using the method of BP neural network model for the main business innovation education quality evaluation for engineering majors, attempt to establish a more mature about college engineering specialty innovative entrepreneurship education quality evaluation system. This will have certain theoretical and practical significance for the development and optimization of innovation and entrepreneurship education in China.

2. Material and method

2.1. Research method

2.1.1. Literature analysis

Through resources such as cnki.com and the library of Chengdu University of technology, we have an in-depth understanding of the progress of innovation and entrepreneurship education at home and abroad and the relevant evaluation system, and summarize the advantages and disadvantages of innovation and entrepreneurship education in Chinese universities.

2.1.2. Questionnaire survey.
Based on the index system of engineering major extracted from Chengdu University of technology, I consulted relevant experts and tutors on campus, prepared questionnaires, conducted surveys on engineering majors, and obtained a real and effective data, so as to determine the weight coefficient of each index.

2.1.3. Empirical analysis
After determine evaluation model, with Chengdu university of technology as case study for the engineering specialty, and for engineering specialty, to calculate weight of the technical indicators related and validity, etc., it is concluded that the Chengdu university of science and engineering professional business innovation in all aspects of the education quality, to analyze the evaluation results, find out the insufficiency, through the empirical study proves the practicability of this study.

2.1.4. BP neural network method.
Compared with traditional data analysis methods, BP neural network has high data adaptability and can be applied to a variety of data types. By improving the BP neural network model, the disadvantages of poor and slow convergence rate can be overcome, so that the model can describe parameter relations and obtain better evaluation results.

2.2. Construction of evaluation index
On the establishment of the quality evaluation index of innovation and entrepreneurship education, Guo Xiaofang and others believe that the quality evaluation of innovation and entrepreneurship education should be composed from the macro level, the teacher level, the student level and the educational environment and form [11]. Zhang Xinze believes that the quality evaluation of innovation and entrepreneurship education is influenced by college environment, teaching design, faculty strength and student performance [12]. Xu hui believes that the quality of innovation and entrepreneurship education is influenced by theoretical knowledge, practical skills, innovative consciousness and creative thinking industry [13]. Gaoming et al. found in their research that innovation and entrepreneurship education should be carried out from the school level, the teacher level, the family level and the social level. Research combined with the level of the teachers and students have mentioned above, the teachers level, this article also think teachers for education in the three elements of the body of the educators, has the leading role, the teachers level is much higher than other factors, the influence of the same, students as the principal part of education according to the elements of the students to innovative entrepreneurship education reflect the effect of the quality of the creative education, the role of this level is also cannot be ignored, in addition, the research of engineering professional, should be mainly focused on the characteristics of the profession itself, as the center, to make the results of the study have credibility, so combining with the practical engineering profession in university, This research will construct the quality evaluation system of innovation and entrepreneurship education for engineering majors in local universities from three aspects: teaching link, professional link and student level. The whole evaluation system is divided into three levels: target level, criterion level and indicator level. Among them, criterion level is divided into three categories, as shown in figure 1.

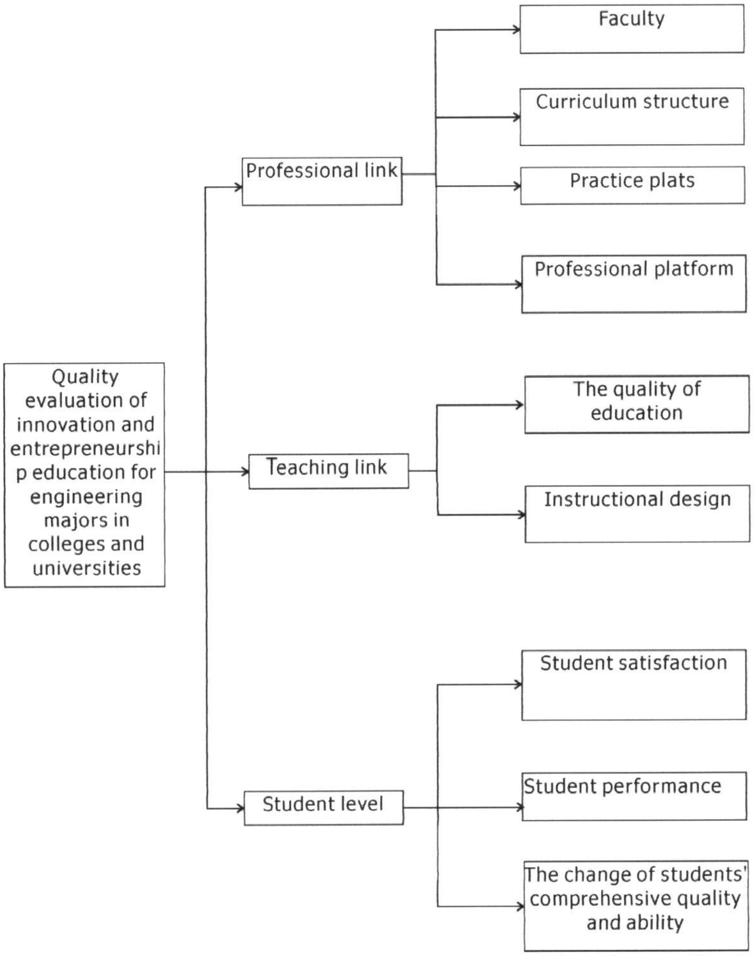

Figure 1. Evaluation index of innovation and entrepreneurship education quality for engineering majors in colleges and universities

2.3. Questionnaire survey and data collection

In Chengdu university of science and engineering students, the author of this paper design the questionnaire survey, the questionnaire (see the appendix) mainly aimed at the Chengdu university of technology institute of nuclear technology and automation engineering, institute of geophysics, school of earth sciences, materials and chemical industry and environment and civil engineering institute of several remarkable characteristics of college students in engineering, professional cover each college is about more than 30 professional students, to ensure the comprehensive and objective of the questionnaire survey, reliability, and representative.

A total of 417 people answered the questionnaire effectively, of which 299 were male, accounting for 71.7%, and 118 were female, accounting for 28.3%. The number of participants in each grade was 1 freshman (0.24%), 23 sophomores (5.52%), 211 junior (50.6%), and 182 senior (43.65%). Since there is only one freshman and the effect of innovation and entrepreneurship education is not outstanding, in order to avoid the resulting errors, it is eliminated. Among the five majors in each college, only one was recovered, which was not representative and was also eliminated. Therefore, the final actual recovery of this questionnaire survey was 411, with a good recovery effect, which could meet the analysis needs. The specific sample statistics are shown in table 1.

Table 1. Statistical data of questionnaire samples

Category 1	quantity	procedure cache percent
schoolboy	299 people	71.7%
schoolgirl	118 people	28.3%
Category 2	quantity	procedure cache percent
sophomore majoring	23 portions	5.52%
junior majoring	211 portions	50.6%
senior majoring	182 portions	43.65%
Category 3	quantity	procedure cache percent
College of nuclear technology and automation engineering	71 portions	17.03%
School of geophysics	27 portions	6.47%
School of GeoSciences	49 portions	11.75%
School of materials and chemistry, chemistry and chemical engineering	30 portions	7.19%
School of environmental and civil engineering	240 portions	57.55%

2.4. Application of BP neural network

The number of nodes in the input layer is determined by the index factors of innovation and entrepreneurship quality evaluation of engineering majors in colleges and universities. It can be known from the previous index construction part that the evaluation system in this study has 31 indicators, so the number of nodes in the input layer of the improved BP neural network model is 31.

The number of nodes in the hidden layer is related to the number of nodes in the input layer and the number of nodes in the output layer. The selection of the number of nodes in the hidden layer usually affects the quality of the selected neural network model. The determination of the number of nodes in the hidden layer generally adopts the following formula:

$$S_2 = \sqrt{S_1 + S_3} + q \tag{1}$$

In the formula, S1 is the number of nodes in the input layer, S3 is the number of nodes in the output layer, S2 is the number of nodes in the desired hidden layer, q is a constant of 1-10, and the number of nodes in the hidden layer is calculated as 8 through trial and error.

The number of nodes in the output layer is the evaluation result. In this paper, the engineering major of Chengdu University of science and technology is taken as an example for evaluation, so the number of nodes in the output layer is 1.

In this paper, analytic hierarchy process (AHP) is used to determine the influence degree of each evaluation index on the upper layer, that is, the weight value. By previous studies, this paper analyzes the role of the relationship between each factor of the target layer, criterion layer and index layer, and then start from the bottom, judge the effect of lower index of the last index size, with 1-9 scaling method to value assignment of index, and through comparing the two, the judging matrix was established and the weight of each index is calculated. The assignment of 1-9 scale method is assumed as follows.

Table 2. Assignment hypothesis of scaling method

Degree of relative importance	determined value
most importantly	9
a great many	7
importance	5
Slightly more important	3
the same importance	1
Slightly unimportant	1/3
unimportance	1/5
trivial	1/7
the least important	1/9
Between two adjacent judgments	2、4、6、8、1/2、1/4、1/6、1/8

By evaluating the importance of each index through the assignment hypothesis table, the weight judgment matrix of target layer, criterion layer and index layer is obtained: G-C, C1-P, C2-P, C3-p. (G is the quality evaluation of innovation and entrepreneurship education, C1 is the professional link, C2 is the teaching link, C3 is the student level, and P is each indicator), which is shown in the following tables.

Table 3. Weight judgment matrix G—C

G	C1	C2	C3	Weight
C1	1	5	3	0.63
C2	/5	1	1/3	0.11
C3	1/3	3	1	0.26

Table 4. Weight judgment matrix C2—P

C2	X14	X15	X16	X17	X18	weight
X14	1	1	1	1/3	1/3	0.11
X15	1	1	1	1/3	1/3	0.11
X16	1	1	1	1/3	1/3	0.11
X17	3	3	3	1	1	0.33
X18	3	3	3	1	1	0.33

Table 5. Output value classification

rank partition	Neural network output
Excellent	1.00-0.90
favorable	0.89-0.80
intermediate	0.79-0.70
pass	0.69-0.60
fail	under 0.59

According to the data collation and normalized processing of the index scale calculated by the evaluation system of innovation and entrepreneurship quality for engineering majors of Chengdu university of technology, the number of nodes in each layer was calculated, the neural network model was established, and MATLAB software was used to carry out neural network training, generate the model, input test data and conduct simulation calculation, simulation results and prediction samples are shown in figure 2 and figure 3.

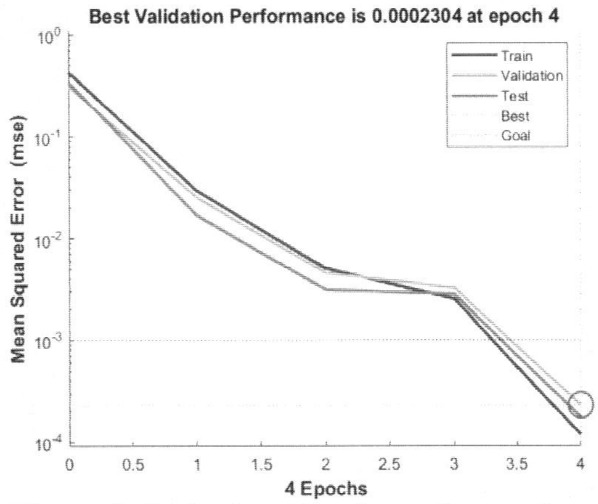

Figure 2. Optimal results of neural network training

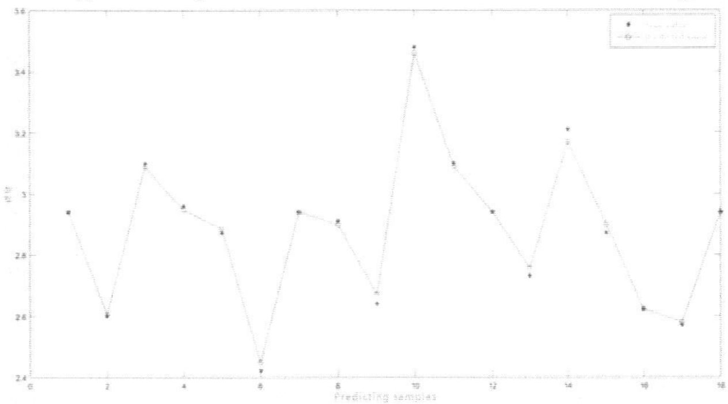

Figure 3. Prediction sample

3. Results and discussion

Through investigation and study and the paper quality evaluation results, we can see that among the 31 indicators, 3 were between 0.89 and 0.8, 9 were between 0.79 and 0.7, 8 were between 0.69 and 0.6, and 11 were below 0.59, in a passing grade, indicates the quality of college engineering specialty innovative entrepreneurial education is still in the elementary level, have greatly improvement space, but also achieved some achievements in some aspects.

3.1. Achievements in innovation and entrepreneurship education for engineering majors

3.1.1. Innovation and entrepreneurship knowledge teaching has a good effect

Through the analysis of the evaluation results of X5=0.85, X14=0.83 and X24=0.83, it can be concluded that the teaching method of innovation and entrepreneurship is popular among students. The unique and innovative teaching method is conducive to increasing students' interest in innovation and entrepreneurship, ensuring their concentration in class, improving the effect of classroom transformation, and thus increasing the participation rate of innovation and entrepreneurship activities.

3.1.2. Achievements in innovation and entrepreneurship improved

X17 = 0.78, with the result of evaluating X22 = 0.75, X29 = 0.70 analysis, preliminary results were obtained in college engineering specialty innovative entrepreneurship education, teachers use their resources to assist and guide the students to participate in the campus and campus platform inside and outside the relevant science and technology competition, exercise the student's innovation ability, can make students better contact with cutting edge innovative undertaking and understanding the dynamic changes of the industry, is conducive to innovation for the future business situation to make objective judgment, increase the success rate.

3.1.3. Awareness of innovation and entrepreneurship has increased

In terms of evaluation indexes, X21=0.73, X25=0.76, X27=0.71 and X30=0.73, innovation and entrepreneurship education of engineering majors in colleges and universities has begun to focus on students' awareness of innovation and entrepreneurship and the ability to cultivate innovation and entrepreneurship, instead of only focusing on hard indicators such as entrepreneurship rate.

3.2. Shortage of innovation and entrepreneurship education in engineering majors in colleges and universities

3.2.1. Lack of teachers for innovation and entrepreneurship education

According to the analysis of evaluation indexes X1=0.64, X2=0.68 and X4=0.41, the proportion of teachers with certain achievements in the field of innovation and entrepreneurship in colleges and universities is relatively small, and the lack of relevant practical experience often leads to the disconnection between theory and practice, hindering the development of innovation and entrepreneurship education in colleges and universities.

3.2.2. Innovation and entrepreneurship knowledge is not closely integrated with professional knowledge

X6 = 0.65 in the work of evaluation index, X9 = 0.28, now most of the existing business entrepreneurship education and professional education section, accounts for a smaller practice courses, students less opportunity to develop itself by actual operation, to meet the needs of students to form a certain innovative entrepreneurial literacy, teaching content and innovative undertaking professional knowledge failed to organic unity together, to reach the purpose of using their own advantages to better innovation entrepreneurship, failed to innovative entrepreneurship education closely associated with students' education, professional knowledge in business entrepreneurship education idea failed to fusion.

3.2.3. The teaching system of innovation and entrepreneurship is not perfect

X18 = 0.64 evaluation indices, X19 = 0.49, college engineering specialty innovative entrepreneurship education teaching system is not perfect, did not establish relevant systems of the building, innovation entrepreneurship courses to learn more courses in the form of, or in the form of lecture lectures on, not set corresponding innovation entrepreneurship course credits, can not cause enough attention to students. There are few extracurricular practical activities, the proportion of class hours of practical courses is small, and the teaching purpose and teaching policy are not clear, which cannot guarantee the internal change of students' innovative and entrepreneurial thinking and ability, which often leads to the problem of treating the symptoms rather than the root cause.

3.2.4. The utilization rate of innovation and entrepreneurship resources is not high

By 8 = 0.40 evaluation indices, X26 = 0.64, engineering major is needed at a certain experimental process, theory or practice experience as engineering specialty in colleges and universities related laboratory quantity is insufficient, the imperfection of the laboratory facilities, not to let students get enough exercise, not effectively stimulate students' innovative entrepreneurial zeal, the innovation of the schools and colleges set up platform for business incubation park utilization rate is not high, not enough publicity, the use of the platform for most of the students' innovation and related policy is not enough to understand, to a certain extent can inhibit students to participate in the activities of innovative entrepreneurial ideas.

4. Conclusion

Starting from the research status of innovation and entrepreneurship education and the evaluation of innovation and entrepreneurship education quality, this paper determines the index system of innovation and entrepreneurship education quality evaluation applicable to the engineering majors in local universities. Through in-depth analysis and research of engineering specialty, the first-level indexes of three levels of engineering specialty are determined, namely, professional link, teaching link and student level. A neural network model for quality evaluation of engineering innovation and entrepreneurship education in universities is established. Through the analysis of the innovative undertaking quality evaluation method, using BP neural network model generalization ability and advantages of nonlinear processing capability index to quantify the problem, on the processing of data, using the AHP to calculate the weight of each index, the result of the weight analysis prove the feasibility of the BP neural network model is established in this paper, for quality assessment of college engineering specialty innovative entrepreneurship education provide new method and train of thought.

Acknowledgement

This work was funded by the Sichuan provincial higher education talent cultivation quality and teaching reform project 2018-2020 (JG2018-485), Chengdu university of technology 2018-2020 higher education personnel training quality and teaching reform project (JG183086) and Chengdu university of technology, local undergraduate university students innovation and entrepreneurship education research base construction project (YJ2017-JD002).

References

[1] Zhang shuo, Xi jinping, at the national conference on ideological and political work in universities, stressed that ideological and political work should be carried out throughout the whole process of education and teaching to create a new situation for the development of higher education in China [N] People's Daily 2016.12.9(1st edition)

[2] Cao duli,Lei jiahui. The development of Chinese university creative education report [M]. Shen: wanjuan publishing company. 2009

[3] Wang fatang, Han xiaohui, Lu qi. The development of innovative education abroad and its enlightenment to Chinese universities [J]. Journal of shandong youth management cadre institute: youth work BBS, 2004(4):74-77.

[4] Tong xiaoling. Research on innovation and entrepreneurship education system in research-oriented universities [D]. Wuhan university of technology

[5] Zhang yiquan, Xu jia. Research review on entrepreneurship education for college students [J]. China electric power education, 2011(16):7-9.

[6] Huang zhaoxin, Wang zhiqiang. On the integration of entrepreneurship education and professional education in universities [J]. Education research, 2013(12):59-67.

[7] Wang zhiqiang. Integration and diversity: the development trend and enlightenment of entrepreneurship education in the eu [J]. Education research, 2014(4):145-151.

[8] Lan lan, Chen youhua. New reform of British educational development [D]. Around southeast asia

[9] Li tongguo. Experience and inspiration of employment guidance for college students in developed countries [J]. Journal of leshan normal university, 2010(6).

[10] Guo xiaofang, hao zhirong, zhang keying. Research on the quality of innovation and entrepreneurship education in shandong liming vocational college of science and technology [D]. 2018

[11] Liu xinze. Evaluation system of innovation and entrepreneurship education quality in

local universities in liaoning province [D].2018
[12] Xu hui. Evaluation criteria for cultivating innovative and entrepreneurial talents in universities [J]. Jiangsu higher education, 2009(6):107-108.
[13] Gao Ming, Li lin. Evaluation index system and promotion factors of college students' innovation and entrepreneurship ability [J]. Journal of chifeng university (natural science edition), 2014(12):183-184.

Journal of Physics: Conference Series **1423** (2019) 012059 doi:10.1088/1742-6596/1423/1/012059

Research on the characteristics of generations of subway vehicles based on artificial neural networks

Fubo Shao[1, 2, 3], **Hui Liu**[3]

[1]The Post-Doctor Station of the CRRC Corporation Limited, Beijing, 100067, China;
[2]State Key Laboratory of Rail Traffic Control and Safety, Beijing Jiaotong University, Beijing 100044, China;
[3]School of Mathematics and Physics, Qingdao University of Science and Technology, Qingdao 266061, China

Corresponding author and e-mail: Fubo Shao, shaofubo@163.com

Abstract. Urban rail transit vehicles have become an important part of public transport due to the outstanding advantages: high speed, large traffic volume, low energy consumption and less land occupation. Metro vehicles have been developed for over one hundred years and there are obvious generational characteristics. In this paper, the generational characteristics of metro vehicles are studied from the top index of metro vehicles. First, the paper synthetically analyses 45 specific indicators of four categories: vehicle transport capacity, safety, comfort and environmental friendliness. And then 14 core indicators are extracted via employing the complex network method. Secondly, the first- and second-generation metro vehicles are analysed based on the 14 core indicators. By employing the artificial neural networks, the relationship between the performance and indicators of metro vehicles is obtained. And then the performance of the next generation metro vehicles can be predicted. Finally, the characteristics of the next generation of metro vehicles are predicted based on the performance of the next generation of metro vehicles and the latest technology. The future developing direction of metro vehicles is pointed and the theoretical basis for the next generation of metro vehicle manufacturing is provided in this paper.

1. Introduction

Urban rail transit is an important part of the urban transportation system. With the acceleration of urbanization, the increasing speed of the total urban population is very fast. Urban rail transit has become the main travel mode of daily commute.

With the emergence of new technologies and the raising of people's requirements, the traditional second-generation metro vehicles can't meet the needs of the society well because of the disadvantages of high operation energy consumption and life cycle cost, insufficient line adaptability, limited passenger carrying capacity, low level of train intelligence, poor comfort, low safety, low level of reliability and availability. Then, the next generate metro vehicles should meet the higher requirements about energy saving, noise reduction, high comfort, high applicability et al.

However, based on the traditional metro vehicle technologies, only the performance improvement and structure improvement of the components cannot systematically solve the shortcomings of the existing metro vehicles, nor meet the various requirements. Then, it is necessary to carry out the top-level technical index research on the new framework of metro vehicle.

The work of this paper is as follows. Firstly, by employing the artificial neural networks and complex network, the index of the traditional first- and second- generation metro vehicles is analyzed and the core characteristic index is extracted. Secondly, the relationship between the vehicle performance and the core indices is given and the performance of the next generate vehicle is predicted via employing artificial neural network. At last, combined with new technologies, the indices of the next generate metro vehicles are given which can be as the guide for the manufacture of the next generate vehicles.

2. Methods

2.1. Complex network

Since Erdös and Rényi began to study the stochastic graph model in 1960s, the theory of complex network [1] has been rising. A complex network or graph is often represented by a pair of tuples (V, e), where V is the node set and e is an edge set. The elements in V are called vertices or nodes and the elements in E are edges in the network. And each edge is related to two nodes in the node set.

The degree k_i of the node i is the number of edges connected to node i. The degree of nodes is defined by the adjacency matrix.

$$k_i = \sum_{j=1}^{n} a_{ij}$$

In the formula, n is the number of network nodes; a_{ij} is the value of ith row and jth column of the adjacency matrix; k_i is the degree of node i which reflects the complexity of the node i. The larger the value of ki, the more nodes connected to node i.

In the complex network [2], the distribution function p (k) is often used to describe the degree distribution of the complex network, where p (k) represents the probability of randomly selecting the nodes whose degree value is k.

2.2. Artificial neural networks

Artificial neural network [3, 4] is an information processing system which aims to imitate the structure and function of human brain. The theory of neural networks has become an interdisciplinary comprehensive frontier subject involving many subjects, such as neurophysiology, cognitive science, mathematical science, psychology, information science, computer science, microelectronics, bioelectronics and optics. In 1890, James, a famous American psychologist, carried out the research on the structure and function of human brain, published the first monograph on the structure and function of human brain, principles of psychology, and made a pioneering research on the basic principles of learning and associative memory. In 1958, Roseblatt, a computer scientist, proposed a neural network structure with three-layer network characteristics, which is called "perceptron" [4]. In 1969 Minsky and Papert published the book perceptron. And in 1982, Hopfield published the famous article "neural network and physical system". In June 1987, the first international neural network academic conference was held in California, the United States, and the International Neural Network Society (INNS) was established, which marks the upsurge of neural network research in the world.

According to the direction of information transmission, neural networks can be divided into two types: feedforward network and feedback network. There are many kinds of neural network learning algorithms. According to the widely used classification method, neural networks learning algorithms can be divided into three categories: supervised learning, unsupervised learning and dead memory learning [5].

Artificial neural network has been applied to almost all fields [6, 7], including pattern recognition, image processing, nonlinear optimization, speech processing, natural language understanding, automatic target recognition and expert system, and good results are achieved.

3. The development of metro vehicles in China

Since London subway, the first subway of the world, was in service on January 10, 1863, the urban rail transit vehicle industry has entered a high-speed development stage. In recent years, the number of metro lines in major cities in the world is more than three times the total length of metro lines built from 1863 to 1963. China's urban rail transit vehicle industry has also developed rapidly since the 1990s. The urban rail transit vehicle industry has made remarkable achievements. The development of urban rail transit vehicles in China can be summarized into two generations and three stages of technological evolution. As shown in Figure 1, the metro vehicle industry can be roughly divided into three stages: initial stage (1960s to early 1980s), stable development stage (mid-1980s-2000) and rapid development stage (from the beginning of the 21st century to now).

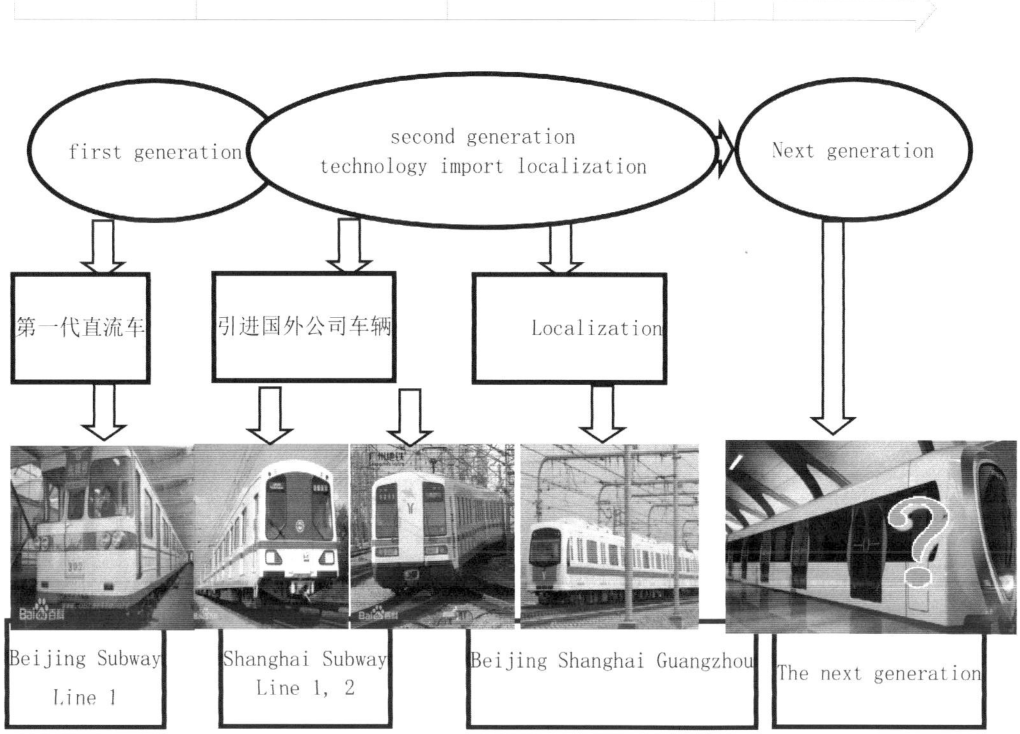

Figure 1. Technological evolution of metro vehicles development in China

4. Generation division of metro vehicles in China

In order to scientifically and objectively analyze the characteristics of the next generation (the third generation) of metro vehicles, the typical characteristics of the first generation and the second generation of metro vehicles is firstly analyzed, and the core universal characteristics between the generation of metro vehicles are extracted which are also as the reference standard for the generation feature division between the second generation and the next generation. And then, the typical characteristics of the next generation of metro vehicles are analyzed.

Specifically, all possible metro vehicle indices are firstly selected from transport capacity, safety, comfort and environmental friendliness aspects. Secondly, according to the correlation between these indices, the complex network model of metro vehicle indices is established, and the core indices of metro vehicles are selected. Thirdly, with the help of neural network theory, the neural network model between metro vehicles indices and vehicle performance is established. Combined with the current metro vehicle index performance, the newest technological development and the shortcomings of the current metro, the next generation metro vehicle indices and characteristics are analyzed and predicted. The specific process is shown in Figure 2.

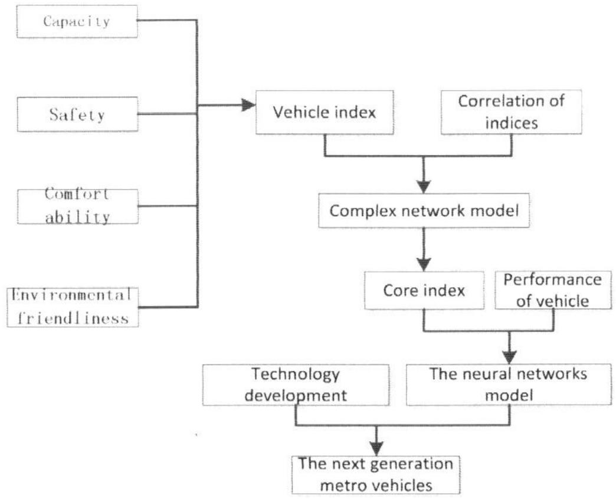

Figure 2. The flow chart of research on characteristics of the next generation metro vehicles

From the four top-level indicators of transport capacity, comfort, safety and environmental friendliness, the indices of metro vehicles are systematic analyzed, and 45 specific indicators of metro vehicles are obtained. The next step work is to extract the core index system of metro vehicles from these 45 indexes.

The 45 indicators are nodes, and according to the correlation between 45 indicators, the edge is generated. Thus, the complex network model is established, which has 45 nodes (indicators) and 466 edges (correlation between indicators). And according to the definition of degree, the degree of 45 indicators are shown in Figure 3.

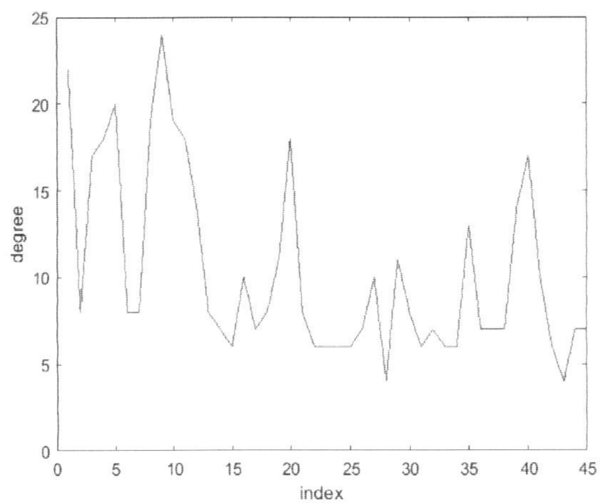

Figure 3. The degree of each index of metro vehicle in complex network model

According to the index degree (index importance) of 45 indices, 14 core indices of metro vehicles are selected which is shown in Table 1.

Table 1. Core indicators of metro vehicles

name	name	name
Operating speed	Critical speed	Passenger compartment noise
accelerated speed	service life	exterior noise
deceleration	failure rate	regenerative braking
Load / axle load	Allowable collision speed	Operational resistance
Train formation	Stationarity	

By employing the neural networks, a two-layer neural network model including the hidden layer and the output layer is established, in which the number of hidden layer neurons is 5, and the specific neural network structure is shown in Figure 4.

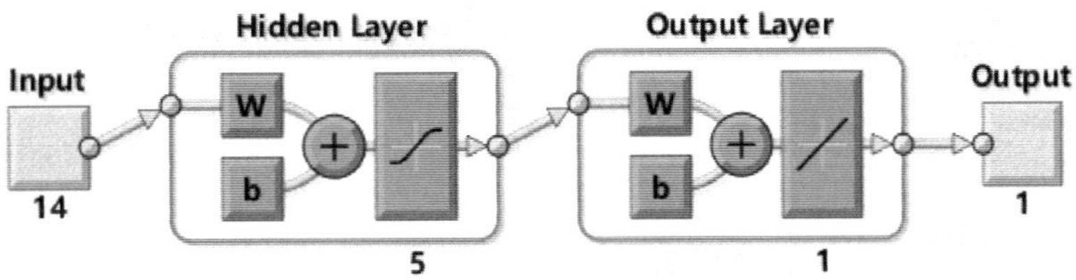

Figure 4. The structure of the neural networks

70% of the data is used for neural network training, 15% of the data is used for neural networks validity test, 15% of the data is used for neural networks test. By employing the neural networks, the neural network model is as follows.

$$y = f(x) = LW\_1 * (IW1\_1 * x + b1) + b2$$

$IW1\_1$ is matrix and it value is shown in Table 2. $b1 = [1.3, 0.7, -0.2, 1.1, -1.5]$, $Lw2_1 = [1.2, 0.3, -0.5, 0.2, -0.2]$, $b2 = -0.6$。

Table 2. The value of the matrix IW1_1.

-0.5	0.4	0.2	-0.2	0.8	-0.9	0.1	0.2	0.0	-0.5	0.0	1.0	0.4	0.7
-0.2	-0.5	-0.2	-0.5	0.2	0.3	0.0	0.3	0.6	0.2	0.8	0.7	0.8	0.7
-0.8	-0.4	0.5	-0.7	-0.3	0.7	0.4	0.1	1.0	0.1	0.5	-0.3	-0.3	0.2
0.5	-0.7	0.2	0.5	0.2	0.1	-0.8	-1.0	-0.2	0.3	0.0	0.2	-0.1	-0.3
-0.5	0.3	-0.5	0.4	0.5	0.6	0.5	0.1	-0.3	0.7	0.5	-0.5	0.4	0.6

Based on the latest technological development, the operation speed, acceleration and deceleration, load / axle load, train formation, service life, failure rate, passenger compartment noise, exterior noise, regenerative braking and stability of the next generation of metro will be greatly improved.

By comparing the characteristic indexes of the current and the next generation of metro vehicles, it can be found that the load / axle load, train formation and regenerative braking of metro vehicles are significantly improved. These three indexes are the specific indexes of the third generation. The characteristics of the next generation of metro vehicles can be summarized as intelligence and energy saving.

5. Conclusions

From the system's view, the top-level indicators of metro vehicles are divided into 45 categories from capacity, safety, comfortable ability and environmental friendliness aspects. With the employment of the complex network model, 14 top-level core indicators are abstracted. Then, with the employment of the neural network model, the neural network model of indices of metro vehicles is established. Combined with the existing rail transit technology, the next generation of metro vehicles is predicted to be characterized by intelligence and energy saving, which are embodied in load / axle load, train formation and regenerative braking. The results can be used as a guide for the future metro vehicle manufacturing.

Acknowledgement
This work was supported by China Postdoctoral Science Foundation (Grant No. 2019M650981).

References

[1] Barabási A. L., Albert R. Emergence of scaling in random networks. Science, 1999, 286(5439): 509-512.
[2] Shao F, Li K. A complex network model for analyzing railway accidents based on the maximal information coefficient, Communications in Theoretical Physics, 2016, 66(4): 459-466.
[3] Hospfield J. J. Neural networks and physical systems with emergent collective computer abilities. Proceedings of National Academy Sciences. 1982, 79(8): 2554-2558.
[4] Rosenblattt F. The perception: A probabilistic model for information storage and organization in the brain. Psychological Review, 1958, 65: 386-408.
[5] Ciganovic N., Beaudry N. J., Renner R. Smooth max-information as one-shot generalization for mutual information. Information Theory, IEEE Transactions on, 2014, 60(3): 1573-1581.
[6] Bjerve S., Doksum K. Correlation curves: measures of association as functions of covariate value. The Annals of Mathematical Statistics, 1993, 21: 890-902.
[7] Narendra K. S., Parthasarathy K. Identification and control of dynamical systems using neural networks. IEEE Transactions on Neural Networks, 1990, 1(1): 4-27.

Journal of Physics: Conference Series **1423** (2019) 012060 doi:10.1088/1742-6596/1423/1/012060

Exploring the results of college English learning under the perspective of artificial intelligence

Ming Cao[1]

[1]School of Humanities and International Education, Xi'an Peihua University, 888 Changning Street, Xi'an China

Corresponding author and e-mail: Ming Cao, 2745359275@qq.com

Abstract. Based on the theory of output hypothesis and the perspective of Artificial Intelligence (AI), the purpose of present study is to verify the result of applying Virtual Reality (VR) experiment. A quantitative method has been used. 69 students as an experimental group and 68 as a control group, by random sampled, have participated this quasi-experiment. After half a year, experimental group has obtained better performance than control group in CET (College English Test). T-test shows significant differences in the result of CET total score ($t=2.05$, $p<0.05$) and listening ($t=2.70$, $p<0.01$) as well as writing ($t=2.32$, $p<0.05$) section score. These results suggest that applying VR method brought positive result for college English learning.

1. Introduction

Turning into 21th century, artificial intelligence (AI) becomes one of the most popular techniques and spreads conspicuously fast. Since education is deeply influenced by technological development, more and more AI equipment has been installed to the university to execute the different tasks. This triggers the discussion of English teaching and learning (Zhu, 2018; Wang, 2019). For Virtual Reality (VR) experiment for English learning, research results are rare. But it is still worth exploring because of its developmental prospect. Weather it effects and how it effects make us pay attention to it. Based on this, the study explores and tries to verify how AI influences college English learning.

2. Theories review

AI refers to how machine imitates the process of human beings' thinking and behaviors so as to allow us to explore the law of intelligent actions of human beings (Winston, 1984). Until now, AI has been experienced two stages and is facing the third one. The first, birth and preliminary exploring period (1950-2003), shows its grow and development; the second is the application and exploring period (2004-2017). During this period of time, many journal articles appeared and the amount of these articles account for 74.71% in total research of AI (Lin, 2019). Nowadays, we move on to the third period --- the peak of development. In this period, many signals show that AI closely relates to education field. For example, Vipkid, Grandescope, Pigai, Mindstorm EV3, AI campus, etc. are popular websites or platforms that have been used for people to solve problems in grading and teaching. While in education filed, it brings about a big challenge for teachers. Traditionally, teachers teach their students in a physical classroom by using textbooks, paper and pens, and evaluate by final examination. Encountering AI, both teachers and students need to explore a new way to teach and learn.

2.1. Output hypothesis
Canadian linguist Merrill Swain (1985) proposed output hypothesis in 1980s. She argued that producing output played very important roles in arousing learners attention and acquiring target language. The output has the function of assumption and self-examination. It promotes "noticing", which means during producing, learners may notice the gap between what they say and what they can say. It serves the process of target language learning not only enhances fluency but also accuracy. After doing a lot of researches and experiments in immersion language teaching, Swain (1995) proposed that output hypothesis had three main functions. (1) noting/triggering function; (2) hypothesis testing function; (3) metalinguistic function. These three functions apply the following aspects: (1) speaking and writing benefits language learners fluency and automaticity; (2) It makes language learner transfer from meaning producing to sentence producing; (3) Speaking and writing the target language make learner detect the hypothesis the learner produces; (4) The output of target language may acquire corresponding feedback(Swain, 1998). Therefore, language learning is a bilateral activity. Focusing on speaking and writing practice has more values than mechanical input only. It cannot be replaced.

2.2. How AI perspective learning contrasts with learning in traditional classroom
Here is the contrast about the characteristics learning in traditional classroom and learning under the perspective of AI. It's easy to understand that both teachers and students change their roles under the perspective of AI. Both teachers and students have rights to talk and discuss.

Table 1. How AI perspective learning contrasts with learning in traditional classrooms.

Items	Learning in Traditional Classroom	AI Perspective Learning
Environment	Teaching in common classrooms	Teaching in classroom with special equipment in special classrooms
Teaching Aids	Using blackboards; chalk; paper; pens	Using blackboards; chalk; paper; pens; Computers; internet; special equipment of AI, etc.
Ways of teaching and learning	Students listen to the teachers only; teachers give homework, check answers, and grade, etc.	Both teachers and students have rights to talk and discuss; Encouraging problem solving; team work; individual study; online/offline learning; projects, etc.
Evaluation	Mid-term/final examination; quiz; test, etc.	Providing process evaluation; peer review; self evaluation; etc.

3. Research design
A quasi-experiment has been designed in this research. 69 students as an experimental group and 68 as a control group, by random sampled, have participated this quasi-experiment. The experimental group has been provided to use the lab once a week during work hours. The lab assistant just taught students how to use the computer and left the students practice by themselves. There was no cellphone website available for them to practice. The control group has been arranged to do a cellphone experiment by providing a web link to make sure they could practice by cellphone after class only. They had no lab practice but they could choose to do it in their spare time. Both groups required to practice after regular class. But during the regular class, teachers set up a report time to check the result of students oral practice for both groups once a week. The teachers made evaluation and provided some comments when they do the report. After half a year, we found the differences.

3.1. Virtual Reality experiment
Under the perspective of AI, the school constructed a language lab for English major students. To improve language output ability, the lab has been designed as a Virtual Reality classroom. In this special classroom, students were allowed to use VR equipment to do language experiment. There were

different topics installed in the computer system and once students wore the VR glasses, they could feel as in the reality according to the topics. When they clicked the button, they would get the task they chose and started their practice. Students could practice for many times via the evaluation from the computer system (online) until they felt that they were confident to present in front of the teacher and students (offline). They could also use their cellphone to be the terminal to practice without wearing glasses at any time just like they were in the lab (online). In fact, there was no feeling of virtual reality. Nevertheless, the process was the same. The procedures of the experiment can be designed as follows (See figure 1).

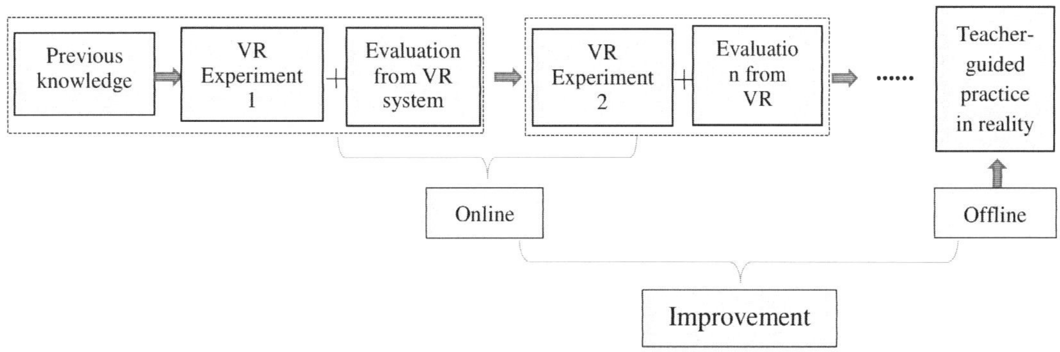

Figure 1. VR experiment procedure.

3.2. Research hypothesis
The hypotheses are as follows. Whether the experimental group will get higher grade than control group from College English Test (CET) or not? If VR has more effects on students English study, on what aspect does it provided the effect?

4. Research results
This part includes a quantitative research results and the analysis of the data. We used T-test and SPSS22.0 to deal with the data. The results are as follows.

4.1. The result of College English Test
We took College English Test (CET) result as an example to analyze students' achievement. Students attended CET voluntarily. Within 57 testers, 45.6% students from experimental group passed the examination, while 30.3% students from control group passed (See table 1). Experimental group shows better performance in total grade as well as in sections of listening and writing. Reading score has a close similarity between two groups. T-test shows significant differences in the result of CET total grade ($t=2.05$, $p<0.05$) and other two sections of listening ($t=2.70$, $p<0.01$) and writing ($t=2.32$, $p<0.05$) (See Table 2).

Table 2. Contrast table

	Experimental group			Control group		
	Numbers attended	Numbers passed	Passing rate (%)	Numbers attended	Numbers passed	Passing rate (%)
Results	57	26	45.6	66	21	30.3

Table 3 T-test of experimental group and Control group

Group	Experimental group(n=57) $\overline{X} \pm SD$	Control group (n=66) $\overline{X} \pm SD$	t
Listening	138.95±25.14	125.47±29.54	**2.70
Reading	142.05±21.40	143.76±22.71	-.43
Writing	136.02±14.14	130.42±12.57	*2.32
Total	417.02±43.73	399.65±49.09	*2.05

(**P<0.01,*P<0.05)

$$t = \frac{(\overline{x}_1 - \overline{x}_2) - (\overline{\mu}_1 - \overline{\mu}_2)}{\sqrt{\frac{s_1^2}{n_1} + \frac{s_2^2}{n_2}}} \quad \text{where} \tag{1}$$

$$df = \frac{(\frac{s_1^2}{n_1} + \frac{s_2^2}{n_2})^2}{\frac{(\frac{s_1^2}{n_1})^2}{n_1 - 1} + \frac{(\frac{s_2^2}{n_2})^2}{n_2 - 1}} \tag{2}$$

4.2. Analysis of the results

Just like Swain (1995) emphasized that output process forces learners to focus on expressions of the target language so that to promote language processing and improve language ability. Experimental group has got better results than control group, which demonstrates there is more effective for VR lab practice than cellphone practice only. There are some reasons need to be paid attention.

4.2.1 VR experiment aroused students interests.

VR practice provided an environment for students just like they attended the real place, so they had more sense of opening their mouth to practice. They may feel that they were in the real setting so that they payed more attention to it. As a result, they learned more words and tried more times.

4.2.2 VR experiment provided a simple feedback.

Although it was pretty rough, it still gave students confidence to practice more. When they heard the sound, they may pay more attention to repeat. As good pronunciation will bring good listening sense, students can easily understand the listening materials from a foreign speakers during CET.

4.2.3 Why writing score is also different?

As we know, writing is closely related to our thinking. It is an integrated effect. As Swain (1998) emphasized both speaking and writing benefits language learners fluency and automaticity, writing is a kind of output activity either, so it did effect. Once students practiced more, they got sense to speak, so they got sense to express their ideas in writing. Good writing needs not only understanding the meaning of the words, but also using the words. VR experiment provided the chance to practice the words. It inevitably provided a positive influence for students' writing.

4.2.4 Other concerns.
Some points may not be related to output hypothesis or AI itself, but students may feel to be more supported than those who were not involved in lab experiment. As control group accepted cellphone experiment, there was no VR environment, so they may feel it is just a common practice. Since no requirements for the practical hours, they may slack themselves. Therefore, they were lack of practice and the CET result is lower than experimental group.

4.2.5 Teachers attitudes may be another factor.
For experimental group, since no cellphone practice carried out, all their practice focused on AI environment so that the teacher may respect students' feelings in different ways. They may ask more student-centered questions, set up more group work during regular class, and provided more chances for students to practice. These perspective of AI may influence students learning output. While for control group, both teachers and students just treated the task as a common assignment, so less importance made them paying less attention to the practice on cellphone. Even if they paid attention to it, they did not get the VR environment. Therefore, it is hard to arouse their interests. As a result, output process did not really happen. For teachers of control group, less influence happened under the perspective of AI.

5. Conclusion
Artificial intelligence brings about a new idea, new concept, and new angle of seeing the world. Although being pretty arguable, for example, lack of knowledge of AI, not considering the feeling of students and teachers, and offending privacy, etc. (Bai, et al, 2019), AI is still an attractive research area which is worth exploring. In current education field, especially for language teaching and learning, we need to treat AI as a practical technology to promote language teaching and reform. Based on the perspective of AI and the theory of output hypothesis, this study just explored a little about the application of VR, introduced the experiment, summarized the result of students' learning and analyzed the effect. It also tried to emphasize the student-centered teaching and learning under the perspective of AI. It acquired that AI did effect the English learning result, but we did not dig out more due to the limitation of the time. There will be another research direction in the future.

Acknowledgement
Thanks to Professor Hao Wang, from Xi'an University of Finance and Economics, who helps me a lot in dealing with the data.
2018 Shaanxi Science of Education "Thirteen-Five" Project "Flipped Class Teaching research on College English Major Based on MOOC Platform", Project No. SGH18H496; 2017 Higher Education Science Project from Shaanxi Association of Higher Education "Research on Countermessures of Improving English Output competence of students in Application-Oriented University Based on Context of Globalization", Project No. XGH17190.

References
[1] Bai, Shehua, Li, Suling, & Ding, Liangxi. Problems and strategies of artificial intelligence on the development of education [J]. Chinese University Science & Technology, 2019(9): 94-96.
[2] Liang, Yingshan. Research on the college English practical system design under the environment of artificial intelligence [J]. Journal of Jiangxi Vocational and Technical College of Electricity, 2019(3): 110-111.
[3] Lin, Xiao-feng &Xie, Kang. The current situation of artificial intelligence and the rational thinking of its educational application [J]. Modern Educational Technology, 2019(8): 12-17.

[4] Swain, M . Communicative competence: Some roles of comprehensible input and comprehensible output in its development [A]. In Gass S & Madden C (eds.). Input in Second Language Acquisition[C]. Rowley, MA: Newbury House, 1985. 235-253.
[5] Swain, M. Focus on form through conscious reflection [A].In Doughty C & Williams J (eds.) Focus on Form in Classroom Second Language Acquisition [C]. New York: Cambridge University Press, 1998: 64-81.
[6] Swain, M. Three functions of output in second language learning [A]. In Cook G. & Seidlhofer B (eds.), Principles and practice in applied linguistics(C). Oxford: Oxford University Press, 1995: 125-144.
[7] Winston, P H. Artificial intelligence [M]. 2nd ed. Boston: Addison-Wesley Longman Publishing. Co., 1984.
[8] Zhu, Yan. Research on college English translation teaching mode under the background of artificial intelligence [J]. Chinese & Foreign Entrepreneurs, 2018(9):179-180.

Journal of Physics: Conference Series **1423** (2019) 012061 doi:10.1088/1742-6596/1423/1/012061

Target extraction of fruit picking robot vision system

Junyan Zhang

Jiangsu Union Technical Institute Suzhou Industrial Park Branch, Suzhou, China

Corresponding author and e-mail: Junyan Zhang, 108645502@qq.com

Abstract. Real-time and accurate target recognition and extraction of crop fruit images in the field is the key technology for picking robot vision system, and the essence of target extraction is image segmentation. When most of the water (vegetable) fruit is in the picking period, the surface color and the background color are quite different, while the surface colors of the same variety are similar, which are reflected in the different distribution characteristics of the fruit surface color and the background color in the color space. According to this characteristic, an image segmentation algorithm based on color space reference table for fruit target extraction of fruit picking robot vision system is proposed. The algorithm first establishes a color space reference table from the fruit sample image, and then uses a method similar to "convolution" to segment the image according to the color space reference table. Compared with other existing methods, the method is based on color information processing, which can remove the background more cleanly. Without segmentation processing and no complicated operation on the background, the method is beneficial to real-time image processing of the robot.

1. Introduction

The rapid development of fruit farming has increased the market demand for orchard machinery. The labor used in the picking operation accounts for 33% to 50% of the labor used in the whole production process. The picking operation is complicated and the seasonality is strong. If manual picking is used, it is not only inefficient, labor-intensive, but also easy to cause fruit damage, which directly affects the quality of the fruit. The use of the picking robot not only improves the picking efficiency, but also reduces the damage rate, saves labor costs, and improves the economic benefits of the fruit growers. Therefore, it is of great significance to improve the mechanization degree of the picking operation. With the decline of agricultural practitioners and the increasing aging trend, the development and utilization of picking robots has enormous economic benefits and broad market prospects, which meet the needs of social development. The normal operation of any kind of fruit picking robot depends on the correct identification of the working object. Therefore, to realize the fruit harvesting of the fruit picking robot, the key is to identify the fruit from the fruit tree and determine the exact spatial position of the fruit so as to be the robot. The movement provides parameters to complete the fruit picking. Therefore, the performance of the visual system will directly affect the fruit picking effect. The performance of the vision system should be evaluated in terms of both accuracy and speed. There are many factors that affect the accuracy and speed of the visual system, such as unstable illumination, wind, fruit and leaves, and fruit growth. At present, the method of relying solely on machine vision to identify fruits has not yet reached commercialization, and it is necessary to improve the algorithm, reduce the use cost, shorten the picking time, improve the machine recognition rate and picking efficiency as soon as possible.

Journal of Physics: Conference Series **1423** (2019) 012061 doi:10.1088/1742-6596/1423/1/012061

2. Problems and analysis of the visual system of fruit picking robot

Through the analysis of the current situation of fruit picking robots at home and abroad in the previous section, it is not difficult to find that the research results on visual direction are relatively rare in our country. Due to the increase in the output of the domestic fruit industry, the corresponding agricultural product harvesting robots are also urgently needed to be developed. There are many common problems in the process of picking robots to consider from the following aspects:

First of all, in the domestic research on fruit picking robots, the recognition rate of picking is not very high. The identification of fruits is generally studied from several aspects: color component difference, gray value threshold, fruit geometric characteristics and so on. However, in the actual picking environment, there is interference of various information, so that the recognition rate is not high. If the identification is based on the way of fruit characteristics, it is necessary to extract a relatively complete boundary, but usually encounters the overlap of fruits or the occlusion of fruits and leaves. At this time, the identification of fruits is more difficult. The fruit is different in color and growth under different growth environments. These differences will make the standard of image segmentation ununified, resulting in reduced adaptability of the picking robot.

In order to meet the requirements of actual picking: to reduce the labor intensity of workers and improve the efficiency of picking, it is necessary to improve the picking speed. This requires not only the high demands on the mechanical system, but also the speed of image processing. This will result in a higher real-time performance in actual work. In addition, in the actual picking environment, the problem of illumination change will be encountered, which brings a lot of inconvenience to the visual system in image processing, and a certain compensation method is needed in image processing.

The research in this paper is mainly aimed at the problems encountered by apple picking robots at work, mainly from the following aspects: (1) Identification and localization of multiple fruits and identification and localization of multiple overlapping fruits are encountered during the picking process. In the case of the arrival, the fruit growth posture and the occlusion often make it impossible to accurately extract the fruit and locate the fruit at the time of picking. This also reduces the picking efficiency of the fruit picking robot. (2) The accuracy of the binocular vision system is also a consideration. When locating the spatial position of the fruit, you need to know the internal and external parameters of the camera. However, when the camera captures images, the lens distortion problem often occurs, which results in a large error in the internal and external parameters of the camera, which indirectly leads to a large positioning error on the fruit. This requires consideration of the camera calibration, improving the distortion problem and improving the calibration accuracy. (3) Exploring accurate image matching algorithms can reduce the matching error rate, which is especially important for picking work.

3. Color space representation system

In the theory of chromatics, the spaces representing colors are RGB, HSV, YCbC, and L* a* b*. This article focuses on the RGB and L* a* b* spaces involved. The RGB color space is based on the principle of three primary colors and is one of the most basic color representation models. It corresponds to a cube in the Cartesian coordinate system, and R, G, and B represent three coordinate axes, representing red, green, and blue. When R, G, and B both take 0 (that is, the origin of the coordinate), it means black; when both take the maximum value, it means white, as shown in Figure 1. The three components are all between 0 and 255. The L* a* b* model is one of the two color representation systems recommended by the International Commission on Illumination for the uniform evaluation of chromatic aberrations. The advantage is that the luminance components are separated, and the spatial distance can be used to represent the difference between the two colors. L* a* b* medium; L* means brightness, its value is 0 to 100; a* means change from red to green, a* is positive for red, negative is green, and value is (-60, 60); b* means yellow to blue change, b* is positive, yellow means negative, blue is the range, and the range is (- 60, 60), as shown in Figure 1. The conversion formula for L* a* b* space and RGB space is

$$L^* = 116(0.299R + 0.587G + 0.114B)^{1/3} - 16$$

$$a^* = 500\left[1.006(0.607R + 0.174G + 0.201B)^{1/3} - (0.299R + 0.587G + 0.114B)^{1/3}\right]$$

$$b^* = 200\left[(0.299R + 0.587G + 0.114B)^{1/3} - 0.846(0.066G + 1.117B)^{1/3}\right]$$

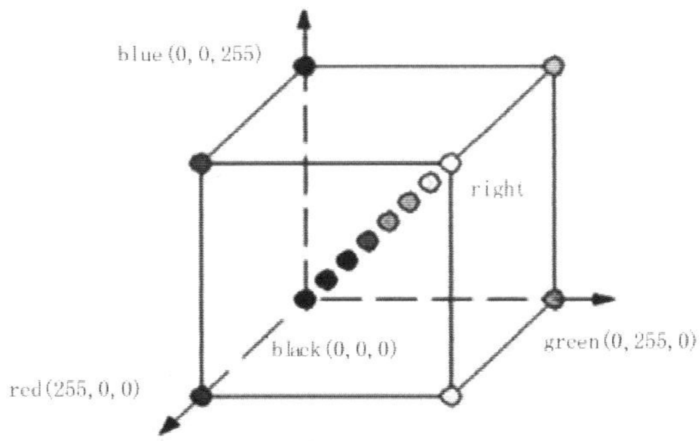

Figure 1. The sketch map of RGB color space

In the L* a* b* color space, a* and b* are luminance-independent components, so the effect of light on the segmentation result in the natural environment can be weakened. These two components can be used to construct the apple sample color space.

4. The role of mathematical morphology algorithm in the process of establishing color space reference table

The fifth step of the color space reference table creation step applies a mathematical morphology operation, which will be described below. The ubiquity of noise in the image determines that the noise obtained from the fourth step also contains the noise of the sample image, which also appears as noise in the result (away from the isolated point of the target region); meanwhile, due to the finite nature of the sample, The sample color distribution area of the result obtained in step 4 is a sparse area in which there are numerous adjacent islands and small holes in the binary image. The purpose of step 5 is to eliminate the noise, fill the small holes and connect the adjacent islands to form a complete area. Expansion and corrosion are the two basic operations of mathematical morphology, let A be the image set and B be the structural element.

The original definition of expansion. The set obtained by expanding B with B is the set of origin positions of B when the displacement of B (the image of B) intersects with at least one non-zero element of A. Improved expansion algorithm. Since there may be a lot of noise in the sample, if the expansion operation actually enlarges the noise according to the original definition of the expansion, the structural element B is taken as a window of 3 × 3, and the origin of the structural element is at the center (Fig. 2). The expansion algorithm is changed to: The set obtained by expanding B with B is the set of origin positions of B when the displacement of B intersects with at least 3 non-zero elements of A. Since the noise is generally far from the target area, there are few non-zero elements around.

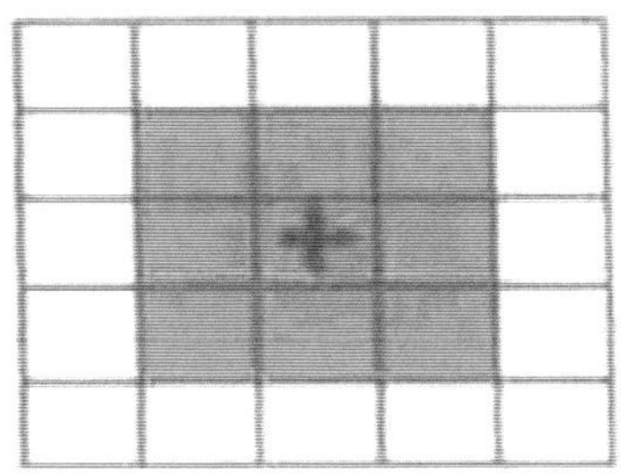

Figure 2. Structure unit B

The original definition of corrosion. The set obtained by etching A with B is the set of origin positions of B when B is completely included in A. Improved corrosion algorithm. Due to the finite nature of the sample, the result obtained from step 4 is a sparse area. If the corrosion operation is performed according to the original definition of corrosion, all the distribution points may be corroded. Therefore, the same as the expansion, the structural element B is 3 × 3 window, the origin of the structural element is at the center. Change the Corrosion Algorithm to: Corrosion with B. The resulting set is the set of origin positions of B when B is displaced by a maximum of 3 zero elements. Also because the noise is generally far from the target area, most of the surrounding is zero elements, so the noise is eroded (filtered), and useful points inside the target area are preserved.

After the expansion and corrosion algorithms are improved, the internal expansion of the target area plays a leading role, and the external corrosion of the target area plays a leading role, and the boundary of the area acts inversely; therefore, a reasonable combination of expansion and corrosion operations can obtain a tightly distributed complete target area. At the same time, noise is eliminated and the boundaries of the region are preserved.

5. Binocular image matching

The dual-purpose stereo matching is based on the fact that two cameras are located at different visual points to capture the same position and then acquire stereoscopic image pairs. The parallax is obtained from the stereo image pair of the target, and finally the three-dimensional geometric information of the target is obtained. In binocular matching, there are many influencing factors in the shooting environment that make the captured image pairs very different, mainly in the following aspects: noise, grayscale and geometric distortion. In the research of binocular stereo matching algorithm, the researchers have proposed a variety of matching algorithms for these problems, but there is still no stable adaptive algorithm. These algorithms are classified according to different constraints: local algorithms and global algorithms. The focus of the global optimization method is on the calculation of parallax, and most of these algorithms are based on the minimum energy. For local optimization, the main consideration is the matching cost calculation and matching cost. Compared with the former, this method is easier to obtain the disparity value, because it is only necessary to determine the disparity value when the matching value is the smallest in the calculation. In actual research, the matching algorithm is generally considered from several aspects:

First, to determine the primitives that can be adapted to the matching object, according to the difference of the matching objects, it is possible to select from the regional aspect, the phase angle or the linear feature in the experiment. The determination of the primitives is crucial to the effectiveness of the experiment. Secondly, in addition to selecting the appropriate primitives, it is also necessary to determine the matching criteria in the features of the image alignment, such as whether the order between the features is consistent, continuous, and not similar or even unique. Finally, based on the above two points, create a real-time effective matching algorithm structure. Through comparison and research, the SSD algorithm is selected as the matching algorithm, which is suitable for the characteristics of the visual system in terms of sensitivity and speed.

The matching primitive has been mentioned in the previous section. Under normal circumstances, it is unrealistic to require the target to match all the image points in the acquired image pair. First, the image is different from the surrounding environment. However, in the process of matching, you can find as many matching points as possible to match. Based on this, it is very important to select the attribute characteristics of the appropriate target. The purpose of selecting a matching primitive is to obtain a corresponding image feature from the image pair. There are three types of primitives that are often chosen: region, feature, and phase match. These three types are suitable for different situations, so when making a match, analyze the feasibility. The regional primitives have the fewest occurrences in the image, but contain the most content, so they have the characteristics of strong object differentiation and high precision when matching. It is the easiest and easy to operate algorithm for the current matching algorithm. The main purpose is to use gray value similarity to correspond to the relationship between image pairs. In the process of algorithm implementation, one of the image pairs is the research object. And the neighborhood of the gray value to be processed is used as the matching template. A similar function is used to evaluate the similarity of the two, and then another pixel is used to find pixels with similar gray areas. However, it can be seen from the above that the algorithm performs region matching according to the gray value, so it is relatively susceptible to external interference, thereby affecting the matching accuracy and increasing the probability of mismatching. However, this can be pre-processed with image pre-processing techniques, which will greatly reduce noise. In the area matching, we also need to choose the appropriate window. The selection of the window size has a great influence on the matching. If it is too large, it is easy to mismatch. If it is too small, the matching is easy to increase the ambiguity and the accuracy is greatly reduced. Feature-based matching the algorithm can limit the search scope to the features of the target, and then extract the matching features in the image pair to complete the matching work. Since the matching is only for feature attributes during matching, the algorithm is dominant in the matching speed. In addition, the algorithm is not easily disturbed by the outside world, and is relatively stable and has high precision. It is also because of this feature of the algorithm that only the sparse disparity map can be extracted at the end of the matching; and the appropriate features are also selected in the matching to obtain the desired result, the selected features should be highly recognized and the feature should be more dense. Phase matching is a matching using the phase alignment of the image, and matching according to the property that the phases of the corresponding points are the same. The matching algorithm has a good matching effect and has good stability for disturbance factors such as noise and distortion.

6. Fruit target extraction and its experimental study

With the fruit surface color space reference table, it is easy to determine whether a pixel is the target pixel. Here, a method for judging whether a pixel is a target pixel in a statistical sense is used: the number of target pixel points is counted in a 5 × 5 neighborhood of each pixel (whether or not it is a target pixel), exceeding In the case of half, the point is considered to be the target pixel, otherwise it is considered to be a non-target pixel. For the interior of the fruit target area, this is itself a kind of "convolution" filtering; for the outside of the fruit target area, this method can remove the background more effectively. Judging whether a pixel is a target pixel actually obtains a binarized image containing the fruit. The ubiquitous noise and the very rich color of the fruit surface make the judgment based on the reference table model may be wrong, so that the fruit target area of the obtained binary image may have a large number of holes, similar to the 4th when the reference table is established in this paper. The resulting binary image; therefore, the fruit image can also be processed using the improved expansion and erosion algorithms mentioned in Table 3 to fill the holes. After the holes are filled, the area of each fruit can be found by the common area marking algorithm, and the circumscribed rectangles of these areas can be obtained, thereby completing the extraction of the fruit target.

Based on the above algorithm, the author selected strawberry, orange and tomato in the HS V, Y CbCr and L*a*b* color spaces for experiments. The results are shown in Tables 1 and 2. Table 1 shows the distribution of fruit surface color samples in the HS, CbCr and a*b* chromaticity spaces, respectively. Table 2 shows the results of the test samples under the HS V, YCbCr and L*a*b* color models.

The background of the fruit picking robot working environment is complex, but basically similar, and the pixels of the background image also have a certain distribution in the color space. In order to better remove the background, the color space reference table of the background is established according to the method of establishing the color space reference table, and the common part of the background color space reference table is subtracted from the fruit color space reference table, so that the segmentation effect is better. The fruit color space reference table. Test results on tomatoes show that this improvement is necessary.

The author has tested the image with a lot of salt and pepper noise and random noise. The results show that the same effect can be achieved without filtering before the target extraction, because the method described in this paper performs noise throughout the image processing. Special treatment, indicating that the method is fast and simple.

Compared with the gray histogram segmentation method, since the method is based on color, more information in the image is utilized, and the background can be removed more cleanly, thereby facilitating further image analysis. Clustering segmentation algorithm and random model-based algorithm are mostly based on more complex mathematical formulas. This method divides the current image by looking up the table. There is no complicated mathematical operation, and the time complexity is linear. Conducive to real-time image processing in robot control. Region-based methods such as region-growth algorithms, segmentation operations are also performed in the background section, while the region-growth algorithm is a recursive algorithm with high time complexity; this method only deals with fruit targets (sometimes including fruit stems) , no processing on the background, saving calculation time, which is conducive to efficient operation of the robot.

After obtaining the color distribution area of the fruit surface, the fruit surface color distribution model can be obtained by fitting the boundary of the region. Compared with the reference table-based method, this method saves memory space (the reference table requires 256 × 256 × 2 B space, if the reference table uses run-length encoding, the actual memory is also very small), but if the boundary is complicated, The calculation is also complicated when the image is divided. Although ellipse can be used to fit the boundary to simplify the calculation when segmenting the image, the assumption that the color distribution is close to the ellipse is not reasonable. In particular, some fruits in the picking period are not fully mature, and the surface color may be distributed in multiple separations. Area. Therefore, compared with the distribution model based method, the reference table based method is not only simpler, but also has a lower misclassification rate.

7. Conclusion

Experiments on the images of strawberries, oranges and tomatoes show that the algorithm based on the fruit color space reference table can quickly target fruit in complex backgrounds under the HS V , Y CbCr and L*a*b* color models. The extraction effect is good, which proves that this algorithm is feasible and efficient.

References

[1] Zhang Tiezhong, Lin Baolong, Gao Rui. Target Extraction of Fruit Picking Robot Vision System [J]. Journal of China Agricultural University, 2004(2).
[2] Zhao Jinying, Zhang Tiezhong, Yang Li. Target Extraction of Tomato Picking Robot Vision System [J]. Transactions of the Chinese Society of Agricultural Machinery, 2006, 37(10): 200-203.
[3] Zhou Jun, Liu Rui, Zhang Gaoyang. Design of Fruit Picking Robot System Based on Stereo Vision [J]. Transactions of the Chinese Society of Agricultural Machinery, 2006, 21(6):158-162.
[4] Wu Yiman. Design of Target Recognition Vision System for Mobile Picking Robot Based on DSP [J]. Agricultural Mechanization Research, 2018.
[5] Wen Zhiqiang, Cai Zixing. Moving Target Extraction Method Based on Mobile Robot Vision System [J]. Computer Engineering, 2007(24):203-205.
[6] ZHANG Guiying, Zhang Xianjie. Target Extraction of Mobile Robot Vision Control System [J]. Manufacturing Automation, 2012(17): 71-72.

Design of human-like behavior learning machine based on neural network FPGA

Ming Tang, Shengli Hao
Xi'an Peihua University, Xi'an, China

Corresponding author and e-mail: Ming Tang, 81408240@qq.com

Abstract. Neural network has been widely used in image recognition, because it can imitate the behavior of biological visual nerve to obtain high recognition accuracy. Aiming at image recognition, this paper studies the machine self-learning system of image recognition results based on neural network FPGA, designs the data sensing and image recognition processing module based on neural network, and studies the key problems of the system self-recognition and machine self-learning process. The image recognition module uses bionics and parallax principle to get the synchronous exposure image of surrounding scenery, reconstruct the three-dimensional shape and position of surrounding objects, and get the information needed by the system design through image processing algorithm. On the basis of this module, a machine self-learning algorithm based on the result of image recognition is designed to build a knowledge base for the machine and guide the machine by modifying the weight and making decision to realize human-like behavior of machine self-learning. The experimental results show that the recognition accuracy of dynamic image recognition in image recognition network is about 80% compared with the corresponding static image recognition, and the time and power consumption are reduced by 24%, 27.8% and 41% respectively.

1. Introduction

Neural network has been widely used in image recognition, because it can imitate the behavior of biological visual nerve to obtain high recognition accuracy. Recently, the rapid growth of modern applications based on deep learning algorithms has further improved research and implementation. In particular, a variety of neural network-based hardware architecture platforms of FPGA have been proposed, which have the advantages of high performance, reconfigurable, rapid development cycle and so on.

In image recognition, the initial state of the computer can only recognize the basic information of the pixels, which is the same as the vision of the organism. The reason why the organism can distinguish objects is the result of the original image processing by the biological nervous system. The key point of image recognition is that image recognition programming is the comprehensive processing of the original image point information. Image recognition usually includes contour recognition, feature recognition, color recognition, material recognition, object recognition and so on Generally, the outline of an object can be obtained from the information of color and brightness. According to the data corresponding to the outline, the content of the outline is determined as what object or what feature, and the judgment of the feature and object can not be separated from the processing of the outline and the corresponding logical data. The feature of material recognition is to recognize according to the degree of reflection of the problem. It can not be separated from the recognition of contour and the judgment of logical data. Therefore, contour recognition is the most important part in image recognition.

Supervised learning begins before learning begins. A sample training set consisting of several known input vectors and corresponding target variables is provided to the neural network tower to adjust the connection weights between neurons by the difference between the given input value and the expected output value and the actual network output value. Unsupervised learning. This learning method only needs to provide input to the neural network, and does not need the expected output value. The neural network Intensive learning. This algorithm does not need to give a clear expected output, but uses an evaluation mechanism to evaluate the quality factor of the output of the neural network corresponding to a given input. The external environment only gives the evaluation results for the output, and improves the system performance by strengthening the award action.

Reconfigurable FPGA has high expansibility and can be applied in many fields such as automation, intelligent control, computer vision system, artificial intelligence, machine learning and so on.

2. System Architecture Design

2.1. System Design Overall Architecture

The system architecture is shown in Figure1. For each part of the system function flow control, image data capture, processing, storage are used separately:

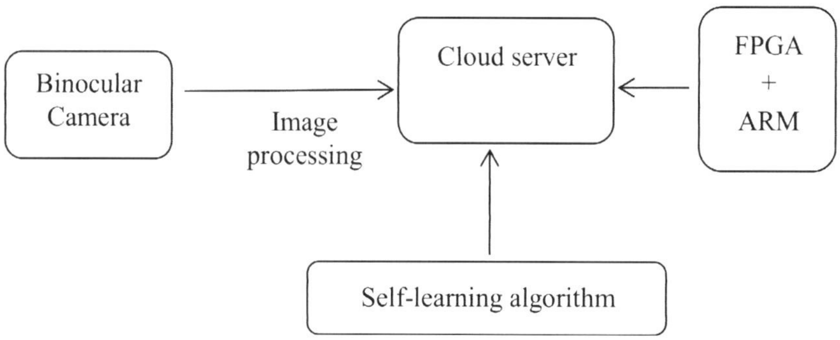

Figure 1. Hardware architecture

2.2. Functional Design of Modules in Embedded System Based on FPGA

2.2.1 Data Sensing and Recognition Module

Data perception and recognition module uses binocular camera. Binocular camera uses bionics and parallax principle to get synchronous exposure image of surrounding scenery by calibrated dual cameras. Based on parallax principle, the three-dimensional geometric information of object can be recovered, and the three-dimensional shape and position of surrounding object can be reconstructed. The acquisition of three-dimensional information is based on the principle of triangulation, that is, a triangle is formed between the image plane of two cameras and the object being measured. Through the position relationship between the two cameras, the three-dimensional dimensions of objects in the common field of view of the two cameras and the three-dimensional coordinates of the feature points of space objects can be obtained.

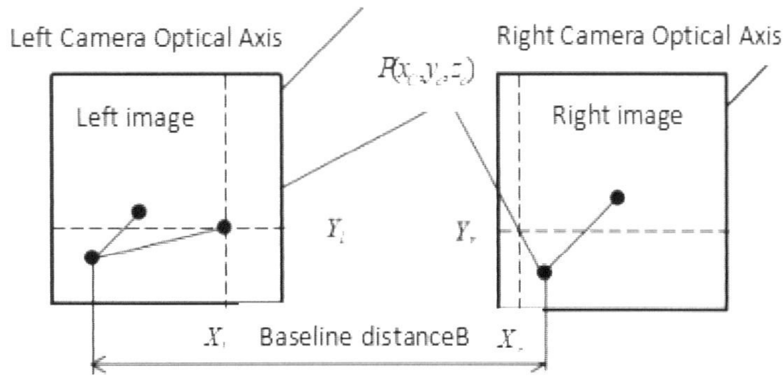

Figure 2. Principle of binocular stereo imaging with horizontal vision

The figure above shows a simple stereo imaging principle of head-up binocular vision. The distance between the projection centers of the two cameras, i.e. the baseline distance B. Two cameras look at the same feature point P of space-time object at the same time, and get the image of point P on "left eye" and "right eye" respectively, their coordinates are respectively $P_l = (x_l, y_l)$; $P_r = (x_r, y_r)$. If the images of two cameras are fixed on the same plane, the Y coordinates of the image coordinates of the feature point P must be the same, that is to say $y_l = y_r = y$. From the triangular geometric relations, the following relations can be obtained.

$$\begin{cases} X_l = f\dfrac{x_c}{z_c} \\ X_r = f\dfrac{(x_c - B)}{z_c} \\ Y = f\dfrac{y_c}{z_c} \end{cases} \tag{1}$$

Then the parallax is as follows $D = X_l - X_r$: From this, the three-dimensional coordinates of feature point P in camera coordinate system can be calculated:

$$\begin{cases} x_c = \dfrac{B.X_L}{D} \\ y_c = \dfrac{B.Y}{D} \\ z_c = \dfrac{B.f}{D} \end{cases} \tag{2}$$

2.2.2 Image Data Processing Module

Image data processing module cameras capture images and output them as analog or digital video signals. The image signals are preprocessed, segmented, extracted and recognized by image processing algorithms (such as filtering, image segmentation, transformation, geometric morphology analysis, etc.). Finally, the information needed for system design is obtained.

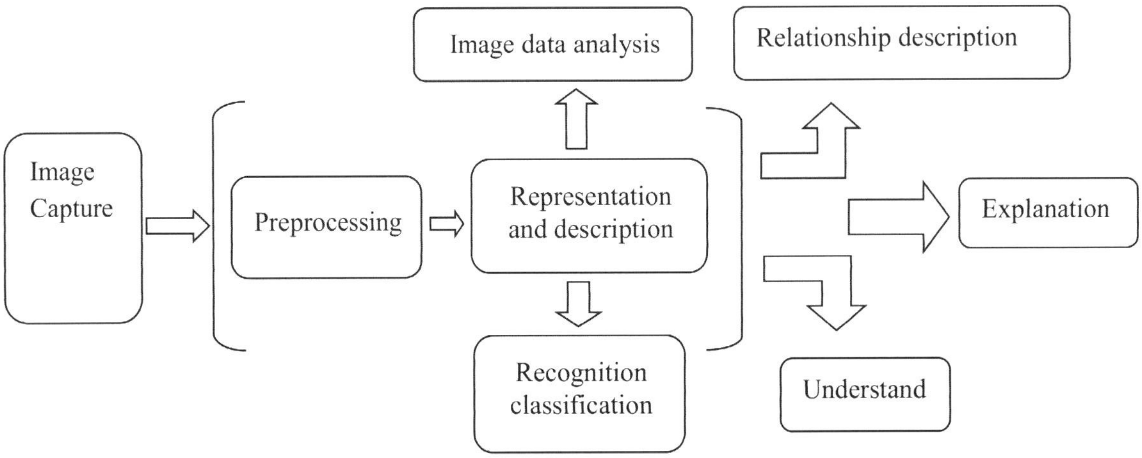

Figure 3. Image processing

2.2.3 Embedded System

The system is designed and implemented based on cloud server scalable FPGA devices, ARM chips and convolution algorithm. The main idea of the system is to divide the neural network algorithm into several basic operations, which can be reconfigured and applied to different projects by compiling different self-learning image processing algorithms. We use FPGA AT3000 series chips (XC3000 series compatible), all memory uses dual-port ARM. Under XILINX development system, the self-learning algorithm circuit is designed with chrysanthemum chain master-slave configuration. After logical simulation, timing simulation, configuration file generation, and then offline experiments.

2.2.4 Self-learning Algorithms

On the basis of open source convolution, reinforcement learning and supervised learning, we add our self-compiled machine self-learning algorithm to build a knowledge base for the machine, and guide the decision-making of the machine by modifying the weights. Finally, the human-like behavior of machine self-learning is realized.

The captured image data is processed locally and sent to the cloud server. After analysis and comparison, some codes are returned:

```
        for (x, y, w, h) in faces:
            cv2.rectangle(frame, (x, y), (x+w, y+h), (0, 255, 0), 2)
        frame_with_text = cv2ImgAddText(frame, tags_str, 0, 0, (255, 255, 255))

        cv2.imshow('my webcam', frame_with_text)

        if cv2.waitKey(1) == 27:
            break  # esc to quit

    # When everything done, release the capture
    cap.release()
    cv2.destroyAllWindows()
```

3. Experiments and Assessment

3.1. Test environment and process

In this paper, the dynamic reconstruction design of image recognition network, data sensing and recognition module uses binocular camera for image capture. The image data processing module outputs the captured image as analog or digital video signal, preprocesses, segmentes, extracts and recognizes the image signal through image processing algorithms (such as filtering, image segmentation, transformation, geometric morphology analysis, etc.); finally obtains the information needed by the system design, and then outputs and inspects it in various forms through the external equipment connected with it. Measurements and responses. The code is written in Python description language, the background server is Aliyun server, and the software environment is Windows 10.

The following code is the image data acquisition and transmission part. The captured image data is processed and sent to the cloud server. After analysis and comparison, the results are returned.

```
def cv2ImgAddText(img, text, left, top, textColor=(0, 255, 0), textSize=20):
    img = Image.fromarray(cv2.cvtColor(img, cv2.COLOR_BGR2RGB))
    draw = ImageDraw.Draw(img)
    fontText = ImageFont.truetype(
        "font/simsun.ttc", textSize, encoding="utf-8")
    draw.text((left, top), text, textColor, font=fontText)
    return cv2.cvtColor(np.asarray(img), cv2.COLOR_RGB2BGR)

cap = cv2.VideoCapture(1)
faceCascade = cv2.CascadeClassifier("haarcascade_frontalface_default.xml")
last_detect = 0
tags_str = "aaaa"
while(True):
    # Capture frame-by-frame
    ret, frame = cap.read()
    if not ret:
        continue
    gray = cv2.cvtColor(frame, cv2.COLOR_BGR2GRAY)
    faces = faceCascade.detectMultiScale(
        gray,
        scaleFactor=1.1,
        minNeighbors=5,
        minSize=(30, 30),
        flags = cv2.CASCADE_SCALE_IMAGE
    )
```

The following code is a single image recognition time measurement in the process of image processing. After many tests, the single image recognition time is set to 4 seconds per time.

```
if time.time() - last_detect >= 4:
    last_detect = time.time()
    try:
        enc_ret, img_data = cv2.imencode('.jpg', frame)
        img_base64_data = base64.b64encode(img_data).decode("utf8")
        ret = api.request_data_plus("https://dtplus-cn-shanghai.data.aliyunc
            "type": 1,
            "content": img_base64_data
        })
        tags = [x['value'] for x in ret['tags']]
```

3.2. Time Power Evaluation

The configuration speed of dynamic reconfiguration of cloud server's FPGA is basically limited by the size of configuration file and the bandwidth of ICAP port. In the design of this paper, four reconfiguration areas are divided, each reconfiguration area has three time-sharing reconfiguration modules, so a complete image recognition network needs a static configuration file and 12 partial configuration files.

Extensible FPGA development boards, ICAP configuration ports with maximum bandwidth of 3.2 Gb/s at 100 MHz maximum clock frequency. With the recognition time of the convolution neural network itself, the total time is 20.3 Ms. Compared with the software implementation based on ARM Cortex A9 processor, as shown in Table 1, there is a great improvement in time and power consumption. Table 1 below shows the time and power consumption of 100 different dynamic and static identifications.

Table 1. Time and Power Consumption of 100 Recognitions for Different Implementation Modes

platform	Realization way	Use time /ms	Power consumption /W
ARM Cortex A9	Software	5120	100
Xilinx.VC-707	Dynamic reconfiguration	1700	76

4. Conclusion

In this paper, for the first time, the dynamic reconfiguration technology of cloud server based on FPGA is applied to the hardware implementation of convolutional neural network, which can effectively solve the problems of low parallel speed, and computing speed cannot meet the real-time requirements of the field, stability, power and size limitations of the chip. It is hoped that in the future, scientific researchers can improve this progress and better promote the development of neural networks and artificial intelligence.

Acknowledgement

This work was supported by Shaanxi Natural Science Fund Project (2018JM7037025).

References

[1] Lu Lu. Study of convolutional neural network and its applications on license plate recognition [D]. Hefei: Hefei University of Technology, 2006.
[2] Shengyu Zhou. Realization of Dynamic Partial Reconfiguration System Based on FPGA [D]. Beijing: Graduate School of Chinese Academy of Sciences, 2007.
[3] Peilong Liu. Research and design of neural network hardware implementation based on FPGA [D]. University of Electronic Science and Technology, 2012.

Journal of Physics: Conference Series **1423** (2019) 012063 doi:10.1088/1742-6596/1423/1/012063

A method for converting entropy to inclusion measure of interval type-2 fuzzy sets

Gao Zheng[1], **Jing Wang**[2]

[1] Department of Electromechanical Management, China Coast Guard Academy, Ningbo, Zhejiang 315801, China
[2] School of Computer, College of Jilin University, Zhuhai, Guangdong 519041, China

Corresponding author and e-mail: GaoZheng, zg197762@163.com

Abstract. As tools for dealing with universal fuzziness in human thinking, traditional fuzzy sets, namely type-1 fuzzy sets, are not able to directly model uncertainties about membership functions, however type-2 fuzzy sets can make up for this deficiency. Interval type-2 fuzzy sets, the simplified cases of general type-2 fuzzy sets, are practical because its membership functions and operations are greatly simplified. The inclusion measure and entropy of fuzzy sets are two important parameters with application prospects, the former describes the degree to which a fuzzy set is included in another, the latter shows the quantity of fuzzy information in a fuzzy set. It is important to make in-depth research on the two measures. In this paper, we mainly discuss the relationship of the inclusion measure and entropy of interval type-2 fuzzy sets proposed by Zheng et al., and present a theorem that converts the entropy to the inclusion measure of interval type-2 fuzzy sets. The conclusion provides the basis for further research.

1. Introduction

As tools for handling the fuzziness of things, fuzzy sets first appeared in 1965, aiming to deal with the universal fuzziness in human thinking [1]. After this, Zadeh, the inventor of fuzzy sets, made a unique approach to propose the concept of "language variable", and used fuzzy rules in the form of "IF-THEN" to quantify knowledge rules in human language [2]. Thus, for the first time, humans acquired the method of converting the knowledge or rules expressed in language into mathematical functions. This laid the foundation for the development of fuzzy system theory. The fuzzy system theory treats an object as a "black box", ignores the mathematical model or physical law inside the "black box", and uses the language to describe the operating experience to form fuzzy rules, then simulates human reasoning processes to implement the management or control of the object [3]. The design of a fuzzy model does not depend on the precise mathematical model of the object. Fuzzy rules are easily understood and accepted by people, even if they do not have advanced theoretical knowledge. Based on fuzzy rules designed by experts, complex nonlinear objects can be handled effectively.

Traditional fuzzy sets, namely type-1 fuzzy sets (T1 FSs), are not able to directly model uncertainties about fuzzy set membership functions because of their completely crisp membership functions [4]. To make up for this deficiency, in 1975, Zadeh [5] improved his theory by defining type-2 fuzzy sets (T2 FSs). T2 FSs use fuzzy values as membership degrees, their membership functions themselves are fuzzy or vague. Thus, they are difficult to understand and use because their membership functions are three-dimensional and their operations are very complex, which limit their development and application. Interval type-2 fuzzy sets (IT2 FSs), also known as interval-valued fuzzy sets, can be considered as simplified cases of general T2 FSs, because their membership functions and operations are greatly simplified [6]. So, most people only use IT2 FSs.

As an important mathematical tool for describing the degree to which a fuzzy set is included in another, the inclusion measure has been paid more and more attention [7]. When Zadeh [1] defined the concept of fuzzy sets, he also discussed the inclusion relationship of two fuzzy sets, this inclusion relationship is clear, which means that a fuzzy set is either contained in another fuzzy set or not. Later, people extended the clear relationship to fuzzy relationship. Thus, the value of this parameter ranges in [0, 1] [8]. In recent years, more and more people have studied the inclusion measures of T2 FSs, especially those of IT2 FSs. In [9], Zeng and Guo gave an axiomatic definition of the inclusion measure of two IT2 FSs. In [7], based on the selected axiomatic definitions, Zheng et al. proposed a computation formula for the inclusion measure of IT2 FSs, gave four theorems that show the relationship between the inclusion measure and similarity measure, and verified their validity through examples. Zhao and Xiao [8] presented a new computation formula of the inclusion measure of IT2 FSs on the basis of an axiom definition, and defined an interval type-2 fuzzy rough set model on the basis of the new inclusion measure to implement the attribute reduction of a continuous domain decision information system. In [10], Zhang and Yang defined a new partial order for typical hesitant fuzzy sets via the disjunctive semantic meaning of a set, based on which the fuzzy inclusion relationship was defined for typical hesitant fuzzy sets.

The entropy, which describes the quantity of the fuzzy information in a fuzzy set [11], is widely used in multi-attribute decision [12] and other fields. So, it is important to made in-depth research on this measure. In 1965, Zadeh [1] first talked about it. Since then, more and more scholars have studied this parameter from different points of view. For example, in 1972, based on Shannon's function, De Luca and Termini [13] introduced a set of properties according with people's intuitive comprehension, and proposed calculation formulas. In [14], based on the distance between a fuzzy set and the centre of the fuzzy unit hypercube, Barini et al. intuitively introduced several new fuzzy entropies. In [15], Zeng and Li introduced a concept of the entropy of IT2 FSs with four properties, gave several calculation formulas, and researched the relationship between the entropy and another important measure, namely the similarity measure. Zheng [16] selected a definition for the entropy of IT2 FSs with four properties, and gave a calculation formula within the limits of those properties. Jiang and Wang [18] presented the constraint condition that is consistent with the intuitionistic facts, gave the axiomatic definition of the entropy fully reflecting the intuition and fuzziness of intuitionistic fuzzy sets, and constructed the improved intuitionistic fuzzy entropy formula according to the entropy axiomatic definition to solve the problem in the entropy ordering theoretically and practically.

This paper mainly discusses the relationship of the inclusion measure and entropy of IT2 FSs presented by Zheng et al. The rest is organized as follows. Section 2 simply reviews the basic knowledge of IT2 FSs, the axiomatic definitions and corresponding computation formulae of the above two parameters of IT2 FSs needed in this research. In Section 3, we give a theorem that converts the entropy to the inclusion measure of IT2 FSs. Finally, we conclude in Section 4.

2. Preliminaries

2.1. IT2 FSs

Roughly speaking, IT2 FSs in common use are simplified versions of general T2 FSs.

Definition 1. An IT2 FS, denoted $\tilde{A}$, can be expressed as (1) [6].

$$\tilde{A} = \int_{x\in X}\int_{u\in J_x} 1/(x,u) \tag{1}$$

where $x \in X$, $u \in J_x \subseteq [0,1]$, $\iint$denotes the union over all admissible x and u , $\int$ can be replaced byΣ indiscrete situations. The upper bound of union of all J_x is called the upper membership function (UMF), the lower bound of union of all J_x is named the lower membership function (LMF).

Suppose that $\overline{\mu}_{\tilde{A}}(x)$ is the UMF of the IT2FS $\tilde{A}$, $\underline{\mu}_{\tilde{A}}(x)$ is the LMF of $\tilde{A}$, $\overline{\mu}_{\tilde{B}}(x)$ is the UMF of another IT2FS $\tilde{B}$, $\underline{\mu}_{\tilde{B}}(x)$ is the LMF of $\tilde{B}$, if $\overline{\mu}_{\tilde{A}}(x)\le\overline{\mu}_{\tilde{B}}(x)$ and $\underline{\mu}_{\tilde{A}}(x)\le\underline{\mu}_{\tilde{B}}(x)$, then we have $\tilde{A}\subseteq\tilde{B}$ [7].

2.2. Inclusion Measure and Entropy of IT2 FSs

In [7], Zheng et al. selectedan axiomatic definition of the inclusion measure of two IT2 FSs, as shown inDefinition 2.

Definition 2.A real function $I: IT2FSs\times IT2FSs\rightarrow[0,1]$ is named the inclusion measure of two IT2 FSs, if I meets the following properties:

$(I1)$ $I(\tilde{A},\tilde{A})=1$;

$(I2)$ $\tilde{A}\subseteq\tilde{B}\Rightarrow I(\tilde{A},\tilde{B})=1$;

$(I3)$ For any $\tilde{A},\tilde{B},\tilde{C}\in IT2FSs$, if $\tilde{A}\subseteq\tilde{B}\subseteq\tilde{C}$, then $I(\tilde{C},\tilde{A})\le I(\tilde{B},\tilde{A})$ and $I(\tilde{C},\tilde{A})\le I(\tilde{C},\tilde{B})$.

Within the limits ofthe three properties in Definition 2, a formula to calculate the inclusion measure ofIT2 FSs $\tilde{A}$ and $\tilde{B}$ was given in [7], as shown in (2).

$$I(\tilde{A},\tilde{B})=\frac{1}{2}\left(\frac{\int_{x\in X}\min\left\{\overline{\mu}_{\tilde{A}}(x),\overline{\mu}_{\tilde{B}}(x)\right\}dx}{\int_{x\in X}\overline{\mu}_{\tilde{A}}(x)dx}+\frac{\int_{x\in X}\min\left\{\underline{\mu}_{\tilde{A}}(x),\underline{\mu}_{\tilde{B}}(x)\right\}dx}{\int_{x\in X}\underline{\mu}_{\tilde{A}}(x)dx}\right) \tag{2}$$

where $\int$ is the integral, and it can be replaced by the summation Σ indiscrete situations.

It can be seen form (2) that UMFs and LMFs play major parts inoperations of IT2 FSs.

In [15], Zeng and Li introduceda concept of the entropy of IT2FSs, as shown inDefinition 3.

Definition 3.A real function $E: IT2FSs\rightarrow[0,1]$ is called an entropy on $IT2FSs$, if E satisfies the followingproperties:

$(E1)$ $E(D)=0$, iff D is a crisp set;

$(E2)$ $E(\tilde{A})=1$, iff $\overline{\mu}_{\tilde{A}}(x)+\underline{\mu}_{\tilde{A}}(x)=1$;

$(E3)$ $E(\tilde{A})\le E(\tilde{B})$ if the fuzziness degree of $\tilde{A}$ is less than that of $\tilde{B}$, that is, $\overline{\mu}_{\tilde{A}}(x)\le\overline{\mu}_{\tilde{B}}(x)$ and $\underline{\mu}_{\tilde{A}}(x)\le\underline{\mu}_{\tilde{B}}(x)$ in the case of $\overline{\mu}_{\tilde{B}}(x)+\underline{\mu}_{\tilde{B}}(x)\le 1$, $\overline{\mu}_{\tilde{A}}(x)\ge\overline{\mu}_{\tilde{B}}(x)$ and $\underline{\mu}_{\tilde{A}}(x)\ge\underline{\mu}_{\tilde{B}}(x)$ in the case of $\overline{\mu}_{\tilde{B}}(x)+\underline{\mu}_{\tilde{B}}(x)\ge 1$;

$(E4)$ $E(\tilde{A})=E(\tilde{A}^c)$.

where $\tilde{A}^c$ denotes the complement of $\tilde{A}$.

Formula (3) was constructed within the limits of four properties in Definition 3 [16].

$$E(\tilde{A})=\frac{\int_{x\in X}\min\left\{\overline{\mu}_{\tilde{A}}(x),1-\underline{\mu}_{\tilde{A}}(x)\right\}dx}{\int_{x\in X}\max\left\{\overline{\mu}_{\tilde{A}}(x),1-\underline{\mu}_{\tilde{A}}(x)\right\}dx} \tag{3}$$

where $\int$ is the integral, and it can be replaced by the summation Σ indiscrete situations.

3. Proposed theorem

Here, we give a theorem that converts the entropy as shown in (3) to the inclusion measure as shown in (2) [16].

For $\tilde{A},\tilde{B}\in IT2FSs$, we define $h(\tilde{A},\tilde{B})\in IT2FSs$,

$$\overline{\mu}_{h(\tilde{A},\tilde{B})}(x)=\max\left(\frac{1+(\overline{\mu}_{\tilde{A}}(x)-\min(\overline{\mu}_{\tilde{A}}(x),\overline{\mu}_{\tilde{B}}(x)))^p}{2},\frac{1+(\underline{\mu}_{\tilde{A}}(x)-\min(\underline{\mu}_{\tilde{A}}(x),\underline{\mu}_{\tilde{B}}(x)))^q}{2}\right),$$

$$\underline{\mu}_{h(\tilde{A},\tilde{B})}(x)=\min\left(\frac{1+(\overline{\mu}_{\tilde{A}}(x)-\min(\overline{\mu}_{\tilde{A}}(x),\overline{\mu}_{\tilde{B}}(x)))^p}{2},\frac{1+(\underline{\mu}_{\tilde{A}}(x)-\min(\underline{\mu}_{\tilde{A}}(x),\underline{\mu}_{\tilde{B}}(x)))^q}{2}\right),$$

where $\overline{\mu}_{h(\tilde{A},\tilde{B})}(x)$ is the UMFof $h(\tilde{A},\tilde{B})$, $\underline{\mu}_{h(\tilde{A},\tilde{B})}(x)$ is the LMFof $h(\tilde{A},\tilde{B})$, $p>0$, $q>0$.

Then, we have the following theorem.

Theorem 1. For $\tilde{A},\tilde{B}\in IT2FSs$, suppose that $E(\tilde{A})$ is the entropy of $\tilde{A}$, then $I(\tilde{A},\tilde{B})=E(h(\tilde{A},\tilde{B}))$ is the inclusion measure of $\tilde{A}$ and $\tilde{B}$.

Proof. $(I1)$ $\overline{\mu}_{h(\tilde{A},\tilde{A})}(x)=\max\left(\frac{1+(\overline{\mu}_{\tilde{A}}(x)-\min(\overline{\mu}_{\tilde{A}}(x),\overline{\mu}_{\tilde{A}}(x)))^p}{2},\frac{1+(\underline{\mu}_{\tilde{A}}(x)-\min(\underline{\mu}_{\tilde{A}}(x),\underline{\mu}_{\tilde{A}}(x)))^q}{2}\right)$

$$=\max\left(\frac{1+(\overline{\mu}_{\tilde{A}}(x)-\overline{\mu}_{\tilde{A}}(x))^p}{2},\frac{1+(\underline{\mu}_{\tilde{A}}(x)-\underline{\mu}_{\tilde{A}}(x))^q}{2}\right)=\frac{1}{2}\text{ and}$$

$$\underline{\mu}_{h(\tilde{A},\tilde{A})}(x)=\min\left(\frac{1+(\overline{\mu}_{\tilde{A}}(x)-\min(\overline{\mu}_{\tilde{A}}(x),\overline{\mu}_{\tilde{A}}(x)))^p}{2},\frac{1+(\underline{\mu}_{\tilde{A}}(x)-\min(\underline{\mu}_{\tilde{A}}(x),\underline{\mu}_{\tilde{A}}(x)))^q}{2}\right)$$

$$=\min\left(\frac{1+(\overline{\mu}_{\tilde{A}}(x)-\overline{\mu}_{\tilde{A}}(x))^p}{2},\frac{1+(\underline{\mu}_{\tilde{A}}(x)-\underline{\mu}_{\tilde{A}}(x))^q}{2}\right)=\frac{1}{2}\Rightarrow\overline{\mu}_{h(\tilde{A},\tilde{A})}(x)+\underline{\mu}_{h(\tilde{A},\tilde{A})}(x)=1$$

$\Rightarrow I(\tilde{A},\tilde{A})=E(h(\tilde{A},\tilde{A}))=1$.

$(I2)$ $\tilde{A}\subseteq\tilde{B}\Rightarrow\overline{\mu}_{\tilde{A}}(x)\le\overline{\mu}_{\tilde{B}}(x)$ and $\underline{\mu}_{\tilde{A}}(x)\le\underline{\mu}_{\tilde{B}}(x)$

$$\Rightarrow\overline{\mu}_{h(\tilde{A},\tilde{B})}(x)=\max\left(\frac{1+(\overline{\mu}_{\tilde{A}}(x)-\min(\overline{\mu}_{\tilde{A}}(x),\overline{\mu}_{\tilde{B}}(x)))^p}{2},\frac{1+(\underline{\mu}_{\tilde{A}}(x)-\min(\underline{\mu}_{\tilde{A}}(x),\underline{\mu}_{\tilde{B}}(x)))^q}{2}\right)$$

$$=\max(\frac{1+(\overline{\mu}_{\tilde{A}}(x)-\overline{\mu}_{A}(x))^p}{2},\frac{1+(\underline{\mu}_{\tilde{A}}(x)-\underline{\mu}_{\tilde{A}}(x))^q}{2})=\frac{1}{2}\text{ and}$$

$$\underline{\mu}_{h(\tilde{A},\tilde{B})}(x)=\min\left(\frac{1+(\overline{\mu}_{\tilde{A}}(x)-\min(\overline{\mu}_{\tilde{A}}(x),\overline{\mu}_{\tilde{B}}(x)))^p}{2},\frac{1+(\underline{\mu}_{\tilde{A}}(x)-\min(\underline{\mu}_{\tilde{A}}(x),\underline{\mu}_{\tilde{B}}(x)))^q}{2}\right)$$

$$=\min(\frac{1+(\overline{\mu}_{\tilde{A}}(x)-\overline{\mu}_{\tilde{A}}(x))^p}{2},\frac{1+(\underline{\mu}_{\tilde{A}}(x)-\underline{\mu}_{\tilde{A}}(x))^q}{2})=\frac{1}{2}$$

$\Rightarrow\overline{\mu}_{h(\tilde{A},\tilde{B})}(x)+\underline{\mu}_{h(\tilde{A},\tilde{B})}(x)=1\Rightarrow I(\tilde{A},\tilde{B})=E(h(\tilde{A},\tilde{B}))=1$.

$(I3)$ $\tilde{A}\subseteq\tilde{B}\subseteq\tilde{C}\Rightarrow\overline{\mu}_{\tilde{A}}(x)\le\overline{\mu}_{\tilde{B}}(x)\le\overline{\mu}_{\tilde{C}}(x)$ and $\underline{\mu}_{\tilde{A}}(x)\le\underline{\mu}_{\tilde{B}}(x)\le\underline{\mu}_{\tilde{C}}(x)\Rightarrow$

$\overline{\mu}_{\tilde{C}}(x)-\overline{\mu}_{\tilde{A}}(x)\ge\overline{\mu}_{\tilde{B}}(x)-\overline{\mu}_{\tilde{A}}(x)$ and $\underline{\mu}_{\tilde{C}}(x)-\underline{\mu}_{\tilde{A}}(x)\ge\underline{\mu}_{\tilde{B}}(x)-\underline{\mu}_{\tilde{A}}(x)\Rightarrow$

$$\overline{\mu}_{h(\tilde{C},\tilde{A})}(x)=\max\left(\frac{1+(\overline{\mu}_{\tilde{C}}(x)-\min(\overline{\mu}_{\tilde{C}}(x),\overline{\mu}_{\tilde{A}}(x)))^p}{2},\frac{1+(\underline{\mu}_{\tilde{C}}(x)-\min(\underline{\mu}_{\tilde{C}}(x),\underline{\mu}_{\tilde{A}}(x)))^q}{2}\right)$$

$$=\max\left(\frac{1+(\overline{\mu}_{\tilde{C}}(x)-\overline{\mu}_{\tilde{A}}(x))^p}{2},\frac{1+(\underline{\mu}_{\tilde{C}}(x)-\underline{\mu}_{\tilde{A}}(x))^q}{2}\right)$$

$$\ge\max\left(\frac{1+(\overline{\mu}_{\tilde{B}}(x)-\overline{\mu}_{\tilde{A}}(x))^p}{2},\frac{1+(\underline{\mu}_{\tilde{B}}(x)-\underline{\mu}_{\tilde{A}}(x))^q}{2}\right)$$

$$=\max\left(\frac{1+(\overline{\mu}_{\tilde{B}}(x)-\min(\overline{\mu}_{\tilde{B}}(x),\overline{\mu}_{\tilde{A}}(x)))^p}{2},\frac{1+(\underline{\mu}_{\tilde{B}}(x)-\min(\underline{\mu}_{\tilde{B}}(x),\underline{\mu}_{\tilde{A}}(x)))^q}{2}\right)=\overline{\mu}_{h(\tilde{B},\tilde{A})}(x)\text{ and}$$

$$\underline{\mu}_{h(\tilde{C},\tilde{A})}(x)=\min\left(\frac{1+(\overline{\mu}_{\tilde{C}}(x)-\min(\overline{\mu}_{\tilde{C}}(x),\overline{\mu}_{\tilde{A}}(x)))^p}{2},\frac{1+(\underline{\mu}_{\tilde{C}}(x)-\min(\underline{\mu}_{\tilde{C}}(x),\underline{\mu}_{\tilde{A}}(x)))^q}{2}\right)$$

$$=\min\left(\frac{1+(\overline{\mu}_{\tilde{C}}(x)-\overline{\mu}_{\tilde{A}}(x))^p}{2},\frac{1+(\underline{\mu}_{\tilde{C}}(x)-\underline{\mu}_{\tilde{A}}(x))^q}{2}\right)$$

$$\geq\min\left(\frac{1+(\overline{\mu}_{\tilde{B}}(x)-\overline{\mu}_{\tilde{A}}(x))^p}{2},\frac{1+(\underline{\mu}_{\tilde{B}}(x)-\underline{\mu}_{\tilde{A}}(x))^q}{2}\right)$$

$$=\min\left(\frac{1+(\overline{\mu}_{\tilde{B}}(x)-\min(\overline{\mu}_{\tilde{B}}(x),\overline{\mu}_{\tilde{A}}(x)))^p}{2},\frac{1+(\underline{\mu}_{\tilde{B}}(x)-\min(\underline{\mu}_{\tilde{B}}(x),\underline{\mu}_{\tilde{A}}(x)))^q}{2}\right)=\underline{\mu}_{h(\tilde{B},\tilde{A})}(x),$$

$\overline{\mu}_{f(\tilde{B},\tilde{A})}(x)+\underline{\mu}_{f(\tilde{B},\tilde{A})}(x)\geq 1\Rightarrow I(\tilde{C},\tilde{A})=E(h(\tilde{C},\tilde{A}))\leq E(h(\tilde{B},\tilde{A}))=I(\tilde{B},\tilde{A})$;

$\overline{\mu}_{\tilde{C}}(x)-\overline{\mu}_{\tilde{A}}(x)\geq\overline{\mu}_{\tilde{C}}(x)-\overline{\mu}_{\tilde{B}}(x)$ and $\underline{\mu}_{\tilde{C}}(x)-\underline{\mu}_{\tilde{A}}(x)\geq\underline{\mu}_{\tilde{C}}(x)-\underline{\mu}_{\tilde{B}}(x)\Rightarrow$

$$\overline{\mu}_{h(\tilde{C},\tilde{A})}(x)=\max\left(\frac{1+(\overline{\mu}_{\tilde{C}}(x)-\min(\overline{\mu}_{\tilde{C}}(x),\overline{\mu}_{\tilde{A}}(x)))^p}{2},\frac{1+(\underline{\mu}_{\tilde{C}}(x)-\min(\underline{\mu}_{\tilde{C}}(x),\underline{\mu}_{\tilde{A}}(x)))^q}{2}\right)$$

$$=\max\left(\frac{1+(\overline{\mu}_{\tilde{C}}(x)-\overline{\mu}_{\tilde{A}}(x))^p}{2},\frac{1+(\underline{\mu}_{\tilde{C}}(x)-\underline{\mu}_{\tilde{A}}(x))^q}{2}\right)$$

$$\geq\max\left(\frac{1+(\overline{\mu}_{\tilde{C}}(x)-\overline{\mu}_{\tilde{B}}(x))^p}{2},\frac{1+(\underline{\mu}_{\tilde{C}}(x)-\underline{\mu}_{\tilde{B}}(x))^q}{2}\right)$$

$$=\max\left(\frac{1+(\overline{\mu}_{\tilde{C}}(x)-\min(\overline{\mu}_{\tilde{C}}(x),\overline{\mu}_{\tilde{B}}(x)))^p}{2},\frac{1+(\underline{\mu}_{\tilde{C}}(x)-\min(\underline{\mu}_{\tilde{C}}(x),\underline{\mu}_{\tilde{B}}(x)))^q}{2}\right)=\overline{\mu}_{h(\tilde{C},\tilde{B})}(x)\text{ and}$$

$$\underline{\mu}_{h(\tilde{C},\tilde{A})}(x)=\min\left(\frac{1+(\overline{\mu}_{\tilde{C}}(x)-\min(\overline{\mu}_{\tilde{C}}(x),\overline{\mu}_{\tilde{A}}(x)))^p}{2},\frac{1+(\underline{\mu}_{\tilde{C}}(x)-\min(\underline{\mu}_{\tilde{C}}(x),\underline{\mu}_{\tilde{A}}(x)))^q}{2}\right)$$

$$=\min\left(\frac{1+(\overline{\mu}_{\tilde{C}}(x)-\overline{\mu}_{\tilde{A}}(x))^p}{2},\frac{1+(\underline{\mu}_{\tilde{C}}(x)-\underline{\mu}_{\tilde{A}}(x))^q}{2}\right)$$

$$\geq\min\left(\frac{1+(\overline{\mu}_{\tilde{C}}(x)-\overline{\mu}_{\tilde{B}}(x))^p}{2},\frac{1+(\underline{\mu}_{\tilde{C}}(x)-\underline{\mu}_{\tilde{B}}(x))^q}{2}\right)$$

$$=\min\left(\frac{1+(\overline{\mu}_{\tilde{C}}(x)-\min(\overline{\mu}_{\tilde{C}}(x),\overline{\mu}_{\tilde{B}}(x)))^p}{2},\frac{1+(\underline{\mu}_{\tilde{C}}(x)-\min(\underline{\mu}_{\tilde{C}}(x),\underline{\mu}_{\tilde{B}}(x)))^q}{2}\right)=\underline{\mu}_{h(\tilde{C},\tilde{B})}(x),$$

$\overline{\mu}_{f(\tilde{C},\tilde{B})}(x)+\underline{\mu}_{f(\tilde{C},\tilde{B})}(x)\geq 1\Rightarrow I(\tilde{C},\tilde{A})=E(h(\tilde{C},\tilde{A}))\leq E(h(\tilde{C},\tilde{B}))=I(\tilde{C},\tilde{B})$.

Hence, we complete the proof of Theorem 1.

4. Conclusion

In this paper, the relationship of the inclusion measure and entropy of IT2 FSs proposed by Zheng et al. is mainly discussed. At first, the development history of T1 FSs, IT2 FSs, inclusion measures and entropies of IT2 FSs are briefly reviewed. Then, the basic knowledge of IT2 FSs, the axiomatic definitions and corresponding computation formulae of the inclusion measure and entropy of IT2 FSs needed in this research are simply thought in retrospect. Finally, the relationship of two measures is discussed and a theorem that converts the entropy to the inclusion measure of IT2 FSs is proposed. The conclusion lays the foundation for further theoretical research and practical application.

Acknowledgement
This work was supported by the General Research Project Foundation of Department of Education of Zhejiang Province (Grant No. Y201737667), the Teaching Quality and Teaching Reform Project Foundation of Department of Education of Guangdong Province (Grant No. 2017009), the Innovative Capacity Development Project Fund of Zhuhai College of Jilin University (Grant No. 2018XJCQ022), the Innovative Entrepreneurship Education Curriculum Development Project Fund of Zhuhai College of Jilin University, the Teacher Education Development Project Fund of Zhuhai College of Jilin University (Grant No. JZ2018JZA11), the Teaching Quality Engineering Construction Project Fund of Zhuhai College of Jilin University in 2019.

References
[1] ZadehL A 1965 *Fuzzy sets*. Information and Controlp 338–353.
[2] Mendel J M 2001*Uncertain Rule-based Fuzzy Logic Systems: Introduction and New Direction* (Upper Saddle River: Prentice Hall).
[3] Fan J F 2017*Fuzzy Control* (Beijing: National Defense Industry Press). (In Chinese)
[4] Mendel J Mand JohnR I B 2002*Type-2 fuzzy sets made simple*. IEEE Transactions on Fuzzy Systemsp117–127.
[5] ZadehL A 1975 *The concept of a linguistic variable and its application to approximate reasoning-1*. Information Sciencesp199–249.
[6] MendelJ M, JohnR Iand LiuF L 2006 *Interval type-2 fuzzy logic systems made simple*. IEEE Transactions on Fuzzy Systemsp808–821.
[7] Zheng G,Xiao J,Jiang Q andZhang Y 2011 *Similarity and inclusion measures between IT2 FSs*. Control and Decisionp861–866.(In Chinese)
[8] Zhao T andXiao J 2013 *Interval type-2 fuzzy rough sets based on inclusion measures*. Acta Automatica Sinicap1714–1721.(In Chinese)
[9] Zeng WYand GuoP 2008 *Normalized distance, similarity measure, inclusion measure and entropy of interval-valued fuzzy sets and their relationship*. Information Sciencesp1334–1342.
[10] Zhang H Yand YangS Y 2016 *Inclusion measure for typical hesitant fuzzy sets, the relative similarity measure and fuzzy entropy*. Soft Computingp1277–1287.
[11] Al-sharhan S, Karray F, Gueaieb W and Basir O2001 *Fuzzy entropy: a brief survey*. Proceedings of IEEE International Conference on Fuzzy Systems (Melbourne) p 1135–1139.
[12] Zhao Y andMao J J 2016 *New type of interval-valued intuitionistic fuzzy entropy and its application*. Computer Engineering and Applicationsp85–89.(In Chinese)
[13] De Luca AandTermini S 1972 *A definition of non-probabilistic entropy in the setting of fuzzy sets theory*. Information and Controlp301–312.
[14] BariniG O, Ngoo L M and Waweru R M 2018 *Some new fuzzy entropy formulae*. International Journal of Computing and Optimizationp 33–38.
[15] Zeng W Y and Li H X 2006 *Relationship between similarity measure and entropy of interval-valued fuzzy sets*. Fuzzy Sets and Systemsp1477–1484.
[16] Zheng G 2011 *Studies on Fuzzy Measure of Type-2 Fuzzy Logic System and its Application to Power Load Forecasting* (Chengdu: Southwest Jiaotong University).(In Chinese)
[17] Jiang D Yand WangY X2018 *A new entropy and its properties based on the improved axiomatic definition of intuitionistic fuzzy entrop*. Mathematical Problems in Engineering p1–6.

Journal of Physics: Conference Series **1423** (2019) 012064 doi:10.1088/1742-6596/1423/1/012064

Data mining in psychological experiments: a research of the relationship between personality, resilience and investment behavior rigidity

Kai Gao[1,*], Xiaoqian Qiao, Zhipeng Mei[1]

[1]Jiangsu Vocational College of Agriculture and Forestry, Zhenjiang, China

[*]Corresponding authors e-mail: Kai Gao, njgaokai@163.com

Abstract. Data mining in psychological experiments is to discover the relationship between other data and psychological data, so as to discover the psychological state of the subjects. Based on the psychological assessment data of students from Jiangsu Vocational College of Agriculture and Forestry during 2017-2019, we applied the association rules in big data processing to calculate the scores of personality, resilience and rigidity of investment behavior of the subjects. The following conclusions are drawn through analysis of promotion degree and correlation degree between values of each attribute. Firstly, pruning lift is of great significance to understand the associated relationship among personality, resilience and investment behavior rigidity. Secondly, the more openness individuals in the Big Five personality, the stronger the investment behavior rigidity. The stronger the neuroticism, the weaker the investment behavior rigidity. The stronger the extroversion, the stronger the investment behavior rigidity. Thirdly, psychological resilience as a mediator affects the impact of openness and neuroticism on the rigidity of investment behavior.

1. Introduction

Data mining in psychological experimental is to discover the relationship between other data and psychological data, so as to discover the psychological state of the subjects. Some scholars applied the association rules of big data processing method to the data processing of psychological assessment, and made valuable exploration [1].

Researchers generally believe that personality traits are importantly related to investment behavior [2-3]. In the literature on the study of personality traits, the most dominant model is the Big Five personality [4]. The Big Five personality model divides personality into five overall dimensions, each of which consists of six levels. Investment behavior rigidity is a behavioral indicator that measures to what extent that investors maintain their investment strategies. It refers to the stability and unchangeable characteristics of the investment behavior in the investment process due to factors such as negative information. In an uncertain situation, the market is full of information with unknown authenticity. Investors can neither obtain complete information related to decision-making, nor can they accurately judge the authenticity of information and the results of decision-making schemes. They can only rely on limited information, according to the logic of process rationality to make decisions. Investors with stronger investment behavior rigidity will maintain more original investment portfolios in investment decisions, while investors with weaker investment behavior rigidity will change more frequently. However, some scholars believe that psychological resilience has an important impact on investment behavior and may affect the relationship between personality and investment behavior, but experimental research on this aspect is lacking.

Based on the psychological assessment data of students from Jiangsu Vocational College of Agriculture and Forestry during 2017-2019, we applied the association rules in big data processing to calculate the scores of personality, resilience and rigidity of investment behavior of the subjects. We measure the responses of different personality types of investors in the context of continuous investment decision-making through behavioral experiments, and analyze the internal relationship of personality, investment behavior rigidity and resilience.

2. Literature Review and Hypotheses

2.1. The Big Five personality

Researchers conducted a lot of studies about the relationship between the Big Five personality and investment behavior. Durand [5] classified investors into different personality types. They pointed out that the stocks of the investments should be consistent with the personality types in order to generate more returns. Evans [6] found that individuals with lower applicability, higher intelligence scores, and lower rigor scores accommodate more losses in financial investments. Oehler [7] used a day's trading data of 80 anonymous traders to find that investors' trading performance was negatively correlated with amenity, neuroticism, responsibility and openness, and positively correlated with extroversion. Mayfield [8] analyzed the data of the traders and traders of the four investment banks in London and found that the trades performed well and were stable and introverted traders and they were highly openness. Akhtar [9] found that the exogenous, openness, and conscientiously high scores of investing individuals have higher returns, while those with high neuroticism and amenity scores have lower financial returns. Wang [10] conducted a questionnaire survey of 1063 investors in seven cities in China, and found that investors have higher knowledge of investment objects, more investment knowledge, and greater possibility of investors' independence and self-efficacy. Equity investors with high extroversion can concentrate their energies and make decisive decisions during investment. They can also eliminate the interference of negative emotions, reasonably analyze and grasp market opportunities and choose rational behaviors. Trading venues are good at getting people to get more information about stock investment decisions. When stocks fall, extroverted individuals may be more rational and decisive, and less affected by sunk cost. Individuals with high appetite are sometimes too skeptical of others' evaluations, and are more susceptible to negative information in the market. It is difficult to adhere to their own ideas in investment decisions. Investors with high agreeableness scores are easy to follow the reaction of most investors in the market. When they see other people buying or selling, they often follow the buying and selling decisions. Individuals with high conscientiousness scores tend to have strong self-discipline, restrain greed when stocks rise, and throw stocks when they fall. Individuals with high neuroticism scores are mainly emotional when operating stocks and are susceptible to incidents. When stock prices change, they often produce extreme emotions and lose their rationality. Investors with high openness are more adventurous, and in the uncertain situation, extroverted investors may have stronger investment behavior rigidity. In addition, the more openness investors are better at getting more information and being able to cope with the complex and volatile stock market.

2.2. Investment behavior rigidity

Investment behavior rigidity refers to the stability and unchangeable characteristics of the investment behavior of the individual in the investment process due to factors such as falling asset prices and negative information. It reflects the persistence behavior of investors in the investment decision in the face of unfavorable information and setbacks. In an uncertain situation, the market is full of information with unknown authenticity. Investors can neither obtain complete information related to decision-making, nor can they accurately judge the authenticity of information and the results of decision-making. They can only rely on limited information, and make decisions according to the logic of process rationality. Investors with stronger investment behavior rigidity will maintain their original investment portfolio more in their investment decisions, while investors with weaker investment behavior rigidity will change more frequently.

Investment behavior rigidity is a behavioral indicator that measures the extent to which investors maintain their investment objectives and strategies in the investment process. When an investor makes an investment decision, due to changes in the market, the surrounding and other constitutional factors, investors will more or less adjust the investment strategy and even the investment target. For some investors, small changes in decision-making situations will cause them to quickly adjust existing decisions, while for others, they may wait and see, and adjust when the situation changes more. Some even realize that the situation has changed, but it will not be adjusted until the investment is over. Therefore, the following assumptions are made:

Hypothesis 1: Investors with higher extroversion personality scores have stronger investment behavior rigidity
Hypothesis 2: Investors with higher agreeableness scores have weaker investment behavior rigidity
Hypothesis 3: Investors with higher conscientiousness scores have weaker investment behavior rigidity
Hypothesis 4: Investors with higher openness score have stronger investment behavior rigidity
Hypothesis 5: Investors with higher neuroticism scores have weaker investment behavior rigidity

2.3. Resilience

Resilience is a good adaptation of individuals to adversity and stress. It means that the "rebound ability" of life stress and frustration has been a hot topic in positive psychology research in recent years. There are three definitions of psychological resilience in the academic world, namely, procedural definition, result definition and quality definition. The procedural definition regards psychological resilience as a dynamic process of development and change, focusing on the process of individuals recovering quickly and successfully responding to major stresses and dangers. The definition of the definition defines the psychological resilience from the development results. It is considered that the psychological resilience is a phenomenon in which the individual's adaptation to the environment and life is still good when the individual is seriously threatened. The definition of quality refers to mental resilience as an individual's ability or quality, which can help individuals to emerge from threats and difficulties. For the composition of psychological resilience, the existing research has not reached an agreement. Borgen and Amundson divided them into three dimensions: personal ability, self-esteem, and interpersonal control. Xing and Sun [11] divided it into self-efficacy, persistence, internal control, and family response. And five dimensions of social networks, Sollie and Kop divided it into five dimensions: goal concentration, emotion control, positive cognition, family support and interpersonal assistance. There are three reasons for the difference in psychological resilience. First, because of the different definitions of psychological resilience, researchers have different internal dimensions of psychological resilience. Second, research by researchers based on different contexts and tasks leads to different understandings of mental resilience. In investment decisions, psychological resilience is reflected in investors' insistence on investment objectives, the control of negative emotions caused by unfavorable situations, and investors' positive perceptions of investment scenarios. We divide the psychological resilience in the investment process into three dimensions: target concentration, emotion control, and positive cognition. In investment decisions, individuals with stronger mental resilience will be more focused on the target, and will not give up easily. The feedback from the outside world can not only properly control the emotions, but also relatively positively interpret the negative information. In addition, outside support also helps such individuals to emerge from the shadow of failure decisions. Therefore, we propose hypothesis 6.

Hypothesis 6: Mental resilience is positively correlated with decision-making rigid behavior, and the stronger the psychological resilience, the higher the rigid behavior coefficient of individuals.

Psychological resilience is closely related to positive personality and personality traits. Studies have found that subjects with higher psychological resilience will show higher due diligence, amenity and openness, and lower neuroticism. Neuroticism is inversely related to psychological elasticity, and high neurotic individuals have poor elasticity recovery; and neurotic negative predicts psychological elasticity. In the study of decision-making behavior, neurotic energy influences investment decision-making behavior through the intermediary role of psychological flexibility. Studies have shown that neuroticism has a negative impact on the generation and development of psychological resilience, which is not conducive to individuals successfully coping with difficulties or frustrations, so neuroticism is a risk factor affecting psychological resilience. Based on this, we propose hypothesis 7.

Hypothesis 7: Mental resilience as a mediator affects the role of neuroticism in investment behavior rigidity.

Based on theoretical assumptions, we construct the research model of this chapter (Figure 1).

3. Methods

3.1. Subjects

The survey was conducted at Jiangsu Vocational College of Agriculture and Forestry. The subjects are 134 students who had finished two courses of "Secondary Investment Theory and Practice" and "Personal Finance" in the past. They came from 21 different majors and had already learned the basic investment knowledge and can understand the basic operations of the investment.

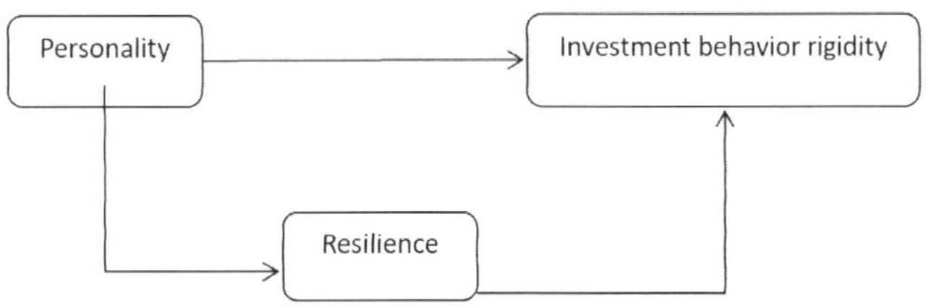

Figure 1. Conceptual model of personality, resilience and investment behavior rigidity

The study was conducted in a random manner. It was divided into two sessions, each of which was divided into two stages. In the first stage, the participants were asked to fill out the Big Five Personality Questionnaire using a computer. After a rest of ten minutes, they take part in the second stage of the experiment under the guiding of the staff to measure investment behavior rigidity. The two stages required the participants to fill in the basic information. In the later statistics, the two-stage data could be combined and analyzed.

In order to obtain effective data and achieve the purpose of practical and effective research results, the screening criteria are set for the data: (1) the questionnaire with obvious answering rules in the questionnaire answer is deleted. (2) If there is a missing item, delete the questionnaire. (3) In the personality questionnaire, the problem of lie detection was set up. Two questions were asked before and after the question of "Always forget to put things back in the right place" and "Lear time and energy spent learning", if the questionnaire option if the absolute value of the difference is greater than 2, the questionnaire is excluded. (4) In order to eliminate the influence of psychological valence on the rigidity of investment behavior, the measurement objects are screened, and the age of the subjects to be analyzed is similar to that of ordinary working-class families. The annual household income is about 10-25 thousand yuan. The monthly living expenses are around 800-1800 yuan. A total of 102 data were finally entered into the analysis.

3.2. Measurements

The measurement of this study involves multiple variables, first of all for the measurement of personality. Most of the existing researches on investment decision-making behavior use the small scale personality measurement questionnaire. Some researchers use 50 questions, and some use 20 questions. Although these scales help the questionnaire to be recycled, the quality of the questions is guaranteed, but the measurement of personality is not perfect. This study used the Big Five Personality Scale revised by Yangzhou University in 2016. The scale was based on the Big Five personality of the International Personality Item Pool and was refined. We used AMOS 20.0 verification validity to remove some items that did not meet the criteria (factor load less than 0.45), and finally included 24 items for extroversion; 24 items for amenity; 24 items for conscientiousness; 24 items for neuroticism Openness includes 24 projects, all of which are scored using a 5-level Likert scale. For the measurement scale of psychological resilience, the more mature ones are the scales compiled by Xing and Sun (2013), Sollie and Kop (2017). In order to respond to investment decision scenarios, we have considered existing scales and built investments around goal focus, emotional control, and positive perception. The psychological resilience scale of decision making.

In order to measure subjects' investment behavior rigidity, we developed an investment decision experiment, which is based on the cards. Subjects are required to perform the guiding operation during the measurement. The measuring tools include: 13 chips (10 chips are priced at 100 yuan, 2 chips are priced at 500 yuan, and 1 chip is priced at 1,000 yuan), which is used to indicate changes in the measured earnings; 20 dedicated measuring cards, each card The back is exactly the same, and the front shows different income changes. Among them, 2 shows a gain of 500 yuan, and 1 shows a gain of 1000 yuan. In addition, there are 17 blanks, indicating a return of 0. Paper and pen are used to record measurement data. The specific measurement process can be expressed in two aspects. (1) The measured object understands the measurement instructions under the guidance of the surveying personnel. The object to be tested needs to be clear of two points of information. First, measurement can help them better understand their investment behavior characteristics, so operating according to their own real ideas will lead to more accurate investment advice. Second, each time a participant draws a card, he needs to pay 100 yuan. There are three kinds of income for each card, which are 0, 500 and 1000, but the probability is unknown. Third, during the evaluation process, the higher the total return obtained by the participants, the better. (2) Start the first experiment. First, the surveyor randomly shuffled 20 cards and placed them on the back with the back facing up. Secondly, the surveyor puts 10 chips with a price of 100 yuan in front of the object to be tested, and informs the object to be tested again: the card can be randomly selected to see if it is a revenue card; each card needs to pay 100 yuan to the surveyor; Obtain a beneficiary card and choose whether to proceed. If the choice is yes, the proceeds marked on the card are invalid, and the object to be tested continues to experiment. If no, the measurement personnel pays the proceeds, and the experiment ends. The first one is opened by the experimenter by default and the receiver is charged 100 yuan. The surveyer records two data: the number of 100-dollar chips saved by the surveyor after the end of the experiment, and the sum S of the final gains of the measured object.

4. Results

4.1. Personality and investment behavior rigidity

We performed an independent sample T test on the ibr score. We grouped the samples according to the extrovert personality type index and its six dimensions, the first group being the low level group and the fourth group being the low level group. In the high-level group, compare the mean values between the groups, and the results are shown in Table 1.

The independent sample T test results showed that there were significant differences in ibr scores for different groups of extroversion levels. The mean value of the low-level group of extroversion was 4.84, and the average of the high-level group was 6.77. There were also significant differences in the ibr score between the two groups of

Journal of Physics: Conference Series **1423** (2019) 012064 doi:10.1088/1742-6596/1423/1/012064

extroversion. The significance levels were 0.05 and 0.01, respectively. This shows that the higher the level of extroversion, the stronger the investment rigid behavior. hypothesis 1 is verified.

The independent sample T test showed that there was no significant difference in the ibr score between the different groups of the level of agreeableness. From the data results, there is no significant correlation between agreeableness and investment behavioral rigidity. hypothesis 2 is not verified. The independent sample T test results showed that there were no significant differences in ibr scores for different groups of due diligence levels. hypothesis 3 is not verified.

The independent sample T test results showed that there were significant differences in the ibr score at the 0.01 level for the different groups of openness levels. The mean for the low level group was 4.04 and the mean for the high level group was 6.77. It indicates that individuals with higher risk are more risk-seeking tendencies in continuous investment decisions. In the investment setbacks, more opportunities are often seen, so it is more willing to risk the original investment and the more rigid the investment behavior. Based on the comprehensive correlation analysis and the independent sample T test results, the openness trait is significantly correlated with the investment behavioral rigidity, and hypothesis 4 is verified.

The independent sample T test results showed that the ibr scores were significantly different at different levels of 0.01 in the different groups of neuroticism. The mean of the low-level group was 7.44, and the mean of the high-level group was 4.52. Based on the comprehensive correlation analysis and the independent sample T test results, the neurotic traits were significantly correlated with the investment behavioral rigidity, hypothesis 5 was verified.

Table 1. Independent sample T-test of personality

	IBR	SD	T	Sig
extroversion (low)	4.84	2.794	-2.784**	0.008
extroversion (high)	6.77	2.122		
Agreeableness (low)	4.84	2.609	-1.728	0.090
Agreeableness (high)	6.00	2.173		
conscientiousness (low)	5.20	2.291	-1.845	0.071
conscientiousness (high)	6.42	2.436		
Openness (low)	4.04	2.557	-4.079**	0.000
Openness (high)	6.77	2.215		
Neuroticism (low)	7.44	2.123	4.594**	0.000
Neuroticism (high)	4.52	2.365		

Table 2. Regression analysis of IBR I

	investment behavior rigidity			
	Model 1		Model 2	
	β	t	β	t
Constant	4.350	2.630	3.602	0.951
Age	-0.037	-0.274	-0.011	-0.094
Aex	0.912	1.738	.948*	2.064
Race	0.152	0.315	0.000	0.001
Education	0.431	1.360	0.118	0.411
Communist	0.850	1.121	0.699	1.050
Cadre	0.488	0.708	-0.106	-0.171
Student Loans	1.559*	2.439	1.619**	2.828
Part-time work	0.333	0.649	0.378	0.836
Experience	-0.036	-0.072	-0.279	-0.625
grade	-0.155	-0.419	-0.156	-0.485
Home income	-0.126	-0.376	-0.070	-0.236
Extroversion			-0.003	-0.137
Agreeableness			0.014	0.466
Conscientiousness			-0.027	-0.932
Openness			0.082**	3.282
Neuroticism			-0.058**	-3.117
F	1.446		3.442	
R2	0.150		0.393	
Adj.R2	0.046		0.279	

4.2. Resilience personality and investment behavior rigidity

First, we use conscientiousness, extroversion, amenity, openness and neuroticism as independent variables, ibr as a dependent variable, adding control variables, and performing hierarchical regression. The regression results are shown in Table 2. In the regression analysis, the coefficient of openness and neuroticism and investment behavioral rigidity is significant at the 0.01 level, which indicates that openness and neuroticism have significant effects on investment rigid behavior. According to the recommendation of Wen Zhonglin et al., we are in the second step of analysis. Excluding the variables such as conscientiousness, extroversion and amenity that are not significant, with openness and neuroticism as independent variables, psychological resilience as dependent variable, and adding control variables such as gender, age and family income status to carry out hierarchical regression. The regression results are shown in Table 2.

The results show that the influence of openness and neuroticism on psychological resilience is significant. Among them, openness has a positive impact on psychological resilience, and neuroticism has a negative impact on psychological resilience. Individuals with more openness are more sensible in facing setbacks or losses, able to vent negative emotions in time, and then actively face life and work, which means that open-ended investors are more likely to emerge from the shadow of setbacks and losses. Mental resilience is stronger than the average investor. Individuals with high levels of neurotic ism often show more negative psychological states in the face of setbacks or losses. They will have different levels of anxiety and depression, and will have impulsive behavior in later life and work, which makes them it is more difficult to get out of the shadow of failure. According to the procedure of mediating role test, openness, neuroticism and psychological resilience are used as independent variables, and investment behavior rigidity is used as the dependent variable, and the control variables such as gender, age and family income status are added to carry out hierarchical regression. The regression results are shown in Table 3.

The results showed that the influence coefficient of openness and neuroticism on psychological resilience was significant at 0.05 level, and the influence coefficient of psychological resilience on investment behavioral rigidity was significant at 0.05 level, and the direction of influence was positive. Individuals with stronger psychological resilience are more likely to face failures and setbacks, and are more likely to emerge from the negative effects of defeat. Failures and setbacks in the investment process are inevitable. Investors with stronger psychological resilience tend to have stronger risk tolerance. Short-term losses will not fundamentally affect their investment direction and plans. Therefore, the more rigid their investment behavior, the hypothesis 2 is verified.

Comparing Tables 2, 3 and 4, after adding the variable of mental resilience, the fitting index of the regression model is increased from 0.296 to 0.358, and the explanatory ability is improved, while the two variables of openness and neuroticism are rigid for investment behavior. The significance of the regression coefficient is reduced, which indicates that psychological resilience plays a partial intermediary role in the influence of openness and neuroticism on the rigidity of investment behavior.

Table 3. Regression analysis of resilience

	Resilience			
	Model 1		Model 2	
	β	t	β	t
Constant	3.332**	6.980	2.881**	2.696
Age	-0.027	-0.707	-0.021	-0.655
Aex	0.093	0.617	0.100	0.775
Race	0.043	0.308	0.030	0.251
Education	0.132	1.447	0.045	0.551
Communist	0.074	0.339	0.008	0.040
Cadre	0.225	1.134	-0.017	-0.095
Student Loans	0.061	0.332	0.061	0.376
Part-time work	0.000	-0.002	0.054	0.420
Experience	-0.026	-0.178	-0.077	-0.608
grade	0.000	0.001	0.005	0.060
Home income	-0.057	-0.587	-0.007	-0.089
Extroversion			0.003	0.405
Agreeableness			-0.012	-1.408
Conscientiousness			0.009	1.054
Openness			0.021**	2.941
Neuroticism			-0.015**	-2.839

Table 3, cont				
F	0.698		3.129	
R2	0.079		0.371	
Adj.R2	-0.034		0.252	

Table 4. Regression analysis of IBR II

	investment behavior rigidity			
	Model 1		Model 2	
	β	t	β	t
Constant	4.350	2.630	0.816	0.214
Age	-0.037	-0.274	0.010	0.088
Aex	0.912	1.738	0.851	1.908
Race	0.152	0.315	-0.029	-0.070
Education	0.431	1.360	0.075	0.268
Communist	0.850	1.121	0.691	1.074
Cadre	0.488	0.708	-0.090	-0.150
Student Loans	1.559*	2.439	1.561**	2.814
Part-time work	0.333	0.649	0.326	0.744
Experience	-0.036	-0.072	-0.205	-0.474
grade	-0.155	-0.419	-0.161	-0.518
Home income	-0.126	-0.376	-0.063	-0.219
Extroversion			-0.006	-0.255
Agreeableness			0.026	0.868
Conscientiousness			-0.035	-1.252
Openness			0.062*	2.441
Neuroticism			-0.043*	-2.313
Resilience			0.967*	2.600
F	1.446		3.856	
R2	0.150		0.438	
Adj.R2	0.393		0.325	

5. Conclusions

In this paper, we developed a concept of investment behavior rigidity, and examined the relationship of investment behavior rigidity, the Big Five personality and resilience. First, the more openness individuals in the Big Five personality, the stronger the investment behavior rigidity. The stronger the neuroticism, the weaker the investment behavior rigidity. The stronger the extroversion, the stronger the investment behavior rigidity. Second, psychological resilience has a significant positive impact on the rigidity of investment behavior. Third, psychological resilience as a mediator affects the impact of openness and neuroticism on the rigidity of investment behavior.

In consideration of stock characteristics, individual investors can increase expected returns through investment behavior rigid measurement tools. Investors with strong investment behaviors can choose small-scale, high-growth technology stocks or stocks of small-scale, low-priced local traditional enterprises, choose long-term holdings and obtain sustained returns. Investors with weak investment behaviors can choose large-scale, high-growth theme stocks, while maintaining good liquidity of assets, while maintaining flexible strategic adjustments which can obtain higher expected returns.

Acknowledgements

This research was financially supported by Jiangsu Philosophy and Social Science Research Project "Research on Users' Repurchase Behavior of Fresh E-commerce Platform under the Background of Rural Revitalization Strategy" (Grant No. 2019SJA1944); Jiangsu Vocational College of Agriculture and Forestry Project "Study on influencing factors and countermeasures of green production behaviors of farmers in the mode of online and offline integration" (Grant No. 2019kj010)

References

[1] European guidelines on cardiovascular disease prevention in clinical practice: executive summary. Fourth Joint Task Force of the European Society of 14 Cardiology and other societies on cardiovascular disease prevention in clinical practice constituted by representatives of nine societies and by invited experts. Eur Heart J,2017,28: 2375-2414.

[2] Durand R, Newby R, Peggs L, Siekierka M. Personality. Journal of Behavioral Finance. 2013; 14(2):116-133.
[3] Evans AM, Revelle W. Survey and behavioral measurements of interpersonal trust. Journal of Research in Personality. 2008; 42(6):1585-1593.
[4] Oehler A, Wendt S, Wedlich F, Horn M. Investors' personality influences investment decisions: Experimental evidence on extraversion and neuroticism. Journal of Behavioral Finance. 2018; 19(1):30-48.
[5] Mayfield C, Perdue G, Wooten K. Investment management and personality type. Financial Services Review. 2008; 17(3):219-236.
[6] Akhtar F, Thyagaraj KS, Das N. The impact of social influence on the relationship between personality traits and perceived investment performance of individual investors. International Journal of Managerial Finance. 2018; 14(1):130-148.
[7] Wang H, Zhang J, Wang L, Liu S. Emotion and investment returns: Situation and personality as moderators in a stock market. Social Behavior and Personality: An International Journal. 2014; 42(4):561-569.
[8] Krishnan R, Beena F. Measurement of Conformity to Behavior Finance Concepts and Association with Individual Personality. IUP Journal of Behavioral Finance. 2009; 6(3/4):25-40.
[9] Borgen WA, Amundson NE, Reuter J. Using Portfolios to Enhance Career Resilience. Journal of Employment Counseling. 2004; 41(2):50-59.
[10] Xing C, Sun J. The role of psychological resilience and positive affect in risky decision-making. International Journal of Psychology. 2013; 48(5):935-943.
[11] Tidball KG, Aktipis A. Feedback enhances greening during disaster recovery: A model of social and ecological processes in neighborhood scale investment. Urban Forestry & Urban Greening. 2018; 34:269-280.

Matrix factorization recommendation algorithms based on knowledge map representation learning[1]

Xuejian Huang[1], Lu Min[2], Gensheng Wang[1], Tian Gan[1], Zhipeng Li[1]

[1]Computer Practice Teaching Center, Jiangxi University of Finance and Economics, Nanchang 330013, China
[2]School of Humanities, Jiangxi University of Finance and Economics, Nanchang 330013, China

Corresponding author and e-mail: Xuejian Huang, 761780598@qq.com

Abstract. The matrix factorization recommendation algorithm does not consider characteristics of the recommendation object itself, resulting in poor recommendation results. Therefore, a matrix factorization recommendation algorithm based on knowledge map representation learning is proposed. Firstly, the recommendation object is represented as a low dimensional semantic vector by using the knowledge map distributed representation learning algorithm. Then the semantic similarity between objects is calculated, and the semantic similarity is incorporated into the objective optimization function of matrix factorization, so that the feature vectors obtained by matrix factorization can also contain semantic knowledge, which makes up for the shortcoming of matrix factorization recommendation algorithm that does not consider characteristics of the recommendation object itself from the semantic perspective. The experimental results show that the improved algorithm has higher accuracy, recall and coverage than the traditional matrix factorization recommendation algorithm.

1. Introduction

Recommendation algorithm is a typical technology to solve the problem of network information overload, which has been widely used in network media, e-commerce, advertising and other commercial fields. At present, recommendation algorithms are divided into three categories according to different recommendation engines: content-based filtering recommendation, collaborative filtering recommendation and hybrid recommendation [1]. Collaborative filtering recommendation algorithm is based on user's historical behavior data and has no domain restrictions. It is the most widely used recommendation algorithm at present. It is mainly divided into User-based CF, Item-based CF, Model-based CF. User-based CF and Item-based CF faces sparse matrix data, it can't achieve good recommendation effect [2]. Model-based CF uses machine learning algorithm to model, which can solve the problem of matrix sparsity to a certain extent [3]. Matrix factorization is a typical algorithm based on model-based collaborative filtering [4].

The matrix factorization recommendation algorithm only uses the user project evaluation matrix without considering other factors, which results in low accuracy of recommendation results. Many experts put forward relevant improvement schemes for this problem. For example, reference [5] proposes a matrix factorization method based on attribute coupling, which combines the attribute information of the project into the matrix factorization model; reference [6] introduces the trust relationship between users to improve the performance of the matrix factorization recommendation algorithm; reference [7] adds additional implicit information (such as browsing, purchase and click History).

It is found that the main method to improve the matrix factorization recommendation algorithm is to add additional relevant information of users or projects. With the development of knowledge map technology, there are a lot of open semantic knowledge data, such as Freebase, OpenKN, DBpedia, BMKG, etc. The knowledge map representation learning algorithm can embed the recommended object into a low-dimensional semantic vector space, so this paper proposes a matrix factorization recommendation

[1]Fund: NSFC funded project (Project No.: 71461012); Science and technology project of Jiangxi Provincial Department of Education (GJJ181550)

algorithm based on knowledge map representation learning to make up for the shortcomings of the matrix factorization recommendation algorithm which does not consider the characteristics of the recommended object itself.

2. Correlation theory

2.1. Matrix factorization recommendation algorithm

Matrix factorization recommendation algorithm (FunkSVD) decomposes two low dimensional user characteristic matrix and item characteristic matrix through the score matrix, and uses these two decomposed matrices to fit the user's score on the project. Matrix factorization is shown in Eq. (1):

$$R \approx U^T V \tag{1}$$

Where R is the user item scoring matrix, $U \in R^{m\times d}$ represents the decomposed user characteristic matrix, $V \in R^{n\times d}$ represents the decomposed project characteristic matrix, m and n represent the number of users and items, and d is the user and item characteristic dimension. The scoring calculation of user i for item j is shown in Eq. (2):

$$R_{i,j} = U_i^T V_j \tag{2}$$

In order to make Eq. (1) fit the real score data, use the idea of linear regression to establish the objective optimization function, as shown in Eq. (3):

$$J = \sum_{i=1}^{m} \sum_{j=1}^{n} I_{i,j} (r_{i,j} - U_i^T v_j)^2 + \lambda(\|U_i\|^2 + \|V_j\|^2) \tag{3}$$

Where $I_{i,j}$ indicates whether user i has scored project j, if so $I_{i,j}$ is 1 otherwise 0, $r_{i,j}$ is the actual score of user i for project j, adding $\lambda(\|U_i\|^2 + \|V_j\|^2)$ to prevent over fitting, λ the regularization parameter. Based on FunkSDV algorithm, an improved Bised MF algorithm is proposed in reference [8]. Bised MF introduces global average sub item, user bias item (difference between user evaluation average score and global average score) and project bias item (difference between project average score and global average score) into objective optimization function Eq. (3). The final objective optimization function is shown in Eq. (4):

$$J = \sum_{i=1}^{m} \sum_{j=1}^{n} I_{i,j} \left(r_{i,j} - U_i^T V_j - \mu - \alpha_i - \beta_j\right)^2 + \lambda\left(\|U_i\|^2 + \|V_j\|^2 + \|\alpha_i\|^2 + \|\beta_j\|^2\right) \tag{4}$$

Where μ is the global average item, α_i is the use i offset item, β_j is project j offset item. The scoring calculation of user i for item j is shown in Eq. (5):

$$R_{i,j} = U_i^T V_j + \mu + \alpha_i + \beta_j \tag{5}$$

2.2. Knowledge map distributed representation learning

The knowledge map uses the triple of "entity-relationship-entity" to describe the relationship between entities and forms a network of knowledge structures through the relationship. Knowledge map distributed representation learning can get semantic vector distributed representation of entities and relationships [9] . Due to its simple parameters and low computational complexity, Transe is the mainstream knowledge map distributed representation learning model, because of its simple parameters, low computational complexity, and remarkable performance on large-scale knowledge map. For each triple (h, r, t) , h and t represent the head entity and the tail entity respectively, and r is the relationship between the head entity and the tail entity, TransE expressed h, t and r as embedding vectors v_h, v_t and v_r. The v_r is the translation between v_h and v_t, also known as translation. The relationship among them is shown in Eq. (6):

$$v_h + v_r \approx v_t \tag{6}$$

The closer the two sides of the equation are, the more likely there is a relationship r between the two entities. The loss function of the TransE model is shown in Eq. (7):

$$f(v_h, v_r, v_t) = \|v_h + v_r - v_t\|_2^2 \tag{7}$$

The total cost function is shown in Eq. (8):

$$L = \sum_{(h,r,t)\in S} \sum_{(h',r,t')\in S'} \max(0, f(v_h, v_r, v_t) - f(v_{h'}, v_r, v_{t'}) + \gamma) \tag{8}$$

Where S is the set of all triples, which is called a positive sample, and S' is the negative sample of set S, that is, to randomly replace the head entity or tail entity of each existing triplet in S, a new triplet is obtained, and the new triplet does not belong to S, and γ is the distance between positive and negative samples.

TransE doesn't distinguish entities under different relationships, and there are shortcomings in dealing with knowledge map of complex relationships. To solve this problem, reference [10] proposes TransR model, which embeds entities and relationships into different spaces, and implements entity representation in corresponding relationship spaces. Its loss function is shown in Eq. (9):

$$f(v_h, v_r, v_t) = \|v_h M_r + v_r - v_t M_r\|_2^2 \tag{9}$$

Where M_r is the matrix of relation r, $v_h M_r$ is the projection of entity vector v_h into relation r space.

3. Fusion recommendation algorithm

The matrix factorization recommendation algorithm only uses the user project evaluation matrix, and does not consider the characteristics of the project itself, resulting in poor recommendation results. This paper proposes a matrix factorization recommendation algorithm based on knowledge map representation learning. The algorithm integrates the semantic similarity between the recommended objects into the objective optimization function of matrix factorization to make up for the defect that the matrix factorization recommendation algorithm does not consider the characteristics of the recommended objects. The algorithm flow is shown in Fig.1:

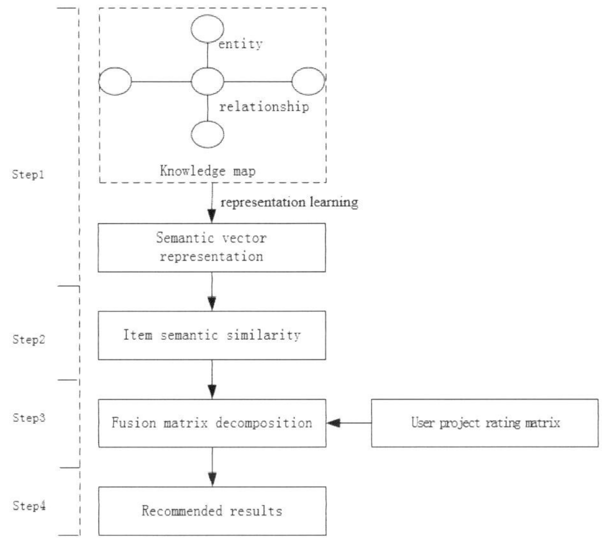

Figure 1. Algorithm flow

The algorithm flow is divided into four steps: semantic vector representation, item semantic similarity calculation, fusion matrix factorization and recommended results.

3.1. Semantic vector representation

According to the knowledge map distributed representation learning algorithm, the vector representation of all entities and relationships in the domain of the recommended object is obtained, and the entity representation of the recommended object is screened out from the entity vector. The vector representation of the recommendation object integrates the entity knowledge related to it in the whole field, so the vector representation contains the context semantic knowledge of the recommendation object. The recommended object entity is represented as a d-dimensional semantic vector, as shown in Eq. (10):

$$I_i = (E_{1i}, E_{2i}, ..., E_{di})^T \tag{10}$$

3.2. Item semantic similarity

The similarity calculation mainly includes cosine similarity, Pearson similarity, Jaccard similarity, log likelihood similarity and Euclidean distance similarity. The loss function of knowledge map distributed representation algorithm is based on Euclidean distance. In order to maintain consistency, the similarity of project semantics is also measured by Euclidean distance, as shown in Eq. (11):

$$d(I_i, I_j) = \sqrt{\sum_{k=1}^{d}(E_{ki} - E_{kj})^2} \tag{11}$$

Adjust it to (0, 1], and the specification calculation is shown in Eq. (12) :

$$sim(i, j) = \frac{1}{1 + d(I_i, I_j)} \tag{12}$$

The larger the value of $sim(i,j)$ is, the closer the semantics of item i and item j are.

3.3. Fusion matrix factorization

Based on the idea that the feature vectors of projects with similar semantics should also be similar, the fusion algorithm integrates the semantic similarity of projects into the objective optimization function Eq. (4) of the Bissed MF matrix factorization. The objective optimization function after fusion is shown in Eq. (13):

$$J=\sum_{i=1}^{m}\sum_{j=1}^{n}I_{i,j}\left(r_{i,j}-U_i^TV_j-\mu-\alpha_i-\beta_j\right)^2+\lambda_1\left(\|U_i\|^2+\|V_j\|^2+\|\alpha_i\|^2+\|\beta_j\|^2\right)+\lambda_2\sum_{i=1}^{n}\left(\frac{\cos(V_j,V_i)+1}{2}-sim(i,j)\right)^2 \quad (13)$$

Eq. (13) adds $\lambda_2\sum_{i=1}^{n}(\frac{\cos(V_j,V_i)+1}{2}-sim(i,j))^2$ on the basis of Eq. (4), to control the consistency of semantic similarity and feature vector similarity. Where $\cos(V_j,V_i)$ is the cosine value of the eigenvectors of items i and j, $sim(i,j)$ is the semantic similarity of items i and j, λ_2 is the fusion coefficient. The calculation of $sim(i,j)$ is shown in Eq. (12).

3.4. Recommended results

The fusion matrix decomposes two low-dimensional user feature matrix and project feature matrix, and uses Eq. (5) to calculate the prediction score. Based on the principle that the higher the prediction score, the more interested the user is, the project with the prediction score greater than a certain threshold is recommended to the user.

4. Experimental Verification and Result Analysis

4.1. Experimental data

In this experiment, movie recommendation is selected as the research object. The experimental data comes from Douban movie review data, which contains 210000 comments of 7800 users on 1500 movies. The number of stars reflects how much users like movies. The star rating is divided into 1-5 stars. In this experiment, 4-5 Stars are labeled as user's favorite movies, and 1-3 stars are user's dislike movies. This experiment uses the latest film knowledge map (BMKG) released by Tsinghua University, which contains more than 720000 movie related entities, 91 attributes and more than 13 million triples. The BMKG integrates the data of Douban film, LinkedMdb and other Chinese and British films. In order to reduce the training time of knowledge map distributed representation learning algorithm, only knowledge related to experimental data is extracted from BMKG.

4.2. Evaluating indicator

In this experiment, *Precision*, *Recall* and *Coverage* are used to measure the performance of the algorithm. Their calculation is shown in Eq. (14), (15) and (16) respectively.

$$\mathrm{Pr}ecision=\frac{TP}{TP+FP} \quad (14)$$

$$\mathrm{Re}call=\frac{TP}{TP+FN} \quad (15)$$

$$Coverage=\frac{N_d}{N} \quad (16)$$

(1)TP、FP、FN are the values in the confusion matrix, as shown in Table 1.

Table 1. Confusion matrix

	Fact =like	Fact =dislike
Predict=like	TP	FP
Predict=dislike	FN	TN

(2)In Eq. (16), N is the number of all films in the experiment, and N_d is the number of films given by the recommended algorithm. The higher the coverage, the better the diversity and novelty of the recommendation results.

4.3. Experimental result analysis

(1) Experimental comparison of different movie entity semantic vector dimensions

In the distributed representation of knowledge map, different dimensions of movie entity vector representation will have a certain impact on the experimental results. Therefore, four groups of comparative experiments are set with dimensions 50, 100, 150 and 200. Other key parameters in the experimental process are shown in Table 2, and the experimental results are shown in Fig.2.

Table 2. Settings of experimental parameters

parameter	value
Regularization parameters λ_1	1e-3
Fusion coefficient λ_2	1
Gradient decline learning rate α	1e-2
Iterations of gradient descent n	300
User and movie feature dimensions d	60

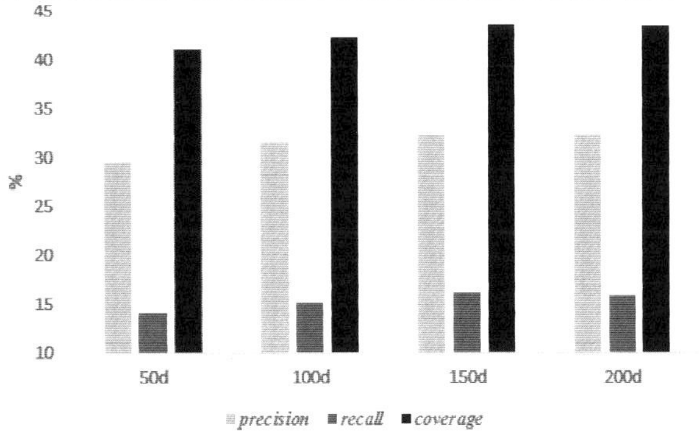

Figure 2. Comparison of experimental results in different dimensions of movie entity semantic vector

From Fig.2, we can see that the accuracy, recall and coverage rate of the algorithm are relatively good when the entity dimension of the distributed knowledge representation algorithm is 150.

(2) Experimental comparison of different users and movie features

The feature dimension d of users and movies needs to be set during matrix factorization. In the experiment, 10 groups (10, 20, 30, 40, 50, 60, 70, 80, 90, 100) were set for comparison. The dimension of movie entity semantic vector was set to 150, and other parameters were consistent with table 2. The experimental results are shown in Fig.3.

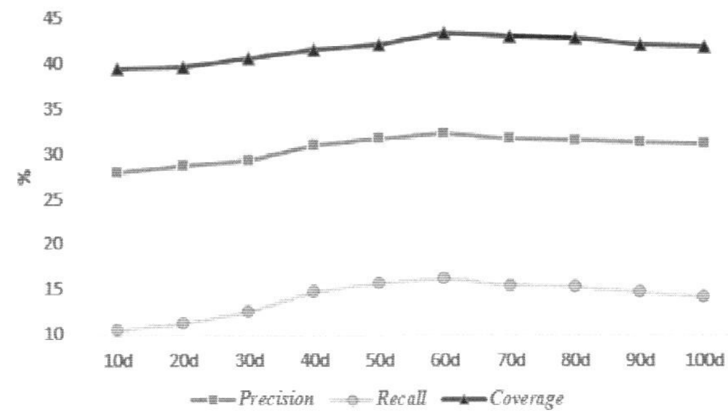

Figure 3. Comparison of experimental results in different user and movie feature dimensions

According to Fig.3, when the user and movie dimensions decomposed by the matrix are 60, the accuracy, recall and coverage of the algorithm are relatively good.

(3) Experimental comparison of different fusion coefficient values

The fusion coefficient λ_2 in Eq. (13) controls the proportion of semantic similarity in the whole algorithm. In this experiment, five kinds of λ_2 values (0, 0.5, 1, 1.5, 2) are set for experimental comparison. The semantic vector dimensions of movie entities are set to 150, the feature dimensions of users and movies are set to 60, and other parameters are consistent with table 1. The experimental results are shown in Fig.4.

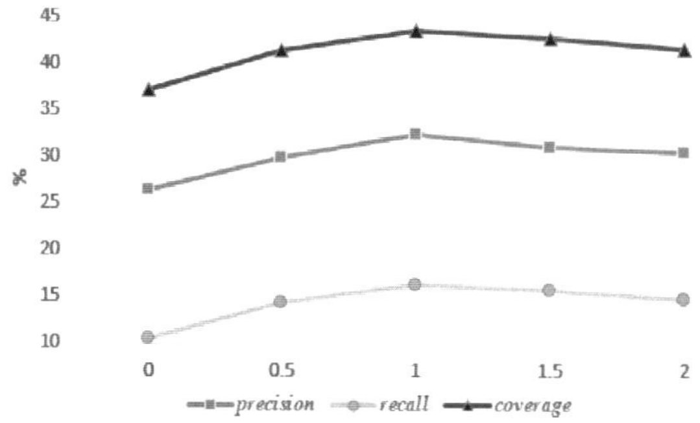

Figure 4. Comparison of experimental results with different fusion rates

When the fusion coefficient is 0, the algorithm in this paper degenerates into the Biased MF matrix factorization recommendation algorithm. When the fusion coefficient is not 0, the semantic similarity of the movie is fused in the Biased MF algorithm. The experimental results show that the fusion algorithm has higher accuracy, recall and coverage than the Biased MF algorithm, and the best result is when the fusion coefficient is 1.

5. Conclusions

The recommendation algorithm based on matrix factorization alleviates the problem of matrix sparsity in collaborative filtering algorithm, but the algorithm only uses the user project evaluation matrix, without considering the additional relevant information of the project, resulting in inaccurate recommendation results. Therefore, this paper proposes a matrix factorization recommendation algorithm based on knowledge map representation learning. The semantic similarity of the project is obtained by knowledge map distributed representation learning algorithm. The semantic similarity is integrated into the objective optimization function of matrix factorization, so that the feature vectors of semantic similar items are similar. The experimental results show that the algorithm is effective. However, the algorithm in this paper also has some shortcomings. On one hand, the algorithm depends on open source knowledge map, which leads to certain domain limitations of the algorithm; on the other hand, when faced with massive data, the efficiency of matrix factorization is low. These are next steps to study.

References

[1] Wu M L, Chang C H, Liu R Z. Integrating content-based filtering with collaborative filtering using co-clustering with augmented matrices [J]. Expert Systems with Applications, 2014,41(6):2754-2761.

[2] Zhang L, Qin T, Teng P Q. An Improved Collaborative Filtering Algorithm Based on User Clustering [J].Information Science, 2014, 32(10):24-27+32. (In Chinese)

[3] Huang L W, Jiang B T, Lv S Y, et al. Survey on Deep Learning Based Recommender Systems [J].Chinese Journal of Computers, 2018, 41(7):1619-1647. (In Chinese)

[4] Bokde D, Girase S, Mukhopadhyay D. Matrix Factorization Model in Collaborative Filtering Algorithms: A Survey [J]. Procedia Computer Science, 2015, 49:136-146.

[5] Yu Y, Wang C, Wang H, et al. Attributes coupling based matrix factorization for item recommendation [J]. Applied Intelligence, 2017, 46(3):521-533.

[6] Forsati R , Mahdavi M , Shamsfard M , et al. Matrix Factorization with Explicit Trust and Distrust Side Information for Improved Social Recommendation[J]. ACM Transactions on Information Systems, 2014, 32(4):1-38.

[7] Shulong C, Yuxing P. Matrix Factorization for Recommendation with Explicit and Implicit Feedback [J]. Knowledge-Based Systems, 2018, 158:109-117.

[8] Bi H L, Zhou W, Lu F Q. Bias based matrix factorization recommender techniques [J]. Application Research of Computers, 2018, 35(10):2928-2931+2964. (In Chinese)

[9] Fan M, Zhou Q, Zheng T F, et al. Distributed representation learning for knowledge graphs with entity descriptions[J]. Pattern Recognition Letters, 2017, 93(01): 31-37.

[10] Lin Y , Liu Z , Sun M , et al. Learning Entity and Relation Embeddings for Knowledge Graph Completion[C]// Twenty-ninth Aaai Conference on Artificial Intelligence. AAAI Press, 2015.

Journal of Physics: Conference Series **1423** (2019) 012066 doi:10.1088/1742-6596/1423/1/012066

Fixed spray type anti-ice deicing control system design

Wenjiang Liu [1], Yunquan Chen [2], Fujin Hou [3], Yucai Jing[4] and Jie Chen[1]

[1] Research Center for Road Safety, Emergency and Disaster Reduction Technology, Shandong Jiaotong University, Jinan 250357, China
[2] Testing Department, Shandong Quanjian Engineering Testing Co., LTD, Jinan 250014, China
[3] Shandong Hi-speed Construction Management Co., LTD, Jinan 250000, China
[4] Shandong Hi-speed Qingdao Highway Co.,LTD, Qingdao 266061, China

Corresponding author and e-mail: Jie Chen, 979656312@qq.com

Abstract. This paper takes intelligent control technology as the core technology, combines with meteorological and road freezing information collection technology and water supply and drainage technology. Firstly, this paper introduces the design of grading subsystem of fixed spray type anti-ice and snow removal control system, and proposes a new type of nozzle suitable for the viaduct at the same time, which can improve spraying effect of the deicing liquid while improving safety guarantee. Then based on analysis of collected on-site road conditions and meteorological actual data and combined with logic control algorithm, a live ice and snow early warning mathematical model is designed to predict the snowy condition of the road surface, and automatically control equipment to spray deicing fluid before or after road icing.. It uses PID control technology to ensure pressure balance inside the spray pipe. The real-time field data is connected to the cloud server with 4G wireless transmission technology, and the existing data and current data are compared and analyzed through the recursive neural network, so as to further accurately predict the ice and snow situation, classify danger level according to national road safety standards, and release it to other platforms. Based on application of actual field, this paper designs a method that can be combined with prevention in advance and post-disposal to eliminate road icing and reduce risks of vehicle driving safety in severe weather.

1. Introduction

With the rapid development of technology, people's concept of de-icing and snowmelt is also undergoing major changes, from the passive form of handling after the occurrence of ice and snow to the active form of prevention in advance. Prevention in advance can reduce the freezing point of road liquid under the condition of water on the road surface or lay a layer of protection on the road surface to prevent snow and ice from directly adhering to the road surface to improve the efficiency of de-icing snow. Therefore, it is an urgent problem to improve the road ice and snow warning and disposal capacity. Nowadays, China mainly uses mechanical vehicles to clean accumulated ice and snow or spread snow melt agent. However, in some remote areas such as long longitudinal slope and other special road sections where snowplow cannot reach quickly, we can only rely on relatively mature fixed spray type anti-ice and snow control system to achieve unmanned automatic disposal of road accumulated ice and snow [1]. Other de-icing methods (conductive abrasion layer, chemical anti-ice coating, thermal conductivity pipe) are still in experimental stage of the road section, and have not been widely promoted.

2. Design of the system hardware structure

With the rapid development of technology and economy, automatic control system has been widely used in all walks of life, including road maintenance. In order to achieve the purpose of timely and rapid automatic disposal of road ice and snow, the main hardware design of this system includes: road and weather detection subsystem, spray subsystem, control subsystem and support subsystem.
Schematic diagram of system structure is shown in Figure1 and Figure2.

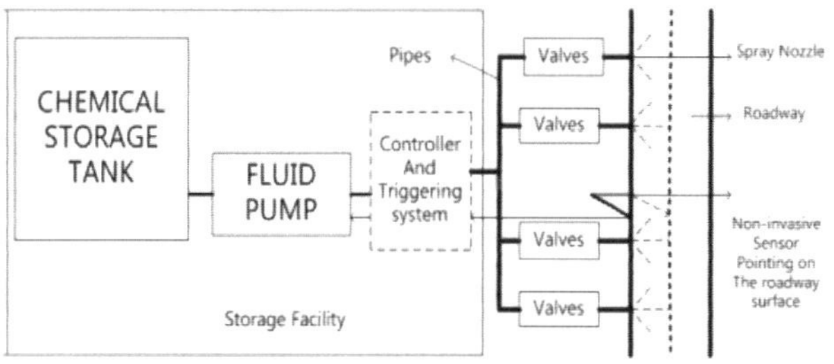

Figure 1. Schematic diagram of system structure

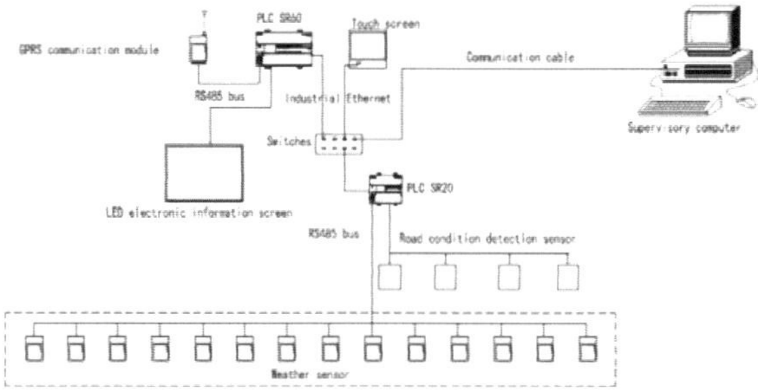

Figure 2. Structure drawing of control system

2.1. Design of road and weather detection subsystem

In practical engineering applications, road and weather sensors should not only be able to accurately measure road and weather data, but also be able to be used under special working conditions.

In order to achieve accurate measurement and cost saving and to facilitate construction,, it is necessary to integrate the measurement sensor used for measured data into a single device, so as to reduce the volume of the measurement instrument and facilitate installation and transportation [2-3]. This system adopts the passive road sensor IRS31-UMB of Lufft and the active road sensor ARS31-PRO as shown in Figure3, which make the measurement of data more accurate and at the same time assist the WS-600 weather sensor as shown in Figure4.

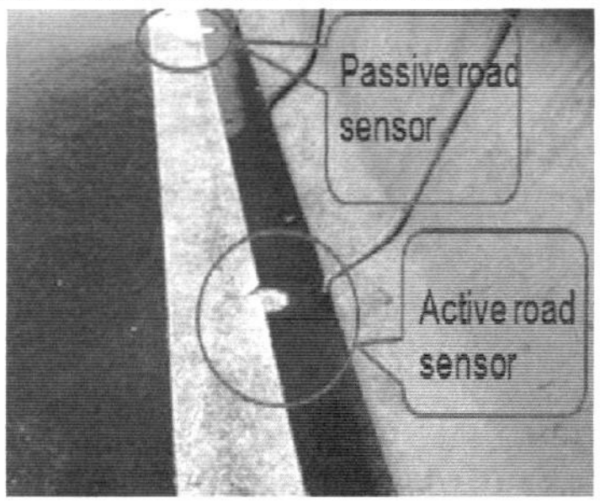

Figure 3. Passive and active sensor

Figure 4. Weather sensor

Lufft road sensors adopt unique capacitance detection technology, microwave radar technology, and combine with multi-parameter fusion algorithm. Through detecting sensors to monitor the state of different materials: capacitance values of the condition of freezing, ice thickness, water, and integrate temperature detection and temperature compensation technology to monitor freezing and ice thickness.

IRS31-UMB passive sensor is an embedded sensor. By using two kinds of shell composition, the surface is hard wear-resistant plastic and others are hard alloy material, which make the sensor package more compact, better durability, and stronger stability. It can detect the freezing point of chloride solution, road temperature, road condition, road water film thickness and other parameters.

ARS31-PRO active road sensors utilize active cooling technology to directly cool the road surface when set conditions are met until icing, and then increase the heat to melt, in the process the real ice temperature can be obtained. This method is not affected by the influence of the snowmelt agent composition, it is the ideal freezing point detection device.

The WS-600 weather sensor measures relative humidity through the capacitive sensor element. Using accurate negative temperature coefficient components (NTC), it measures the temperature. 24 KHz Doppler radar is adopted to perceive each drop and each snowflake. At the same time, precipitation types are distinguished according to different falling speed of raindrops and snowflakes, and the amount of precipitation and precipitation intensity are calculated by using the falling speed and size of raindrops (snowflakes). Ultrasonic transmitter receiver is used to detect wind speed and direction.

2.2. Execution part: the design of spray subsystem

2.2.1. Main supply circuit design

The main liquid supply loop is divided into two parts with the shunt as the node. The front end is composed of liquid storage tank, manual ball valve, electric valve, filter screen, flow meter and pipe connector. The back end is composed of water pump, pressure gauge, check valve, buffer pressure tank, relief valve and connecting pipe fitting. The small pressure tank can absorb the sudden increase of pressure in the pipeline to prevent "water hammer effect", which increases the pressure stability of the pipeline, and prevents the damage caused by high pressure to the pipeline.

When selecting the main material, the main material shall be determined according to the physical and chemical characteristics of the working liquid of the system and the working pressure of the system. In general, the main material in the pump station shall give priority to using the metallic materials and when metallic materials are not applicable, consider non-metallic materials; steel pipes are preferred for metallic materials, and non-ferrous metals are considered later. Due to short distance of the main liquid supply circuit, the corrosion resistant stainless steel material is generally selected. The material of liquid storage tank can choose polyethylene.

2.2.2. Design of spray circuit

Spray main pipe consists of main infusion pipe, solenoid valve, spray water hose, nozzle and its accessories. In order to be suitable for the rolling and bending road trend of construction road section and make installation convenient and standardize production and processing, the main infusion pipe had better adopt the pressure-resistant rubber hose which is easy to bend.

Among them, fish and bird sprinkler head and fan-shaped sprinkler head are widely used in de-icing system. The spraying distance of fish and bird sprinkler head is relatively long, but there are spray blind areas, which leads to uneven proportion of the sprayed road solution, and reduces the effect of road de-icing. At the same time, the protruding protrusion of the fish and bird sprinkler head will also cause a safety hazard to the vehicle. The fan-shaped sprinkler head orifice channel is large and smooth. Although the blocking phenomenon is reduced, the spray solution is relatively wasted in practical application. At the same time, its spray distance at the same height of the same pressure is much shorter than that of the fish and bird sprinkler head, and the spray coverage area is relatively small.

In order to improve the safety of sprinklers to work, improve the utilization rate of deicing medium and improve the sprinkler spray distance, improve the system running timeliness, we design a new type of nozzle, which is compact and has a smooth surface that can be embedded in the concrete barrier basting reduces the safety impact on moving vehicle. The water outlet is composed of a plurality of orifices of different sizes. The upper large hole ensures the spray distance. The lower holes ensure uniform spraying. The inner arc-shaped curved surface reduces the water hammer effect of the nozzle to further increase the spraying distance and uniformity of spraying. The object of the new nozzle is shown in Figure 5.

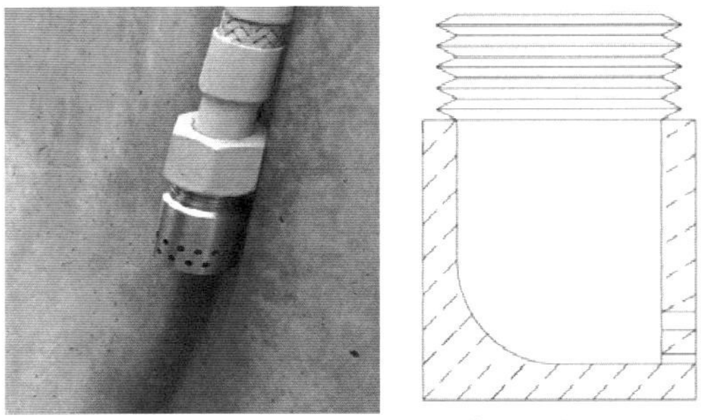

Figure 5. New type of nozzle

2.2.3. Design of solenoid valve controller
In order to improve system construction efficiency, modular production and cost saving, solenoid valve controller is designed by us. In addition to the basic I/O port controller, the terminal design is divided into power inlet and outlet terminal, signal inlet and outlet terminal and solenoid valve power supply terminal. In construction as long as the wiring is correct, it can be quickly connected. The power supply adopts safe DC 24V.RS-485 interface is adopted for data transmission. And they are sealed in a water drain box.

Among them, the connecting party of the main controller and the solenoid valve controller had better adopt the "hand-in-hand" topological structure. Other topological structures are easy to generate reflected signals and affect communication quality. At the same time, the topological structure reduces the distance between the equipment and the main line and further reduces interference. 485 repeaters should be used if the distance between the main controller and the solenoid valve controller is long. The communication cable used must be twisted pair with shielding layer above double strands, which has strong anti-interference ability and is more conducive to differential mode of communication adopted by 485 communication. Meanwhile, all equipment should have good grounding to prevent outdoor lightning surge interference [3].

2.3. Design of control subsystem
In this system, the main body of the control subsystem is a set of relatively large electrical cabinets. It is divided into two parts: one is responsible for connecting the distribution area of external power supply lines, and the other is responsible for controlling the control area of external equipment. The touch screen, control knob and threading hole are embedded on the door panel corresponding to the electrical cabinet.

In general, the industrial control system can use the main controller of the single chip microcomputer and the peripheral circuit. The main controller of the programmable logic controller, the subsidiary module, and the industrial control computer are the core of the main controller.

In the spray type active anti-ice deicing system, the single chip microcomputer needs to complete the external circuit by itself. By displaying drive circuit, pulsing width modulation circuit, D/A conversion circuit, serial communication circuit, control relay circuit, etc., to complete the overall control function of the system.

In the spray type active anti-ice deicing system, only the model of PLC (Programmable Logic Controller) and its affiliated modules need to be selected and connected with other hardware devices in the system by using power cord and signal cable. In general, there is no need to design peripheral circuits by us. Logical control program is used to control the operation of serial ports or I/O ports, so as to realize data detection and processing and spraying of the system [4].

In the spray type active anti-ice deicing system, IPC also only needs to use power cord and signal cable to connect with other hardware devices in the system, and generally does not need to design the peripheral circuit by itself. VB and other software programming can be used to control serial communication, data detection and processing and spraying of the system.

In the design and construction of spray type active anti-ice deicing system, the form of the main controller can be selected according to installation and construction environment of the system and the requirements of using. Generally speaking, PLC is widely used in the spray type active anti-ice deicing system because it does not need too much self-designed circuit in the process of using, which makes it easy to have strong reliability and can adapt to complex and harsh working environment [5].

3. Design of control algorithm

3.1. Spray decision control

In order to realize early warning and forecast accurately and simply, mathematical models are often established by using the relationship between various kinds of data in practical applications, and mathematical formulas which are easy to understand and operate are formed. In the prediction of road ice coagulation and snow cover, a mathematical model can be established after analyzing and extracting summarized laws of road ice coagulation mechanism, and the abstract experimental phenomena and principles can be converted into a concise mathematical formula that is easy to be applied in the early warning and control system. For example, the formula to predict the time of freezing snow and the amount of spray.

Formula of road water freezing time prediction:

$$t_{jb} = \frac{(T_L - Q)}{V_L} \tag{1}$$

Formula of the amount of spraying time required by a single nozzle to spray deicing liquid:

$$X = \frac{L_L \bullet D_P \bullet D_L \bullet \left(\frac{T_L - T_S}{Q_C + Q}\right)}{100000 \bullet np \bullet xp} \tag{2}$$

Time required for road surface snow prediction:

$$t_{jx} = \frac{\frac{1}{V_L} + \frac{1}{V_Q} + V_F}{S_X} \tag{3}$$

Formula of the amount of spraying time required by a single nozzle to spray deicing liquid on road surface with snow:

$$X = \frac{L_L \bullet D_P \bullet S_{XL} \bullet F}{1000np \bullet nx} \tag{4}$$

Among them: t_{jb}: Predicted freezing time; T_L: The road surface temperature; Q: Real-time freezing point; V_L: Road temperature drop rate; D_L: The water film thickness; D_P: The spray width; L_L: Laying length; Q_C: The freezing point of a deicing liquid in a storage tank; T_S: The set temperature difference; np: The number of nozzles; xp: Spray amount per unit time of nozzle; t_{jx}: Forecast snow cover time; V_Q: Air temperature drop rate; V_F: The wind speed; T_D: Atmospheric temperature; S_X: Snow rate; F: Snow melting efficiency of deicing liquid.

For example: on December 27, 2018, the detection data of the automatic de-icing and snow removal system on er-huan south road in Jinan city are shown in Table 1.

Table 1. Automatic spray system test data

Testing time	Road surface temperature	Freezing point temperature	Water film thickness	Coefficient of friction	Traffic conditions	Air temperature	Wind speed	Precipitation	Precipitation type	Line pressure	Instantaneous pipeline flow	Deicing liquid level
2018-12-27 13:05	0.53	0.00	22.21	0.81	10.00	-3.43	4.74	341.68	0.00	0.02	-0.00	46.38
2018-12-27 13:10	0.47	0.00	15.55	0.82	10.00	2	1.77	341.68	0.00	0.02	-0.00	46.41
2018-12-27 13:15	0.47	0.00	13.33	0.82	10.00	4.04	2.75	341.68	70.00	0.02	0.00	46.41
2018-12-27 13:20	0.49	0.00	13.33	0.82	10.00	-3.99	2.75	341.74	70.00	0.02	-0.00	46.42
2018-12-27 13:25	0.63	0.00	13.33	0.82	10.00	-4.07	1.97	341.89	70.00	0.02	0.00	46.41
2018-12-27 13:30	0.31	0.00	11.11	0.33	10.00	-4.08	3.99	341.94	70.00	0.48	1.2	44.38
2018-12-27 13:35	0.12	0.00	22.21	0.54	40.00	-4.32	0.00	341.99	70.00	0.53	0.7	43.44
2018-12-27 13:40	-0.07	0.00	119.9	0.67	25.00	-4.08	2.71	342.07	70.00	0.02	0.00	43.42
2018-12-27 13:45	-0.24	0.00	198.8	0.62	25.00	-4.19	2.22	342.15	70.00	0.02	-0.00	43.38
2018/12/27 13:50	-0.33	-8.77	102.7	0.77	25.00	-3.79	2.35	342.27	70.00	0.02	-0.00	43.41
2018/12/27 13:55	-0.49	-8.77	83.97	0.78	25.00	-3.97	1.87	342.35	70.00	0.02	0.00	43.28
2018/12/27 14:00	-0.71	-8.77	57.21	0.81	25.00	-4.11	2.13	342.41	70.00	0.02	-0.00	43.47

As can be seen from Table 1, the storage time interval between the same data type is 5 minutes. A total of 12 pieces of data were selected for a total of 1 hour. In the table, as time goes by, the road surface temperature drops slightly. However, after the snowfall, the detected road surface temperature rises due to heat released from snowfall and the snow will cover the surface of the road sensor. After the spraying starts, the snow on the road starts to melt and absorb heat, and the road surface temperature starts to drop again. Curve of road temperature change is shown in Figure 6.

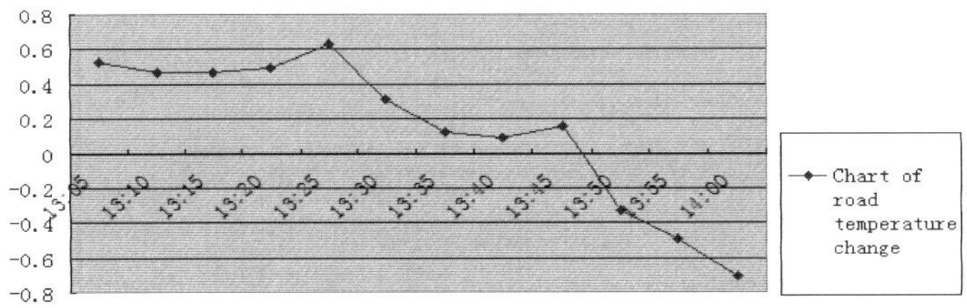

Figure 6. Chart of road temperature change

In the early stage of snowfall, the water film thickness of road surface is kept within the safe range. At this time, the change trend of atmospheric temperature and road surface is similar, and there will be a temperature rise phenomenon. The change curve of atmospheric temperature is shown in Figure 7.

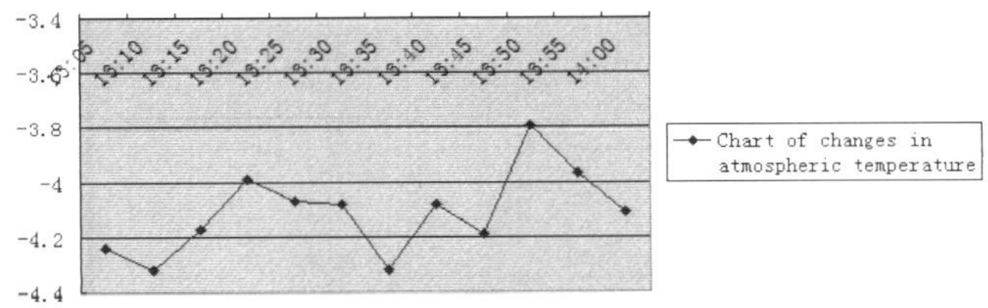

Figure 7. Chart of changes in atmospheric temperature

The snow cover time predicted by the system early warning model at 13:20 is 16.34 minutes, the difference between the time of 13:35 and the time of the actual sensor to detect snow in the outlet area is less than 2 minutes. The change curve of road condition is shown in Figure 8.

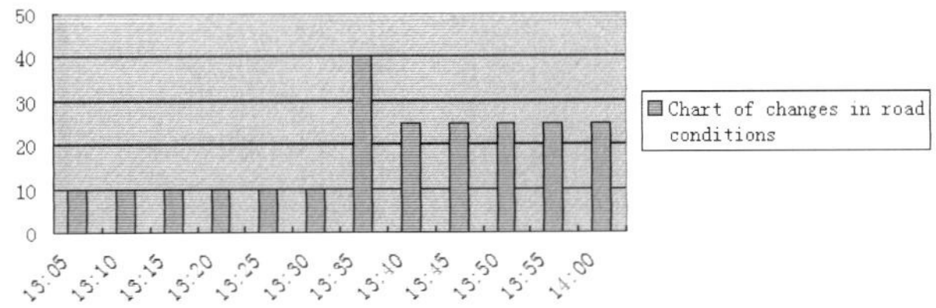

Figure 8. Chart of changes in road conditions

In order to improve the efficiency of anti-ice and snow removal, the system will enter spraying state five minutes before the warning time to eliminate snow and ice hazards in advance. It can be seen from the change of the friction coefficient of road surface, the friction coefficient of road surface with thick snow will decrease rapidly when the road surface have snow, and the friction coefficient of road surface will gradually reach the safe range after the start of spraying as shown in Figure 9.

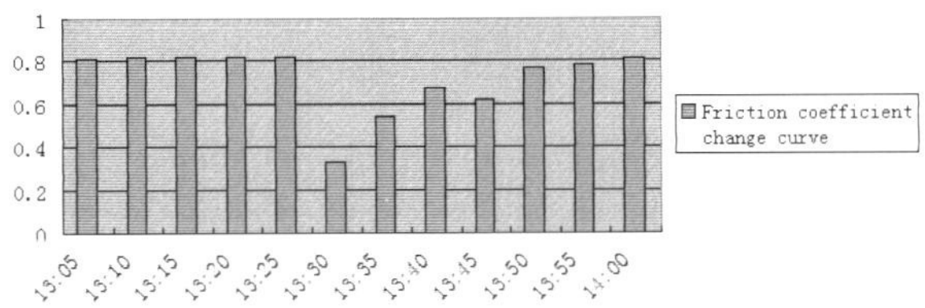

Figure 9. Friction coefficient change curve

3.2. Design of pressure control

In this system, PLC executes PID function block by timing (according to sampling time) and takes the relatively ideal proportional-integral-differential value obtained through repeated experiments as the basic fixed set value according to PID operation rules. When the spraying is started, a fixed pressure value will be given for each nozzle to achieve the spraying effect when the system is running, and then the real-time pressure value detected by the pressure sensor in the pipeline will be used as the feedback value, and the voltage signal obtained after calculation is the output value. The output signal is connected with the corresponding input port of the frequency converter by cable and converted into controllable frequency (range 0~50 Hz), so as to achieve the purpose of constant pressure spraying.

However, in practical projects, the pipe wall and valve will cause losses to the pressure of solution in the pipe with the extension and increase of the number of pipes, the spray pressure away from the nozzle of the pump will decrease with the extension of the distance between the pipes under the condition of constant pump outlet pressure, making the spray effect impossible. Therefore, in the design of pipeline pressure control, the sprinkler head within each distance corresponds to different running frequency of water pump, and the end sprinkler head corresponds to the highest running frequency of water pump. In order to facilitate the control, the nozzles closest to the water pump are taken as NO.1, and the nozzles number increases successively with the extension of the distance. When designing the program, the nozzle number is used as the trigger signal to change the set pressure value of the input PID module in real time [6-7]. PID control output simulation is shown in Figure 10.

At the initial stage when the water pump started, the pipeline pressure gradually rises. After the control and adjustment of PID, the pipeline pressure is basically maintained within reasonable range of the set value of 0.5Mpa.

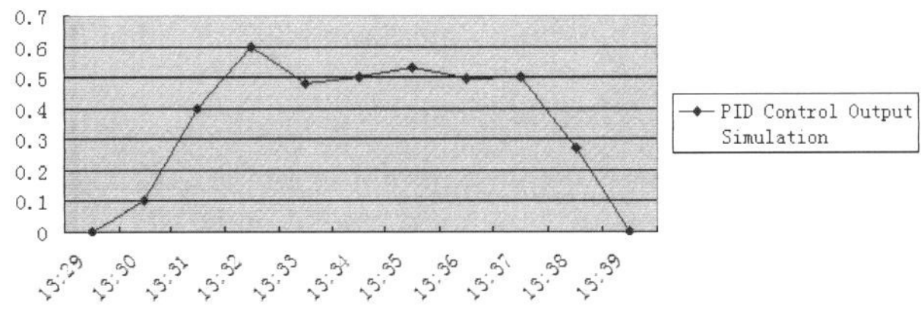

Figure 10. PID control output simulation

3.3. Road ice and snow warning

To solve this problem, "road safety warning and disposal platform" is designed by using cloud computing on the basis of intelligent anti-icing and snow removal control system, and 4G network is used to realize remote data docking. The cloud platform is embedded with icing snow warning function, and the recursive neural network icing snow warning model is used to combine the received local road meteorological data information with the local overall weather forecast information to predict the remaining occurrence time(greater than 5min) of icing or snow on the special road surface. At the same time with the combination of the controller in real-time decision tree of early warning function, danger level: high risk, medium risk, slight risk and no risk are divided. When there is no danger, the system does not execute action processing. In the case of slight danger, the system issues the speed limit signal to remind the driver and gives the reasonable driving speed. In the case of medium risk, the system turns on the spraying mechanism on the basis of the light dangerous action and sprays a small amount of deicing liquid for prevention. At high risk, the amount of deicing liquid would be increased to ensure the anti-ice ability of the road.

Using the cloud for big data analysis and advance the ability to learn quickly, and the host controller real-time logic control function of different risk measures the deicing snow removal function more detailed accurate. Rational use of the dosage of the deicing fluid can promote efficiency of road deicing snow melting, further efficiently solve icy roads of snow and prevent road problems of "black ice" and secondary icing [8].

4. Effect verification

Through the verification of the result of actual project, in winter when the road surface has accumulated water or snowfall due to rainfall, etc.,, when it is predicted that the snow accumulation time is less than 10 minutes, the spray will be automatically started, add a layer of "protective outerwear" before the road surface is frozen or snowy.. After the road surface temperature dropped below zero degrees Celsius, the spray of the deicing fluid surface due to the increase of the thickness of water film is a decrease in the road surface friction coefficient, but generally no less than 0.45.The vehicle is not easy to appear skid sideslip phenomenon. The surface of the non-spray deicing fluid has already formed ice, and the friction coefficient of the road surface reaches 0.1. The vehicle is prone to have accidents under such road conditions. This system can respond quickly and prevent road icing in time. Below shows laying the overpass ramp of this system, five o 'clock in the morning it began to snow, after 20 minutes the system automatically started to spray. Road of floating snow began to melt the outflow while other pavements still had snow. The snow continues. Freezing point system detected at half past seven in the morning the road solution and pavement temperature difference is less than the set value and the system opened a second spray, while other roads had up to two centimetres of snow, the sprayed road had no snow. The road maintenance department and environmental protection after other pavement, freezing point system detected the road solution maintained in a safe range from start. At this time, the system can detect the ice and snow condition of the road in real time to prevent the secondary icing phenomenon. Spray scene effect diagram are shown in Figure 11.

Figure 11. Experimental effect

Acknowledgement
This work is supported by the Key Research and Development Foundation Project of Shandong under Grant 2017GSF220008; Science and Technology Project of Shandong Provincial Transportation Department under Grant 2017B33.

References

[1] Ren yi, Yuan tongsen, wan zhi, Qin min. Overview of the current situation of highway ice-prevention and de-icing technology at home and abroad [J]. Hunan traffic science and technology, 2014, 40(2): 71-75.
[2] Ou yan, Pu xiang, Zhou xuchi, Lu ye, Ren yi, sun guai. Research progress of pavement ice monitoring technology [J]. Highway, 2013(4): 191-195.
[3] Qu zhi, Liang jia, Zheng min. Referring to the ice monitoring system of xihan expressway-talking about the construction method of road ice monitoring system in shaanxi province [J]. Straits science and technology & industry, 2016(10): 95-96.
[4] Anburaj Muthumani, Laura Fay, Michelle Akin, Shaowei Wang, Jing Gong, Xianming Shi. Correlating lab and field tests for evaluation of deicing and anti-icing chemicals: A review of potential approaches [J]. Cold Regions Science and Technology, 2014, 97: 21 32
[5] Fu zhen, Peng dan dan, sun guo qiang, qin juze. Application of fixed automatic spray snow and ice control technology in North America and Europe [J]. Highway engineering, 214, 39(4): 64-68.
[6] Wu lipu, Zhang long, Zheng changbing. Remote monitoring system for intelligent flyover snowmelt spraying [J]. Microcomputer and application, 2015, 34(2): 69-71.
[7] Tao Ma, Lei Geng, Xunhao Ding, Deyu Zhang, Xiaoming Huang. Experimental study of deicing asphalt mixture with anti-icing additives [J]. Construction and Building Materials, 2016, 127: 653-662.
[8] Wu chuanhong. Analysis on severe weather warning system of expressway [J]. China traffic informatization, 2016(2): 94-96.

The analysis method and application of automobile driving direction deviation

Peijiang Chen[1], Mei Zhang[2], Liying Wu[1]

[1] School of Mechanical and Vehicle Engineering, Linyi University, Linyi, Shandong, China
[2] Archives, Linyi University, Linyi, Shandong, China

Corresponding author and e-mail: Peijiang Chen, chenpeijiang@163.com

Abstract. When the automobile body deviates, being one of the hidden dangers of road traffic safety, it affects not only the automobile operation stability, but also the accuracy of headlight detection results. The driving directions analyzing method based on binocular stereo vision was presented. The shape center of the front license plate was extracted as automobile body feature point based on mathematical morphology and Hough transform. The sequence of the feature points was fitted into a straight line, which was the driving direction of the automobile. The measurement error of the headlight beam irradiation was corrected. The experiment results show that that this method can effectively detect the deviation angle of the automobile driving direction, and improve the accuracy of headlight detection.

1. Introduction

When the automobile body is deviated, there is a certain angle between the automobile body and the driving direction. The deviation can affect the operational stability, and it is a potential problem for road safety. In addition, the body deviation can also affect the accuracy of the headlight deflection testing results [1]. When the headlights are detected, the lead rail of the detector should be aligned with the automobile body, otherwise it can cause measurement errors for beam irradiation direction. If the automobile body is deviated, the irradiation direction of the headlamp is inconsistent with the driving direction, then the measurement error is increased, resulting low pass rate and poor reproducibility of headlight detection.

Thus, correcting the measurement results of the automobile headlights is very necessary. Wu Peidi [2] uses the laser ranging technology to measure the angle of the automobile body skew, which is used to correct the detecting data of the headlight testing. Lin Huiying [3] uses he binocular stereo vision technology to measure the deviation angle of automobile axis deviation. These methods are difficult to achieve real automatic detection.

This study combines the characteristics of digital image processing technology and binocular stereo vision technology; the visual images of the automobile are captured in the driving process. Extracting the body longitudinal axis, detecting the actual driving direction and analyzing the skew angle of automobile body, the measurement results are applied to the error correction of headlight detection, to improve the accuracy of measurement data.

2. The Driving Directions Analyzing Method Based on Binocular Stereo Vision

In the measurement scheme of automobile driving direction based on the binocular stereo vision technology, a number of frames of visual image of automobile are captured and processed. The outline of front license plate is extracted, the form center of front license plate is analyzed, and the automobile body feature points are obtained. The feature point sequence is fitted to get the driving direction and the positional deviation between the actual driving direction and ideal lead line is analyzed.

The process of analyzing the driving direction of an automobile based on the binocular stereo vision technology is as follows.

(1) According to the camera calibration results and the automobile position, the general location of the front license plate is estimated for the left and right images of each frame of the vehicle images.

(2) The front license plate of the automobile is positioned roughly by using the mathematical morphology. The contour of front license plate of the automobile is extracted quickly and the front license plate is positioned precisely based on the Hough transform with aspect ratio constraint and angle constraint.

(3) The area of front license plate is analyzed to extract the shape center point, which can be seen as the automobile body feature point.

(4) The automobile body feature points of left and right images of each frame are taken three-dimensional reconstruction to calculate their coordinates in the world coordinate system, and to give a sequence of feature points of automobile body.

(5) The sequence of feature points is fitted to get a straight line, which is the ultimate driving direction of the automobile.

2.1. Feature point detection of automobile body

For each image, the Hough transform is used to detect outer contour of the front license plate [4]. The shape center of front license plate is extracted as the feature point of automobile body.

Suppose that in the rectangular coordinates of image space, there is a straight line through the point (x, y), with slope of a, intercept of b. Since the slope may be infinite, it can be expressed with normal form as shown in formula (1).

$$\rho = x\cos\theta + y\sin\theta \quad (1)$$

Where, ρ is the distance from the origin to the normal line of the straight line, θ is the angle between x-axis and the normal line [5].

After the Hough transform, it can be shown with the point (ρ, θ) in parameter space, that is, the problem of line detection in the image space is converted into the corresponding points detection in the parameter space. The detection task can be completed by feature point voting in parameter space.

Based on the detection accuracy, the parameter space is discretized and divided into a number of accumulator units $A\,[\theta_i, \rho_j]$.

For each non-background image point, by traveling through all the values of θ_i, the corresponding values of ρ_j are calculated, and the accumulator unit A $[\theta_i, \rho_j]$ is added 1.

If the value of an accumulator unit is local maximum value point, and is greater than a given threshold, the corresponding linear parameters (θ_i, ρ_j) can represent the straight lines detected by the Hough transform [6].

When the front surface of automobile is parallel to the camera plane, after the Hough transform for the front license plate image of an automobile, its horizontal edges and vertical edges can generate two peaks at $\theta = 90°$, $\theta = 0°$ respectively in the parameter space.

The distance between the two former points is about r times the distance between the latter. Therefore, the ratio r can be used to eliminate the interference of corresponding peak points of the not plate frames [7].

According to the four peak points selected in the Hough transform domain, the corresponding straight lines are gotten in the original image space, which forms a rectangular, that is the front license plate.

2.2. Analysis of automobile driving direction
The body feature points of the left and right image of each frame are taken three-dimensional reconstruction, getting the coordinates in the world coordinate system. The sequence of feature points on the body image are fitted to form a straight line, which is the traveling direction of the automobile.

In this project, the task is to analyse the horizontal driving direction, and correct the measurement data of the horizontal irradiation direction of the headlight test. Therefore, the deviation between the fitted line and the ground plane cannot be considered, and the three-dimensional linear fitting is changed into a two-dimensional linear fitting, which can reduce the computation amount and improve the real time of detection.

The most commonly used method for linear fitting of is the least squares method. Assuming that there are n measured discrete data points, in the given class of functions, if there is a minimum value among the quadratic sums of distance between the demand function and given points $\sum_{i=1}^{n}(f(x_i)-y_i)^2$, then the function $f(x)$ is the least squares solution, that is the function expression of the fitted straight line [8].

This fitted straight line is the driving direction of the tested automobile.

3. Headlight detection error correction based on driving direction
In order to verify the measurement effect of the automobile driving direction, it is applied into the error correction of the automobile headlight detection. In this application, it needs to determine the actual driving direction of the automobile and analyze the deviation between the driving directions and the ideal position, and then correct the measurement data error of the headlight beam irradiation direction.

The error correction principle based on driving directions is given, as shown in Fig. 1.

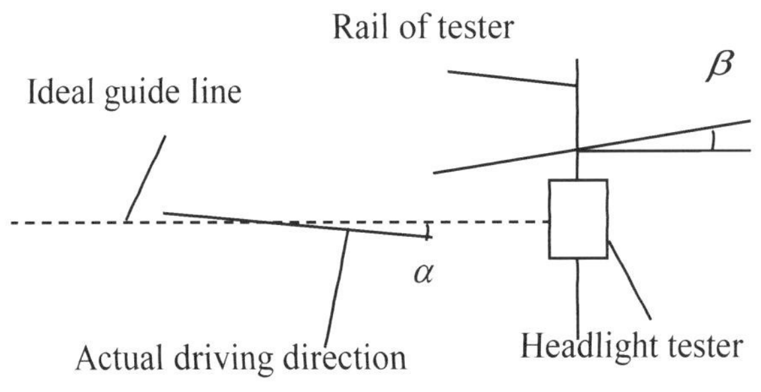

Figure 1. The Error Correction Principle of Headlight Detection Based on Driving Direction

Assuming the declination angle between the actual driving direction and ideal guide line is α, the offset of horizontal beam irradiation of headlight measured by the headlamp tester is β, then the actual beam irradiation direction can be gotten, as shown in formula (2).

$$\theta=\beta-\alpha \tag{2}$$

Where, making the ideal guide line of the test field as the reference basis, positive α means that the actual driving direction is to left, or the negative α is to right; similarly, if the measurement data and correcting results of the headlight detection are positive, the direction are to left, and vice versa.

In the experiment, the automobile goes to the headlight detection position about 5 meters away. In the travelling process, a sequence of visual images are captured.

The automobile body feature points of these images are extracted. According to the above method, the equation of the fitted straight line is gotten, that is $z=-120y+12781$. The angle can be gotten, then $\alpha=-0.4775°$.

The tested automobile parks in the position for headlight detection. By the headlight tester, the measurement data of the horizontal beam irradiation are gotten, as shown in Table 1.

According to Eq.2, the measured data of the horizontal beam irradiation direction of the headlight can be corrected, the results are shown in Table 1.

Table 1. The testing and data correction of the automobile headlight

Detection Item	High beam		Lower beam	
	Left headlight	Right headlight	Left headlight	Right headlight
Horizontal deviation (mm/10m)	-134	-426	12	52
Correction Result (mm/10m)	-50	-342	96	136

According to the standard of headlight detection, the right deviation angle of the high light beam of the right headlight exceeds requirements. After error correction, the detection result can meet the testing requirements.

In order to further verify the method, the parking position of the same car is changed, the horizontal beam irradiation direction of the headlamp is measured and error corrected. On the basis of the automobile driving direction, the measured data of the headlight detection can be error corrected. The maximum difference of the horizontal beam irradiation direction is significantly reduced. The experimental results show that the consistency of headlight detection after error correction is significantly improved.

4. Conclusions

In this project, several analyzing methods of automobile driving direction are analyzed, and their advantages and disadvantages are summarized. The morphological operations and Hough transform technique are used; a method of automobile driving direction based on binocular stereo vision technology is proposed, and it is applied to correct the measurement error in headlight detection. The method is non-contact measurement, and it has better real-time and can be engaged with automobile testing station.

Acknowledgement

This study is sponsored by the Science and Technology Planning Project of Higher Education of Shandong Province under Grant J16LB69, and the Doctoral Scientific Research Startup Foundation of Linyi University under Grant LYDX2016BS081, People's Republic of China.

References

[1] K. F. Sokmen, E. Pulat, N. Yamankaradeniz, and S. Coskun 2014 Thermal Computations of Temperature Distribution and Bulb Heat Transfer in an Automobile Headlamp. Heat and Mass Transfer, vol. 50, issue 2, pp. 199-210.

[2] P. Wu 2009 Study on Correction Method of Parking Error in Automobile Headlamp Testing, Changchun, China: Jilin University.

[3] H. Lin 2008 Study on Testing System of the Position Deviation of Vehicle-body and Axle Based on Stereo Vision, Jilin University, Changchun, China.

[4] Robert A. Mclaughlin 1998 Randomized Hough Transform: Improved Ellipse Detection with Comparison, Pattern Recognition Letters, vol. 19, issue 3-4, pp. 299-305.

[5] D. Fan, H.i Bi, and L. Wang, 2012 Implementation of Efficient Line Detection with Oriented Hough Transform, International Conference on Audio, Language and Image Processing, Shanghai, China. Pp. 45-48.

[6] S. Guo, W. Zhai, Q.Tang, and Y. Zhu 2012 Combining the Hough Transform and an Improved Least Squares Method for Line Detection. Computer Science, vol. 39, issue 4, pp. 196-200.

[7] W. Jiang, and S. Zheng 2010 Research OH Vehicle License Plate Recognition System Based on Hough Transform and Neural Network. ,kmmal of Anqing Teachers College (Natural Science Edition), vol. 16, issue 2, pp. 52-55.

[8] Y. Zheng, G. Wang, B. He, and X. Lin 2011 PET Bottle Cap Inspection Method Based on Linear Fitting Algorithm. Application Research of Computer, vol. 28, issue 11, pp. 4398-4400.

Density-dependent quantized kernel least mean square with desired smoothing

Qing Lin[1], Yulei Huang[1] and Yan Li[1]

[1] Intelligent Science and Information Engineering College, Xi'an Peihua University, Xi'an 710125, China

Corresponding author and e-mail: Qing Lin, 530610232@qq.com

Abstract. Kernel least mean square algorithm is a simple and effective adaptive algorithm, but it is constrained by its infinite network size. Many schemes have been proposed to reduce the size of the network, but the distribution of input data and the desire output are seldomly considered together. In this paper, we present a density correlation quantization scheme in input space and a method of smoothing the desired output by using the average value of the neighbourhood in output, and call it the DQKLMS-S algorithm. The good performance of DQKLMS-S is verified by experiments.

1. Introduction

The kernel method uses the Mercer kernel. In most cases, the rich expressiveness of kernel Gaussian function in mapping low-dimensional input space to high-dimensional feature space has made great achievements in batch kernel learning, such as support vector machine [1,2], kernel principal component analysis [3], online learning kernel (OKL) [4]. OKL incrementally learns the target function, especially when the data arrives in sequence.

Kernel adaptive filter [5] is one of the most famous OKL, including the kernel least mean square (KLMS) [6-7]. However, a common problem with kernel adaptive filters is that the network size grows as new data arrives. The growing network size presents computing and memory challenges. More importantly, overfitting can also occur if the network size is too large, making the algorithm too sensitive to outliers. Increasing the sparsity of the kernel adaptive filter is helpful to solve the complex calculation and overfitting problems.

Up to now, in order to restrain the growth of network size, various sparse technologies have been applied to the kernel adaptive filter. These technologies use sparse standards and accept only important input data as the new centers. Typical sparsity criteria include novelty criteria [8], predictive variance criteria [9], surprise criteria [10], and approximate linear correlation criteria [11]. A novel quantization scheme proposed [12] limits the network size and applies to the quantized kernel adaptive filter. The quantization algorithm greatly reduces the network size and even has better performance than KLMS.

In addition to network size, the representations of the input data and the output also affect generalization performance. In QKLMS, the distribution of quantization centers is approximately uniform in the input field. In fact, the distribution of the input data is often not uniform distributed. In order to improve the learning performance, the quantization technique should quantify the input data according to the distribution of the input data. Another problem in QKLMS is that the expected output of the quantified data is considered equal to the output of the nearest center. In many cases, the difference between the outputs of the neighborhood may be very large, especially when the underlying system is disturbed by impulsive noise. In order to deal with the above two problems, we consider a density correlation quantization scheme in input space and a method of smoothing the desired output by using the average value of the neighborhood in output, and call it the DQKLMS-S algorithm.

The rest of this article is organized as follows. The second part briefly reviews QKLMS. The new DQKLMS-S is proposed in section 3. In the fourth part, simulations on chaotic time series prediction are presented. Finally, we conclude the paper in the section 5.

2. Quantized kernel least mean square

The purpose of the adaptive filter is to learn a continuous input/output mapping $d(i)=f(\boldsymbol{u}(i))$ based on the input/output sequence. $\boldsymbol{u}(i)\in \mathbb{U}$ is an m-dimensional input vector. The output is assumed to be one-dimensional $d(i)$.

The least mean square (LMS) [13] is an incremental linear algorithm that updates weight vectors using the following steps

$$\begin{cases} f_0=\mathbf{0} \\ e(i)=d(i)-f_{i-1}(\boldsymbol{u}(i)) \\ f_i=f_{i-1}+\eta e(i)\boldsymbol{u}(i) \end{cases} \tag{1}$$

where $e(i)$ is the prediction error, η represents the step size, and the f_i denotes the learned mapping at iteration i. The output of LMS in iteration i is $d^*(i)=f_{i-1}(\boldsymbol{u}(i))=w(i)^T\boldsymbol{u}(i)$. Since LMS is a linear combination of input samples, LMS will fail if the mapping between inputs and outputs is non-linear.

In order to overcome the limitation of linearity while maintaining the simplicity of LMS, the input $\boldsymbol{u}(i)$ is converted to the high-dimensional feature space [6] by the kernel induction mapping $\varphi(\boldsymbol{u}(i))$.

If the Mercer condition is satisfied in the feature space, the inner product can be expressed by positive definite kernel function, which is expressed as

$$\kappa(\boldsymbol{u},\boldsymbol{u}')=\exp\left(-\frac{\|\boldsymbol{u}-\boldsymbol{u}'\|^2}{2\sigma^2}\right) \tag{2}$$

where σ denotes the kernel size.

Similar to the learning rules in the original input space, KLMS in the feature space can be expressed as

$$\begin{cases} f_0=\mathbf{0} \\ e(i)=d(i)-f_{i-1}(\boldsymbol{u}(i)) \\ f_i=f_{i-1}+\eta e(i)\kappa(\boldsymbol{u}(i),\cdot) \end{cases} \tag{3}$$

As can be seen from the above equation, one input $\boldsymbol{u}(i)$ corresponds to one coefficient $\eta e(i)$, resulting in the KLMS network being a growing radial-basis function network with the same size as the training set, which makes the calculation burden and memory demand very large.

In order to restrain the growth of KLMS network, QKLMS uses a simple online vector quantization method to quantify the input space. The learning rule of QKLMS is expressed as

$$\begin{cases} f_0=\mathbf{0} \\ e(i)=d(i)-f_{i-1}(\boldsymbol{u}(i)) \\ f_i=f_{i-1}+\eta e(i)\kappa(Q[\boldsymbol{u}(i)],\cdot) \end{cases} \tag{4}$$

where $Q[\square]$ is the quantization operator in the input space, the output of $Q[\boldsymbol{u}(i)]$ is either the $\boldsymbol{u}(i)$ itself or the nearest center.

$$Q[\boldsymbol{u}(i)] = \begin{cases} \boldsymbol{C}_{j^*}(i-1) & dis(\boldsymbol{u}(i), \boldsymbol{C}(i-1)) \le \varepsilon \\ \boldsymbol{u}(i) & dis(\boldsymbol{u}(i), \boldsymbol{C}(i-1)) > \varepsilon \end{cases} \tag{5}$$

where $\boldsymbol{C}(i)$ represents the center set, $dis(\boldsymbol{u}(i), \boldsymbol{C}(i-1))$ is the distance between $\boldsymbol{u}(i)$ and $\boldsymbol{C}(i-1)$, and ε denotes the quantization size

$$dis(\boldsymbol{u}(i), \boldsymbol{C}(i-1)) = \min_{1 \le j \le size(\boldsymbol{C}(i-1))} \| \boldsymbol{u}(i) - \boldsymbol{C}_j(i-1) \| \tag{6}$$

where $\boldsymbol{C}_j(i-1)$ is the j-th element of the center set $\boldsymbol{C}(i)$, and

$$j^* = \arg\min_{1 \le j \le size(\boldsymbol{C}(i-1))} \| \boldsymbol{u}(i) - \boldsymbol{C}_j(i-1) \| \tag{7}$$

If $dis(\boldsymbol{u}(i), \boldsymbol{C}(i-1)) > \varepsilon$, a new center is added into the center set $\boldsymbol{C}(i) = \{\boldsymbol{C}(i-1), \boldsymbol{u}(i)\}$, $\boldsymbol{\alpha}(i) = [\boldsymbol{\alpha}(i-1), \eta e(i)]$, if not, the coefficient of the closest center will be updated $\boldsymbol{\alpha}_{j^*}(i) = \boldsymbol{\alpha}_{j^*}(i-1) + \eta e(i)$.

Due to the fixed quantization threshold, the distribution of the center in the input data area in the dictionary is approximately consistent. Uniform distribution means that the quantization output of dense region is the same as that of sparse region, which cannot reflect the distribution of input data. What's more, in QKLMS, the expected output of the quantified data is considered equal to the output of the nearest center. In many cases, the difference between the outputs of the neighborhood may be very large, especially when the underlying system is disturbed by impulsive noise. In order to deal with the above two problems, we consider a density correlation quantization scheme in input space and a method of smoothing the desired output by using the average value of the neighborhood in output.

3. Density-dependent quantized kernel least mean square with desired smoothing

In this section, we discuss a new algorithm based on QKLMS, which is improved in processing input quantization and smoothing output results. On this basis, a density correlation quantization scheme and a method of smoothing the desired output by using the average value of the neighborhood are proposed into QKLMS, and an input density correlation quantization kernel least mean square with desired smoothing (DQKLMS-S) can be expressed as

$$\begin{cases} f_0 = \mathbf{0} \\ e(i) = \mathbf{D}_{j^*}(i) - f_{i-1}(\boldsymbol{u}(i)) \\ f_i = f_{i-1} + \eta e(i) \kappa(Q[\boldsymbol{u}(i)], \cdot) \end{cases} \tag{8}$$

The output of $Q[\boldsymbol{u}(i)]$ is either the $\boldsymbol{u}(i)$ itself or the nearest center

$$Q[\boldsymbol{u}(i)] = \begin{cases} \boldsymbol{C}_{j^*}(i-1) & dis(\boldsymbol{u}(i), \boldsymbol{C}(i-1)) \le \varepsilon \\ \boldsymbol{u}(i) & dis(\boldsymbol{u}(i), \boldsymbol{C}(i-1)) > \varepsilon \end{cases} \tag{9}$$

We introduce tow vectors $\mathbf{N}$ and $\mathbf{D}$. $\mathbf{N}$ is used to record the number of input data quantizing into each center. $\mathbf{D}$ denotes the smoothed desired output. If $dis(\boldsymbol{u}(i), \boldsymbol{C}(i-1)) > \varepsilon_{j^*}$, we add the input as a new center: $\boldsymbol{C}(i)=\{\boldsymbol{C}(i-1), \boldsymbol{u}(i)\}$, $\boldsymbol{\varepsilon}=[\boldsymbol{\varepsilon},\varepsilon_0]$, $\boldsymbol{\alpha}(i)=[\boldsymbol{\alpha}(i-1), \eta e(i)]$, $\mathbf{L}=[\mathbf{L},1]$, $\mathbf{N}(i)=[\mathbf{N}(i-1),1]$, $\mathbf{D}(i)=[\mathbf{D}(i-1),d(i)]$. If $\frac{\boldsymbol{\varepsilon}_{j^*}}{2}<dis(\boldsymbol{u}(i),\boldsymbol{C}(i-1))<\boldsymbol{\varepsilon}_{j^*}$ and $\mathbf{L}_{j^*}>\delta$, we update the quantization dictionary $\boldsymbol{C}(i)=\{\boldsymbol{C}(i-1), \boldsymbol{u}(i)\}$, $\boldsymbol{\alpha}(i)=[\boldsymbol{\alpha}(i-1), \eta e(i)]$ and alter the quantization threshold $\boldsymbol{\varepsilon}_{j^*}=\frac{\boldsymbol{\varepsilon}_{j^*}}{2}$, $\boldsymbol{\varepsilon}=[\boldsymbol{\varepsilon},\boldsymbol{\varepsilon}_{j^*}]$ and $\mathbf{L}_{j^*}=1$, $\mathbf{L}=[\mathbf{L},1]$, $\mathbf{N}(i)=[\mathbf{N}(i-1),1]$, $\mathbf{D}(i)=[\mathbf{D}(i-1),d(i)]$. Otherwise, the quantization center set remains unchanged $\boldsymbol{C}(i)=\boldsymbol{C}(i-1)$, and the effect of this input is added into the coefficient of the closest center: $\boldsymbol{\alpha}_{j^*}(i)=\boldsymbol{\alpha}_{j^*}(i-1)+\eta e(i)$, $\mathbf{L}_{j^*}=\mathbf{L}_{j^*}+1$. We also update as follows: $\mathbf{N}_{j^*}(i)=\mathbf{N}_{j^*}(i-1)+1$, and the smoothed desired output can be calculated as [5]

$$\mathbf{D}_{j^*}(i)=\frac{\mathbf{N}_{j^*}(i-1)\mathbf{D}_{j^*}(i-1)+d(i)}{\mathbf{N}_{j^*}(i)} \tag{10}$$

4. Simulation results

In this section, The Mackey-Glass (MG) chaotic time series prediction is considered. The time series is generated by a delay ordinary differential equations and is expressed as follows [14]:

$$\frac{dx(t)}{dt}=-bx(t)+\frac{ax(t-\tau)}{1+x(t-\tau)^{10}} \tag{11}$$

with $b=0.1$, $a=0.2$ and $\tau=30$. The sampling period of the signal is 6 seconds. The goal of MG chaotic time series prediction is to use the previous points to predict the present value. 1000 samples are served as the training data, and 100 samples are served as the test data. The time embedding dimension is 7, that is the $\boldsymbol{u}(i)=[x(i-1),x(i-2),...,x(i-7)]^T$ is selected as the input vector to predict $x(i)$. During the simulation, the training data are interfered by additive noise. We consider two kinds of noise: Gaussian noise with mean of 0 and variance of 0.04; mixed Gaussian noise with density

$$0.95\times N(0,0.04)+0.05\times N(0,1) \tag{12}$$

where $N(\mu,\sigma^2)$ represents the Gaussian density.

We compare the performance of QKLMS and DQKLMS-S. A Gaussian kernel of width 1.0 is selected as the Mercer kernel. The step sizes of these two algorithms are selected to obtain almost the same initial convergence rate. The quantization size is set as $\varepsilon=0.2$. Simulation results averaged over 1000 independent Monte Carlo runs. Average learning curves and network growth curves are shown in Fig. 1 and Fig. 2, respectively. The test MSEs in the final iteration is shown in Table I. It can be clearly seen that whether Gaussian noise or mixed Gaussian noise, the DQKLMS-S has better performance than QKLMS. In this example, the network size is reduced to about 200, which is very small compared to the original network size (1000).

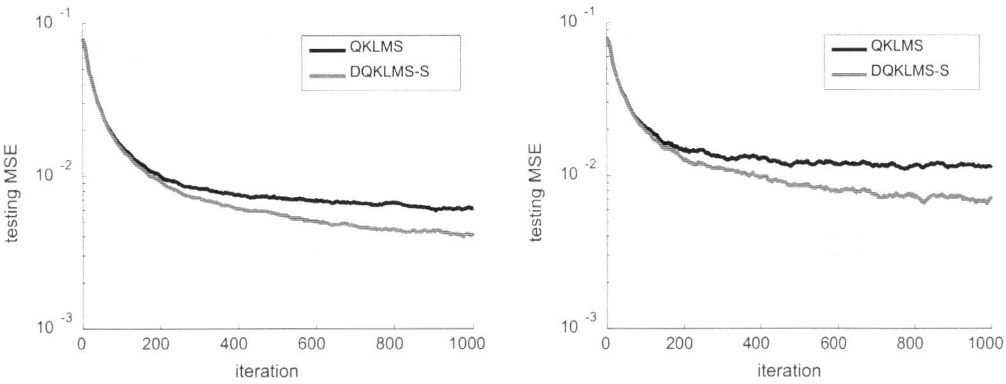

(a) Gaussian (b) Mix-Gaussian

Figure 1. Learning curves with different noise

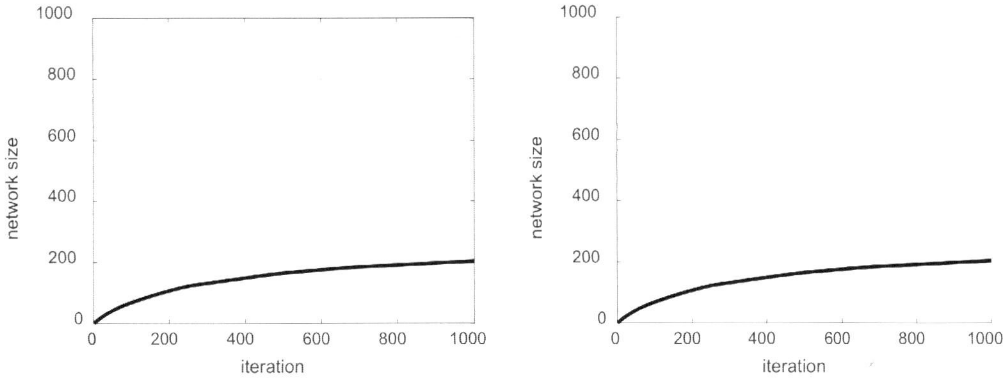

(a) Gaussian (b) Mix-Gaussian

Figure 2. Network growth curves with different noise

Table 1. Testing MSE at final iteration.

Algorithms	Testing MSE (Gaussian noise)	Testing MSE (mix-Gaussians noise)
QKLMS	$0.0061177 \pm 6.4794 \times 10^{-5}$	0.0116 ± 0.00016736
DQKLMS-S	$0.0041969 \pm 1.0447 \times 10^{-5}$	0.0070029 ± 0.00021841

5. Conclusions

In this paper, for improving the learning performance of the QKLMS, a density correlation quantization scheme in input space and a method of smoothing the desired output by using the average value of the neighborhood in output have been introduced into QKLMS to form a new algorithm, called DQKLMS-S. In the new algorithm, the distribution of quantization centers is consistent with the distribution of input data, and the output is smoothed by the desired signals output. Simulation results show that the algorithm is effective.

Acknowledgement

Fund Program: University-level Scientific Research Program of Xi'an Peihua University in 2019, "Research on Intelligent Tourism Service Platform Based on 5G Network" (No. PHKT19033)

References

[1] Lima M D, Costa N L and Barbosa R. Improvements on least squares twin multi-class classification support vector machine. Neurocomputing, 313(3):196-205, 2018.
[2] Miao X, Liu Y, Zhao H and Li C. Distributed Online One-Class Support Vector Machine for Anomaly Detection Over Networks. IEEE Transactions on Cybernetics, 49(4):1-14, 2019.
[3] Fan J and Chow T W S. Exactly Robust Kernel Principal Component Analysis. IEEE Transactions on Neural Networks and Learning Systems, 1-13, 2019.
[4] Ding S, Mirza B, Lin Z, Cao J, Lai X and Tam V. Kernel based online learning for imbalance multiclass classification. Neurocomputing, 277(14):139-148, 2018.
[5] Liu W, Principe J C, and Haykin S. Kernel adaptive filtering: a comprehensive introduction, volume 57. John Wiley & Sons, 2011.
[6] Liu W, Pokharel P P and Principe J C. The kernel least-mean-square algorithm. Signal Processing, IEEE Transactions on, 56(2):543–554, 2008.
[7] Bouboulis P and Theodoridis S. Extension of wirtinger's calculus to reproducing kernel hilbert spaces and the complex kernel lms. Signal Processing, IEEE Transactions on, 59(3):964–978, 2011.
[8] Platt J. A resource-allocating network for function interpolation. Neural computation, 3(2):213–225, 1991.
[9] Csato L and Opper M. Sparse on-line gaussian processes. Neural computation, 14(3):641–668, 2002.
[10] Liu W, Park I and Principe J C. An information theoretic approach of designing sparse kernel adaptive filters. Neural Networks, IEEE Transactions on, 20(12):1950–1961, 2009.
[11] Engel Y, Mannor S and Meir R. The kernel recursive least-squares algorithm. Signal Processing, IEEE Transactions on, 52(8):2275–2285, 2004.
[12] Chen B, Zhao S, Zhu P and Principe J C. Quantized kernel least mean square algorithm. Neural Networks and Learning Systems, IEEE Transactions on, 23(1):22–32, 2012.
[13] Haykin S and Widrow B. Least-mean-square adaptive filters, volume 31. John Wiley & Sons, 2003.
[14] Xu X, Qu H, Zhao J, Yang X and Chen B. Quantised kernel least mean square with desired signal smoothing. Electronics Letters, 51(18):1457-1459, 2015.

AUTHOR INDEX

AUTHOR INDEX